Contents

List of Contributors vii

Preface ix

Acknowledgments xiii

Contents of Volume II xv

THE AIR FORCE RESEARCH PROGRAM—A NEW LOOK
Colonel Ronald A. Bena 1

THE BAYESIAN AND NONPARAMETRIC APPROACH TO RELIABILITY STUDIES: A SURVEY OF RECENT WORK
I. N. Shimi and C. P. Tsokos 5

BAYESIAN LIFE TESTING USING THE TOTAL Q ON TEST
William S. Jewell 49

A BAYESIAN APPROACH FOR TESTING INCREASING FAILURE RATE
Robert H. Lochner and A. P. Basu 67

NONPARAMETRIC ESTIMATION OF DISTRIBUTION FUNCTIONS
Myles Hollander and Ramesh M. Korwar 85

A CLASS OF DISTRIBUTIONS USEFUL IN LIFE TESTING AND RELIABILITY WITH APPLICATIONS TO NONPARAMETRIC TESTING
Z. Govindarajulu 109

A GENERALIZED WILCOXON MANN-WHITNEY STATISTIC WITH SOME APPLICATIONS IN RELIABILITY
A. P. Basu 131

THE ASYMPTOTIC THEORY OF EXTREME ORDER STATISTICS
Janos Galambos 151

ESTIMATING AND FORECASTING FAILURE-RATE PROCESSES BY MEANS OF THE KALMAN FILTER
H. F. Martz, Jr., K. Campbell, and H. T. Davis 165

SHOCK AND WEAR MODELS AND MARKOV ADDITIVE PROCESSES
Erhan Çinlar 193

THE MAINTENANCE OF SYSTEMS GOVERNED BY SEMI-MARKOV SHOCK MODELS
Richard M. Feldman 215

BOUNDS FOR THE DISTRIBUTIONS AND HAZARD GRADIENTS OF MULTIVARIATE RANDOM MINIMUMS
Moshe Shaked 227

THE ROLE OF THE POISSON DISTRIBUTION IN APPROXIMATING SYSTEM RELIABILITY OF k-OUT-OF-n STRUCTURES
R. J. Serfling 243

CONVERTING DEPENDENT MODELS INTO INDEPENDENT ONES, WITH APPLICATIONS IN RELIABILITY
N. Langberg, F. Proschan, and A. J. Quinzi 259

STRUCTURAL INFERENCE ON RELIABILITY IN A LOGNORMAL MODEL
Danny Dyer 277

UNIFORMLY MOST POWERFUL UNBIASED TESTS FOR THE PARAMETERS OF THE GAMMA DISTRIBUTION
Max Engelhardt and Lee J. Bain 307

UNBALANCED RANDOM-EFFECT MODEL WITH UNEQUAL GROUP VARIANCES: ESTIMATION OF DESIGN PARAMETERS ALLOWING FOR SCREENED BATCHES
K. W. Fertig and N. R. Mann 315

SOME STOCHASTIC CHARACTERIZATIONS OF MULTIVARIATE SURVIVAL
William B. Buchanan and Nozer D. Singpurwalla 329

A FAMILY OF BIVARIATE LIFE DISTRIBUTIONS
Henry W. Block 349

TECHNIQUES FOR ANALYZING MULTIVARIATE FAILURE DATA
Richard E. Barlow and Frank Proschan 373

OPTIMAL REPLACEMENT POLICIES FOR DEVICES SUBJECT TO A GAMMA WEAR PROCESS
Mohamed Abdel-Hameed 397

SOME LIMIT THEOREMS IN STANDBY REDUNDANCY WITH RENEWALS
Z. Khalil 413

SOME ASPECTS OF THE USE OF THE MANN-FERTIG STATISTIC TO OBTAIN CONFIDENCE INTERVAL ESTIMATES FOR THE THRESHOLD PARAMETER OF THE WEIBULL
Paul N. Somerville 423

OPPORTUNISTIC REPLACEMENT POLICIES
Davinder P. S. Sethi 433

ON SELECTING WHICH OF k POPULATIONS EXCEED A STANDARD
Ray E. Schafer 449

ON GENERALIZED MULTIVARIATE GAMMA TYPE DISTRIBUTIONS AND THEIR APPLICATIONS IN RELIABILITY
P. R. Krishnaiah 475

ON SOME OPTIMAL SAMPLING PROCEDURES FOR SELECTION PROBLEMS
Shanti S. Gupta and Deng-Yuan Huang 495

CLASSIFICATION RULES FOR EXPONENTIAL POPULATIONS: TWO PARAMETER CASE
A. P. Basu and A. K. Gupta 507

A BIVARIATE EXPONENTIAL MODEL WITH APPLICATIONS TO RELIABILITY AND COMPUTER GENERATION OF RANDOM VARIABLES
D. S. Friday and G. P. Patil 527

The Theory and Applications of Reliability

With Emphasis on Bayesian and Nonparametric Methods

VOLUME I

ACADEMIC PRESS RAPID MANUSCRIPT REPRODUCTION

The papers in this volume were presented at the Conference on the Theory and Applications of Reliability with Emphasis on Bayesian and Nonparametric Methods held at the University of South Florida, Tampa, Florida, December 15–18, 1975. The Conference was jointly sponsored by the United States Air Force, Air Force Office of Scientific Research (AFOSR), and the University of South Florida, Tampa, Florida.

The Theory and Applications of Reliability

With Emphasis on Bayesian and Nonparametric Methods

VOLUME I

EDITED BY

CHRIS P. TSOKOS
Department of Mathematics
University of South Florida
Tampa, Florida

I. N. SHIMI
Air Force Office of Scientific Research (AFOSR)
Bolling Air Force Base
Washington, D.C.

ACADEMIC PRESS, INC. New York San Francisco London 1977

A Subsidiary of Harcourt Brace Jovanovich, Publishers

ACADEMIC PRESS, INC.
111 Fifth Avenue, New York, New York 10003

United Kingdom Edition published by
ACADEMIC PRESS, INC. (LONDON) LTD.
24/28 Oval Road, London NW1

Library of Congress Cataloging in Publication Data

Conference on the Theory and Applications of Reliability with Emphasis on Bayesian and Nonparametric Methods, University of South Florida, 1975.
The theory and applications of reliability with emphasis on Bayesian and nonparametric methods.

"Papers . . . presented at the Conference on the Theory and Applications of Reliability with Emphasis on Bayesian and Nonparametric Methods held at the University of South Florida, Tampa, Florida, December 15–18, 1975 . . . jointly sponsored by the United States Air Force, Air Force Office of Scientific Research (AFOSR), and the University of South Florida, Tampa, Florida."
1. Reliability (Engineering) – Mathematical models–Congresses. 2. Bayesian statistical decision theory – Congresses. I. Tsokos, Chris P. II. Shimi, I. N. III. United States. Air Force. Office of Scientific Research. IV. Florida. University of South Florida, Tampa. V. Title.
TS173.C7 1975 620′.004′5 76-57988
ISBN 0–12–702101–9 (v. 1)

PRINTED IN THE UNITED STATES OF AMERICA

List of Contributors

Mohamed Abdel-Hameed Department of Mathematics, University of North Carolina, Charlotte, North Carolina 28200

Lee J. Bain Department of Mathematics, University of Missouri–Rolla, Rolla, Missouri 65401

Richard E. Barlow Department of Operations Research, University of California at Berkeley, Berkeley, California 94700

A. P. Basu Department of Statistics, University of Missouri–Columbia, 222 Mathematics Science Building, Columbia, Missouri 65201

Ronald A. Bena Air Force Office of Scientific Research, Bolling Air Force Base, Washington, D. C. 20052

Henry W. Block Department of Mathematics, University of Pittsburgh, Pittsburgh, Pennsylvania 15261

William B. Buchanan Operations Reevaluation Group, Center for Naval Analysis, 1401 Wilson Blvd., Arlington, Virginia 22209

K. Campbell Statistical Service Group, University of California, Los Alamos Scientific Laboratory, Post Office Box 166, Los Alamos, New Mexico 87544

Erhan Çinlar Department of Industrial Engineering and Management Sciences, Northwestern University, Evanston, Illinois 60201

H. T. Davis Statistical Service Group, University of California, Los Alamos Scientific Laboratory, Post Office Box 166, Los Alamos, New Mexico 8754

Danny Dyer Department of Mathematics, University of Texas at Arlington, Arlington, Texas 76019

Max Engelhardt Department of Mathematics, University of Missouri–Rolla, Rolla, Missouri 65401

Richard M. Feldman Department of Industrial Engineering, Texas A & M University, College Station, Texas 77843

K. W. Fertig Rocketdyne Rockwell Intl. Corp., 6633 Canoga Avenue, Canoga Park, California 91304

D. S. Friday Department of Statistics, Pennsylvania State University, 219 Pond Laboratory, University Park, Pennsylvania 16802

Janos Galambos College of Liberal Arts, Temple University, Philadelphia, Pennsylvania 19122

Z. Govindarajulu Department of Statistics, University of Kentucky, Lexington, Kentucky 40506

A. K. Gupta Department of Statistics, University of Michigan, Ann Arbor, Michigan 48104

Shanti S. Gupta Department of Statistics, Mathematics Science Building, Purdue University, West Lafayette, Indiana 47907

Myles Hollander Department of Statistics and Statistical Consulting Center, Florida State University, Tallahassee, Florida 32306

Deng-Yuan Huang Department of Statistics, Mathematics Science Building, Purdue University, West Lafayette, Indiana 47907

William S. Jewell Operations Research, University of California, Berkeley, California 94720

Z. Khalil Department of Mathematics, Concordia University, 1455 de Maisonneuve Blvd. W., Montreal, Quebec, H3G 7M8

Ramesh M. Korwar Department of Mathematics and Statistics, University of Massachusetts, Amherst, Massachusetts 01002

P. R. Krishnaiah Aerospace Research Laboratories, Wright-Patterson Air Force Base, Ohio 45433

N. Langberg Department of Statistics, Florida State University, Tallahassee, Florida 32306

Robert H. Lochner Department of Mathematics and Statistics, Marquette University, Milwaukee, Wisconsin 53233

N. R. Mann Rocketdyne Rockwell Intl. Corp., 6633 Canoga Avenue, Canoga Park, California 91304

H. F. Martz, Jr. Statistical Service Group, University of California, Los Alamos Scientific Laboratory, Post Office Box 166, Los Alamos, New Mexico 87544

G. P. Patil Department of Statistics, Pennsylvania State University, 219 Pond Laboratory, University Park, Pennsylvania 16802

Frank Proschan Department of Statistics, Florida State University, Tallahassee, Florida 32306

A. J. Quinzi Department of Statistics, Florida State University, Tallahassee, Florida 32306

Ray E. Schafer Hughes Aircraft Company, Fullerton, California 92634

R. J. Serfling Department of Statistics, Florida State University, Tallahassee, Florida 32306

Davinder P. S. Sethi Operations Research Center, 3113 Etcheverry Hall, University of California at Berkeley, Berkeley, California 94720

Moshe Shaked Department of Mathematics and Statistics, University of New Mexico, Albuquerque, New Mexico 87131

I. N. Shimi AFOSR/NM, Building 410, Bolling Air Force Base, Washington, D.C. 20332

Nozer D. Singpurwalla Department of Operations Research, The George Washington University, Washington, D. C. 20006

Paul N. Somerville Florida Technological University, Box 25000, Orlando, Florida 32816

C. P. Tsokos Department of Mathematics, University of South Florida, Tampa, Florida 33620

Preface

Life testing in reliability studies in the past two decades has received a substantial amount of interest from theoreticians as well as reliability engineers. Their concern was a product of the increased complexity and sophistication in electronic, mechanical, and structural systems that came into existence very rapidly during this period. In the early 1950s scientists began to explore the field of parametric life testing under the assumption of an *exponential* time-to-failure distribution. These studies produced a series of papers that were to influence future work in the theory and practice of reliability estimation. Shortly thereafter came other failure distributions such as the *gamma* and *Weibull* models. The *Poisson* distribution has been considered as the failure model in situations in which the number of failures that occur in a given time period behaves as a random variable. Recently, textbooks and monographs have been written that give an excellent approach to the complete theoretical structure of most classical probability distributions as failure models.

In the ensuing years, the problem of reliability and life parameter estimation for a specified failure distribution received wide attention. A partial list of selective research efforts in this area is given here.

Most of the reliability studies were initiated and structured from purely theoretical behavior. The problems that the practicing reliability engineer encountered were not well documented in the structuring of these models. This lack of communication between the model building theoretician and practicing reliability scientist still exists in our present efforts in the subject area.

In the past six or seven years there has been a considerable amount of interest in the Bayesian and nonparametric approach to reliability studies. A Bayesian analysis implies the exploitation of suitable prior information in association with Bayes' theorem. It rests on the notion that a parameter within a failure model is not merely an unknown fixed quantity but rather a random variable that is characterized by some probability distribution. In the area of life testing, it is indeed realistic to assume that a life parameter is stochastically dynamic. This assertion is strongly supported by the fact that the complexity of mechanical, electronic, or structural systems is likely to cause undetected component interactions resulting in unpredictable fluctuation of the life parameter.

To a reliability engineer, a Bayesian approach would seem appealing because it provides a way for the formulation of a distributional form for an unknown parameter based on the prior convictions or information available to him. Recently, Drake[1] gave an excellent account for the use of Bayesian statistics in reliability problems. Drake points out,

[1] Bayesian Statistics for the Reliability Engineer, *Proc. 1968 Annu. Symp. on Reliability,* pp. 315–320, 1966.

> The (Bayesian) realizes . . . that his selection of a prior (distribution) to express his present state of knowledge will necessarily be somewhat arbitrary. But he greatly appreciates this opportunity to make his entire assumptive structure clear to the world . . .

A further argument of the Bayesian approach to reliability is given by Evans,[2] who questions and comments:

> Why should an engineer not use his engineering judgment and prior knowledge about the parameters in the statistical distribution he has picked? For example, if it is the mean time between failures (MTBF) of an exponential distribution that must be evaluated from some tests, he undoubtedly has some idea of what the value will turn out to be. If he does not, he is about to get fired. Then he can get a much better idea about the true MTBF by combining some test results with his prior knowledge.

Furthermore, from a practical point of view, Feduccia[3] states that, using reliability prediction and its associated measure of uncertainty, the coefficient of variation as the "prior" in a Bayesian expression makes sense because:

(1) Reliability predictions are a requirement in nearly every military electronic equipment development program.
(2) Reliability prediction techniques are based on the pooled and organized experience of countless individuals and organizations. Thus, when one uses these techniques, he benefits from this accumulated experience and does not have to guess, using engineering judgment, or rely on his own limited experience.
(3) Reliability predictions have been shown to provide reasonable estimates of equipment operational performance.
(4) The consumer and producer would be more likely to agree on the use of λ_ρ (prior predicted failure rate of the equipment), the predicted equipment failure rate, as prior evidence than most other types of prior information.

Finally Crellin[4] and other scientists have given additional eloquent presentations of the basic philosophy and fruitfulness of the Bayesian approach to reliability.

The nonparametric approach analyzes the reliability of components and systems under the assumption of a time-to-failure distribution with a wide defining property rather than a specific parametric class of probability distributions. To the practicing reliability engineer these nonparametric classes are easier to assume and justify. Some basic examples of these classes are the increasing failure rate, increasing failure rate average, and new better than used distributions. A large number of very practical and powerful results have been recorded from assuming such types of distributions, yet much more is needed to be done to fully utilize the power of these classes.

[2] Prior knowledge, engineers versus statisticians, *IEEE Trans. Reliability* **18**, Editorial.
[3] A Bayesian/Classical Approach to Reliability Demonstration, RADC-TR-70-72, Tech. Rep., pp. 1-23, 1970.
[4] The philosophy and mathematics of Bayes equation, *IEEE Trans. Reliability* **R-21**, 131-135.

A significant amount of progress has been made in both Bayesian and nonparametric approaches to reliability in recent years. However, there are still many important problems both from the theoretical and practical points of view that need to be studied. Moreover, from the practitioner's view there are still some very basic and indeed important questions to be considered with respect to the implementation of existing theory. For example: How to test whether the life distribution is increasing failure rate, increasing failure rate average, or new better than used type distributions? When should one employ a Bayesian approach to his reliability problem? How should one structure or identify the prior probability distribution? What about computational feasibility? What size of random sample is necessary for Bayesian reliability analysis?

The aim of the conference on The Theory and Applications of Reliability with Emphasis on Bayesian and Nonparametric Methods was to bring together practicing reliability scientists and model building theoreticians so that they might exchange information, discuss open problems like the ones stated above, and set directions on the subject areas of mutual interest. To attempt to achieve these objectives the conference was structured so as to have (1) technical presentations, (2) a clinical session, (3) round table discussions.

The technical sessions were devoted primarily to recent original findings on the theory and applications of reliability with emphasis on Bayesian and nonparametric methods. The clinical session was structured to provide practicing reliability scientists the opportunity to ask technical questions concerning actual problems that they were encountering. Three panels of experts in Bayesian approach to reliability, nonparametric approach to reliability, and empirical Bayes approach to reliability led the round table discussions. The aims of the panels were to discuss the present status of each of the areas and propose new directions in both theory and applications in the subject areas.

The proceedings of the conference have been divided into two volumes. The first volume contains the theoretical presentations which would be of value to both applied and theoretical scientists working in the subject area. The second volume consists of the more practical oriented papers of the conference which would be useful to practicing reliability scientists. Both volumes contain the articles by Col. Ronald A. Bena, entitled, "The Air Force Research Program—A New Look" and a survey by the editors, "The Bayesian and Nonparametric Approach to Reliability Studies: A Survey of Recent Work."

Acknowledgments

The editors wish to thank all of the individuals who, in addition to the speakers, contributed to the success of the conference. A special thanks is due to the following participants for their kind assistance:

ADVISORY COMMITTEE

MR. CURTIS AASEN Braddock, Dunn, and McDonald Corporation
DR. RALPH A. EVANS Durham, North Carolina
MR. JEROME E. HOROWITZ Laurence G. Hanscom Field
DEAN E. KOPP University of South Florida
MR. J. J. NARESKY Rome Air Development Center
DR. FRANK PROSCHAN Florida State University
DR. I. SHIMI Air Force Office of Scientific Research

SESSION CHAIRPERSONS

MR. CURTIS AASEN Braddock, Dunn, and McDonald Corporation
DR. JAMES J. HIGGINS University of South Florida
DR. CARL F. KOSSACK University of Georgia
DR. V. LAKSHMIKANTHAM University of Texas at Arlington
DR. W. J. PADGETT University of South Carolina
DR. I. SHIMI Air Force Office of Scientific Research
DR. K. "TED" WALLENIUS Clemson University

CLINICAL SESSION

MR. DAVID F. BARBER Rome Air Development Center
MR. JEROME KLION Rome Air Development Center

PANELS: COMMENTATOR—R. A. EVANS

BAYESIAN APPROACH TO RELIABILITY

DR. DAVE MASTRAN Arthur Young and Company
DR. RAY E. SCHAFER Hughes Aircraft Company
DR. SHELEMYAHU ZACKS Case Western Reserve University

EMPIRICAL BAYES APPROACH TO RELIABILITY

DR. G. K. BENNETT University of South Florida
DR. G. C. CANAVOS Virginia Commonwealth University
DR. HARRY F. MARTZ Los Alamos Scientific Laboratory

NONPARAMETRIC APPROACH TO RELIABILITY

DR. RICHARD E. BARLOW University of California at Berkeley
DR. MYLES HOLLANDER Florida State University
DR. ERNEST SCHEUER California State University

We would also like to express our appreciation to Col. Ronald A. Bena, Deputy Director of the United States Air Force, Office of Scientific Research, and Dr. Charles Polk, Head of the Electrical Science and Analysis Section, National Science Foundation, for their special addresses to the conference. For their kind assistance during the progress of the conference we wish to thank Dr. A. N. V. Rao, Dr. Stephen Smeach, Dr. Truett Smith, Mr. Steven Bean, Mr. Gary Chao, Mr. James Chen, Mr. Curtis Church, Ms. Chris Deans, Mr. Steve Howell, Mr. Robert Jernigan, Ms. Diane Rotsell, Ms. Rebecca Watson, Ms. Cathy Welch, and Mr. Richard Welch. Our special thanks to Dr. James J. Higgins for his kind assistance in the preparation of the conference and the proceedings. Finally, we wish to gratefully acknowledge the excellent help that we received from Ms. Sherrie L. Dominguez during the preparation, progress, typing, and editing of the proceedings.

Chris P. Tsokos
I. N. Shimi

Contents of Volume II

THE AIR FORCE RESEARCH PROGRAM—A NEW LOOK
Colonel Ronald A. Bena

THE BAYESIAN AND NONPARAMETRIC APPROACH TO RELIABILITY STUDIES: A SURVEY OF RECENT WORK
I. N. Shimi and C. P. Tsokos

BAYES: IN THEORY & PRACTICE
Ralph A. Evans

BAYES ESTIMATION OF THE RELIABILITY OF SERIES AND PARALLEL SYSTEMS OF INDEPENDENT EXPONENTIAL COMPONENTS
Shelemyahu Zacks

COMPARISON OF BAYESIAN ESTIMATES OF FAILURE INTENSITY FOR FITTED PRIORS OF LIFE DATA
J. J. Higgins and C. P. Tsokos

AN ANALYSIS REGARDING THE DETERMINATION OF BAYESIAN CONFIDENCE LIMITS FOR THE RELIABILITY OF DISTRIBUTION-FREE PARALLEL SUBSYSTEMS
Donald R. Smith

BAYESIAN CONFIDENCE BOUNDS FOR THE POISSON FAILURE MODEL
A. Papadopoulos and A. N. V. Rao

BAYESIAN ESTIMATION OF SYSTEM RELIABILITY FOR WEIBULL DISTRIBUTION USING MONTE CARLO SIMULATION
Satish J. Kamat

BAYES ESTIMATION OF RELIABILITY FOR THE LOGNORMAL FAILURE MODEL
W. J. Padgett and Chris P. Tsokos

MODIFIED BAYESIAN PROCEDURES IN RELIABILITY TESTING
Bennet P. Lientz

ROBUSTNESS AND THE PRIOR DISTRIBUTION IN THE BAYESIAN ESTIMATION
George C. Canavos

BASIC CONCEPTS OF EMPIRICAL BAYES METHODS WITH SOME RESULTS FOR THE WEIBULL DISTRIBUTION
G. Kemble Bennett

BAYES AND EMPIRICAL BAYES POINT AND INTERVAL ESTIMATION OF RELIABILITY FOR THE WEIBULL MODEL
H. F. Martz, Jr. and M. G. Lian

A NEW APPROACH TO THE NONPARAMETRIC TESTS OF EXPONENTIAL DISTRIBUTION WITH UNKNOWN PARAMETERS
Y. H. Wang and Stella A. Chang

APPLICATION OF NONPARAMETRIC METHODS IN THE STATISTICAL AND ECONOMIC ANALYSIS OF WARRANTIES
Wallace R. Blischke and Ernest M. Scheuer

A SURVEY OF STATISTICAL METHODS IN SYSTEMS RELIABILITY USING BERNOULLI SAMPLING OF COMPONENTS
Bernard Harris

ESTIMATION OF A MULTI-PARAMETER FAILURE RATE USING A GENERAL REPLACEMENT TESTING PLAN
L. H. Crow and I. N. Shimi

ANALYSIS OF VARIANCE FOR WEIBULL POPULATIONS
John I. McCool

ACCELERATED LIFE TESTS USING THE POWER LAW MODEL FOR THE WEIBULL DISTRIBUTION
Nozer D. Singpurwalla and Faiz A. Al-Khayyal

RELIABILITY MODELS WITH POSITIVE MEMORY DERIVED FROM THE MEAN RESIDUAL LIFE FUNCTION
Eginhard J. Muth

LEAST SQUARES ANALYSIS OF ACCELERATED LIFE TESTS FOR THE POWER RULE AND ARRHENIUS MODELS
Henry D. Kahn

RELIABILITY ANALYSIS OF REPARABLE THREE-STATE DEVICE MARKOV MODELS
Balbir Singh and C. L. Proctor

RELIABILITY ESTIMATION UNDER COMPETING CAUSES OF FAILURE
M. M. Desu and Subhash C. Narula

NUMERICAL COMPARISON OF SOME TRUNCATED SEQUENTIAL TEST PLANS FOR THE MEAN OF AN EXPONENTIAL DISTRIBUTION
Joseph A. Yahav and Lawrence S. Young

A MONTE CARLO COMPARISON OF THE METHOD-OF-MOMENTS TO THE MAXIMUM-LIKELIHOOD ESTIMATES OF WEIBULL PARAMETERS FOR CAS DATA
Frank Saylor

CONFIDENCE LIMITS FOR SYSTEM RELIABILITY WHEN TESTING TAKES PLACE AT THE COMPONENT LEVEL
Robert G. Dostal and Louis M. Iannuzzelli

ESTIMATION OF NON-LINEAR MODELS AND SPECIFICATIONS OF THE ERRORS
E. G. Charatsis

THE AIR FORCE RESEARCH PROGRAM - A NEW LOOK

by COLONEL RONALD A. BENA
Department of the Air Force

The Air Force research program which had been subjected to reduction in recent years under the pressure of decreasing DOD budgets and high costs of operational hardware was bolstered by a firm policy letter from the then Secretary of the Air Force, John L. McLucas, in October 1974. His points were: (1) Research is a fundamentally important part of the overall Air Force Research and Development program and the preservation of the quality of that program is of utmost importance; (2) Research funding should be protected from competition from development and production programs; (3) the FY 76 funding should be monitored or increased from the FY 75 level and future research budgets should be maintained at a level no lower than that percentage represented by the FY 75 level; (4) The emphasis of Air Force basic research should be shifted from predominantly in-house activities to outside university support.

Earlier in a letter to the Secretary of Defense, Senator Barry Goldwater expressed serious concern for the apparent erosion of basic research in the department. He indicated that while the Defense RDT&E appropriation increased from $7,730 million in FY 69 to $8,333 million in FY 74, the Defense Research Sciences budget dropped from $330 million to $276 million during the same period. He observed with concern that the majority of the reduction had been in contractual and grant programs. The Senator wrote that with respect to research, the DOD was turning inward and appeared to be shifting toward a predominantly in-house posture.

With the McLucas guidance and also the recommendation from a special study by the Air Force Systems Command (AFSC) on the utilization of Air Force laboratories, AFSC has repostured the management of its research program. The new structure, which is expected by AFSC to increase the research payoff, incorporates the key recommendations of the Secretary of the Air Force and should ease the concerns of Senator Goldwater. The principal managerial change was that the Director of the Air Force Office of Scientific Research (AFOSR) was designated as the single manager of the Air Force research program.

Under the new arrangement, the Director of AFOSR is responsible for the overall planning, management, implementation and control of a unified Air Force basic research program. AFOSR still reports to the Director of Science and Technology at AFSC Headquarters at Andrews AFB, Maryland. This change of organizational command lines places research into a separate and distinct category from development where it will not compete with development for its funds.

Additionally AFOSR assumes the task of interrelating research with other Department of Defense and government agencies as well as developing an integrated Air Force research plan. Tight coordination of research planning will be of mutual benefit to all agencies concerned and will assure that unwarranted duplication of effort does not exist.

In order to shore up the eroding university support, the ratio of in-house to out-house research is being changed from the FY 76 ratio of 40-60. This new program will devote 70 percent of its funding to an extramural contract/grant program with universities and industry. The remaining 30 percent will be dedicated to in-house programs at Air Force laboratories. The majority of the extramural program will be monitored directly by AFOSR; some programs, however, will be monitored by the respective laboratories.

Other major changes were the decentralization of the in-house program and the trend toward "full spectrum" (basic research through development) in the Air Force laboratories. The Aerospace Research Laboratories (ARL) at Wright-Patterson AFB Ohio were disestablished. Most of the ARL programs and science and engineering personnel were integrated into a new super organization, the Air Force Wright Aeronautical Laboratories. This organization was formed by pulling together the Air Force Aero-Propulsion Laboratory, and the Air Force Materials Laboratory.

The Air Force Cambridge Research Laboratories (AFCRL) which were a major component of the Air Force research system, have been divided into two groups. Both, however, will continue to remain at L G Hanscom AFB in Bedford, Massachusetts. One group has become a Detachment of the Rome Air Development Center (RADC). The remaining personnel have been restructured into an exploratory development laboratory - The Air Force Geophysics Laboratory (AFGL).

In a move to quantify the role and magnitude of research in the full spectrum of research and development in the laboratories, the AFSC Commander established that in FY 77 approximately 7% of the total manpower of the Air Force laboratories would be devoted to basic research. This figure will vary among the nine Air Force laboratories, but it is anticipated that all laboratories will have both in-house and contractual research programs which will be closely related to the technology needs and objectives of the individual laboratories. This in-house and contractual effort at the laboratories is an integral part of the overall single manager Air Force research program and is coupled with the contract and grant program at AFOSR.

Most of the Air Force research program will be designed to respond to the technology needs and objectives of the individual laboratories; however, a smaller portion of the program will provide for the exploitation of new scientific opportunities which have not yet been identified.

The new management responsibilities for AFOSR also include the assignment of two detachments to its organizational structure. The Frank J. Seiler Research Laboratory at the Air Force Academy, Colorado Springs, Colorado, and the European Office of Aerospace Research and Development, London, England, now report to AFOSR instead of HQ AFSC. Their missions have not been changed.

The Air Force laboratories have now been repostured to address research as well as development. They have thus become working partners with AFOSR conducting basic research tasks in areas important to their missions. This joint undertaking marks a significant point in the course of Air Force research.

THE BAYESIAN AND NONPARAMETRIC APPROACH TO RELIABILITY STUDIES: A SURVEY OF RECENT WORK

by I. Shimi
U. S. Air Force Office of Scientific Research
and C. P. Tsokos
University of South Florida

Abstract. The aim of the paper is a survey of recent work in the Bayesian and nonparametric approaches to reliability studies. Our presentation consists of four parts. The basic philosophy and preliminary concepts on reliability are given in section one. In section two we review some of the recent findings in the Bayesian approach to reliability. The empirical Bayes developments on the subject are given in section three. Finally, section four contains some of the fundamental thinking and recent developments in the nonparametric approach to reliability.

1.0 Introduction. The study of reliability has long been considered to be of major importance. Clearly, a knowledge of the failure behavior of a component can lead to savings in its costs of production and maintenance and, in some cases, to the preservation of human life.

Little observation is required, however, to determine that mechanical failures tend to occur quite randomly. Such an observation immediately leads to a conclusion that studies concerning reliability should be made from a probabilistic or statistical viewpoint. Within the field of reliability theory, such an approach is universally used.

The probabilistic failure behavior of a component or system is generally specified in terms of a function, say $f(t)$, known as the failure density. In fact, $f(t)$ is a probability density defined on the positive real axis (failures do not occur before the beginning of use), specifying the instantaneous probability of failure at time t, and written $f(t;\theta_1, \theta_2,\ldots,\theta_n)$, where

θ_i, $i=1,2,\ldots,n$ are the parameters of f. It can also be written $f(t|\underline{\theta})$, where $\underline{\theta} = (\theta_1,\theta_2,\ldots,\theta_n)$. A failure model, then consists of the specification of the functional form of f and the values of its parameters.

Corresponding to the density $f(t)$ is its distribution function,

$$F(t) = \int_0^t f(x)dx$$

which is the cumulative probability of failure up to time t. We are interested, however, in the probability that a component or system has not failed at time t. This function,

$$R(t) = 1 - F(t),$$

is known as the reliability function. Another function often considered is the hazard rate,

$$h(t) = \frac{f(t)}{F(t)},$$

which is just the failure rate conditioned on past survival. We can also define the expected time until failure assuming survival to time t given by

$$M(t) = \frac{1}{F(t)} \int_0^\infty xf(t+x)dx$$

and known as the mean residual life. Models specified and described in this way are called continuous failure models.

An alternative formulation is to consider a number of identical components or systems being tested up to a fixed time, say $t = t_o$, known as the mission length or mission time. In such a case, the reliability at time $t = t_o$ is a constant $R_{t_o} = R(t_o)$ and is a parameter of the probability distribution of the number of successes or failures in the test. Suppose n items are tested and k of them survive the mission time. Then we have

$$p(k) = p(k;n,R_{t_o}).$$

We call such a formulation a discrete failure model. In some cases, such models are specified in terms of the mean component life,

$$M_o = M(0),$$

in which case R_{t_o} is a function of the parameter M_o.

The choice of a functional form for $f(t)$ or $p(k)$ in the above models can be done in several ways. One can, for purposes of theoretical investigation, arbitrarily choose a classical, well-known probability density. Alternatively, a density can be derived from known physical properties of the component or system of interest. We note, however, e.g., see Mann, Schafer, and Singpurwalla [68], chapter 4, or Barlow and Proschan [8], that almost all derived distributions turn out to have classical forms. It is also possible to fit a density to existing data, but such a procedure presupposes some idea of the general form of the failure density. Thus we shall be concerned with models whose failure rate or probability of success obeys a known distribution.

Let us now consider several distributions commonly used in formulating discrete models. Perhaps the simplest is the binomial,

$$p(k;n,R_{t_o}) = \binom{n}{k} R_{t_o}^{k} (1-R_{t_o})^{n-k}, k=0,1,2,\ldots,n,$$

which characterizes the testing of n identical item to time t_o with a component reliability of R_{t_o} for that length of time. If n is very large and R_{t_o} is nearly equal to 1, then we may characterize the number of failures by the Poisson density,

$$p(k;\theta) = \frac{e^{-\theta}\theta^k}{k!}, \quad k \; 0,1,2,\ldots,$$

with θ given approximately by

$$\theta = n(1 - R_{t_o}).$$

An alternative use of the Poisson is to treat the hazard rate as a constant,

$$h(t) = h,$$

in which case the number of failures is characterized by a Poisson density whose parameter is proportional to the length of testing time. That is,

$$\theta = ht_o,$$

In either of the above cases, the reliability at time t_o is given by

$$R_{t_o} = e^{-\theta} .$$

Other distributions which have been used for discrete models include the geometric, hypergeometric, and negative-binomial.

Among the distributions used for continuous models, the most general is probably the Weibull [99] given by

$$f(t;\alpha,\beta,\gamma) = \frac{\beta}{\alpha} (t-\gamma)^{\beta-1} \exp -\left\{\frac{(t-\gamma)^{\beta}}{\alpha}\right\} ,$$

$$\gamma, \beta > 0, \gamma \geq 0, \qquad 0 \leq t < \infty$$

in which γ is the guarantee time (almost always set equal to zero), β is the shape parameter, and γ is the scale parameter. This density is most useful since it has the exponential $(\beta = 1)$ and Rayleigh $(\beta = 2)$ densities as special cases. We shall henceforth assume that γ equals zero. For the resulting two-parameter distribution, the reliability and hazard rate are given by

$$\bar{F}(t) = \exp -\left\{\frac{t^{\beta}}{\alpha}\right\}, \quad o \le t < \infty, \; o < \alpha, \; \beta$$

and

$$h(t) = \frac{\beta}{\alpha} t^{\beta-1}, \quad o \le t < \infty, \; o < \alpha, \; \beta.$$

Other distributions used in continuous failure models include the extreme-value,

$$f(t;\alpha,\gamma) = \left\{\frac{1}{\alpha} \exp \frac{1}{\alpha} (t-\gamma) - \exp\left[-\frac{1}{\alpha}(t-\gamma)\right]\right\}, \quad o \le t < \infty$$

$$\alpha > 0, \; -\infty < \gamma < \infty,$$

with location parameter γ and scale parameter α, the log-normal,

$$f(t;\alpha,\gamma) = \frac{1}{\sqrt{2\pi}\,\alpha t} \exp\left\{-\frac{1}{2}\left(\frac{\ln t-\gamma}{\alpha}\right)^2\right\}, \quad o \le t < \infty$$

$$\alpha > 0, \; -\infty < \gamma < \infty,$$

and the gamma,

$$f(t;\alpha,\beta) = \frac{t^{\beta-1}}{\alpha^{\beta}\Gamma(\beta)} \exp\left\{-\frac{t}{\alpha}\right\}, \quad o \le t < \infty, \quad o < \alpha, \; \beta.$$

Some authors have approached the study of reliability by classifying systems according to the qualitative behavior of their failure models. For example, Proschan [84] defines a system to be New Better than Used (NBU) if

$$\bar{F}(t_1 + t_2) \le \bar{F}(t_1)\bar{F}(t_2) \quad \text{for all } t_1, t_2 \ge o .$$

Hence, the replacement of an old component increases the reliability of the system. Similarly, he defines a New Better than Used in Expectation (NBUE) to be one for which

$$M(t) \le M(0) \quad \text{for all} \quad t \ge 0$$

such that $\bar{F}(t) > 0$, that is, for which the mean residual life of an old component is less than that of a new one. The duals of these, NWU (New Worse than Used) and NWUE, are easily defined. In addition, a failure rate or model is said to have Increasing Failure Rate (IFR) if $\bar{F}(x+t)/\bar{F}(t)$ is a decreasing function in t for all $x > 0$, Decreasing Mean Residual Life (DMRL) if $M(t)$ is decreasing in t, and Increasing Failure Rate Average (IFRA) if $\bar{F}(t)^{1/t}$ is decreasing for $t > 0$. In precisely the same way as previously, the duals, DFR, IMRL, and DFRA, can be defined. Haines and Singupurwalla [49], have defined a distribution to have Decreasing Percentile Residual Life (DPRL) if, for some α in $(0,1)$, the $(100\ \alpha)^{th}$ percentile of the residual life at time t is a decreasing function for $t \geq 0$.

Marshall and Proschan [74], have compared the NBU and NBUE systems under various maintenance policies, replacement at age, block replacement, replacement at failure only, by comparing the expected numbers of failures and replacements of functioning (still surviving) components. Ross [89], demonstrated that the time until the first failure of a coherent (built up by series and parallel circuits) systems of components with exponential failure rates has an NBU distribution.

Esary, Marshall, and Proschan [42], considered the behavior of systems subject to a Poisson shock process. They obtained results showing that if a system has a discrete NBU (NBUE) probability to failure is also NBU (NBUE). A-Hameed and Proschan [1], [2] have extended these results to more general shock processes.

Marshall and Proschan [74] and Esary, Marshall, and Proschan [41], have studied the question of which of the properties, NBU, NBUE, NWU, and NWUE, are preserved under various reliability operations, including formation of coherent systems, convolution (the distributions of sums of independent lifetimes), arbitrary mixture,

$$f(t) = \sum p_n f_n(t) \quad \text{and} \quad \sum p_n = 1$$

and mixtures of distributions that do not cross (F_1 and F_2 do not cross if there does not exist $t_1, t_2 \geq 0$ such that

$$F_1(t_1)-F_2(t_1) < 0$$

and

$$F_1(t_2)-F_2(t_2) > 0 .$$

Marshall and Proschan [74], [75] and Haines and Singpurwalla [48] have obtained bounds for $F(t)$ and $R(t)$ for NBU and NBUE failure models and series or parallel systems of independent NBU or NBUE components. An example of the bounds obtained is that an NBUE system satisfies $F(t) \leq t/M(t)$ for $t \leq M(t)$. For a series or parallel system of NBUE components:

$$M_s(0) \geq \left(\sum_{i=1}^{n} M_i(0)^{-1}\right)^{-1}$$

and

$$M_p(0) \leq \int_0^{\infty} [1 - \prod_{i=1}^{n} (1-e^{-t/M_i(0)})]dt.$$

These bounds are, in fact, obtained when the components have exponential failure distributions.

Most investigators, however, are interested in obtaining some type of point estimates or confidence bounds for the parameters of a failure model whose functional form has been specified and for which some life test data are available. There are four basic approaches to such problems: <u>classical</u> or <u>frequentist</u>, <u>standard Bayes</u>, <u>empirical Bayes</u> and <u>nonparametric</u>.

When complete life test data are available for properly chosen samples of components obeying a given failure law, the techniques for deriving classical unbiased, minimum variance (best), or consistent estimators are well established and can be found in any good references of statistics (e.g., Kendall and Stuart [57].) Unfortunately, cost and time limitations often lead to a severe censoring of data in one of two ways. The first (Type I censoring) occurs when the life test is stopped after a specific length of time and the second (Type II) occurs when the test stops after a specific number of failures have been observed. The classical approach is not of primary interest to us in the present study. However, we will briefly list some results for the construction of classical estimators based on censored samples drawn from probability models of interest to those in the field of reliability.

Epstein and Sobel [39], developed a maximum likelihood estimator for θ in a sample drawn from an exponential density and subject to Type II censoring. If n items are tested and m failure times, $t_1,\ldots,t_m$, are observed, then

$$\hat{\theta} = \frac{1}{m}\,(t_1+t_2+\ldots+t_m+(n-m)\,t_m)$$

is a maximum likelihood, unbiased, minimum variance estimator based on a sufficient statistic for θ. The results are generalized to the class of distributions of the form:

$$f(t;\ \theta) = \frac{g'(t)}{\theta}\exp\left\{-\frac{g(t)}{\theta}\right\},$$

in which case

$$\hat{\theta} = \frac{g(t_1)+g(t_2)+\ldots+g(t_m)+(n-m)\,g(t_m)}{m}$$

They later extended their results to an exponential density with a location parameter,

$$f(t;\theta,\gamma) = \frac{1}{\theta}\exp\left\{-\frac{(t-\gamma)}{\theta}\right\},\quad 0 \le t < \infty,\ t \ge \gamma.$$

Bartholomew [14], examined the same situation subject to Type I censoring and derived the sampling distribution the maximum likelihood estimator

$$\hat{\theta} = \frac{1}{k}\sum_{i=1}^{n}\,[a_i\,t_i + (1-a_i)T],$$

where k is the number of failures observed before the end of testing at time T and $a_i = 1$ if t_i, the failure time of the ith item, is less than T and $a_i = 0$, otherwise. He suggested the use of an alternate estimate

$$\hat{\theta}' = \frac{T}{-\ln(1-\frac{k}{n})}$$

which is simple and highly efficient.
Basu [13], gave minimum variance unbiased estimators of the reliability for Type II censored data in the one and two parameter exponential models. They are

$$\hat{R}(t) = \left\{ 1 - \frac{1}{\sum_{i=1}^{m} t_m + (n-m)t_m} \right\}^{m-1}$$

for the single parameter exponential and

$$\hat{R}(t) = \frac{n-1}{n}\left\{ 1 - \frac{t - t_1}{\sum_{i=1}^{m} t_m + (n-m)t_m + (n-1)t_1} \right\}^{m-2}$$

for the exponential with location parameter. These results generalize to the Weibull failure model in which case t^{β} has an exponential density. Sarkar [90], Lieberman and Ross [61], and Mann and Grubbs [67] have examined the problem of constructing system confidence bounds for series (cascade) systems whose subsystems have exponential failure rates for which Type II censored data is available.

Cohen [29], considered maximum likelihood estimators for the two-parameter Weibull model for censored data. He pointed out that the two types of censoring were different only in that the time of termination, t_T, was equal to T for Type I censoring and equal to t_m for Type II. Harter and Moore [51], studied maximum likelihood estimation from Type II data for the three-parameter gamma and Weibull models. In both papers, the iterative solution of a set of equations is required and the computation necessary may be quite large. Johns and Lieberman [55], developed exact asymptotically efficient confidence bounds for the two-parameter Weibull model with Type II censored data. The method, however, required the use of an extensive set of tables. Mann [65], developed another method, also requiring special tables, of using best linear invariant (BLI) estimators for the reliability in the Weibull model with Type II data. She demonstrated that the resulting estimators had a smaller expected loss than the best linear unbiased (BLU) estimators. D'Agostino [33] gave a modification of the technique used by Johns and Lieberman which did not require the use of special tables. Bain [3], obtained a best linear unbiased estimator of the scale parameter of an extreme-value distribution using Type II censored data. The results are applicable to the Weibull modell since the logarithms of Weibull deviates obey an extreme-value probability density law.

The remainder of this presentation will be devoted to a consideration fo the standard Bayes, empirical Bayes, and non-parametric approaches to reliability studies.

2.0 The Bayesian Approach To Reliability. In the Bayes approach, the parameter to be estimated, say θ in the failure model $f(t;\theta)$, is presumed to have associated with it a weighting function, $g(\theta)$, which may or may not represent a probability density. If it does not, it is called an improper prior. The use of such a prior function follows naturally either from the assumption that θ is, in fact, a random variable or from the desire to more efficiently use data already available. Drake [35], Evans [43], Feduccia and Klion [44], and Crelin [31] have, among others, presented the arguments which are in favor of the application of Bayesian techniques to reliability problems.

The investigator who seeks to use such methods usually has available the functional forms of $f(t;\theta)$ and $g(\theta)$ and a set of data, $\underline{t} = (t_1, t_2, \ldots, t_n)$, drawn from the distribution f. Generally, he wishes to obtain an estimate, $\hat{\theta}(\underline{t})$, or a confidence interval, $(\hat{\theta}_L(\underline{t}), \hat{\theta}_U(\underline{t}))$, which is functionally determined by the information in his possession. If the density, f, is treated as a conditional probability $(f(t;\theta) \equiv f(t|\theta))$ then the joint density may be obtained by

$$f(t,\theta) = f(t;\theta)g(\theta).$$

Hence, the posterior density of θ given t is given by

$$h(\theta_o|t) = \frac{f(t;\theta_o)g(\theta_o)}{\int_\Omega f(t;\theta)g(\theta)d\theta}, \quad \theta\varepsilon\Omega, \text{ a well-defined set,}$$

which is a result of Bayes Law. If we wish to condition θ on $\underline{t}$ then we have

$$h(\theta_o|\underline{t}) = \frac{\ell(\underline{t};\theta_o)g(\theta_o)}{\int_\Omega \ell(\underline{t};\theta)g(\theta)d\theta}, \quad \theta \varepsilon \Omega$$

where $\ell(\underline{t};\theta)$ is the likelihood function of $\underline{t}$. For independent sample failure times,

$$\ell(\underline{t};\theta) = f(t_1;\theta)f(t_2;\theta)\ldots f(t_n;\theta).$$

The problem that now arises is what point in the distribution of θ given $\underline{t}$ we shall use as our estimate. This is dealt with by the specification of a <u>loss function</u>, $L(\hat{\theta},\theta)$, which gives the investigator's measure of the importance of the error resulting from the specification of $\hat{\theta}$ as the estimate when, in fact, θ is the true value.

The specification of $L(\hat{\theta},\theta)$ would seem to be entirely arbitrary. However, there is a well-known loss function which yields a simple answer to our estimation problem. That is, the <u>squared error</u> or <u>quadratic</u> loss function,

$$L(\hat{\theta},\theta) = (\hat{\theta}-\theta)^2 .$$

The motivation for such a loss function is that as long as the error is reasonable, the losses are of the same magnitude for both high and low estimates. The loss becomes large when the estimate is grossly off the true value. It also offers nice analytical tractibility. With this type of loss function, the Bayes estimator, defined as taht value of $\hat{\theta}$ which minimizes the quantity

$$E(L|\underline{t}) = \int_{\Omega} L(\hat{\theta},\theta)h(\theta|\underline{t})d\theta, \quad \theta\varepsilon\Omega$$

is given by

$$\hat{\theta} = \int_{\Omega} \theta h(\theta|\underline{t})d\theta,$$

that is, the mean of the posterior density. Other loss functions have been considered by some authors.

Canfield [27], in a paper dealing with the estimation of reliability in the exponential failure model, uses the assumption that over-estimation of $R(t)$ is worse than underestimation to justify a loss function of the form

$$L(\hat{R},R) = \begin{cases} k_1 \left(\frac{\hat{R}}{R} - 1\right)^2 & \text{if } \hat{R} \leq R \\ k_1 \left(\frac{\hat{R}}{R} - 1\right)^2 + k_2 \left(\frac{\hat{R}}{R} - 1\right) & \text{if } \hat{R} > R \end{cases}$$

He uses a beta prior

$$g(R;p,q) = \frac{1}{\beta(p,q)} R^{p-1} (1-R)^{q-1}, \quad R\varepsilon(0,1),$$

which is a natural conjugate of the exponential model (the posterior density is also beta). He concludes that when no prior information is available ($p=q=1$ implying g is uniform) and a symmetric loss is used ($k_2 = 0$), then the resulting estimator is the minimum variance unbiased estimator of the reliability,

$$R(t) = e^{-\theta t}$$

The remainder of this section will be a description of some recent results in the field of reliability which have been obtain obtained using standard Bayesian methods.

Springer and Thompson [94], consider a series system of components or subsystems, each of which exhibits a binomial failure model, and for each of which the prior density is of the beta form (the natural conjugate to the binomial). They construct Bayesian confidence bounds for the product of the binomial parameters (the product of the reliabilities is the reliability of the system) by deriving the posterior density of the product using Mellin transform techniques. Papadapoulos and Tsokos [80], derived estimates for the reliability of a binomial failure model when the prior density is either uniform or beta. In addition, the first two moments of the estimators were given and the use of game theory techniques for estimation were investigated.

Pugh [85], examined the applicability of Bayesian methods to the development of confidence bounds for the reliability of an exponential failure model with uniform prior. For a life test until K failures and a mission time t_m he obtains the result

$$P[R>R_o] = 1 - \sum_{\nu=o}^{k} \frac{R^{(t_k/t_m)+1}}{\nu!} [-\ln R^{(t_R/t_m)+1}]^{\nu} .$$

El-Sayyad [36], considered a large number of estimators for the parameter of an exponential density assuming a prior of the form

$$g(\theta) = \theta^{a-1} e^{-b\theta}, \quad o \leq \theta \quad \infty \quad , o < a, b.$$

If the loss function is of the form

$$L(\hat{\theta},\theta) = \theta^{\alpha} \; (\hat{\theta}^{\beta} - \theta^{\beta})^2 \; ,$$

then the estimate of θ is

$$\hat{\theta} = \frac{1}{b+S}\left\{\frac{\Gamma(a+n+\alpha+\beta)}{\Gamma(a+n+\alpha)}\right\}^{1/\beta}$$

where $S = \sum^{n} t_k$ is the sum of the failure times from a sample of size n. If, however, the loss function is

$$L(\hat{\theta},\theta) = (\ln\hat{\theta} - \ln\theta)^2 ,$$

then

$$\hat{\theta} = \frac{\phi(a+n)}{b+S}$$

where

$$\phi(n) = \exp\left\{\frac{d}{dn} \ln \Gamma(n)\right\} \xrightarrow[n\to\infty]{} (\ln n - \frac{n}{2})$$

Bhattacharya [18], made a Bayesian analysis of the exponential model for data subject to either type of censoring. He considered the uniform, inverted gamma, and exponential densities as prior for θ. It was also observed that if the improper prior, $g(\theta) = 1/\theta^2$, is used, then the resulting Bayes estimate coincides with the minimum variance unbiased estimate (MVUE) of Epstein and Sobel [37]. Finally, a proof was given showing that this is the only prior for which this is true.

Springer and Thompson [94], [95], [96] have developed Bayesian confidence bounds for series and parallel systems of components with exponential failure times and for series systems some of whose components are known to have exponential failure rates and for some of which no failure rate information is known. They use a prior of the form

$$g(R) = \frac{(\beta_o+1)^{r_o+1}}{\Gamma(r_o+1)} R^{\beta_o} \left(\ln \frac{1}{R}\right)^{r_o}, \quad R = e^{-\frac{t_m}{\theta}} ,$$

and utilizing Mellin transform techniques to derive the posterior density of the system reliability

$$R_s = R_1 R_2 \ldots R_n$$

or

$$R_p = 1 - \prod_{i=1}^{n} (1-R_i).$$

Basically, their results involve a fast way of writing down the partial fraction expansion for the Mellin transform of the system posterior density of R. Fertig [48], attempted to determine the optimal component priors for the above priors with optimal defined in the sense that the resulting confidence intervals are as close as possible to the uniformly most accurate (UMA) unbiased confidence intervals, see Ferguson [47], and Lehmann [59]. He concluded that the prior used by Springer and Thompson was the best when prior data was available, but that no prior was optimal when no prior information existed $(\beta_o = r_o = 0)$. Mann [66], used Monte Carlo computer techniques to obtain the values of r_o and β_o for each component so that the resulting system prior yielded Bayesian confidence intervals that were optimal in the same sense. Results similar to Fertig's were obtained showing that no choice exists, independent of previous data, which results in optimal confidence intervals.

El-Sayyad [36], applied a method of sequential testing proposed by Lindley [64], to the exponential model. The method begins with a measure of information suggested by Shannon which estimates the amount of knowledge about θ which we have in $g(\theta)$:

$$I = \int_\Omega g(\theta) \ \ln g(\theta) d\theta$$

When additional data $\underline{x} = (x_1, x_2, \ldots, x_n)$ has been obtained we have a new amount of information,

$$I_n = \int_\Omega h(\theta|\underline{x}) \ \ln h(\theta|\underline{x}) \ d\theta,$$

which uses the posterior density. We may then treat the posterior as prior and obtain more information. The method says, in effect, collect information until I_n reaches a certain value. Furthermore, the method is generalizable by choosing a monotone function $\phi(\theta)$ of θ and defining

$$I_\phi = I_\theta + \int_\Omega g(\theta) \ln \frac{d\phi}{d\theta} \, d\theta.$$

Clearly, the choice of ϕ will have a strong effect on the form of the posterior from which we finally make a Bayesian inference. El-Sayyad uses

$$g(\theta) = \frac{b^a}{\Gamma(a)} \theta^{a-1} e^{-b\theta}, \quad o \leq \theta < \infty, \ o < a, b$$

for the prior of θ, form which

$$I_\theta = a \ln b - \ln \Gamma(a) - a + (a-1)[\phi(a) - \ln b]$$

$$\simeq \frac{1}{2} \ln \frac{b^2}{2\Pi a} - \frac{a+1}{2a}.$$

The ϕ here is the derivative of $\ln \Gamma(a)$, as before. He then considers several forms of $\phi(\theta)$: $\phi(\theta) = \sqrt{\theta}$, for which

$$I_\phi \simeq \frac{1}{2} \ln b + \ln 2 - \ln \sqrt{2\Pi} - \frac{1}{2} + \frac{1}{4a};$$

$\phi(\theta) = \ln \theta$, for which

$$I_\phi \simeq \frac{1}{2} \ln \frac{a}{2\Pi} - \frac{1}{2};$$

$\phi(\theta) = 1/\theta$, for which

$$I_\phi = I_\theta + 2E(\ln\theta)$$

$$\simeq \frac{1}{2} \ln \left(\frac{a}{2\Pi b^2}\right) - \frac{(a+1)a}{2}.$$

Deely, Tierney, and Zimmer [34], have considered the use of the "Maximum Entropy Principle" (MEP) in choosing a prior for the binomial and exponential models. The principle defines the optimal choice to be that prior which, within the constraints of available data, maximizes the entropy of the prior density of

$$S_\theta = -I_\theta,$$

that is, which minimizes the presumed measure of information available about θ. They introduce the notion of a "least favorable distribution." If G is a family of prior distributions, then G_o is least favorable over G if its minimum expected loss (the loss expected for the Bayes estimator) is greater than that of any other member of G. Under the assumption of a quadratic loss, it is shown that a least favorable distribution may be a better choice of prior than the MEP choice.

The difficulties arising from the MEP were approached in a different manner by Jaynes [54], who suggested that the principle could be generalized by consideration of the measure,

$$S_\phi = - \int_\Omega g(\theta) \ln [\frac{g(\theta)}{\phi(\theta)}] \, d\theta,$$

where ϕ is, as El-Sayyad [36], noted, an appropriate monotone function of θ. He further suggested that the choice of $g(\theta)$ should be made using transformation group techniques. Essentially, such an approach says to choose a g in such a way that changes in the parameters of $g(\theta)$ do not change the entropy measure. For example, if θ_1, is a location parameter and θ_2 is a scale parameter, then we seek a prior $g(\theta_1,\theta_2)$ such that

$$g(\theta_1,\theta_2) = a\, g\, (\theta_1 + b,\, a\theta_2).$$

The solution is of the form

$$g(\theta_1,\theta_2) - \frac{\text{constant}}{\theta_2} .$$

In a similar way, it is argued that the binomial model leads to the equation

$$\theta(1-\theta)g'(\theta) = (2\theta-1)g(\theta),$$

whose solution is

$$g(\theta) = \frac{\text{constant}}{\theta(1-\theta)}$$

Harris and Singpurwalla [50] dealt with the application of Bayesian estimation methods to failure models with hazard rates of the form

$$h(t) = \theta\alpha t^{\alpha-1}, \ 0 \leq t < \infty, \ 0 < \theta, \ \alpha$$

when complete samples are available. This hazard rate corresponds to the Weibull model and becomes the exponential when $\alpha = 1$. They derived estimators for θ in the exponential and Weibull cases with the uniform, two-point, and gamma priors. They also investigated the estimation of α where θ is known but obtained a solution only for the two-point prior for α. Soland [92] determined a Bayesian acceptance sampling procedure for θ when α is known for the Weibull model and data subject to Type II censoring (m failures out of n items tested simultaneously), using a gamma prior. Later, Soland [93], extended the results to the case in which both α and θ are unknown. Bury [23], developed a Bayesian decision process for the hazard rate of a Weibull failure model with a conjugate prior. The results permit the optimization of sequential, simple item sampling schemes, such as might be used with high cost equipment.

Canavos and Tsokos [27], constructed a Bayesian analysis of the scale and shape parameters and the reliability function of the Weibull model for the uniform, exponential, and inverted gamma prior densities. The results are a direct generalization of the work of Bhattacharya in which the one-parameter exponential model was concerned. The two parameter case was analyzed under the assumption that the parameters were independently distributed. The estimates and their variance for Type II censored date were obtained for the parameters and the reliability function, $R(t) = e^{-t^{\alpha}/\theta}$, and were compared with the corresponding minimum variance unbiased or maximum likelihood estimator by an analysis of the relative mean squared errors (MSE). The Bayesian estimates were found to have uniformly smaller MSE. Tsokos [98], continued this work by studying the effect of the choice of a wrong prior on the MSE of the estimators by use of Monte Carlo simulation. Papadopoulos and Tsokos [79] derived upper and lower Bayes confidence bounds for scale parameter and lower bounds for the reliability function of the **Weibull** model for the inverted gamma and uniform priors. The effect of a wrong choice between the two priors was also studied. It was found that a wrong choice still yielded results with a lower MSR than the classical maximum likelihood (ML) estimator.

Tsokos and Rao [97], studied the robustness of the Bayes estimators of the scale parameter, reliability function, and hazard function of the Weibull model with respect to a wrong choice of the prior distribution of the scale parameter. The average MSE of the resulting estimates were compared with those for the natural conjugate (inverted gamma) prior with the parameters of the priors chosen so as to make their first two moments equal. The beta, Poisson, inverted gamma, truncated normal, log-normal, and extreme-value priors were studied. It was concluded that the effects of higher prior moments on the results make it necessary for the investigator to be very careful in his choice of the prior.

Papadopoulos and Tsokos [81] have extended the work of Soland [92] by deriving a Bayes reliability estimator for the case in which both the scale and shape parameters are unknown. The result was compared with the MLE for the reliability using numerical computer simulation techniques.

3.0 The Empirical Bayes Approach To Reliability. The empirical Bayes approach to estimation, as first proposed and later developed by Robbins [86], [87], is an extension of the standard Bayesian technique in that it uses prior information, usually a set of previous estimates, to construct a prior distribution rather than specifying an arbitrary density function.

Assume that a set of samples of failure times with their associated classical estimators of the parameter distribution are available, say

$$(t_1, \hat{\theta}_1), (\underline{t}_2, \hat{\theta}_2), \ldots, (\underline{t}_k, \hat{\theta}_k).$$

Assume also that the prior distribution, $G(\theta)$, and the failure model, $f(t|\theta)$, are known. Then the Bayes solution for a quadratic loss can be written

$$w_g(\underline{t}) = \frac{\int_\Omega \theta \ell(\underline{t}|\theta) g(\theta) d\theta}{\int_\Omega \ell(\underline{t}|\theta) g(t) d\theta}$$

$$= \int_\Omega \theta h(\theta|\underline{t}) d\theta,$$

where Ω is a well defined set and $\ell(\underline{t}|\theta)$ is the likelihood function of the sample $\underline{t} = (t_1, t_2, \ldots, t_n)$ and $h(\theta|\underline{t})$ is the posterior probability density of θ.

If a sufficient estimator exists for θ, then, using the Fisher-Neyman factorization theorem, we may write the likelihood function as

$$\ell(\underline{t}|\theta) = q^*(\hat{\theta}(\underline{t})|\theta) y(\underline{t}),$$

where $\hat{\theta}(\underline{t})$ is the sufficient estimator, q^* is its conditional probability density function, and y is a function of $\underline{t}$ alone. Thus, $w_g(\underline{t})$ may be rewritten as

$$w_g(\underline{t}) = \frac{\int_\Omega \theta q^*(\hat{\theta}|\theta) g(\theta) d\theta}{\int_\Omega q^*(\hat{\theta}|\theta) g(\theta) d\theta}.$$

The usual approach is to construct a sequence of such Bayes solutions $w_1(\underline{t}), w_2(\underline{t}), \ldots, w_k(\underline{t})$, each depending on an increasing amount of previous information. We call this sequence of estimators consistent if

$$\lim_{k\to\infty} w_k(\underline{t}) = w_G(\underline{t})$$

with probability 1. If the condition,

$$\lim_{k\to\infty} R[G, w_k(\underline{t})] = R[G, w_G(\underline{t})],$$

holds with probability 1 where $R[\cdot,\cdot]$ is the Bayes risk, then the estimates $w_k(\underline{t})$ are said to be asymptotically optimal relative to $G(\theta)$.

Two approaches exist for the construction of the estimates $w_k(\underline{t})$. One seeks to determine a value for $w_k(\underline{t})$ without first approximating the density $g(\theta)$. Robbins [86], [87] was able to provide a set of consistent estimators which, when the loss function was bounded in the parameter, were also asymptotically optimal.

Samuel [91] developed a set of optimal empirical Bayes rules, based on the estimators considered by Robbins, by which a simple hypothesis could be tested. Using a limit process, he extended the applicability of these rules to the testing of one- and two-sided composite hypotheses about the parameter of a discrete failure model. The resulting rules were specified for the Poisson, geometric, negative binomial, and binomial models. A further limiting process extended the rules to the testing of hypotheses concerning the parameters of a continuous distribution which is of the exponential family.

The second approach to the development of empirical Bayes rules requires the approximation of the prior distribution of the parameter, $g(\theta)$. This method was first introduced by Maritz [69], [70], [71], who suggested the use of a step function to approximate $g(\theta)$. In such a case, the smoothed empirical Bayes estimator, $w_k^*(\underline{t})$, is given by

$$w_k^*(\underline{t}) = \frac{\sum_{j=1}^{k} \theta_j q^*(\hat{\theta}_k(\underline{t})|\theta_j)}{\sum_{j=1}^{k} q^*(\hat{\theta}_k(\underline{t})|\theta_j)} .$$

He investigated the resulting mean square errors obtained by use of the classical best (minimum variance) simple empirical Bayes (Robbins type), and smooth empirical Bayes estimators and determined that the simple empirical Bayes Approach required too much prior information to have a MSE as small as the best estimator, whereas the smooth empirical Bayes estimator required very little. Similar results were obtained by Lemon and Krutchkoff [61], who demostrated that, for a discrete true prior, the smooth empirical Bayes estimates were superior to the best estimator in general when the resulting MSE's were compared for small samples. Lemon [62] extended this conclusion to several continuous densities, including the exponential and normal probability distributions.

Bennett [15] and Bennett and Martz [16], [17] suggested a further improvement by the use of a continuously smooth approximation to the prior, $g(\theta)$. The resulting continuously smooth empirical Bayes estimator is given by

$$w_k'(\underline{t}) = \frac{\sum_{j=1}^{k} \int_{\Omega} \theta q^*(\hat{\theta}_k(\underline{t})|\theta) g_k(\theta)d\theta}{\sum_{j=1}^{k} \int_{\Omega} q^*(\hat{\theta}_k(\underline{t})|\theta) g_k(\theta)d\theta} .$$

Their approximation to $g(\theta)$ was based on the class of consistent estimators of a probability density function which were proposed by Parzen [82]. In essence, these approximation are weighted averages of the form

$$g_k(\hat{\theta}) = \frac{1}{k\delta(k)} \sum_{j=1}^{k} V \frac{\hat{\theta}_k - \hat{\theta}_{k,j}}{\delta(k)} ,$$

where

$$\lim_{k \to \infty} \delta(k) = 0,$$

$$\lim_{k \to \infty} k\delta(k) = \infty,$$

$$\int |V(y)| dy < + \infty,$$

and

$$\int |V(y) dy = 1.$$

The weighting function, V, is usually even as required by Parzen and has many possible forms. Two of the most often used are the exponential,

$$V(y) = \tfrac{1}{2} e^{-|y|} ,$$

and the trigonometric form,

$$V(y) = \frac{1}{2\pi} \frac{\sin^2(y/2)}{(y/2)} .$$

Bennett and Martz, using Monte Carlo procedures, demonstrated that these estimators, when applied to a Poisson failure model, have a smaller MSE than the conventional minimum variance estimators. Canavos [26], also using Monte Carlo techniques and a Poisson model, obtained similar results with respect to conventional maximum likelihood estimators.

Canavos and Tsokos [24], developed empirical Bayes estimators for the scale parameter and reliability function of the two-parameter Weibull failure model with a shape-parameter that is assumed known. The estimator for the scale parameter which was developed was a continuously smoothed type with the prior approximated by a Parzen-type expression using the exponential weighting function and a previous sequence of sufficient estimates.

Canavos and Tsokos [25] developed an empirical Bayes estimate for the random scale parameter in a gamma failure model with known shape parameter. The technique used was similar to that of their work on the Weibull model. A Monte Carlo simulation was carried out to compare the mean square errors (MSE's) of the classical Bayes, empirical Bayes, and conventional minimum variance unbiased estimators. The classical Bayes estimators were found to have the smallest MSE and, for moderate or large sample sizes, the empirical Bayes estimators had a smaller MSE than the conventional ones. The main advantage of the empirical Bayes approach, of course, is the lack of the necessity of specifying the form of the true prior. Nichols and Tsokos [77] have obtained consistent sequences of empirical Bayes point estimators for four broad families of probability distribution.

Couture and Martz [30] used continuously smoothed empirical Bayes estimation procedures for the determination of the stochastic scale parameters of a two-parameter Weibull distribution with known shape parameter and for the simultaneous determination of stochastic scale and shape parameters. Their approximation of the prior was based on a sequence of previously available maximum likelihood estimates of the parameter. It was shown, using Monte Carlo simulation, that the resulting estimators, for a true prior in the Pearson family of distributions, yielded a much smaller MSE than the conventional maximum likelihood estimators.

4.0 The Nonparametric Approach. A useful approach in reliability studies, developed in the last ten years or so, is to classify the life distributions for components or systems according to the qualitative behavior of their failure modes. This approach leads, in many cases, to a more practical, easier to assume and justify, reliability model. These are what have been called the nonparametric models in reliability studies. The most rewarding feature of these nonparametric models is that we do not lose much, in terms of answering the typical questions of concern in reliability studies, as compared to the more restrictive parametric models. Besides the obvious inherent advantages of the nonparametric methods, we also have the nice property of relative insensitivity to outliers in the data, superior power properties for a wide class of alternative distributions, and using the test statistics we can obtain estimators and distribution free confidence intervals for the population parameters of interest.

The present **mainstream activities in nonparametric reliability** studies started as a result of the classification of life distributions according to the monotonicity properties of the failure rate, the average failure rate, the mean residual life, the conditional survival probability of a used unit compared to the survival probability of a new unit, and the conditional mean residual life for a used unit compared to the mean residual life for a new unit.

The first systematic treatment of properties of distributions with monotone failure rate was given by Barlow, Marshall, and Proschan [4]. Another paper that presents a number of different notions of aging is by Bryson and Siddiqui [22]. What we would like to present in this section is the definitions of these classes and some of the implications and properties of these classes.

(1) The life distribution for the reliability function $\bar{F}(t) = 1-F(t)$ is said to have increasing failure rate (IFR) if

$$\bar{F}(x|t) \overset{\text{def}}{=} \frac{\bar{F}(x+t)}{\bar{F}(t)}$$

is decreasing in t for all $x > 0$.

An equivalent definition is, F is IFR if and only if $\log \bar{F}$ is concave. If F is absolutely continuous and f is a density for F, then F is IFR is equivalent to

$$h(t) = \frac{f(t)}{\bar{F}(t)}$$

increasing in t. This notion of increasing failure rate is equivalent to stating that the residual life length of an unfailed item of age t is stochastically decreasing in t. Thus IFR distributions correspond to a class of distributions which are affected adversely by aging. If it is supposed that the aging is beneficial to the device, we are led to consider the class of decreasing failure rate (DFR) distributions. This class is defined in terms of $\bar{F}(x\ t)$ increasing in $t \geq 0$ for all $x > 0$.

It can be very easily shown that the Weibull distribution with shape parameter $\alpha \geq 1$ is an IFR distribution, and a DFR distribution for $0 < \alpha < 1$. Similarly for the gamma distribution with density

$$g(t) = \beta^{\alpha} t^{\alpha-1} \rho^{-\beta t}/\Gamma(\alpha),$$

for $t \geq 0$, where $\alpha, \beta > 0$, it can be shown that this is an IFR distribution if $\alpha \geq 1$ and a DFR distribution if $0 < \alpha \leq 1$. The log normal distribution is also an IFR distribution.

Some of the most useful properties of IFR distributions will now be given. For references see Barlow and Proschan [10].

a) If F is IFR and $F(x) < 1$, then F is absolutely continuous on $(-\infty, x)$, that is, there exists a density for F on $(-\infty, x)$. There might exist a jump in F at the right hand end point of its interval of support when this interval is finite. This property was used effectively by Crow and Shimi [32].

b) If a density function f is a Polya frequency function of order 2 (PF_2) then the corresponding distribution F is IFR. See Karlin [56] for a definition of PF_2.

c) Coherent systems of IFR components need not have IFR distributions. This was first shown in Esary and Proschan [40].

This last property of IFR distributions leads to the search for the smallest class of distributions which gives the life distribution of coherent systems of IFR components. This class is defined as follows.

(2) A life distribution F is said to be increasing failure rate average (IFRA) if

$$[\bar{F}(t)]^{1/t}$$

is decreasing in $t > 0$.

This is equivalent to saying that F is IFRA if and only if

$$\overline{F}(\alpha t) \geq \overline{F}^{\alpha}(t) \quad \text{for} \quad 0 < \alpha < 1 \quad \text{and} \quad t \geq 0.$$

It is also equivalent to saying F is IFRA if $-\log \overline{F}(t)$ is increasing in $t \geq 0$.

It turns out that IFR distributions form a subclass of the IFRA distributions. Also, coherent systems of IFRA components have an IFRA distribution, Ross [88]. The IFRA class of distributions is the smallest class containing the exponential distributions, closed under formation of coherent systems and taking limits in distribution. These properties and some other useful characterizations of IFRA distributions were given in Birnbaum, Esay and Marshall [21]. This class has been subsequently investigated by Barlow and Marshall [9], and Barlow [12].

Beside the fact that IFRA life distributions arise naturally when coherent systems are formed from components with independent IFR distributions, it has also been shown that the IFRA distributions also arise naturally when one considers cumulative damage shock models. These shock models can be described as follows. Shocks occur in time according to some process, each independently causing random damage to a device, the damages accumulate until a certain tolerance level is exceeded, causing the device to fail. The time of failure was shown to be a random variable with an IFRA distribution. For details see Esary, Marshall, and Proschan [42]. They assumed that the shocks occur according to a homogeneous Poisson process and discovered several properties for the survival probability $\overline{F}$ from properties of the sequence of probabilities giving the probability of surviving k shocks. This has been generalized to the case of a nonhomogeneous Poisson process by A-Hameed and Proschan [7], and to the case of a sequence of shocks occurring randomly in time according to a nonstationary pure birth process by A-Hameed and Proschan [2].

Other classes of distributions are defined next.

(3) F is a decreasing mean residual life (DMRL) distribution if

$$\int_0^\infty \overline{F}(x+t)\,dx/\overline{F}(t)$$

is decreasing in t.

This is equivalent to saying that the residual life of an unfailed item of age t has a mean that is decreasing in t.

(4) F is a new better than used (NBU) distribution if

$$\overline{F}(x)\overline{F}(t) \geq \overline{F}(t + x)$$

for all $x, t \geq 0$.

This is equivalent to saying that the life length of a new item is stochastically greater than the residual life length of an unfailed item of age t.

(5) F is a new better than used in expectation (NBUE) distribution if

$$\int_0^\infty \overline{F}(x)\,dx \geq \int_0^\infty \overline{F}(t + x)\,dx/\overline{F}(t) \quad \text{for all} \quad t \geq 0.$$

This is equivalent to saying that the expected life length of a new item is greater than the expected residual life of an unfailed item of age t.

The following implications are easily checked (Bryson and Siddiqui [22]:

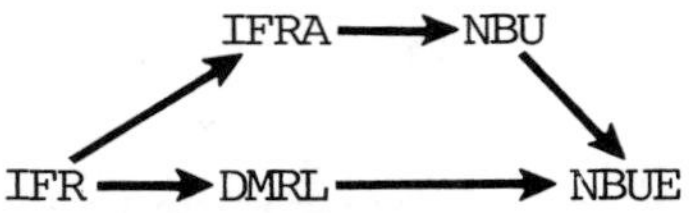

No additional implications exist among these classes of distributions.

If in definitions 1-5 above we reverse the direction of the inequality and make whatever was said to increase decrease and vice versa we obtain new classes of distributions characterized by the properties implied by their names. These are:

(1) decreasing failure rate IDFR) distributions

(2) decreasing failure rate average (DFRA) distributions

(3) increasing mean residual life (IMRL) distributions

(4) new worse than used (NWU) distributions

(5) new worse than used in expectation (NWUE) distributions.

As we noted before, the formation of coherent systems of independent IFRA components preserved the IFRA property, yet if the components were IFR the coherent system need not have been IFR. This lead to considering which of the above classes of distributions is preserved under the different reliability operations. The reliability operations of interest are: (a) Formation of coherent systems, (b) Addition of life lengths, (c) Mixture of distributions.

The following table gives a summary of the results concerning preservation of life distributions under the appropriate reliability operations.

	REALIABILITY OPERATIONS		
LIFE DISTRIBUTION CLASS	FORMULATION OF COHERENT SYSTEMS	ADDITION OF LIFE DISTRIBUTIONS	MIXTURE OF DISTRIBUTIONS
IFR	Not preserved	Preserved	Not preserved
IFRA	Preserved	**Preserved**	Not preserved
DMRL	Not known	Not known	Not known
NBU	Preserved	Preserved	Not preserved
NBUE	Not preserved	Preserved	Not preserved
DFR	Not preserved	Not preserved	Preserved
DFRA	Not preserved	Not preserved	Preserved
IMRL	Not known	Not known	Not known
NWU	Not preserved	Not preserved	Not preserved
NWUE	Not preserved	Not preserved	Not preserved

The results related to the preservation properties of IFR and DFR under the appropriate reliability operations were presented for the first time in Barlow, Marshall and Proschan [4]. For preservation under mixtures for DFR and DFRA the reader is referred to Esary, Marshall and Proschan [41]. The NBU, NWU, NBUE and NWUE classes are discussed and their applications in maintenance policies are described in Marshall and Proschan [74].

For the different bounds on reliability and other parameters, assuming the underlying life distribution is IFR, IFRA, DFR or DFRA and assuming one moment or one percentile to be known, the reader is referred to Barlow and Marshall [5], Barlow [6], Barlow and Marshall [7] and [9]. For the preservation of NBU under formation of coherent structures see Esary, Marshall and Proschan [41]. All these results are now presented in the excellent new book by Barlow and Proschan [10].

Now we turn our attention to inference problems related to the above classes. Thy hypothesis testing procedures will be discussed first and then some of the estimation problems will be presented. We notice that, until very recently, much less attention has been devoted to statistical inference for the above classes as compared to the work related to their probabilistic properties.

The IFR class has received the greatest amount of study from the inferential point of view. Although this class was around for some time the first attempt in terms of testing a hypothesis appeared in (1967). Testing for constant versus monotone increasing failure rate based on the ranks of the normalized spacing between the ordered observations was considered by Proschan and Pyke [83]. The asymptotic normality of these statistics was proved for fixed alternatives, and they computed the ratio of the efficacies of one of their rank tests to the best statistics for the Gamma and Weibull alternatives. Bickel and Doksum [20] proved that the asymptotic normality proved by Proschan and Pyke for fixed alternative also holds for sequences of alternatives F_{θ_n} that approach the H_0 distribution $1 - \exp(-\lambda t),\ t \geq 0$, as $n \to \infty$. They also proved that the above mentioned ratios of efficacies are in fact Pitman efficiencies and that the test statistic $T_1 = \Sigma i R_i$, where $R_1, \ldots, R_n$ are the ranks of the normalized spacings, is asymptotically equivalent to the Proschan-Pyke statistic. By determing the Pitman efficiency of T_1 with respect to $T_2 = \Sigma i \log [1-R/(n+1)]$ for all sequences of alternatives F_{θ_n}, they proved that T_1 is asymptotically inadmissible. This paper also investigated properties of the power of tests based on statistics that are linear in the normalized spacings when λ is known and when it is unknown. Unbiasedness was shown for tests that are monotone in the normalized spacings. Comparisons with the likelihood ration tests presented by Barlow [12] were made using Monte Carlo power estimates.

Let

$$X_{(1)} \leq \cdots \leq X_{(n)}$$

be the order statistics of the sample, and let

$$D_i = (n-i+1)\ (X_{(i)} - X_{(i-i)}),\ 1 \leq i \leq n$$

be the rank of D_i among $D_i, \ldots, D_n$. If the null hypothesis to be tested is that F is negative exponential with unknown parameter, a level α tests based on statistics of the form

$$\sum_{i=1}^{n} h_n(X_i/\ X_i)$$

(see Moran [76]), linear functions of $(D/\Sigma D_i, \ldots, D_n/\Sigma D_i)$ (see Barlow [12]), Lewis [13], locally most powerful tests based on $(R_1,\ldots,R_n)$ (see Bickel and Doksum [20]), and linear functions of

$$[-\log(1-R_1/(n+1)),\ \ldots,\ -\log\ (1-R_n/(n+1))]$$

were shown by Bickel [19] to be similar tests for the null hypothesis; under some regularity conditions each one of these classes contains a test asymptotically equivalent to the asymptotically most powerful similar test. In Bickel and

Doksum [20] it was shown that when the parameter of the null hypothesis is known, the last three classes do not contain asymptotically most powerful tests. As a consequence the ranks are not asymptotically sufficient.

Barlow [12] investigated the problem of testing the null hypothesis of constant fialure rate versus monotone failure rate. The test statistic was based on the likelihood **ratio statistic.** In this paper unbiased tests at all significance levels were given for the problems H_o: $F\varepsilon\{IFRA\}$ versus H_1: $F\varepsilon\{DFRA\}$, H_o: $F\varepsilon$ {class of exponential distributions with possible truncation on the right}$\equiv F_o$ versus H_1: $F\varepsilon\{IFR\}-F_o$, H_o: $F\varepsilon\{DFR\}$ versus H_1: $F\varepsilon\{IFR\}$, H_o: $F\varepsilon\{\text{exponential}\}$ versus H_1: $F\varepsilon\{IFR\}$, and various other combinations of DFR AND IFR alternatives. The main tool for these test was the MLE for IFR, DFR distributions given by Marshall and Proschan [72].

The first attempt to test the hypothesis H_o: F is exponential with unspecified scale parameter versus H_1: F is NBU (and not exponential) is given in Hollander and Proschan [52]. The test statistic used was motivated by the parameter

$$\gamma(F) \overset{\text{def}}{=} \iint [\bar{F}(x)\bar{F}(y) - \bar{F}(x+y)]dF(x)dF(y),$$

which is zero when F is exponential, and nonnegative when F is NBU. The test statistic used is the U-statistic which is asymptotically equivalent to $\gamma(F_n)$, where F_n is the empirical distribution. The statistic is unbiased, asymptotically normal, and consistent. Since no other test statistic existed against NBU alternatives, they determined the asymptotic relative efficiency of their test statistic relative to statistics used against IFR alternatives. Since the IFR class of distributions is contained in the NBU class, it is expected that the efficiencies, under IFR alternatives, to select the IFR tests. They showed that for a class of NBU alternatives the NBU test out-performs the IFR tests. The small sample null distribution of the U-statistic used was considered and exact probabilities were computed in special cases. Some useful tables were also developed in this paper. This paper was followed by a second paper by Hollander and Proschan [53], in which they developed tests for decreasing mean residual life (DMRL) and new better than used in expectation (NBUE) alternatives. The NBUE statistic derived was shown to be equivalent to the total time on test (TTOT) statistic, which was considered to be a test statistic for exponentiality versus IFR or IFRA alternatives, the larger class of NBUE alternatives. Thus, significantly large values of the TTOT statistic, which used to indicate acceptance of an IFR and IFRA alternative, should now be used to accept an NBUE alternative. The DMRL statistic derived is new, and critical constants and a large sample approximation are obtained. Consistency and symptotic relative efficiency of the DMRL and NBUE tests are investigated.

The estimation problems for the above classes will now be presented. Marshall and Proschan [72] (see also Grenander [46]) considered the maximum likelihood estimation for life-distributions with monotone failure rate over the support of the distribution functions. They considered data arising from a testing plan which does not allow censoring, time-truncation or replacements. They proved that for this testing plan the MLE for life-distributions with increasing failure rate is strongly consistent.

Crow and Shimi [32] considered the maximum likelihood estimation of life-distributions with monotone failure rate over the interval $[0,T)$, where T is a fixed positive real number, and no other assumptions about the distribution or its failure rate are given outside that interval. The following renewal type testing plan is used, which allows for time-truncation and replacement. At time zero, the beginning of the testing, n new items form the population to be tested are put on test. When an item fails it is instantaneously replaced with another new item from the same population and at time T all testing is stopped. The maximum likelihood estimates of the distribution function and its failure rate over [0,T) are given and shown to be uniformly strongly consistent as n tends to infinity. Since the class of distributions under consideration, IFR on (0,T), is a nonparametric family of distributions for which there exists no sigma-finite measure relative to which all the measures induced by the class are absolutely continuous, the usual concept of maximum likelihood estimation cannot be applied. The general definition of MLE due to Kiefer and Wolfowitz [58] is used in this paper to determine the MLE of the life-time distribution over [0,T).

In a recent report Marshall and Proschan [75] proved that the MLE for IFRA distributions is inconsistent. We note that MLE yields strongly consistent estimators in the subclass of IFR distributions and in the class of all distributions, but not in the class of IFRA distributions. In the DFRA case the maximum likelihood estimation procedure fails, that is, no MLE exists for the DFRA distributions.

There are no estimators with desirable properties for the NBU or the DMRL classes of distributions.

REFERENCES

[1] A-Hameed, M. S. and F. Proschan, Nonstationary Shock Models, *Stoch. Proc. applic. 1:* (1973), pp. 383-404.

[2] ______________________________, Shock Models with Underlying Birth Process, *J. Appl. Prob. 12:* (1975), pp. 18-28.

[3] Bain, L. J., Inferences Based on Censored Sampling from the Weibull or Extreme-Value Distribution, *Technometrics*, Vol. 14, No. 3, (1972), pp. 693-702.

[4] Barlow, R. E., A. Q. Marshall and F. Proschan, Properties of Probability Distributions with Monotone Hazard Rate, *Ann. Math. Statist. 34:* (1963), pp. 375-389.

[5] _____________ and A. W. Marshall, Bounds for Distributions with Monotone Hazard Rate, I and II, *Ann. Math. Statist. 35:*

(1964), pp. 1234-1274.

[6] Barlow, R. E., Bounds on Integrals with Applications to Reliability Problems, *Ann. Math. Statist.* 36: (1965), pp. 565-574.

[7] ____________ and A. W. Marshall, Tables of Bounds for Distributions with Monotone Hazard Rate, *J. Amer. Statist. Assoc.* 60: (1965), pp. 872-890.

[8] ____________ and F. Proschan, Mathematical Theory of Reliability, John Wiley and Sons, Inc. New York, New York, (1965).

[9] ____________ and A. W. Marshall, Bounds on Interval Probabilities for Restricted Families of Distributions, *Proc-ceedings of the Fifth Berkeley Symposium on Mathematical Statistics and Probability*. III: (1967), pp. 229-257.

[10] ____________ and F. Proschan, Statistical Theory of Reliability and Life Testing, Probability Models, John Wiley and Sons, New York, (1965).

[11] ____________________________, Mathematical Theory of Reliability, John Wiley and Sons, New York, (1965).

[12] ____________, Likelihood Ratio Tests for Restricted Families of Probability Distributions, *Ann. Math. Statist.* 39: (1968), pp. 547-560.

[13] Basu, A. P., Estimates of Reliability for Some Distributions Useful in Life Testing, *Technometrics*, Vol. 6, (1964), pp. 215-219.

[14] Bartholomew, D. J., The sampling Distribution of an Estimate Arising in Life Testing, *Technometrics*, Vol. 5, No. 3, (1963), pp. 361-374.

[15] Bennett, G. K., Smooth Empirical Bayes Estimation with Applications to the Weibull Distribution, NASA Technical Memorandum, X-58048, (1971).

[16] Bennett, G. K. and H. F. Martz, Jr., A continuous Empirical Bayes Smoothing Technique, *Biometrika*, Vol. 59, (1972).

[17] Bennett, G. K. and H. F. Martz, An Empirical Bayes Estimation for the Scale Parameter of the Two-Parameter Weibull Distribution, *Naval Research Logistics Quarterly*, Vol. 20, No. 3, (1973), pp. 387-393.

[18] Bhattacharya, S. K., Bayesian Approach to Life Testing and Reliability Estimation, *Journ. Amer. Statistics Assoc.*, Vol. 62, (1967), pp. 48-62.

[19] Bickel, P. J., Tests for Monotone Failure Rate II, *Ann. Math. Statist. 40*: (1969), pp. 1250-1260.

[20] ____________ and K. A. Doksum, Tests for Monotone Failure Rate Based on Normalized Spacings, *Ann. Math. Statist. 40*: (1969), pp. 1216-1235.

[21] Birnbaum, Z. W., J. D. Esary and A. W. Marshall, Stochastic Characterization of Wearout for Components and Systems, *Ann. Math. Statist. 37*: (1966), pp. 816-825.

[22] Bryson, M. C. and M. M. Siddiqui, Some Criteria for Aging, *J. Amer. Statist. Assoc. 64*: (1969), pp. 1472-1783.

[23] Bury, K. V., On the Reliability Analysis of a Two-Parameter Weibull Process, *INFOR*, Vol. 110, (1972), pp. 129-139.

[24] Canavos, G. C. and C. P. Tsokos, Ordinary and Empirical Bayes Approach to Estimation of Reliability in the Weibull Life Testing Model, *Proc. of Sixteenth Conference on the Design of Experiments in Army Research, Development and Testing*, Fort Lee, Virginia, (1970), pp. 379-392.

[25] ______________________________, A Study of an Ordinary and Empirical Bayes Approach to Estimation of Reliability in the Gamma Life Testing Model, *Proc. of IEEE Symposium of Reliability*, (1971).

[26] ______________, An Empirical Bayes Approach for the Poisson Life Distribution, *IEEE Transactions on Reliability*, Vol. R-22 No. 2, (1973).

[27] Canavos, G. C. and C. P. Tsokos, Bayesian Estimation of Life Parameters in the Weibull Distribution, *Journal of Operations Research*, Vol. 21, No. 3, (1973), pp. 755-763.

[28] Canfield, R. V., A Bayesian Approach to Reliability Estimation Using a Loss Function, *IEEE Transactions on Reliability* Vol. R-19, No. 1, (1970), pp. 13-16.

[29] Cohen, A. C., Maximum Likelihood Estimation in the Weibull Distribution Based on Complete and Censored Samples, *Technometrics*, Vol. 7, No. 4, (1965), pp. 579-588.

[30] Couture, D. J. and H. F. Martz, Jr., Empirical Bayes Estimation in the Weibull Distribution, *IEEE Transactions on Reliability*, Vol. R-21, No. 2, (1972), pp. 75-83.

[31] Crelin, G. L., The Philosophy and Mathematics of Bayes Equation, *IEEE Transactions on Reliability*, Vol. R-21, No. 3, (1972), pp. 131-135.

[32] Crow, L. H. and I. N. Shimi, Maximum Likelihood Estimation of Life-Distributions from Renewal Testing, *Ann. Math. Statist. 43:* (1972), pp. 1827-1838.

[33] D'Agostino, R. D., Linear Estimation of the Weibull Parameters, *Technometrics*, Vol. 13, No. 1, (1971), pp. 171-182.

[34] Deely, J. J., M. S. Tierney and W. J. Zimmer, On the Usefulness of the Maximum Entropy Principle in the Bayesian Estimation of Reliability, *IEEE Transactions on Reliability* Vol. R-19, No. 3, (1970), pp. 110-115.

[35] Drake, A. W., Bayesian Statistics for the Reliability Engineer, *Proc. 1966 Annual Symposium on Reliability*, (1966), pp. 315-320.

[36] El-Sayyad, G.M., Estimation of the Parameter of an Exponential Distribution, *JRSS*, Series B, Vol. 29, (1967), pp. 525-532.

[37] Epstein, B. and M. Sobel, Life Testing, *Journ. Amer. Statist. Assoc.*, Vol. 48, (1953), pp. 486-502.

[38] ______________________, Some Theorem Relevant to Life Testing From an Exponential Distribution, *Annals Math.*

Statist., Vol. 25, (1961), pp. 373-381.

[39] Epstein, B. and M. Sobel, Sequential Life Tests in the Exponential Case, *Ann. Math. Statist.*, Vol. 26, (1955), pp. 183-189.

[40] Esary, J. D. and F. Proschan, Relationship Between System Failure Rate and Component Failure Rate, *Technometrics* 5: (1963), pp. 183-189.

[41] ___________, A. W. Marshall and F. Proschan, Some Reliability Applications of the Hazard Transform, *SIAM J. Appl. Math. 18*: (1970), pp. 849-860.

[42] ____, Shock Models and War Processes, *Ann. Probability 1*: (1973), pp. 627-649.

[43] Evans, R. A., Prior Knowledge, Engineers Versus Statisticians, *IEEE Transactions on Reliability*, Vol. 1, Editorial, (1969).

[44] Feducia, J. A. and J. Klion, How Accurate Are Reliability Predictions, RADC, (1968).

[45] _____________, A Bayesian/Classical Approach to Reliability Demonstration, RADC-TR-70-72, Technical Report, (1970), pp. 1-23, (Rome Air Development Center).

[46] Grenander, V., On the Theory of Mortality Measurement, II, Skand. Akuarietidskr. 39: (1956), pp. 125-153.

[47] Ferguson, T. S., Mathematical Statistics, A Decision Theoretic Approach, Academic Press, New York, (1967).

[48] Fertig, K. W., Bayesian Prior Distributions for Systems with Exponential Failure-Time Data, *Annals of Math. Stat.* Vol. 43, No. 5, (1972), pp. 1441-1448.

[49] Haines, A. L. and N. D. Singpurwalla, Some Contributions to Stochastic Characterization of Wear, Reliability and Biometry, ed. by F. Proschan and R. J. Serfling, *SIAM Publication* (1974).

[50] Harris, C. M. and N. D. Singpurwalla, Life Distributions Derived from Stochastic Hazard Functions, *IEEE Transactions on Reliability*, Vol. R-17, No. 2, (1968), pp. 70-79.

[51] Harter, H. and A. H. Moore, Maximum-Likelihood Estimaton of the Parameters of Gamma and Weibull Populations from Complete and from Censored Samples, *Technometrics*, Vo. 7, No. 4, (1965), pp. 639-643.

[52] Hollander, M. and F. Proschan, Testing Whether New is Better than Used, *Ann. Math. Statist. 43*: (1972), pp. 1136-1146.

[53] ________________, Testing for the Mean Residual Life, *Biometrika*, (to appear), (1975).

[54] Jaynes, E. T., Prior Probabilities, *IEEE Trans. Syst. Sci. Cybernetics*, Vol. SSC-4, No. 3, (1968), pp. 227-241.

[55] Johns, M. V. and G. J. Lieberman, An Exact Asymptotically Efficient Confidence Bound for Reliability in the Case of the Weibull Distribution, *Technometrics*, Vol. 8, No. 1, (1966), pp. 135-175.

[56] Karlin, S., Total Positivity, Vol. I. Stanford University Press, Stanford, California, (1968).

[57] Kendall, M. G. and A. Stuart, Vol. I, Vol. II, Vol. III, The Advanced Theory of Statistics, Hafner Publishing Corp., New York, (1952), (1960), (1966).

[58] Kiefer, J. and J. Wolfowitz Consistency of the Maximum Likelihood Estimator in the Presence of Infinitely Many Incidental Parameters, *Ann. Math. Statist. 27*: (1956), pp. 887-906.

[59] Lehmann, E. L., Testing Statistical Hypothesis, John Wiley and Sons, New York, (1957).

[60] Lemon, G. H., An Empirical Bayes Approach to Reliability, *IEEE Trans. on Reliability*, Vol. R-21, No. 3, (1972), pp. 155-158.

[61] ________ and R. Krutchkoff, An Empirical Bayes Smoothing Technique, *Biometrika*, Vol. 56, (1969), pp. 361-365.

[62] Lieberman, G. J. and S. M. Ross, Confidence Intervals for Independent Exponential Series Systems, *Jour. Amer. Statist. Assoc.*, Vol. 66, (1971) pp. 837-840.

[63] Lewis, P. A. W., Some Results on Tests for Poisson Processes, *Biometrika*,52, (1965), pp. 67-77.

[64] Lindley, D. V., Sequential sampling: Two decision problems with linear losses for binomial and normal random variables, *Biometrika*, 52, (1961), pp. 507-532.

[65] Lock, H. W. and Savits, Th., "The IFRA Closure Problem", *Annuals of Probability*, to appear (1976).

[66] Mann, N. R., Tables for Obtaining the Best Linear Invariant Estimates of Parameters of the Weibull Distribution, *Technometrics*, Vol. 9, No. 4, (1967), pp. 629-645.

[67] __________, Computer-Aided Selection of Prior Distribution for Generating Monte Carlo Confidence Bounds on System Reliability, *Nav. Research Log. Quarterly*, Vol. 17, (1970), pp. 41-54.

[68] __________ and F. E. Grubbs, Approximately Optimum Confidence Bounds on Series System Reliability for Exponential Time to Failure Data, *Biometrika*, Vol. 59, No. 1, (1972), pp. 191-204.

[69] __________, R. E. Schafer and N. D. Singpurwalla, Methods for Statistical Analysis of Reliability and Life Data, John Wiley and Sons, New York, (1974).

[70] Maritz, J. S., Smooth Empirical Bayes Estimation for One-Parameter Discrete Distributions, *Biometrika*, Vol. 53, (1966), pp. 417-429.

[71] ____________, Smooth Empirical Bayes Estimation for One-Parameter Discrete Distributions, *Biometrika*, Vol. 54, (1967), pp. 17-29.

[72] ____________, Smooth Empirical Bayes Estimation for Continuous Distributions, *Biometrika* , Vol. 54, (1967), pp. 435-450.

[73] Marshall, A. W. and F. Proschan, Maximum Likelihood Estimation for Distributions with Monotone Failure Rate, *Ann. Math. Statist. 36:* (1965), pp. 69-77.

[74] ____________________________, Mean Life of Series and Parallel Systems, *J. App'l. Prob.*, Vol. 7, (1970), pp.

165-174.

[75] Marshall, A. W. and F. Proschan, Classes of Distributions Applicable in Replacement, with Renewal Theory Implications, *Proceedings of the 6th Berkeley Symposium on Mathematical Statistics and Probability*, Vol I: (1972) pp. 395-415, University of California Press, Berkeley, California.

[76] ______________________________, Inconsistency of Maximum Likelihood Estimator of Distributions Having Increasing Failure Rate Average, FSU Statistics Report No. M350, (1975).

[77] Moran, P. A. P., The Random Division of an Interval; Part II, *J. Roy. Statist. Soc. Ser.* Vol. 13: (1951), pp. 147-150.

[78] Nichols, W. and C. P. Tsokos, Empirical Bayes Point Estimation in a Family of Probability Distributions, *International Statistical Institute*, Vol. 40, No. 2, (1971).

[79] ___________________________, Empirical Bayes Point Estimation in a Family of Probability Distributions, *Int. Stat. Rev.*, Vol. 40, No. 2, (1972), pp. 147-151.

[80] Papadopoulos, A. and C. P. Tsokos, Bayesian Confidence Bounds on the Random Scale Parameter and Reliability Functions of the Weibull Failure Model with Computer Simulations, *IEEE Transactions on Reliability*, (1975).

[81] ________________________________, Bayesian Reliability Estimates of the Binomial Failure Model, *JUSE* Vol. 21, No. 1, (1974), pp. 9-26.

[82] ________________________________, Bayesian Analysis of the Weibull Failure Model with Unknown Scale and Shape Parameters, *Statistica*, (1976).

[83] Parzen, E., On Estimation of a Probability Density Function and Mode, *Ann. Math. Statist.*, Vol. 33, (1962), pp. 1056-1076.

[84] Proschan, F. and R. Pyke, Tests for Monotone Failure Rate, *Proceedings of the 5th Berkeley Symposium on Mathematical Statistics and Probability*, *III* (1967), pp. 293-312, University of California Press, Berkeley, California.

[85] Proschan, F., Recent Research on Classes of Life Distributions Useful in Maintenance Modeling, FSU Stat. Report M291, (1974).

[86] Pugh, E. L., The Best Estimate of Reliability in the Exponential Case, *Opns. Res.*, Vol. II, (1963), pp. 157-163.

[87] Robbins, H., An Empirical Bayes Approach to Statistics, *Proc. Third Berkeley Sympos.*, *I*, (1955), pp. 157-163.

[88] Robbins, H. The Empirical Bayes Approach to Statistical Decision Problems, *Ann. Math. Statist.*, Vol. 35, (1964), pp. 1-20.

[89] Ross, S. M., Introduction to Probability Models, Academic Press, New York, (1972).

[90] ___________, On Time to First Failure In Multicomponent Exponential Reliability Systems, In preparation, (1974).

[91] Sarkar, T. K., An Exact Lower Confidence Bound for the Reliability of a Series System Where Each Component has an Exponential Time to Failure Distribution, *Technometrics*, Vol. 13, No. 3, (1971), pp. 535-546.

[92] Samuel, E., An Empirical Bayes Approach to the Testing of Certain Parametric Hypotheses, *Ann. Math. Statist.*, Vol. 34, (1963), pp. 1370-1385.

[93] Soland, R. M., Bayesian Analysis of the Weibull Process with Unknown Scale Parameters and Its Applications to Acceptance Sampling, *IEEE Transactions on Reliability*, Vol. 17, (1968), pp. 84-90.

[94] _____________, Bayesian Analysis of the Weibull Process with Unknown Scale and Shape Parameters, *IEEE Transactions on Reliability*, Vol. R-18, No. 4, (1969), pp. 181-184.

[95] Springer, M. D. and W. E. Thompson, Bayesian Confidence Limits for the Product of N Binomial Parameters, *Biometrika*, Vol. 53, pp. 611-613.

[96] _________________________________, Bayesian Confidence Limits for the Reliability of Cascade Exponential Subsystems, *IEEE Trans. on Reliability*, Vol. R-16, No. 2,

(1967), pp. 86-89.

[97] Springer, M. D. and W. E. Thompson, Bayesian Confidence Limits for Reliability of Redundant Systems When Tests are Terminated at First Failure, *Technometrics*, Vol. 10, No. 1, (1968), pp. 29-36.

[98] Tsokos, C. P. and A. N. V. Rao, Robustness Studies in Bayesian Developments, *Proceedings of the 20th Conference on the Design of Experiments in Army Research Development and Testing*, ARO, 75-2, (1974), pp. 273-302.

[99] ____________, Bayesian Approach to Reliability Using the Weibull Distribution with Unknown Parameters with Simulation, *JUSE*, Vol. 19, No. 4, (1973).

[100] Weibull, W., A Statistical Distribution Function of Wide Applicability, *J. Appl. Mech.*, 18, 293, (1951).

BAYESIAN LIFE TESTING USING THE TOTAL Q ON TEST

by WILLIAM S. JEWELL
University of California, Berkeley

Abstract. Suppose the basic shape of the cumulative failure (hazard) function has been identified for a certain component, and that an unknown parameter θ for a new production run of similar components is to be estimated. In particular, suppose that the failure function is of *proportional* type, $R(x) = \theta Q(x)$, where Q is the known shape function, and that θ is sampled from a prior gamma density. By using a new statistic, called the *total Q on test* (TQT), it is possible to perform Bayesian updating during a variety of lifetime testing programs in a manner similar to total time on test plots. This statistic can also be used with complete lifetime data, extending over several product runs, to identify the failure form Q, and to estimate the gamma hyperparameters. Extensions include the use of several TQT statistics to estimate the relative strength of competing hazard functions.

1.0 Introduction. Suppose that the random lifetime, $\tilde{x}$, of a certain component type has a distribution function $P(x \mid \theta)$, where θ is an unknown parameter that may vary from one production run to another, but is assumed constant within each run. We assume a prior distribution of interbatch variation, $P(\theta)$, is available.

There are two main types of estimation problems within the Bayesian framework:

(1) *Model identification*, in which the forms of $P(x \mid \theta)$ and $P(\theta)$ are to be inferred from previous production runs;

(2) *Parameter estimation*, in which selected components from a given batch are operated under normal or accelerated test conditions to infer the particular value of θ for

that batch.

These two problems are interrelated, since the choice of model is often made to simplify parameter estimation, and the success of Bayesian updating depends upon how close the assumed models are to the true ones. However, a third factor which influences both of these tasks, and which is often overlooked, is the test design.

In the following, we shall examine Bayesian estimation under normal, but incomplete test conditions. We first show that a convenient modeling family is the proportional hazard class (with generalization to competing hazard families), and that, with a certain natural conjugate prior, parameter estimation is easily carried out through a test statistic we call "total Q on test," a generalization of the concept of "total time on test." Furthermore, model identification can be simplified by using special total Q on test plots of the sample data.

2.0 Incomplete Life Tests. The classical manner in which lifetime distributions are inferred is to place N components, assumed to have the same values of θ, in a test environment which matches normal operating conditions for T hours, or until all components have all failed, if earlier. Normally, the economics of testing are such that the (random) number which have failed, $\tilde{C}(T)$, is much less than N. (See e.g., [1] where this scheme is called a truncated data test, and [5], where it is a Type I censored test). The outcomes of the incomplete test can be described in terms of a data set D of the ordered random lifetimes $\{\tilde{x}_{i:N}\}$ and the random $\tilde{C}(T)$ as follows:

$$D = \left\{ \left[\tilde{x}_{1:N} = x_1\right] \leq \left[\tilde{x}_{2:N} = x_2\right] \leq \cdots \leq \left[\tilde{x}_{C:N} = x_C\right] \leq T; \quad (2.1) \right.$$

$$\left. \tilde{x}_{\tilde{C}+1:N} > T; \ \ldots \ \tilde{x}_{N:N} > T \right\} .$$

If $p(x \mid \theta)$ and $P^c(x \mid \theta)$ are the density and complementary (tail) distribution, respectively, corresponding to $P(x \mid \theta)$, then the likelihood of the data set D is:

$$L(D \mid \theta) = p(x_1, x_2, \ldots, x_N; C(T) \mid \theta)$$

$$= \frac{N!}{(N-C)!} \prod_{i=1}^{C} p(x_i \mid \theta) \left[P^c(T \mid \theta) \right]^{N-C} \prod_{i=C+1}^{N} \delta(x_i - T) \quad (2.2)$$

$$(x_1 \leq x_2 \leq \cdots \leq x_C \leq T)$$

where unit impulse (degenerate) densities $\delta(x)$ are used for convenience so that $\sum_{C=0}^{N} \iint L(D \mid \theta) dx_1 dx_2 \ldots dx_N = 1$.

An alternate testing scheme (called a censored data test in [1] and a Type II censored test in [5]), fixes the number of failures, C, in advance, and thus the duration of the test is a random variable, $\tilde{T}(C) = \tilde{x}_{C:N}$. The likelihood (2.2) becomes a density

$$L(D \mid \theta) = p(x_1, x_2, \ldots, x_N \mid \theta)$$

$$= \frac{N!}{(N-C)!} \prod_{i=1}^{C} p(x_i \mid \theta) \left[P^c(x_C \mid \theta) \right]^{N-C} \prod_{i=C+1}^{N} \delta(x_i - x_C). \quad (2.3)$$

$$(x_1 \leq x_2 \leq \cdots \leq x_C)$$

with $\iint L(D \mid \theta) dx_1, dx_2 \ldots dx_N = 1$.

3.0 Proportional Hazard Family: Total Q On Test and Maximum Likelihood Estimators. Although (2.2) and (2.3) could, in principle, be calculated for any underlying distribution, it is desirable to pick a family which is both useful from a modeling point of view, and simple computationally. The Koopman-Pitman-Darmois exponential family has extremely useful properties in regular Bayesian forecasting [2] [4]; but, it has no convenient integral form and does not seem useful in life testing applications.

However, every complementary distribution function can be written $P^c(x \mid \theta) = \exp(-R(x \mid \theta))$, where $R(x \mid \theta)$ is the *cumulative failure (hazard) function*, given θ; a convenient way to parametrize this form is to suppose that $R(x \mid \theta) = \theta Q(x)$. Since $Q(x)$ is a monotone nondecreasing function which contain all of the failure shape information (thus preserving the properties of IFR, IFRA, etc. [1] for all θ), we could call it the *unit-or prototype-failure function*, and refer to the density and cumulative distributions:

$$P^c(x \mid \theta) = e^{-\theta Q(x)} \quad ; \quad p(x \mid \theta) = \theta q(x) e^{-\theta Q(x)} \quad ; \tag{3.1}$$

as the *proportional hazard family* $(q(x) = dQ(x)/dx)$. We note the following properties:

(a) The family includes the exponential, Weibull with given shape parameter, and Gumbel distributions;

(b) A random lifetime, $\tilde{x}$, with distribution (3.1) and θ an integer can be interpreted as $\tilde{x} = \min(\tilde{y}_1, \tilde{y}_2, \ldots, \tilde{y}_\theta)$, where $\Pr\{\tilde{y}_i > y\} = \exp(-Q(y))$, for all $i = 1,2,\ldots,\theta$, which suggests a certain physical failure model for this family;

(c) It *may* be reasonable that production of different lots under varying conditions maintains the same shape for the hazard function and changes only the relative intensity of failures.

Most importantly, this family greatly simplifies estimation in life-testing, since the likelihood (2.2) becomes

$$p(x_1, x_2, \ldots, x_N ; C \mid \theta)$$

$$= \frac{N!}{(N-C)!} \prod_{i=1}^{C} q(x_i) \prod_{i=C+1}^{N} \delta(x_i - T) \left\{ \theta^C e^{-\theta [TQT(x_1, x_2, \ldots, x_C)]} \right\}, \tag{3.2}$$

$$(x_1 \le x_2 \le \cdots \le x_C)$$

where we call

$$TQT(x_1, x_2, \ldots, x_C ; C = C(T)) = \sum_{i=1}^{C} Q(x_i) + (N-C)Q(T) \quad (3.3)$$

the *total Q on test*, a statistic of varying dimensions that is a natural generalization of the total time on test concept for the exponential density, $Q(x) = x$ [1]. Corresponding to (2.3) and the second testing scheme, we obtain a similar, but fixed-dimensional statistic:

$$TQT(x_1, x_2, \ldots, x_C) = \sum_{i=1}^{C} Q(x_i) + (N-C)Q(x_C), \quad (3.4)$$

and a likelihood similar to (3.2).

Since only the term in braces in (3.2) is a function of θ, it is easy to show that the maximum likelihood estimators of $\tilde{\theta}$ and $1/\tilde{\theta}$ are:

$$\hat{\theta}^{-1} = MLE\{(\tilde{\theta})^{-1}\} = \frac{TQT(D)}{C}; \quad \hat{\theta} = MLE\{\tilde{\theta}\} = \frac{C}{TQT(D)} \quad (3.5)$$

which coincides with known results for the exponential density under the two tests [1] [5].

4.0 Other Life Tests. A variety of other life testing schemes also use appropriate generalizations of the total Q on test statistic. For example, if N units are placed on test, but a set of fixed, but possibly different test durations $\{T_i\}$ are given for all units, then we observe $\tilde{y}_i = \min\{\tilde{x}_i, T_i\}$. The likelihood of this unordered data is:

$$p(y_1, y_2, \ldots, y_N ; C \mid \theta)$$

$$= \prod_{i\in C} q(y_i) \prod_{j\notin C} \delta(y_j - T_j) \left\{ \theta^C e^{-TQT(y_1, y_2, \ldots, y_N ; C)} \right\} \quad (4.1)$$

$$(0 \le y_i \le \infty)$$

where $C = \{i \mid \tilde{y}_i < T_i , i = 1,2, \ldots, N\}$ and $C = ||C||$. The total Q on test is

$$TQT(y_1,y_2, \ldots, y_N ; C) = \sum_{i=1}^{N} Q(y_i) = \sum_{i\varepsilon C} Q(x_i) + \sum_{j\not\varepsilon C} Q(T_j). \quad (4.2)$$

This is called testing with withdrawals in [1].

Another testing procedure might be called renewal testing. Suppose a component is started on a single test stand at time zero; after the first failure, a new item is started; and so on, until the procedure terminates at a fixed time T, with the $(\tilde{C} + 1)^{st}$ item still operative. The likelihood associated with $D = \left\{\tilde{x}_1 = x_1, \tilde{x}_2 = x_2, \ldots, \tilde{x}_C = x_C ; \tilde{x}_{\tilde{C}+1} > T - \sum_{i=1}^{C} x_i\right\}$ is just:

$$p(x_1,x_2, \ldots, x_C ; C \mid \theta) = \prod_{i=1}^{C} q(x_i)\left\{\theta^C e^{-TQT(x_1,x_2,\ldots,x_C;C)}\right\} \quad (4.3)$$

over the simplex $0 \le \sum_{i=1}^{C} x_i \le T$, with total Q on test

$$TQT(x_1,x_2, \ldots, x_C ; C) = \sum_{i=1}^{C} Q(x_i) + Q\left(T - \sum_{i=1}^{C} x_i\right). \quad (4.4)$$

The generalization to stopping with the Cth failure, multiple test stands, and other combined censoring-truncating-withdrawal-renewal test schemes should now be obvious.

In fact, with the proportional hazard family it is clear that, no matter what the test set-up, one needs only to monitor the actual (complete or incomplete) lifetimes $\{y_i\}$ for each item, and the (fixed or stochastic) number , C, of completed tests. Then $TQT = \sum_{\text{all } i} Q(y_i)$ and (3.5) and the Bayesian formulae of the next section always hold.

5.0 Bayesian Updating Under Life Testing; Mixed Distributions. In a Bayesian formulation, we do not use (3.5) directly because we have prior information on the possible values of θ, perhaps through cumulative data on previous production runs, or, if one is so persuaded, from sampling expert opinion. To simplify the application of Bayes' law, it is convenient to pick a *natural conjugate prior*, whose form matches that in the braces in (3.2); this is just the gamma density

$$p(\theta) = \frac{Q_o(Q_o\theta)^{C_o-1} e^{-\theta Q_o}}{\Gamma(C_o)} \qquad (0 \le \theta \le \infty) \tag{5.1}$$

with hyperparameters C_o, Q_o. The usefulness of (5.1) in modeling unimodal densities over $[0,\infty)$ is well-known. It is easy to see that Bayes' law then gives a posterior-to-data density of θ, $p(\theta \mid D)$, which is also gamma, but with updated hyperparameters:

$$\begin{aligned} C_o &\leftarrow C_o + C , \\ Q_o &\leftarrow Q_o + TQT(D) . \end{aligned} \tag{5.2}$$

In place of the classical estimators (3.5) we have the posterior-to-data expectations

$$E\{(\tilde{\theta})^{-1} \mid D\} = (1 - Z_1)E\{\tilde{\theta}^{-1}\} + Z_1\hat{\theta}^{-1}(D) \tag{5.3}$$

$$[E\{\tilde{\theta} \mid D\}]^{-1} = (1 - Z_2)[E\{\tilde{\theta}\}]^{-1} + Z_2[\hat{\theta}(D)]^{-1} \tag{5.4}$$

with "credibility factors"

$$Z_1 = C/(C_o - 1 + C) \ ; \ Z_2 = C/(C_o + C) . \tag{5.5}$$

This means that if the test gives a large number of complete observations, relative to C_o, then the Bayesian and MLE estimators coincide. However, for relatively incomplete tests, more weight is given the prior expectations, $E\{\tilde{\theta}^{-1}\}$ or $E\{\tilde{\theta}\}$. Linear mixing formulae of this type are well known in Bayesian forecasting [2] [3] [4].

It is also of interest to examine the form of the *mixed density* of lifetimes, obtained by averaging over many batches. From (3.1) and (5.1)

$$P^C(x) = EP^C(x \mid \theta) = \{1 + [Q(x)/Q_o]\}^{-C_o} ;$$
$$p(x) = [C_o q(x) \,/\, Q_o]\{1 + [Q(x)/Q_o]\}^{-C_o - 1} ; \tag{5.6}$$

which is a generalization of a shifted Pareto. We see that if the prototype failure function is Gumbel, we get exponential tails (for large x) in the mixed distribution, while if the underlying **failures** are Weibull, we get the "more dangerous" algebraic tails. If life test data D are drawn from a batch with fixed $\theta = \theta_1$, it follows that the forecast distribution for the remaining components from this batch, $P^C(x \mid D)$, is of form (5.6) with updated parameters (5.2). Furthermore, one can show that if the test data set D is large enough, $P^C(x \mid D)$ approaches $\exp\{-\theta_1 Q(x)\}$, the true distribution for this batch, with probability one.

Generally, the failure function of the mixed distribution increases less rapidly than that of the prototype hazard:

$$R_{MIXED}(x) = -\ln[P^C(x)] = C_o \ln[1 + (Q(x)/Q_o)] \tag{5.7}$$

and may be DFR even if $q(x)$ is increasing. (5.7) should not be confused with the average behavior of the individual failure functions:

$$R(x) = ER(x \mid \tilde{\theta}) = C_o Q(x) \; Q_o \; , \tag{5.8}$$

which always has the shape $Q(x)$.

6.0 Total Q On Test Transform. We turn now to the problem of identifying the prototype failure function, $Q(x)$, and the hyperparameters C_o, Q_o. First, let us examine the effect of plotting TQT with a Q which is not the necessarily correct one.

Paralleling Barlow and Campo [1], we define the *total Q on test transform* (TQTT) of any distribution F with respect to Q as:

$$H^{-1}(t) = H^{-1}_{F/Q}(t) = \int_0^{F^{-1}(t)} [1 - F(u)]dQ(u) \quad t \in [0,1]. \qquad (6.1)$$

This awkward notation is used because $H_{F/Q}$ (the inverse of $H^{-1}_{F/Q}$) is a distribution on $[0,\mu]$, where

$$\mu = H^{-1}(1) = \int_0^{F^{-1}(1)} F^c(u)dQ(u) \leq \infty, \qquad (6.2)$$

and is the mean of F if $Q(x) = x$, the usual total on test transform.

Now if $F^c(x) = \exp\{-\theta Q(x)\}$, it is easy to verify that

$$H^{-1}(t) = t/\theta, \quad t \in [0,1] \qquad (6.3)$$

while, for a general failure function, $F^c(x) = \exp\{-R_F(x)\}$,

$$\left.\frac{dH^{-1}(t)}{dt}\right|_{x = F^{-1}(t)} = \frac{q(x)}{r_F(x)} \qquad (6.4)$$

where $r_F(x) = dR_F(x)/dx$.

Thus, if we could plot the TQTT for any Q, we could isolate those areas where $r(x)$ was larger [smaller] than $q(x)$ as those regions where the TQTT has slope less [greater] than unity. Constants of proportionality can be eliminated by plotting the scaled total Q on test transform (STQTT), $H^{-1}(t)/H^{-1}(1)$, thus giving us a continuous increasing function on the unit square which intersects $(0,0)$ and $(1,1)$. And the closer $Q(x)$ is to the form of $R(x)$, the closer the STQTT is to the straight line (obtained for the exponential distribution with the conventional total time on test transform).

The empirical usefulness of this for identification can be seen from the following. Suppose we have observed a censored life test for a sample of N items from the proportional hazard family with fixed, but unknown θ, say $F(x) = P(x \mid \theta)$, and there are C failures. The empirical lifetime distribution, $F_C(x)$, can be obtained from the ordered complete samples $\{x_{1:N} \le x_{2:N} \cdots x_{C:N}\}$

$$F_C(x) = \begin{cases} 0 & 0 \le x < x_{1:N} \\ i/C & x_{i:N} \le x < x_{i+1:N} \\ 1 & x_{C:N} \le x \end{cases} ; \tag{6.5}$$

however this is unlikely to be of direct use because of the small size of C.

Suppose we plot instead the *empirical Q on test ratio*

$$\mathrm{EQTR}(i \mid D) = \frac{\mathrm{TQT}(x_{1:N}, x_{2:N}, \ldots, x_{i:N})}{\mathrm{TQT}(x_{1:N}, x_{2:N}, \ldots, x_{C:N})} \tag{6.6}$$

versus $(i \mid C)$ $(i = 1,2, \ldots, C)$, where TQT is the statistic defined in (3.4). Since

$$\frac{1}{N}\,\mathrm{TQT}(x_{1:N}, x_{2:N}, \ldots, x_{C:N}) = \int_0^{F_C^{-1}(C/N)} [1 - F_C(u)]\,dQ(u), \tag{6.7}$$

one can show, following [1], that

$$\mathrm{EQTR}(i \mid D) \to \frac{H^{-1}(t)}{H^{-1}(u)} \quad (0 \le t \le u \le 1) \tag{6.8}$$

uniformly as $N \to \infty$, and $i/N \to t$, $C/N \to u$. In other words, the empirical Q on test ratio tends to look like the scaled total Q on test transform as the number of data points increases, and thus can be used for model identification merely by trying different Q functions, until a satisfactory straight line approximation is obtained. Naturally, for any finite sample, there will be fluctuations in the ratio (6.6); Barlow and Campo [1] give results for the exponential case using total time on test, which should also be useful once an approximate form Q has been found.

Perhaps it is appropriate at this point to compare our procedure for nonexponential model identification with that set forth in Barlow and Campo, assuming always there is sufficient data. In their approach, they would plot empirical ratios corresponding to the scaled total time on test (e.g. (6.6) and (3.4) with $Q(x) = x$); departure from the exponential would show up as a departure from a straight line. Special characteristics, such as IFR, IFRA, NBU, etc. would be apparent from inspection, and then through the use of families of transparent overlays on the unit square, the final distribution would be chosen visually.

In our approach, we propose to try successive forms for Q, plotting the empirical ratio (6.6) each time. Regions of departure from a straight line, together with (6.4), are supposed to suggest new empirical modifications in Q, which are carried out until the EQTR is apparently a straight line.

7.0 Example. Figures 1 through 5 show various empirical Q on test ratios, based on 107 samples for the right rear brake of a D9G-66A caterpillar tractor given by Barlow and Campo [1]. Three prototype failure functions, $Q(x) = x^1$, $x^{1.5}$, and x^2 are given in each plot, from top to bottom respectively.

Figures 1, 2, and 3 illustrate the effect of varying the number of complete samples, with Figure 1 showing the original 107 lifetimes, and 2 and 3 showing the curves with only 54 and 27 lifetimes, selected at random. Clearly the variability increases with decreasing data, although $Q(x) = x^{1.5}$ is always the best of the functions chosen. Figure 1 suggests, via (6.4), that a slight modification of the failure function between 0-0.05 and 0.65-1.00 would give a better fit.

Figures 4 and 5 show the effect of incomplete life data by censoring the original 107 samples at the 54th and 27th lifetimes, respectively. Again, the model comparison is remarkably consistent, even with small duration tests, although the discrimination (separation between the plots) is less with censored data.

8.0 Data From Several Batches; Estimation of Hyperparameters and Model Identification. The use of (6.6) to identify Q depends on having sufficient complete lifetimes for a fixed value of θ. However, a more usual situation would be that we have a moderate number of samples from several production runs, each with different values of θ.

If we pool the data, then the resulting mixed distribution will be given by (5.6), from which we see that it is difficult to extricate the identification of Q from the estimation of C_o and Q_o. In fact, the TQTT of $P(x)$ with respect to the correct $Q(x)$ is *not* a straight line, but is

$$H^{-1}(t) = \frac{Q_o}{C_o - 1}\left\{1 - (1-t)^{C_o - 1/C_o}\right\} \tag{8.1}$$

which is straight only for C_o large. A normalized plot of this function is shown in Figure 6.

However, maximum likelihood estimates $\hat{C}_o$, $\hat{Q}_o$ can be reliably obtained from a large amount M of pooled data, given Q. First, $\hat{Q}_o$ is found so that the following two sums are equal,

$$\sum(\hat{Q}_o) = \sum_{i=1}^{M}\left[\frac{Q(x_i)}{\hat{Q}_o + Q(x_i)}\right] = \sum_{i=1}^{M} \ln\left[1 + \frac{Q(x_i)}{\hat{Q}_o}\right], \tag{8.2}$$

and then:

$$\hat{C}_o = M/\sum(\hat{Q}_o). \tag{8.3}$$

There remains the problem of identifying Q. We suggest the following:

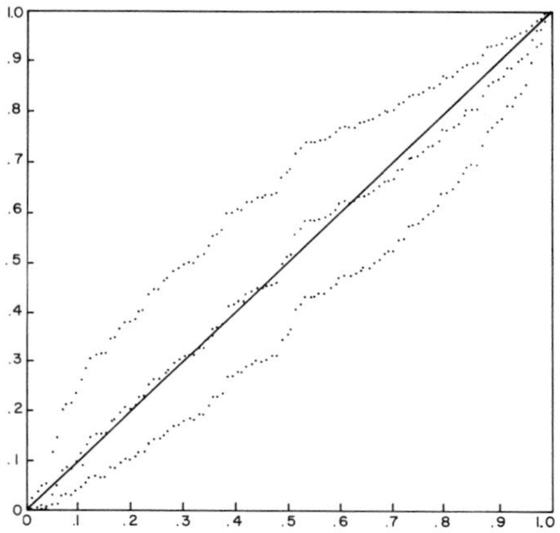

FIGURE 1: Empirical Q on test ratio. 107 complete samples, right rear brake [Barlow and Campo]. $Q(x) = x^1$, $x^{1.5}$, and x^2.

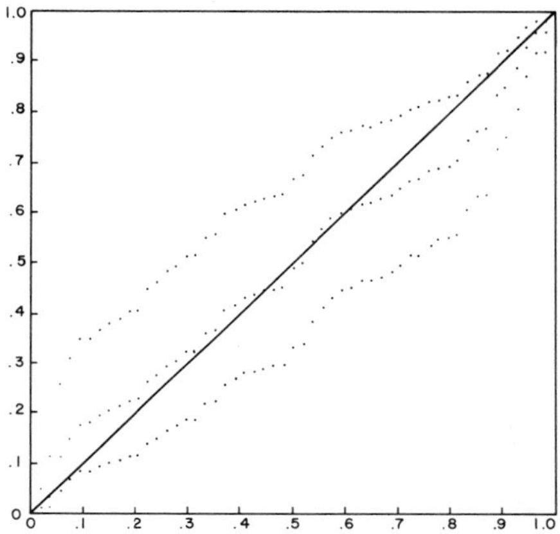

FIGURE 2: Empirical Q on test ratio. 54 complete samples, right rear brake [Barlow and Campo]. $Q(x) = x^1$, $x^{1.5}$, and x^2.

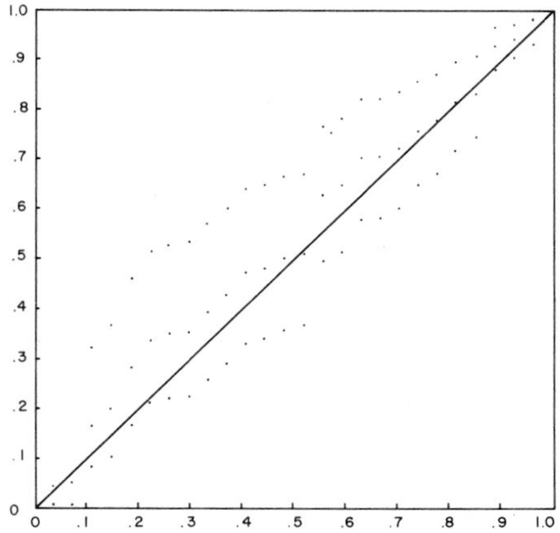

FIGURE 3: Empirical Q on test ratio. 27 complete samples, right rear brake [Barlow and Campo]. $Q(x) = x^1$, $x^{1.5}$, and x^2.

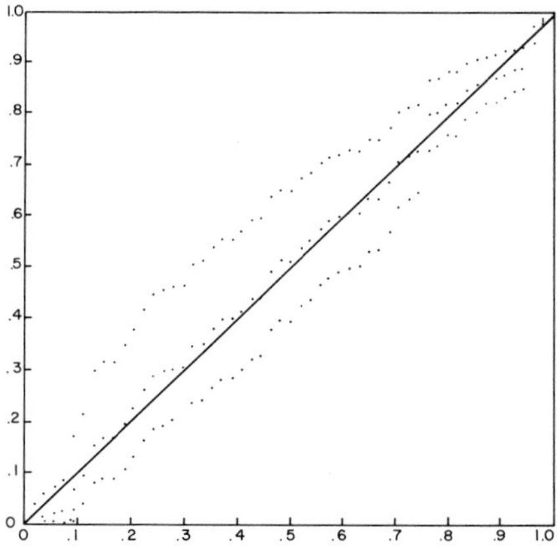

FIGURE 4: Empirical Q on test ratio. 107 samples, censored at 54th sample, right rear brake [Barlow and Campo]. $Q(x) = x^1$, $x^{1.5}$, and x^2.

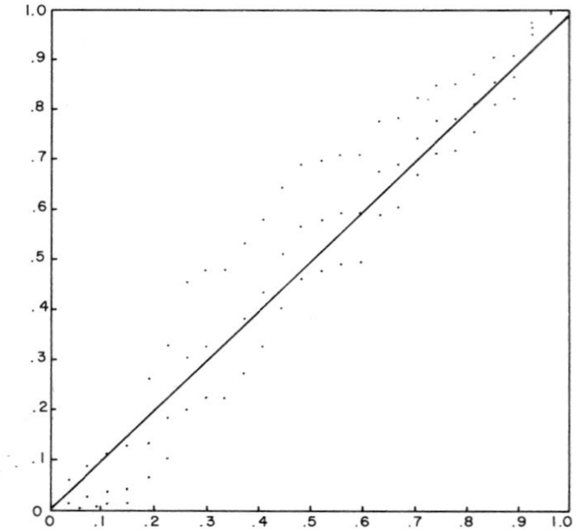

FIGURE 5: Empirical Q on test ratio. 107 samples, censored at 27th sample, right rear brake [Barlow and Campo]. $Q(x) = x^1$, $x^{1.5}$, and x^2.

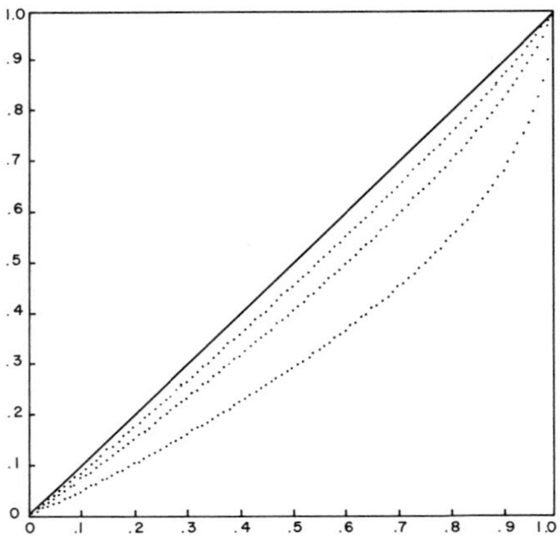

FIGURE 6: Normalized total Q on test transform for the gamma-mixed proportional hazard family with the correct Q(x), for values of the hyperparameter $C_o = 2, 4, 8$.

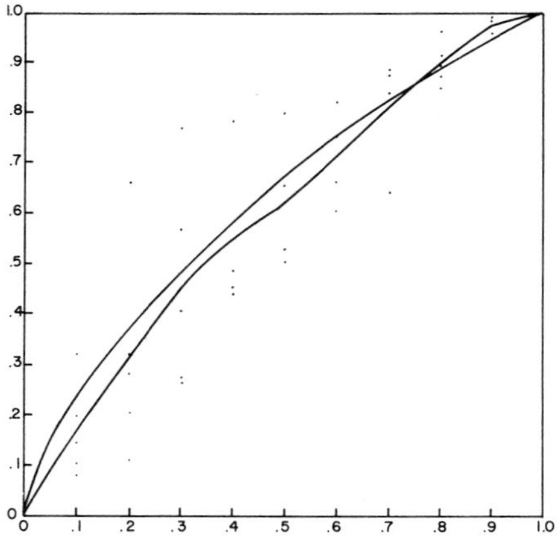

FIGURE 7: Simulated TQT data from proportional hazard family with $R(Q^{-1}(x)) = x^{1.5}$. Five runs, 10 complete samples from each run, comparing raw data, average curve, and theoretical curve.

(1) For each value of θ (production run), plot the empirical ratio (6.5) for the appropriate number of completed lifetimes;

(2) Connect these points by straight lines, or other reasonable curves;

(3) For various values of the abscissa, find the average value of the ordinate over all runs.

Even though the individual curves may vary because of the few samples per batch, the average STQTT should more nearly reflect the theoretical transform curve. One then varies the form of Q, as before, until a satisfactory fit is obtained.

Figure 7 shows this procedure for 5 runs, with 10 complete samples in each run, when $R(Q^{-1}(x)) = x^{1.5}$. The raw data is shown unconnected, the average values over all runs for $t = 0(0.1)1.0$ are connected by straight line segments, while the theoretical STQTT is the curvilinear line. Limited computational experience with other simulated data gives similar results, surprisingly smooth when averaged over all runs.

9.0 Extensions. A natural extension to (3.1) is to consider the *competing hazard family*

$$P^c(x \mid \theta) = \exp\left[-\sum_{i=1}^{m} \theta_i Q_i(x)\right] ;$$

$$p(x \mid \theta) = \left[\sum_{i=1}^{m} \theta_i q_i(x)\right] \exp\left[-\sum_{i=1}^{m} \theta_i Q_i(x)\right] . \tag{9.1}$$

Here the Q_i, and their derivatives, q_i, are prototype failure constituents which are competing in an unknown mixture specified by parameters $\theta_1, \theta_2, \ldots, \theta_m$, subject to a known m-dimensional prior density. For instance, with $m = 2$, one constituent might be random, $q_1(x) = 1$, and the other IFRA, $Q_2(x)/x \uparrow$ in x. Unfortunately, the likelihood is now more complex; corresponding to (3.2), we have for $m = 2$

$$p(x_1, x_2, \ldots, x_N ; C \mid \theta_1, \theta_2) = \frac{N!}{(N-C)!} \prod_{i=C+1}^{N} \delta(x_i - T). \tag{9.2}$$

$$\left\{ \left\{ \prod_{i=1}^{C} [\theta_1 q_1(x_i) + \theta_2 q_2(x_i)] \right\} e^{-\theta_1 TQ_1T(D) - \theta_2 TQ_2T(D)} \right\}$$

where TQ_1T and TQ_2T refer to the two possible test statistics. In place of θ^C, we have C terms corresponding to the possible powers of θ_1 and θ_2. The term in inner braces is proportional to

$$\sum_{i=1}^{C} p_i(D)\, \theta_1^i \theta_2^{C-1}$$

where

$$p_i(D) = K \cdot \prod_{\substack{\text{permutations} \\ \text{of } x_1, x_2, \ldots, x_C}} \overbrace{q_1(x_{j_1}) q_1(x_{j_2}) \cdots q_1(x_{j_i})}^{i \text{ terms}} \overbrace{q_2(x_{j_{i+1}}) \cdots q_2(x_{j_C})}^{(C-i) \text{ terms}} \tag{9.3}$$

and K is adjusted so that $\sum p_i(D) = 1$. The complexity for general m is easily visualized.

A convenient prior consist of independent gammas:

$$p(\theta_1, \theta_2 \mid C_{o1}, Q_{o1}; C_{o2}, Q_{o2}) \propto \left(\theta^{C_{o1}-1} e^{-Q_{o1}\theta_1}\right) \left(\theta^{C_{o2}-1} e^{-Q_{o2}\theta_2}\right); \tag{9.4}$$

we see that the posterior-to-data density is a mixture of C terms:

$$p(\theta_1, \theta_2 \mid D) = \sum_{i=1}^{C} p_i(D) \cdot p(\theta_1, \theta_2 \mid C_{o1}+i, Q_{o1}+TQ_1T(D); C_{o2}+TQ_2T(D)) \tag{9.5}$$

from which, for example, the updated means can be obtained, e.g.,

$$E\{\tilde{\theta}_1 \mid D\} = \sum_{i=1}^{C} p_i(D) \frac{C_{o1} + i}{C_{o1}[E\{\tilde{\theta}_1\}]^{-1} + TQ_1T(D)} , \tag{9.6}$$

Although these Bayesian calculations are more complicated than before, they are easily computerized for $m = 2$ or 3.

Another possible extension is to a *competing lifetime model family*

$$P^c(x \mid \theta) = \sum_{i=1}^{m} \pi_i \exp[-\theta_i Q_i(x)] ;$$

$$p(x \mid \theta) = \sum_{i=1}^{m} \pi_i \theta_i q_i(x) \exp[-\theta_i Q_i(x)]. \tag{9.7}$$

Here the $\pi_i \geq 0$ $(i = 1,2, \ldots, m)$, $\sum_{j=1}^{m} \pi_j = 1$, are random variables, with, say, a Dirichlet prior distribution, and the θ_i may be fixed or subject to a joint or independent prior. What happens, in general, is that, as data is received, the posterior estimates of the π_i change (along with the θ_i), thus tending to select the "correct" model. Unfortunately, even for $m = 2$, this approach leads to terms in which varying numbers of the Q_i are summed, eliminating the usefulness of the TQT statistic.

However, it is possible to extend the class of priors somewhat by using a *competing prior model family*

$$p(\theta) = \sum_{i=1}^{m} \pi_i p_i(\theta) \; ; \; P^c(x \mid \theta) = \exp\{-\theta Q(x)\} ; \tag{9.8}$$

where the π_i are either fixed, or have their own prior. Here the size of the computation remains fixed, and the TQT changes the weights π_i as a function of the data. For instance, if the $p_i(\theta) = p(\theta \mid C_{oi}, Q_{oi})$ are gamma, and the initial π_i are fixed, then posterior to the data we have from (3.2):

$$\pi_i(D) = K \cdot \pi_i \; \frac{\Gamma(C_{oi} + C)}{\Gamma(C_{oi})} \; \frac{Q_{oi}^{C_{oi}}}{[Q_{oi} + TQT(D)]^{C_{oi}+C}} , \qquad (9.9)$$

where K is chosen so that $\sum \pi_i(D) = 1$. Of course, this approach requires much more effort in estimating the hyperparameters.

Acknowledgement. Computations and plotting were ably programmed by Jan Kwiatkowski.

REFERENCES

[1] Barlow R. E. and R. A. Campo, "Total Time on Test Processes and Applications to Failure Data Analysis," ORC 75-8, Operations Research Center, University of California, Berkeley, (1975).

[2] Jewell, W. S., "Credible Means are Exact Bayesian for Simple Exponential Families," ORC 73-21, Operations Research Center, University of California, Berkeley, (October 1973), *Astin Bulletin*, Vol. VIII, Part 1, (1974), pp. 77-90.

[3] Jewell, W. S., "Regularity Conditions for Exact Credibility," ORC 74-22, Operations Research Center, University of California, Berkeley, (July 1974), *Astin Bulletin*, Vol. VIII, Part 3, (1975), pp. 336-341.

[4] Jewell, W. S., "Exact Multidimensional Credibility," ORC 74-14, Operations Research Center, University of California, Berkeley, (May 1974), Mitteilungen der Vereinigung schweizerischer Versicherungsmathematiker, Band 74, Heft 2, (1974), pp. 193-214.

[5] Mann, N. R. , R. E. Schafer and N. D. Singpurwalla, Methods for Statistical Analysis of Reliability and Life Data, Wiley, New York, (1974).

A BAYESIAN APPROACH FOR TESTING INCREASING FAILURE RATE*

by ROBERT H. LOCHNER
Marquette University

and A. P. BASU
University of Missouri - Columbia

Abstract. In situations where a statistical analysis is carried out on life-testing data, it is often useful to study the failure rate of the population. The failure rate, q(t,x), of a population over the time interval (t,t+x) is defined as the ratio of the probability that a member selected at random from the population will die in the time interval (t,t+x), given that the member is alive at time t, divided by x. It has been found, both by theoretical and practical investigators, that many of the probability distributions useful in life-testing have either a monotone increasing or monotone decreasing (or constant) failure rate. That is, q(t,x) is a monotone function of t for fixed x. Several tests have been developed by others to test if a population has monotone failure rate. In this article inference for q(t,x) is developed from a Bayesian viewpoint. The failure rate for a future observation is obtained based on an existing data set. Bounds are obtained for the posterior probability that a distribution has increasing (decreasing) failure rate. The methods can be used with random samples or multiply censored samples.

1.0 Introduction. Consider a population of random life times having cumulative distribution function F(t) where $F(0^-) = 0$. Then the failure rate for the population over the interval (t,t+x) is defined as

$$q(t,x) = \frac{F(t+x)-F(t)}{x(1-F(t))} \text{ for } 0 < x,\ 0 < t,\ F(t) < 1. \qquad (1.1)$$

* Research supported by the Air Force Office of Scientific Research under Grant No. AFOSR-75-2795.

If $F(x)$ has a density function $f(x)$, then the instantaneous failure rate, or failure rate function, is defined as

$$q(t) = \lim_{x \to 0^+} q(t,x) = f(t)/(1-F(t)) \text{ for } 0 < t, \quad F(t) < 1. \tag{1.2}$$

$F(t)$ is said to have increasing failure rate (IFR) or decreasing failure rate (DFR) if $q(t,x)$ is a monotone increasing or decreasing function of t for all $x > 0$.

Several authors have developed estimators or hypothesis testing procedures for $q(t)$ or $q(t,x)$. Epstein [10] presented several tests for constant failure rate versus various alternatives. Proschan and Pyke [18] derived a test for constant failure rate versus IFR. Exact and asymptotic properties of the test were considered. Barlow and Proschan [4] developed an extensive theory of monotone failure rates, but most of their results are structural probabilistic rather than inferential. Barlow and Proschan [5] and Hollander and Proschan [12] considered tests for conditions which contain the increasing failure rate condition. Properties of tests for increasing failure rate have been considered by Barlow [3], Bickel [6], and Bickel and Doksum [7], and others. All of these tests for failure rate have been based on the Neyman-Pearson formulation for hypothesis testing.

Only a few papers have used any type of Bayesian consideration for failure rates (See Aldrich and Morton [1], for example).

In this article the posterior failure rate over a given time intejval is obtained and bounds are given for the probability a distribution is IFR (or DFR). The methods can be used with multiply censored or grouped sample data as well as with random samples. The sampling model is given in Section 2, the probability bounds in Section 4, and the posterior failure rate in Section 5. Selection of prior parameters is discussed in Section 6. Section 7 contains illustrative examples.

2.0 Sampling Model. Let $X_1, X_2, \ldots, X_n$ be a random sample from a population of life times with cdf $F(t)$ where $F(0) = 0$. F can be either discrete or continuous. Assume that the ith sample unit is put on life test for fixed time $\tau_i (i = 1,2,\ldots,n)$. Then an observation on X_i is available if $X_i \leq \tau_i$. Let t^* be some arbitrarily selected number at least as large as any of the finite τ_i's. We partition the interval $(0,t^*)$ into a set of k subintervals $\{(t_{j-1}, t_j), j = 1,2,\ldots,k\}$ where $0 = t_o < t_1 <\ldots< t_k = t^*$ and each finite τ_i is equal to some t_j. (The τ_i's are not necessarily all distinct). Then the set of distinct finite τ_i's is a subset of $\{t_1, t_2, \ldots, t_k\}$.

We next define s_j as the number of sample units which fail in the time interval $(t_{j-1},t_j](j = 1,2,\ldots,k)$ and define r_j as the number of units which are removed from testing at time t_j without having failed during the test $(j = 1,2,\ldots,k-1)$. Let N_j be the number of units having fixed test times $(\tau_i$'s) equal to $t_j(j = 1,2,\ldots,k-1)$. We also let r_k equal the number of units removed from test at time t_k or observed to fail after time t_k and let N_k equal the number of units having test times, τ_i, greater than or equal to t_k. Then

$$\sum_{j=1}^{k} (s_j + r_j) = \sum_{j=1}^{k} N_j = n.$$

3.0 Preliminary Results.

Let

$$p_j = F(t_j) - F(t_{j-1}) ,$$

$$\underline{p} = (p_1,\ldots,p_k)$$

and $$\underline{s} = (s_1,\ldots,s_k; r_1,\ldots,r_k).$$

It is shown in Lochner and Basu [14] that $\underline{s}$ is a sufficient statistic for $\underline{p}$. The exact and asymptotic probability functions for $\underline{s}$ are derived in that report.

We will assume throughout that $\underline{p}$ has, a priori, a Dirichlet density function $f(\underline{p})$ over

$$\rho = \{\underline{p}\colon 0 \le p_j \le 1, j = 1,2,\ldots,k, \sum_{j=1}^{k} p_j \le 1\}. \tag{3.1}$$

where

$$f(\underline{p})\, d\underline{p} \propto p_1^{v_1-1} \cdots p_k^{v_k-1} (1-p_1 - \cdots - p_k)^{v_{k+1}-1}\, dp_1 \cdots dp_k\ .$$

Selection of appropriate values for $v_1,\ldots,v_{k+1}$ is considered in Section 6. For now, we simply require that either the v_j's be all positive or all equal to zero.

4.0 Bounds on IFR Probabilities.

Let $\underline{u} = (u_1, \ldots, u_k)$ where

$$u_1 = 1 - p_1 \tag{4.1}$$

$$u_j = (1-p_1 - \cdots - p_j)/(1-p_1 - \cdots - p_{j-1}), (j=2,3,\ldots,k).$$

Then

$$u_j = 1-q(t_{j-1},t_j-t_{j-1})(t_j-t_{j-1}) \tag{4.2}$$

where $q(t_{j-1},t_j-t_{j-1})$ is the failure rate over the interval (t_{j-1},t_j). (See equation 1.1).

It follows from equation (3.1) that $u_1, u_2, \ldots, u_k$ are, a priori, independent beta variables with density functions $f_j(u|\underline{v})$, $j = 1,2,\ldots,k$, where

$$f_j(u|\underline{v})\ du \propto u^{w_j-1}(1-u)^{v_j-1}\ du\quad 0 \le u \le 1,$$

$w_j = v_{j+1} + \cdots + v_{k+1}$ and either $v_j > 0$ for all j or $v_j = 0$ for all j. It can also be shown that the u_j's are, a posteriori, independent beta variables with density functions given by

$$f_j(u|\underline{s},\underline{v}) = B^{-1}(\alpha_j,\beta_j)u^{\alpha_j-1}(1-u)^{\beta_j-1}, \quad 0 \leq u \leq 1 \tag{4.3}$$

where

$$\alpha_j = \sum_{i=j+1}^{k} (s_i + r_i + v_i) + r_j + v_{k+1},$$

$$\beta_j = s_j + v_j,$$

and

$$B^{-1}(\alpha,\beta) = \Gamma(\alpha+\beta)/(\Gamma(\alpha)\cdot\Gamma(\beta)).$$

(See Lochner and Basu [14]). We assume throughout that α_j and β_j are positive for all j.

Suppose $F(t)$ has an increasing failure rate. Then if we require that the t_i's be equally spaced (that is, $t_j=jt^*/k, j = 1,2,\ldots,k$), it is clear from equation (4.2) that $u_1,u_2,\ldots,u_k$ for a decreasing sequence from left to right. The converse is not necessarily true, however. That is $u_1 > u_2 > \ldots > u_k$ does not imply that $F(t)$ is IFR. Hence

$$\text{Prob }(F(t) \text{ is IFR}|\underline{s},\underline{v}) < \text{Prob}(u_1 < u_2 < \ldots < u_k|\underline{s},\underline{v})$$

where

$$\text{Prob}(u_1>u_2>\ldots>u_k|\underline{s},\underline{v}) = \int_0^1\int_0^{u_1}\cdots\int_0^{u_{k-1}} \prod_{j=1}^{k} \{B^{-1}(\alpha_j,\beta_j) u_j^{\alpha_j-1}(1-u_j)^{\beta_j-1}\}du_k\ldots du_1 \tag{4.4}$$

If the β_j's are integers (which would be the case whenever the v_j's are all integers) then

$$\text{Prob}(u_1>u_2>\ldots>u_k|\underline{s},\underline{v}) = [\prod_{j=1}^{k} B(\alpha_j,\beta_j)]^{-1} \sum_{j_k=0}^{\beta_k-1} \cdots \sum_{j_1=0}^{\beta_1-1} \tag{4.5}$$

$$\prod_{i=1}^{k} \left[\binom{\beta_i - 1}{j_i} (-1)^{j_i} \left(\sum_{d=i}^{k} (\alpha_d + j_d) \right)^{-1} \right]$$

Equations (4.4) and (4.5) give an upper bound for the posterior probability that $F(t)$ is IFR. However, if k is large we would expect $\mathrm{Prob}(u_1 > u_2 > \dots > u_k | \underline{s}, \underline{v})$ to be approximately equal to Prob $(F(t)$ is IFR$|\underline{s},\underline{v})$. To obtain an upper bound for the posterior probability that $F(t)$ is DFR we compute $\mathrm{Prob}(u_k > u_{k-1} > \dots > u_1 | \underline{s}, \underline{v})$. This can be calculated using equations (4.4) or (4.5) by inverting the notation (that is, replacing (α_j, β_j) by $(\alpha_{k+1-j}, \beta_{k+1-j})$, $j = 1,2,\dots,k)$.
Paralleling the IFR case, we have that
Prob$(F(t)$ is DFR$|\underline{s},\underline{v}) < \mathrm{Prob}(u_k > u_{k-1} > \dots > u_1 | \underline{s}, \underline{v})$.

5.0 Probability Distribution for Future Samples. Jeffreys [13] and others have used Bayesian analysis to predict the behavior of future samples based on information from an existing sample. That method will be used here to obtain a joint posterior time to failure cdf for a future sample. As a special case (sample size = 1) a population posterior time to failure cdf, or a "predictive distribution" of time to failure (and hence an estimate of reliability) will be obtained. A conditional posterior probability function for $\underline{p}$, which can be regarded as a posterior failure rate, is also given in this section.

Result 5.1. Let $\underline{s}$, $\underline{p}$, and $\underline{u}$, be as defined before. Let $\underline{X} = (X_1,\dots,X_c)$ be the (unknown) times to failure of a sample to be tested at a future date. (We assume that all units in this future sample will be tested for at least time t^*.) Let Y_j denote the value i for which $t_{i-1} < X_j \le t_i$ holds, and let $\underline{Y} = (Y_1,\dots,Y_c)$. Then the posterior probability function for $\underline{Y}$ given $\underline{s}$ is

$$\text{Prob}(\underline{y}|\underline{s}) = \prod_{i=1}^{k} B\left[\sum_{j=1}^{c}\sum_{m=i+1}^{k+1}\delta_{mj}+\alpha_i,\ \sum_{j=1}^{c}\delta_{ij}+\beta_i\right]/B(\alpha_i,\beta_i). \qquad (5.1)$$

where α_i and β_i are defined in Section 4 and

$$\delta_{ij} = \begin{cases} 1, & \text{if } t_{i-1} < X_j \le t_i \\ 0, & \text{if } X_j \notin (t_{i-1}, t_i]. \end{cases}$$

Proof. See Appendix.

As a special case of equation (5.1), if $c = 1$ (i.e., the future sample is of size one), then

$$\text{Prob}(y_1|\underline{s}) = \prod_{i=1}^{k} B\left(\sum_{m=i+1}^{k}\delta_{m1} + \alpha_i,\ \delta_{i1} + \beta_i\right)/B(\alpha_i,\beta_i)$$

and so

$$\text{Prob}\{t_{i-1} < X_1 \le t_i|\underline{s}\} = \prod_{j=1}^{i-1} (\alpha_j/(\alpha_j+\beta_j))\ \ \beta_i/(\alpha_i+\beta_i)$$
$$= \gamma_1\ldots\gamma_{i-1}\ (1-\gamma_i) \qquad (5.2)$$

where

$$\gamma_j = \alpha_j/(\alpha_j + \beta_j).$$

Similarly, for $c = 1$,

$$\text{Prob}\{X_1 \le t_i|\underline{s}\} = \sum_{j=1}^{i} \gamma_1\ldots\gamma_{j-1}\ (1-\gamma_j) \qquad (5.3)$$

$$= 1 - \gamma_1\ldots\gamma_i, \quad i = 1,\ldots,k.$$

Using equations (5.2) and (5.3) we can immediately obtain the posterior failure rate over any of the intervals $(t_{i-1},t_i]$:

$$q(t_{i-1},t_i-t_{i-1}) = \text{Prob}(t_{i-1}<X_1\le t_i|t_{i-1}<X_1)/(t_i-t_{i-1})$$
$$= \frac{\text{Prob}(t_{i-1}<x_1\le t_i)}{(t_i-t_{i-1})\text{Prob}(t_{i-1}<X_1)} \qquad (5.4)$$
$$= (1-\gamma_i)/(t_i-t_{i-1}).$$

6.0 Selection of Prior Parameters. In Section 3 we assigned a Dirichlet prior density to $\underline{p}$ without giving any explanation for this action. The Dirichlet distribution has been used as a prior distribution for multinomial variables by several authors. (See, for example, Novick and Grizzle [17], Altham [2] and Lochner and Basu [15]). Conner and Mosimann [8] proposed a generalized Dirichlet distribution which lochner [16] applied to life testing situations. Ferguson [11] and others have considered Dirichlet processes in which for any partitioning $(A_1,\ldots,A_m)$ of the sample space of the aprent population, $(P(A_1),\ldots,P(A_m))$ has a Dirichlet distribution, where $P(A_i) = \int_{A_i} f(x)dx$. The reason most often cited for using a Dirichlet prior is that it is the conjugate prior for the multinomial distribution. The Dirichlet distribution and the generalized Dirichlet distribution are conjugate priors for the probability function given by equation (A.3).

In selecting values for the prior distribution parameter **vector** $\underline{v}$, a distinction should be made between the situations where extensive prior information is available and where the experimenter is in a state of relative ignorance. If extensive prior information is available, the only problem is how to incorporate that information into the prior density. For a Dirichlet distribution,

$$E(p_i) = v_i/(v_1 + \ldots + v_{k+1}) \tag{6.1}$$

and

$$Var(p_i) = v_i(v_1+\ldots+v_{k+1}-v_i)/((v_1+\ldots+v_{k+1})^2(v_1+\ldots+v_{k+1}+1)).$$

Suppose we have prior estimates, $\hat{p}_i$, of the expected values of the p_i's, where $\hat{p}_1+\ldots+\hat{p}_{k+1} = 1$. If we set $v_i = \eta\hat{p}_i$ for some $\eta > 0$ then, a priori, $E(p_i) = \hat{p}_i$

and

$$Var(p_i) = \hat{p}_i\ (1-\hat{p}_i)/(\eta+1). \tag{6.2}$$

Since $E(p_i)$ is independent of η but $Var(p_i)$ is monotone decreasing in η, it would be reasonable to let η be a measure of our "confidence" in our prior estimates $\hat{p}_1,\ldots,\hat{p}_{k+1}$.

Suppose we have very little prior information about the population being considered, and have little "confidence" in our estimates $\hat{p}_1,\ldots,\hat{p}_{k+1}$. To reflect this situation we would want to make $Var(p_i)$ large. This can be accomplished by making η small. It would be tempting to set $\eta = 0$ to reflect no prior information, but then $f(p_i) \to \infty$ as $p_i \to 0$ or 1, which is generally unsatisfactory. If $\eta > 0$, then $f(p_i) \to 0$ as $p_i \to 0$ or 1, a much better situation. The problems caused by setting $\eta = 0$ are discussed further by Raiffa and Schlaifer [19], pages 63-65, for the case $k = 1$.

To pursue the interpretation of η further, suppose no prior information was available and the parameter vector $\underline{v} = (v_1^*,\ldots,v_{k+1}^*)$ represents a "noninformative" prior for $\underline{p}$. Then $(v_1^* +\ldots+ v_{k+1}^*)$ would be close to zero to guarantee large prior variances for the p_i's. Suppose further that it was decided to run a life test on N_k^* units for time t_k, and use the resulting posterior density for $\underline{p}$ as the prior density function for $\underline{p}$ in a future experiment. Let $\underline{s}^* = (s_1^*,\ldots,s_k^*,r_k^*)$ denote the data vector obtained from the N_k^* observations $(r_k^* = N_k^* - s_1^* -\ldots- s_k^*$ and $r_1^* = \ldots = r_{k-1}^* = 0)$. Then the density function for $\underline{p}$ given $\underline{s}^*$ is again Dirichlet with parameter vector $v = (v_1,\ldots,v_{k+1}) = (s_1^* + v_1^*,\ldots,s_k^* + v_k^*, r_k^* + v_{k+1}^*)$. We then have $E(p_i|\underline{s}^*) = (s_i^*+v_i^*)/(N_k^*+v_1^*+\ldots+ v_{k+1}^*) = \hat{p}_i$, say. Since $v_1^* +\ldots+ v_{k+1}^*$ is close to zero, $E(p_i|\underline{s}^*)$ is approximately equal to s_i^*/N_k^*, the maximum likelihood estimator of p_i in a sampling theory context. If we now set $v_i = \eta\hat{p}_i$, then $\eta = N_k^* + v_1^* + \ldots + v_{k+1}^* \doteq N_k^*$. Hence, in the situation where prior information is available, η can represent an estimate of the number of sample units which would need to be tested in order to arrive at the state of knowledge in which an experimenter finds himself at the start of his life testing experiment. This method of determining a prior parameter value based on an estimate of the sample size needed to attain the amount of information which is a **priori available was first** proposed by Draper and Guttman [9].

7.0 Illustrative Example. Six samples were obtained using a pseudo-random number generator. All samples were from Weibull distributions with mean life of 100 hours. That is, the parent populations had distribution functions of the form

$$F(x) = 1 - \exp\{-(x/\theta)^{\beta}\}, \quad x \geq 0, \; \beta > 0, \tag{7.1}$$

where

$$\theta = 100/\Gamma(1/\beta+1).$$

For the first two samples the shape parameter, β, was 2. For the second two samples β was 1 (exponential distribution). For the last two samples β was 0.5. For Weibull distributions, $\beta > 1$ implies IFR and $\beta < 1$ implies DFR. In generating our samples we set $t_0 = 0$, $t_1 = 50$, $t_2 = 100$ and $t_3 = t^* = 150$. For the first, third and fifth samples $N_1 = N_2 = N_3 = 6$. For the second, fourth and sixth samples $N_1 = N_2 = 5$, $N_3 = 10$. The sample data obtained is summarized in Table 1.

Table 1: Sample Data

Sample number	, Shape parameter	Sample sizes N_1	N_2	N_3	$\underline{s} = (s_1,s_2,s_3;\ r_1,r_2,r_3)$
1	2	6	6	6	(3,2,4; 5,4,0)
2	2	5	5	10	(5,3,4; 3,3,2)
3	1	6	6	6	(10,2,1; 2,2,1)
4	1	5	5	10	(7,2,1; 4,3,3)
5	.5	6	6	6	(9,1,1; 2,3,2)
6	.5	5	5	10	(11,2,1; 2,1,3)

Upper bounds on the posterior probabilities that each data set came from IFR or DFR populations were calculated using equation (4.5). One minus the sum of these two bounds provided a lower bound on the probability the parent population was neither IFR or DFR. Two different prior parameter vectors were used with each data set: $\underline{v}_A = (4,2,1,1)$ and $\underline{v}_B = (0,0,0,0)$. $\underline{v}_A$ represents a prior belief that the failure rate is constant. It has the smallest integer values possible for the v_i's while still maintaining a constant prior failure rate. $\underline{v}_B$ denotes either no prior information or an extreme prior, depending on your point of view (see Section 6). The resulting prior and posterior probabilities are given in Table 2. Note that there is little difference in probability bounds for the different values of $\underline{v}$. Although no conclusions can be drawn based on this limited data, it appears that the procedure is reasonably robust under variation in $\underline{v}$. This is rather surprising when one considers that $\underline{v}_A$ represents prior information equivalent to $N_k^* = 8$ observations (see Section 6) and all the samples generated here contain 20 or fewer observations. In Table 2, the major changes between prior and posterior probability bounds occur in the IFR and DFR columns. This is to be expected. With a prior assumption of constant failure rate $(\underline{v}_A)$ or "no" prior assumption $(\underline{v}_B)$, and with sample sizes of only 18 or 20, it is not surprising that there is no strong rejection of the "neither" possibility.

Table 2: Bounds on probability a distribution is IFR, DFR, or neither

Sample number	$\underline{v} = (4,2,1,1)$ P(IFR)	P(DFR)	P(neither)	$\underline{v} = (0,0,0,0)$ P(IFR)	P(DFR)	P(neither)
Prior	.17	.17	.65	(not computable)*		
1	.53	.00	.47	(not computable)*		
2	.41	.01	.58	.45	.00	.54
3	.09	.23	.68	.06	.24	.69
4	.08	.27	.64	.05	.33	.62
5	.03	.25	.72	.01	.23	.73
6	.02	.43	.54	.01	.50	.49

* Probabilities not computable because one or more of the distribution parameters were zero.

In Table 3 conditional posterior probabilities of failure are given for each interval. That is, for a future sample observation X,

$$P(t_{i-1} < X \leq t_i \mid t_{i-1} < X; \underline{s},\underline{v})$$

$$= (t_i - t_{i-1})\, q\, (t_{i-1}, t_i - t_{i-1})$$

$$= \beta_i/(\alpha_i + \beta_i)$$

is calculated for $i = 1, 2, 3$ based on each of the six samples generated, using both $\underline{v}_A$ and $\underline{v}_B$. When $\underline{v} = \underline{v}_B$ the posterior probabilities are the sampling theory maximum likelihood estimates of these probabilities. As was the case with Table 2, the differences between posterior probabilities based on $\underline{v}_A$ and $\underline{v}_B$ are slight. Using $\underline{v}_A$ tends to "smooth out" the posterior probabilities since $\underline{v}_A$ represents prior information that the probabilities are equal.

Table 3: Posterior conditional probability of failure

Sample number	$P(t_{i-1} < X \leq t_i \mid t_{i-1} < X;\ \underline{s},\underline{v})$					
	$\underline{v} = (4,2,1,1)$			$\underline{v} = (0,0,0,0)$		
	i=1	2	3	i=1	2	3
Prior	.50	.50	.50	(not computable)		
1	.27	.29	.83	.17	.20	1.00
2	.32	.31	.62	.25	.25	.67
3	.54	.40	.50	.56	.33	.50
4	.39	.31	.33	.35	.22	.25
5	.50	.27	.40	.50	.14	.33
6	.54	.36	.33	.55	.29	.25

The examples given here are based on small samples. This was not just for convenience, but was in fact necessary. Although the methods presented here are valid for any size sample, equation (4.5) presents serious computational difficulties for sample sizes much larger than those used here. The problem is that equation (4.5) is an alternating series. Accuracy beyond the capability of most computers is required to successfully evaluate equation (4.5) for moderate or large sample sizes. No suitable approximation to equation (4.5) has been found so far. The methods of Section 5 do not suffer from this accuracy problem, and may be easily used with any size sample.

8.0 Appendix. Proof of Result 5.1:

$$\text{Prob}(\underline{y},\underline{s}) = \text{Prob}\{t_{y_j-1} < X_j \leq t_{y_j},\ j=1,\ldots,c, \text{ and } \underline{s} \text{ is observed}\}$$

$$= \int_\rho f(\underline{y},\underline{s},\underline{p})\,d\underline{p}$$

$$= \int_\rho f(\underline{y}|\underline{s},\underline{p})\,f(\underline{s}|\underline{p})\,f(\underline{p})\,d\underline{p} \tag{A.1}$$

where

$$f(\underline{y}|\underline{s},\underline{p}) = f(\underline{y}|\underline{p})$$

$$= \prod_{i=1}^{k} p_i^{\sum_{j=1}^{c} \delta_{ij}} (1-p_1-\ldots-p_k)^{\sum_{j=1}^{c} \delta_{k+1,j}} \tag{A.2}$$

It is shown in Lochner and Basu [14] that

$$f(\underline{s}|\underline{p}) = C(\underline{s}) \prod_{i=1}^{k} p_i^{s_i}(1-p_1-\ldots-p_i)^{r_i} \text{ for } \underline{s} \in \Omega, \text{ where} \tag{A.3}$$

$$C(\underline{s}) = \prod_{i=1}^{k} \binom{N_i}{r_i} \binom{s_1+\ldots+s_i+r_1+\ldots+r_{i-1}-N_1-\ldots-N_{i-1}}{s_i} \quad \text{and}$$

$$\Omega = \{\underline{s}: \{0 \le r_i \le N_i, 0 \le s_i \le N_i+\ldots+N_k-s_{i+1}-\ldots-s_k),$$

$$\sum_{j=1}^{i} N_j-r_j \le \sum_{j=1}^{i} s_j, \ \sum_{j=1}^{k} (N_j-r_j) = \sum_{j=1}^{k} s_j, \ i=1,\ldots,k\}.$$

Hence using equations (A.1), (A.2), (A,3) and (3.1) we have

$$\text{Prob}(\underline{y},\underline{s}) \propto \int_\rho \left[\prod_{i=1}^{k} p_i^{\sum_{j=1}^{c} \delta_{ij}} \right] \cdot \tag{A.4}$$

$$\cdot (1-p_1-\ldots-p_k)^{\sum_{j=1}^{c} \delta_{k+1,j}} \quad C(\underline{s}) \prod_{i=1}^{k} p_i^{s_i+v_i-1} (1-p_1-\ldots-p_i)^{r_i}$$

$$\cdot (1-p_1-\ldots-p_k)^{v_{k+1}-1} d\underline{p}$$

Making the variable change defined in equation (4.1) we obtain

$$\text{Prob}(\underline{y},\underline{s}) \propto C(\underline{s}) \int_0^1 \cdots \int_0^1 \prod_{i=1}^{k} u_i^{\alpha_i + \sum_{m=i+1}^{k+1} \sum_{j=1}^{c} \delta_{mj}} \cdot (1-u_i)^{\sum_{j=1}^{c} \delta_{ij}+\beta_i-1}$$

$$du_1 \ldots du_k$$

$$\propto C(s) \prod_{i=1}^{k} B\left[\sum_{j=1}^{c}\sum_{m=i+1}^{k+1} \delta_{mj}+\alpha_i, \sum_{j=1}^{c} \delta_{ij}+\beta_i\right] . \qquad (A.5)$$

It is easily shown that

$$f(\underline{s}) = \int_0^1 \cdots \int_0^1 f(\underline{s}|\underline{u})f(\underline{u})\ du_1 \ldots du_k \qquad (A.6)$$

$$\propto C(\underline{s}) \prod_{i=1}^{k} B(\alpha_i, \beta_i).$$

Substituting equations (A.5) and (A.6) into the right side of

$$\text{Prob}(\underline{y}|\underline{s}) = f(\underline{y},\underline{s})/f(\underline{s})$$

completes the proof.

Acknowledgement. We are grateful to the Marquette University Computer Services Division for computing time.

REFERENCES

[1] Aldrich, C. A. and Morton, T. E., Bayesian modification of the failure rate function. *J. Amer. Statist. Assoc.*, *68*, (1973), pp. 483-484.

[2] Altham, P. M. E., Exact Bayesian analysis of a 2x2 contingency table and Fisher's "exact" significance test. *J. R. Statist. Soc. B*, *31*, (1969), pp. 261-269.

[3] Barlow, R. E., Likelihood ratio tests for restricted families of probability distributions. *Annals Math. Statist.*, *39*, (1969), pp. 547-560.

[4] Barlow, R. E. and Proschan, F., Mathematical Theory of Reliability. Wiley, New York, (1965).

[5] ______________________, A note on tests for monotone failure rate based on incomplete data. *Annals Math. Statist.*, *40*, (1969), pp. 595-600.

[6] Bickel, P. J., Tests for monotone failure rate II. *Annals Math. Statist.*, *40*, (1969), pp. 1250-1260.

[7] ____________ and Doksum, K., Tests on monotone failure rate based on normalized spacings. *Annals Math. Statist.*, *40*, (1960), pp. 1216-1235.

[8] Conner, R. J. and Mosimann, J. E., Concepts of independence for proportions with a generalization of the Dirichlet distribution. *J. Amer. Statist. Assoc.*, *64*, (1969), pp. 194-206.

[9] Draper, N. R. and Guttman, I., The value of prior information. T. R. 135, Dept. of Statist., Univ. of Wisconsin, (1967).

[10] Epstein, B., Tests for the validity of the assumption that the underlying distribution of life is exponential. *Technometrics*, *2*, (1960), pp. 83-101 and 167-183.

[11] Ferguson, T. S., A Bayesian analysis of some nonparametric problems. *Ann. Statist.*, 1, (1973), pp. 209-230.

[12] Hollander, M. and Proschan, F., Testing whether new is better than used. *Annals. Math. Statist.*, *43*, (1972), pp. 1136-1146.

[13] Jeffreys, H., Theory of Probability, 3rd Ed., Clarendon Press, Oxford, (1961).

[14] Lochner, R. H. and Basu, A. P., Bayesian analysis of time truncated samples. Tech. Report No. 201, Department of Statistics, University of Wisconsin, (1969).

[15] ____________________________, Bayesian analysis of the two-sample problem with incomplete data. *J. Amer. Statist. Assoc.*, *67*, (1972), pp. 432-438.

[16] ____________, A generalized Dirichlet distribution in Bayesian life testing. *J. Royal Statist. Soc. B*, *37*, (1975), pp. 103-113.

[17] Novick, M. R. and Grizzle, J. E., A Bayesian approach to the analysis of data from clinical trials. *J. Amer. Statist. Assoc.*, *60*, (1965), pp. 81-96.

[18] Proschan, F. and Pyke, R., Tests for monotone failure rate. *Proc. Fifth Berkeley Symp. Math. Statist. Prob.*, *3*, (1967), pp. 293-312.

[19] Raiffa, H. and Schlaifer, R., Applied Statistical Decision Theory. Harvard University Graduate School of Business Administration, Boston, (1961).

NONPARAMETRIC ESTIMATION OF DISTRIBUTION FUNCTIONS*

by MYLES HOLLANDER
The Florida State University
and RAMESH M. KORWAR
University of Massachusetts

Abstract. Consider the nonparametric estimation of n + 1 distribution functions $F_1, F_2, \ldots, F_{n+1}$, on the basis of samples $\underline{X}_i = (X_{i1}, \ldots, X_{im_i})$ from F_i, $i = 1, \ldots, n + 1$. In an empirical Bayes context where $\underline{X}_1, \ldots, \underline{X}_n$ were the "past" samples, Korwar and Hollander [14] used Ferguson's [9] Dirichlet process to propose an estimator of F_{n+1} that utilized all n + 1 samples. Their estimator was developed for the equal sample sizes case $m_1 = m_2 = \ldots = m_{n+1}$. Here the Korwar-Hollander estimator is generalized to the unequal sample sizes case, and exploited for simultaneous estimation of $F_1, F_2, \ldots, F_{n+1}$. We establish a necessary and sufficient condition for the proposed estimators to be better (riskwise) than using, for each distribution, the corresponding empirical distribution function as an estimator. One sufficient condition, easily obtained from our necessary and sufficient condition, is $n \cdot \min(m_1, m_2, \ldots, m_{n+1}) > \max(m_1, m_2, \ldots, m_{n+1})$.

1.0 *Introduction.* For the problem of estimating a distribution function F, on the basis of a sample $X_1, \ldots, X_m$ from F, the empirical distribution function $\hat{F}_m$ is known to be an excellent

* Research sponsored by the Air Force Office of Scientific Research, AFSC, USAF, under Grant AFOSR-74-2581B. (Invited paper for Air Force Office of Scientific Research Conference on The Theory and Applications of Reliability).

nonparametric estimator of F. The Glivenko-Cantelli theorem states that, as $m \to \infty$, $\sup_{-\infty<t<\infty} |F(t) - \hat{F}_m(t)| \to 0$ a.s. Minimaxity results for $\hat{F}_m$, under various loss functions, can be found in Aggarwal [1], Dvoretzky, Kiefer and Wolfowitz [8], and Phadia [18]. Furthermore, when there is no restriction on F, that is, when F is assumed only to be a member of the class of all distributions, $\hat{F}_m$ is the maximum likelihood estimator (MLE) of F.

Some authors restrict the class to which F can belong, still keeping it nonparametric. Grenander [12] obtained the MLE for F when F is assumed to be a life distribution [$F(t) = 0$ for $t < 0$] having an increasing failure rate (IFR) [F is a member of the IFR class if $\bar{F}(x + t)/\bar{F}(t)$ is decreasing in t whenever $x > 0$, where $\bar{F} = 1 - F$]. Marshall and Proschan [16] showed that the MLE for an IFR distribution is strongly consistent. Marshall and Proschan [17] derived the MLE for the increasing failure rate average class (IFRA) [a life distribution F is a member of the IFRA class if $-t^{-1}\ell n\, \bar{F}(t)$ is increasing in $t > 0$] and proved it is not strongly consistent; the estimator does not converge a.s. to the true F but converges a.s. to another function.

Another nonparametric approach is to define a prior distribution P on the class Ω (say) of distributions F and then choose F to minimize the average risk $\int_\Omega \{\int L(F, \hat{F})\, dF\} dP(F)$; such an F is of course a Bayes estimator with respect to the prior P. This approach has recently been advocated by Ferguson [9] who presented a class of prior distributions called Dirichlet process priors. Whereas other prior distributions on can be defined [cf. Kraft [15], Dubins and Freedman [7], Ferguson [10] and additional references in those papers], Ferguson's stochastic process is particularly useful because: (i) It is tractable - if a random sample $X_1, \ldots, X_m$ is taken from a distribution function F that is a random sample function of a Dirichlet process P, then the posterior distribution of P given $X_1, \ldots, X_m$ is also a Dirichlet process and is manageable analytically (for example, risks of estimators can be computed); (ii) The class of priors is rich enough to incorporate a wide range, and varying degrees, of prior information; (iii) The process can be parametrized in a manner that enables direct incorporation and interpretation of prior information. These advantages have been emphasized by Ferguson [9], [10] and Antoniak [2].

Ferguson's nonparametric Bayes estimator of F is as follows. [For Dirichlet process definitions see Section 2, and for more details, see Ferguson's [9] paper.] To estimate F, Ferguson assumes $X_1, \ldots, X_m$ is a sample from a distribution F which is chosen by a Dirichlet process P. The parameter $\alpha(\cdot)$, of the process, is assumed to be σ-additive. Let the parameter and the action spaces be the set of all the distributions on $(\mathcal{R},\mathcal{B})$, the measurable space of the real line $\mathcal{R}$ and the σ-field of Borel subsets of $\mathbf{R}$. Take the loss function to be

$$L(F, \hat{F}) = \int_{\mathcal{R}} (F(t) - \hat{F}(t))^2 \, dW(t), \tag{1.1}$$

where W is a given finite measure on $(\mathcal{R},\mathcal{B})$ (a weight function) and $\hat{F}$ is an estimator of F. When P is a Dirichlet process on $(\mathcal{R},\mathcal{B})$ with parameter α, Ferguson's Bayes estimator of F is given by

$$\tilde{F}(t) = pF_0(t) + (1 - p)\hat{F}(t), \tag{1.2}$$

where

$$p = \alpha(\mathcal{R})/(\alpha(\mathcal{R}) + m), \tag{1.3}$$

$$F_0(t) = \alpha((-\infty, t])/\alpha(\mathcal{R}), \tag{1.4}$$

and $\hat{F}(t) = (\#\text{ of } X_i\text{'s} \leq t)/m$ is the empirical distribution function of the sample $X_1, \ldots, X_m$. The Bayes estimator given by (1.2) is a weighted average of F_0, the prior "guess" at F, and the sample distribution function $\hat{F}$. Ferguson notes this brings out the role of $\alpha(R)$ as a measure of faith in the prior. (More general classes of nonparametric Bayesian estimators of F, which include $\tilde{F}$ as a special case, are considered by Doksum [6], Ferguson [10], and Goldstein [11].)

Motivated by Ferguson's Bayes estimator, Korwar and Hollander [14] proposed an empirical Bayes estimator of the distribution function which required less prior information about $\alpha(\cdot)$. Only $\alpha(R)$ need be specified. They considered the following model. Let $(F_i, \underline{X}_i)$, i = 1, 2, ... be a sequence of pairs of independent random elements. The F's are random probability measures which have a common prior distribution given by a Dirichlet process on (R,B). The parameter $\alpha(\cdot)$ of the process is assumed to be σ-additive. Given $F_i = F'$ (say), $\underline{X}_i = (X_{i1}, \ldots, X_{im_i})$ is a random sample of size m_i from F'. For the problem of estimating F_{n+1} on the basis of $\underline{X}_1, \ldots, \underline{X}_{n+1}$, Korwar and Hollander proposed a sequence $G = \{G_{n+1}\}$ of estimators which, for the loss function given by (1.1), was shown to be asymptotically optimal in the sense of Robbins [20].

The Korwar-Hollander sequence $\{G_{n+1}\}$ was developed in the equal sample sizes case $m_1 = m_2 = \ldots = m$. In this paper G_{n+1} is generalized to be applicable in situations where the sample sizes are unequal. The proposed sequence of estimators is, for n = 1, 2, ...,

$$H_{n+1}(t) = p_{n+1}\sum_{i=1}^{n}\hat{F}_i(t)/n + (1 - p_{n+1})\hat{F}_{n+1}(t), \tag{1.5}$$

where

$$p_{n+1} = \alpha(R)/(\alpha(R) + m_{n+1}), \tag{1.6}$$

and $\hat{F}_i$ is the empirical distribution function of $\underline{X}_i$, i = 1, ..., n+1.

Section 3 establishes an optimal property of H_{n+1} which is similar to, but slightly weaker than, asymptotic optimality. In Section 3 we also compare the performance of the estimator H_{n+1} with that of the sample distribution function $\hat{F}_{n+1}$ based on $\underline{X}_{n+1}$. We show that the inequality

$$\frac{1}{m_{n+1}} > \frac{\alpha(R) \sum_{i=1}^{n} m_i^{-1} + n}{n^2\{\alpha(R) + m_{n+1}\}} \tag{1.7}$$

is a necessary and sufficient condition for the Bayes risk of $\hat{F}_{n+1}$, with respect to the Dirichlet process prior, to be larger than the overall expected loss using H_{n+1}. Furthermore, inequality (1.8)

$$n \cdot \min(m_1, m_2, \ldots, m_{n+1}) > \max(m_1, m_2, \ldots, m_{n+1}) \tag{1.8}$$

is a sufficient condition for H_{n+1} to be better than $\hat{F}_{n+1}$. Another sufficient condition is

$$(2n - 1) \cdot \min(\alpha(R), m_1, m_2, \ldots, m_{n+1})$$
$$> \max(\alpha(R), m_1, m_2, \ldots, m_{n+1}). \tag{1.9}$$

In Section 4 we show how the proposed estimators can be used for the problem of simultaneously estimating n+1 distribution functions. The estimators are computed for some data of Proschan [19] on the times of successive failures of the air conditioning systems of Boeing 720 jet airplanes.

2.0 *Dirichlet Process Preliminaries*. In this section we present some basic definitions and results that pertain to Ferguson's Dirichlet process. See Ferguson [9] for a more comprehensive coverage.

Definition 2.1. Let $Z_1, \ldots, Z_k$ be independent random variables with Z_j having a gamma distribution with shape parameter $\alpha_j \geq 0$

and scale parameter 1, $j = 1, \ldots, k$. Let $\alpha_j > 0$ for some j. The Dirichlet distribution with parameter $(\alpha_1, \ldots, \alpha_k)$, denoted by $\mathcal{D}(\alpha_1, \ldots, \alpha_k)$, is defined as the distribution of

$(Y_1, \ldots, Y_k)$, where $Y_j = Z_j / \sum_{i=1}^{k} Z_i$, $j = 1, \ldots, k$.

This distribution is always singular with respect to Lebesgue measure on k-dimensional Euclidean space. Also, if any $\alpha_i = 0$, the corresponding Y_i is degenerate at 0. However, if $\alpha_i > 0$ for all $i = 1, \ldots, k$, the (k-1)-dimensional distribution of $(Y_1, \ldots, Y_{k-1})$ has density, with respect to Lebesgue measure on the (k-1)-dimensional Euclidean space, given by

$$f(y_1, \ldots, y_{k-1} | \alpha_1, \ldots, \alpha_k) \tag{2.1}$$

$$= \frac{\Gamma(\alpha_1 + \ldots + \alpha_k)}{\Gamma(\alpha_1) \ldots \Gamma(\alpha_k)} \left(\prod_{i=1}^{k-1} y_i^{\alpha_i - 1} \right) \left(1 - \sum_{i=1}^{k-1} y_i\right)^{\alpha_k - 1} I_S(y_1, \ldots, y_{k-1}),$$

where S is the simplex

$$S = \{(y_1, \ldots, y_{k-1}) : y_i \geq 0, \sum_{i=1}^{k-1} y_i \leq 1\}.$$

For $k = 2$, (2.1) becomes the density of a Beta distribution, $\mathcal{Be}(\alpha_1, \alpha_2)$. Note that the condition $\alpha_i > 0$ for some $i = 1, \ldots, k$, is required in Definition 2.1 so that $\sum_{i=1}^{k} Z_i$ is not degenerate at 0.

Let (X, A) be a measurable space. Ferguson defined the following stochastic process $\{P(A), A \in A\}$.

Definition 2.2. (Ferguson, [9]). Let (X, A) be a measurable space. Let α be a non-null finite measure (nonnegative and finitely additive) on (X, A). Then P is a Dirichlet process on (X, A) with parameter α if for every $k = 1, 2, \ldots$, and measurable partition $(B_1, \ldots, B_k)$ of X, the distribution of

$(P(B_1), \ldots, P(B_k))$ is Dirichlet with parameter $(\alpha(B_1), \ldots, \alpha(B_k))$.

If F is chosen by a Dirichlet process, then F is discrete with probability one [See Ferguson [9], Berk and Savage [3], Blackwell [4], and Blackwell and MacQueen [5]].

A sample from a Dirichlet process is next defined.

Definition 2.3. (Ferguson, [9]). The $\mathcal{X}$-valued random variables $X_1, \ldots, X_m$ constitute a sample of size m from a Dirichlet process P on $(\mathcal{X}, \mathcal{A})$ with parameter α if for any $\ell = 1, 2, \ldots$ and measurable sets $A_1, \ldots, A_\ell, C_1, \ldots, C_m$, $Q\{X_1 \in C_1, \ldots,$

$$X_m \in C_m | P(A_1), \ldots, P(A_\ell), P(C_1), \ldots, P(C_m)\} = \prod_{i=1}^{m} P(C_i) \text{ a.s.},$$

where Q denotes probability.

Roughly speaking, we may view a sample of size m from a Dirichlet process as follows. The process chooses a random distribution F, say, and then given F, $X_1, \ldots, X_m$ is a random sample from F.

Theorem 2.4 gives the posterior distribution of a Dirichlet process P, given a sample $X_1, \ldots, X_m$ from the process.

Theorem 2.4. (Ferguson, [9]). Let P be a Dirichlet process on $(\mathcal{X}, \mathcal{A})$ with parameter α, and let $X_1, \ldots, X_m$ be a sample of size m from P. Then the conditional distribution of P given $X_1, \ldots, X_m$ is a Dirichlet process on $(\mathcal{X}, \mathcal{A})$ with parameter $\beta = \alpha + \sum_{i=1}^{m} \delta_{X_i}$, where, for $x \in \mathcal{X}$, $A \in \mathcal{A}$, $\delta_x(A) = 1$ if $x \in A$, 0 otherwise.

Theorem 2.5 is a generalization of Ferguson's [9] Proposition 4.

Theorem 2.5. (Korwar and Hollander, [14]). Let P be a Dirichlet process on $(\mathcal{R}, \mathcal{B})$ with parameter α and let $X_1, \ldots, X_m$ be a sample of size m from P. Then

$$Q\{X_1 \le x_1, \ldots, X_m \le x_m\}$$
$$= \{\alpha(A_{x_{(1)}})\ldots(\alpha(A_{x_{(m)}}) + m-1)\}/\{\alpha(R)\ldots(\alpha(R) + m-1)\},$$

where $x_{(1)} \le \cdots \le x_{(m)}$ is an arrangement of $x_1, \ldots, x_m$ in increasing order of magnitude, $A_x = (-\infty, x]$, and Q denotes probability.

3.0 *Properties of H_{n+1}*. In this section we consider properties of H_{n+1}. First we will show that the difference between the risk of the Bayes estimator of F_{n+1} (based on $\underline{X}_{n+1}$) and the overall expected loss using H_{n+1} converges to zero as $n \to \infty$. Then in a comparison with $\hat{F}_{n+1}$ (the sample distribution function based on $\underline{X}_{n+1}$), we prove H_{n+1} does better than $\hat{F}_{n+1}$ riskwise, if and only if inequality (1.7) is satisfied.

In our empirical Bayes framework, Ferguson's Bayes estimator (1.2) of $\mathbf{F}$ based on $\underline{X}_{n+1}$ is

$$\tilde{F}_{m_{n+1}}(t) = p_{n+1}F_0(t) + (1 - p_{n+1})\hat{F}_{n+1}(t), \qquad (3.1)$$

where p_{n+1}, $F_0(t)$ are respectively given by (1.6) and (1.4), and where $\hat{F}_{n+1}$ is the sample distribution function of $\underline{X}_{n+1}$. The Bayes risks $R_n(\alpha)$ and $R(H_{n+1}, \alpha)$ of the estimators (3.1) and (1.5), respectively, with respect to the Dirichlet prior, are

$$R_n(\alpha) \overset{\text{def.}}{\equiv} R(\tilde{F}_{m_{n+1}}, \alpha)$$
$$= E_{\underline{X}_{n+1}}[\int \{E_{F(t)|\underline{X}_{n+1}}(F(t) - \tilde{F}_{m_{n+1}}(t))^2\}\, dW(t)], \qquad (3.2)$$

and

$$R(H_{n+1}, \alpha)$$
$$= E_{\underline{X}_{n+1}}[\int \{E_{F(t)|\underline{X}_{n+1}}(F(t) - H_{n+1}(t))^2\}\, dW(t)]. \qquad (3.3)$$

Let $R_{n+1}(H, \alpha)$ be the expectation of $R(H_{n+1}, \alpha)$ with respect to $\underline{X}_i$, $i = 1, \ldots, n$ (the past observations).

Theorem 3.1. Let $\alpha(R)$ be known. Then the sequence $\{H_{n+1}\}$ has the optimal property that

$$\lim_{n\to\infty}[R_{n+1}(H, \alpha) - R_n(\alpha)] = 0. \tag{3.4}$$

Our proof of Theorem 3.1 uses Lemma 2.5 of Korwar and Hollander [14] restated here as Lemma 3.2.

Lemma 3.2. Let the hypotheses be those of Theorem 2.5 and let $F(t) = P((-\infty, t])$ and $\hat{F}(t)$ be the sample distribution function of $\underline{X} = (X_1, \ldots, X_m)$. Then for each $t \in R$,

$$E(F(t)|\underline{X}) = \tilde{F}_m(t), \tag{3.5}$$

$$E(F^2(t)|\underline{X}) = \tilde{F}_m(t)(\tilde{F}_m(t)\beta(R) + 1)/(\beta(R) + 1), \tag{3.6}$$

$$E(\hat{F}(t)) = F_0(t), \tag{3.7}$$

and

$$E(\hat{F}^2(t)) = F_0(t)/m + [(m - 1)F_0(t)\{F_0(t)\alpha(R) + 1\} /\{m(\alpha(R) + 1)\}], \tag{3.8}$$

where $\tilde{F}_m(t)$ is the analog of (3.1) for $\underline{X}$, $F_0(t)$ is given by (1.4), and $\beta(R) = \alpha(R) + m$.

Proof of Theorem 3.1. The proof is almost identical to the proof of Theorem 2.4 of Korwar and Hollander [14], insofar as the computation of $R_{n+1}(H, \alpha)$ and $R_n(\alpha)$ is concerned. We give the details here for the sake of completeness. Using (3.5) of Lemma (3.2), we can rewrite $R(H_{n+1}, \alpha)$ as

$$R(H_{n+1}, \alpha) = R(\tilde{F}_{m_{n+1}}, \alpha)$$

$$+ E_{\underline{X}_{n+1}}\left[\int \{E_{F(t)|\underline{X}_{n+1}}(\tilde{F}_{m_{n+1}}(t) - H_{n+1}(t))^2\}\, dW(t)\right], \tag{3.9}$$

where $\tilde{F}_{m_{n+1}}(t)$ is given by (3.1). Now,

$$\tilde{F}_{m_{n+1}}(t) - H_{n+1}(t) = p_{n+1}\{F_0(t) - \sum_{i=1}^{n} \hat{F}_i(t)/n\}, \quad (3.10)$$

and this is independent of $\underline{X}_{n+1}$ and $F(t)$. Thus,

$$R(H_{n+1}, \alpha) = R(\tilde{F}_{m_{n+1}}, \alpha) + \int (\tilde{F}_{m_{n+1}}(t) - H_{n+1}(t))^2 \, dW(t). \quad (3.11)$$

It now follows by (3.11), and the definition of $R_{n+1}(H, \alpha)$, that

$$R_{n+1}(H, \alpha) = R(\tilde{F}_{m_{n+1}}, \alpha) + \int \{E_{\underline{X}_1, \ldots, \underline{X}_n} (\tilde{F}_{m_{n+1}}(t) - H_{n+1}(t))^2\} dW(t). \quad (3.12)$$

The first term on the right of (3.12) is given by (2.19) of Korwar and Hollander [14] with m replaced by m_{n+1}:

$$R_n(\alpha) = [\alpha(R)/\{(\alpha(R) + 1)(\alpha(R) + m_{n+1})\}] \cdot \int F_0(t)(1 - F_0(t)) \, dW(t). \quad (3.13)$$

We will next evaluate the second term on the right of (3.12). We have by (3.10),

$$E_{\underline{X}_1, \ldots, \underline{X}_n} (\tilde{F}_{m_{n+1}}(t) - H_{n+1}(t))^2 \quad (3.14)$$

$$= p_{n+1}^2 E_{\underline{X}_1, \ldots, \underline{X}_n} (F_0^2(t) - 2F_0(t) \sum_{i=1}^{n} \hat{F}_i(t)/n + \sum_{i,i'=1}^{n} \hat{F}_i(t)\hat{F}_{i'}(t)/n^2).$$

Now use Lemma 3.2 to evaluate the right side of (3.14). Note that $\hat{F}_i$ depends only on $\underline{X}_i$, and $\hat{F}_i$, $\hat{F}_{i'}$, are independent when $i \neq i'$. Hence (3.14), after simplification, becomes

$$E_{\underline{X}_1, \ldots, \underline{X}_n} (\tilde{F}_{m_{n+1}}(t) - H_{n+1}(t))^2 \tag{3.15}$$

$$= p_{n+1}^2 \{(\alpha(R) \sum_{i=1}^{n} m_i^{-1} + n)/(n^2(\alpha(R) + 1))\} F_0(t)(1 - F_0(t)).$$

Now we have by (1.6), (3.12), (3.13) and (3.15), after simplification, that

$$R_{n+1}(H, \alpha)$$

$$= [1 + \alpha(R)(\alpha(R) \sum_{i=1}^{n} m_i^{-1} + n)/\{n^2(\alpha(R) + m_{n+1})\}] R_n(\alpha). \tag{3.16}$$

It follows from (3.13), (3.16) and the fact $\alpha(R)/(\alpha(R) + m_{n+1}) < 1$, $m_i \geq 1$ for each i, that

$$0 \leq R_{n+1}(H, \alpha) - R_n(\alpha) < (1/n) \int F_0(t)(1 - F_0(t))\, dW(t),$$

which yields (3.4). ||

We now turn to the comparison of H_{n+1} with the sample distribution function $\hat{F}_{n+1}$. Theorem 3.3 shows that inequality (1.7) is a necessary and sufficient condition for the empirical Bayes estimator H_{n+1} to be better than the sample distribution function $\hat{F}_{n+1}$.

Theorem 3.3. Let $\alpha(R)$ be known. Let $\hat{F}_{n+1}$ be the sample distribution function based on $\underline{X}_{n+1}$. Then inequality (1.7) is a necessary and sufficient condition for $R(\hat{F}_{n+1}, \alpha)$, the Bayes risk of $\hat{F}_{n+1}$ with respect to the Dirichlet prior, to be larger than $R_{n+1}(H, \alpha)$, the overall expected loss of H_{n+1}.

Proof. From (3.3) of Korwar and Hollander [14], replacing m by m_{n+1}, we have

$$R(\hat{F}_{n+1}, \alpha) = [1 + \alpha(R)/m_{n+1}] R_n(\alpha), \tag{3.17}$$

where $R_n(\alpha)$ is given by (3.13). Comparing (3.17) with (3.16) yields

$R_n(\hat{F}_{n+1}, \alpha) > R_{n+1}(H, \alpha)$ if and only if

$$\frac{1}{m_{n+1}} > \frac{\alpha(R) \sum_{i=1}^{n} m_i^{-1} + n}{n^2\{\alpha(R) + m_{n+1}\}} . \quad ||$$

Corollary 3.4. A sufficient condition for $R(\hat{F}_{n+1}, \alpha)$ to be greater than $R_{n+1}(H, \alpha)$ is

$$n \cdot \min(m_1, m_2, \ldots, m_{n+1}) > \max(m_1, m_2, \ldots, m_{n+1}).$$

Proof. Direct algebra shows inequality (1.7) is equivalent to $c_n > 0$ where

$$c_n = n/m_{n+1} - \sum_{i=1}^{n} m_i^{-1}/n + (n - 1)/\alpha(R). \tag{3.18}$$

Letting $M_1 = \max(m_1, \ldots, m_{n+1})$ and $M_2 = \min(m_1, \ldots, m_{n+1})$, we note

$$c_n \geq n/M_1 - 1/M_2.$$

Thus $c_n > 0$ if $nM_2 > M_1$. ||

Remark 3.5. If all the sample sizes $m_1, \ldots, m_{n+1}$ are equal, then inequality (1.8) reduces to $n > 1$, the condition given in Theorem 3.1 of Korwar and Hollander [14].

Corollary 3.6. A sufficient condition for $R(\hat{F}_{n+1}, \alpha)$ to be greater than $R_{n+1}(H, \alpha)$ is

$$(2n - 1)\min(\alpha(R), m_1, m_2, \ldots, m_{n+1})$$
$$> \max(\alpha(R), m_1, m_2, \ldots, m_{n+1}).$$

Proof. Recall inequality (1.7) is equivalent to $c_n > 0$, where c_n is given by (3.18). Let $M_1' = \max(\alpha(R), m_1, \ldots, m_{n+1})$ and $M_2' = \min(\alpha(R), m_1, \ldots, m_{n+1})$. Then $c_n \geq \frac{n}{M_1'} - \frac{1}{M_2'} + \frac{(n-1)}{M_1'}$, and thus $c_n > 0$ if $(2n - 1)M_2' > M_1'$. ||

4.0 *Simultaneous Estimation of Distribution Functions*. Note that by interchanging the roles of samples 1 and n+1, the H estimator defined by (1.5) becomes an estimator of F_1 based on $\underline{X}_1$ and the "past" samples $\underline{X}_2, \ldots, \underline{X}_{n+1}$. More generally, an estimator of F_j based on all the samples is

$$H_j(t) = p_j \sum_{\substack{i=1 \\ i \neq j}}^{n+1} \hat{F}_i(t)/n + (1 - p_j)\hat{F}_j(t), \qquad j = 1, 2, \ldots, n+1 \tag{4.1}$$

where

$$p_j = \alpha(R)/(\alpha(R) + m_j), \tag{4.2}$$

and $\hat{F}_i(t)$ is the empirical distribution function of the sample $\underline{X}_i$.

We now illustrate the use of the estimators defined by (4.1). Table I, adapted from Proschan [19], gives the intervals between successive failures of the air conditioning systems of three 720 jet airplanes. (Proschan's data set contains information on thirteen 720's but, for simplicity, we have restricted our consideration to three planes.)

TABLE I

Intervals Between Failures of Air Conditioning Systems

PLANE

7912	7913	7914
23	97	50
261	51	44
87	11	102
7	4	72
120	141	22
14	18	39
62	142	3
47	68	15
225	77	197
71	80	188
246	1	79
21	16	88
42	106	46
20	206	5
5	82	5
12	54	36
120	31	22
11	216	139
3	46	210
14	111	97
71	39	30
11	63	23
14	18	13
11	191	14
16	18	
90	163	
1	24	
16		
52		
95		

Source: F. Proschan [19]

For the Table I data, $n + 1 = 3$, $m_1 = 30$, $m_2 = 27$, and $m_3 = 24$. Note that in this case inequality (1.8) is satisfied. To use (4.1) we must specify $\alpha(R)$. Ferguson [9], Goldstein [11], Korwar and Hollander [14], and condition (1.9) of this paper give various justifications for interpreting $\alpha(R)$ as the "prior sample size" of the process. Note from (4.1) that as $\alpha(R)$ decreases, the estimator H_j of F_j puts more weight on the observations from the jth sample, and less weight on the observations from the other samples. We consider, for example, the case where $\alpha(R)$ is specified to be 7. Then from (4.2) we have $p_1 = 7/(7 + 30) = .19$, $p_2 = 7/(7 + 27) = .21$, $p_3 = 7/(7 + 24) = .23$, so that

$$H_1(t) = .19(\hat{F}_2(t) + \hat{F}_3(t))/2 + .81(\hat{F}_1(t)),$$

$$H_2(t) = .21(\hat{F}_1(t) + \hat{F}_3(t))/2 + .79(\hat{F}_2(t)),$$

$$H_3(t) = .23(\hat{F}_1(t) + \hat{F}_2(t))/2 + .77(\hat{F}_3(t)).$$

Figure 1a contains plots of $\hat{F}_1$ and H_1, Figure 1b plots of $\hat{F}_2$ and H_2, and Figure 1c plots of $\hat{F}_3$ and H_3.

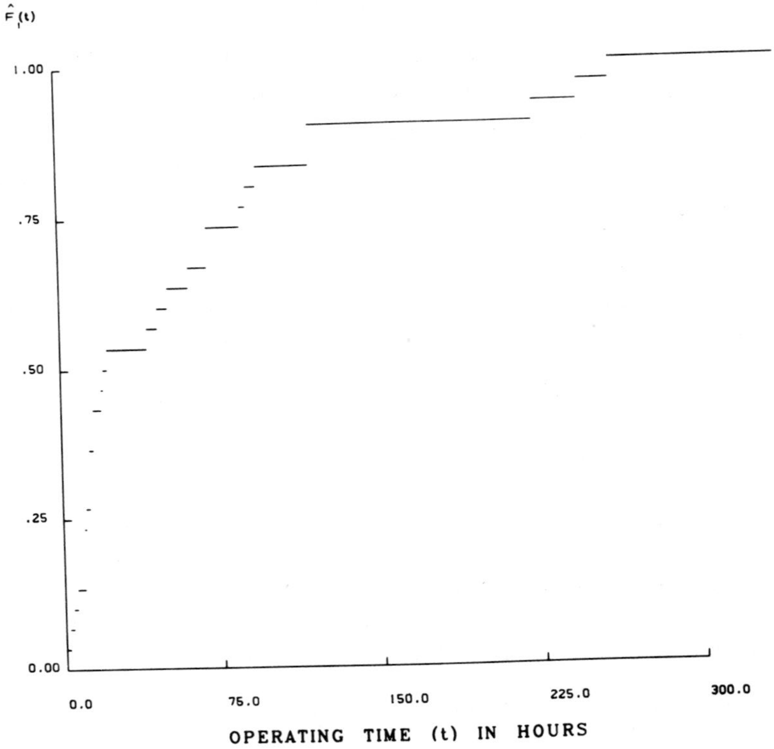

Figure 1a. Plots of $\hat{F}_1$ and H_1

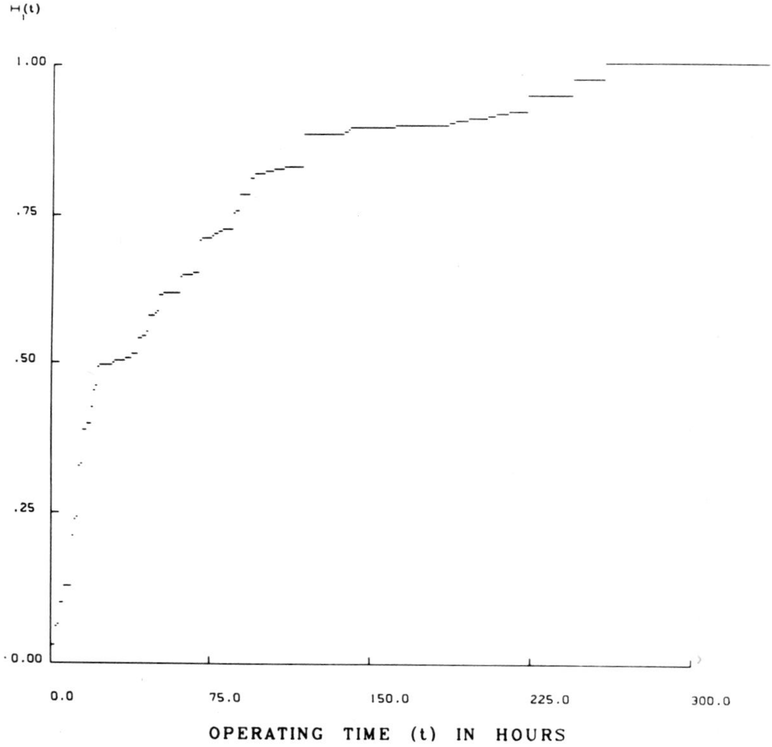

Figure 1a. Plots of $\hat{F}_1$ and H_1 continued

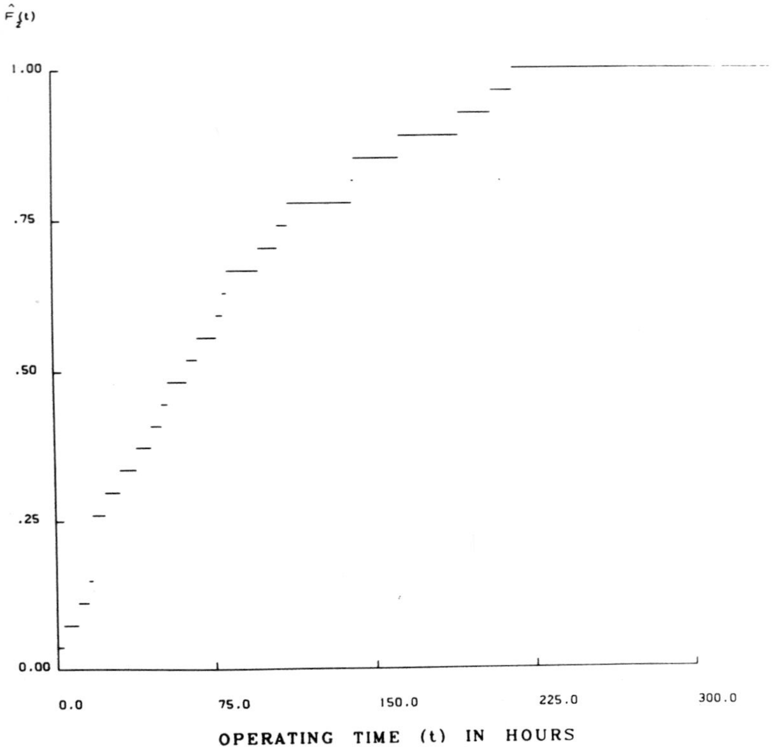

Figure 1b. Plots of $\hat{F}_2$ and H_2

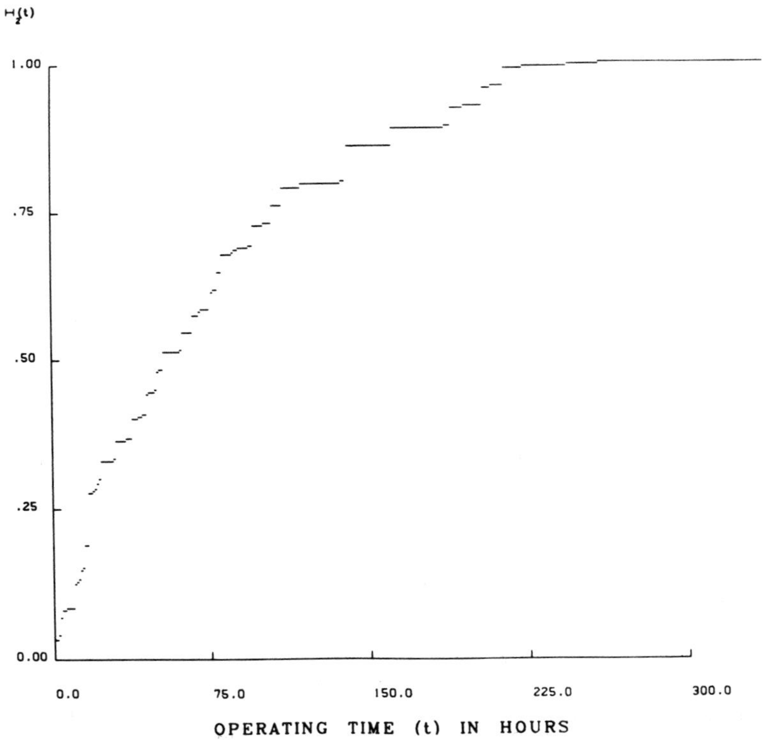

Figure 1b. Plots of $\hat{F}_2$ and H_2 continued

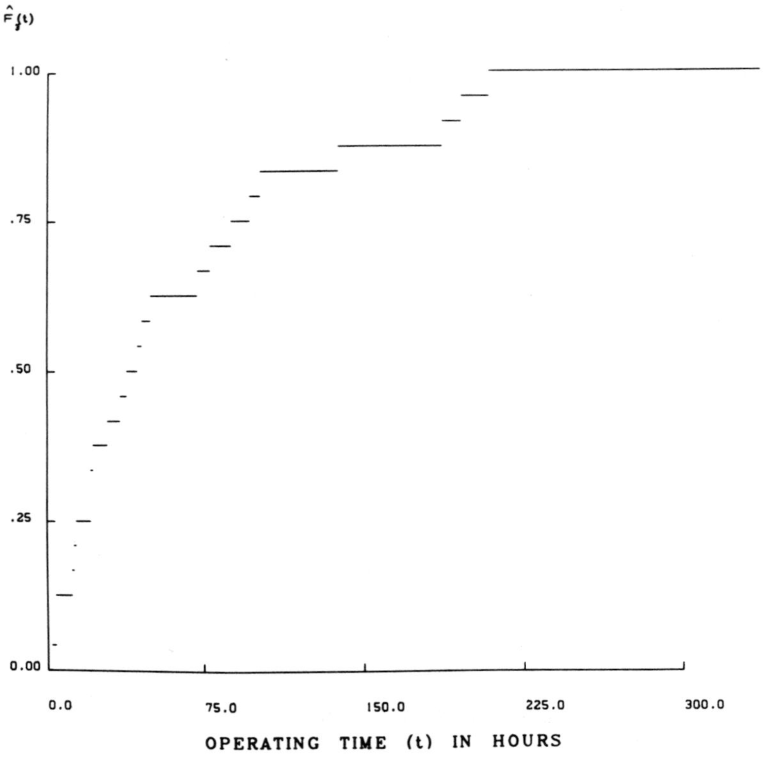

Figure 1c. Plots of $\hat{F}_3$ and H_3

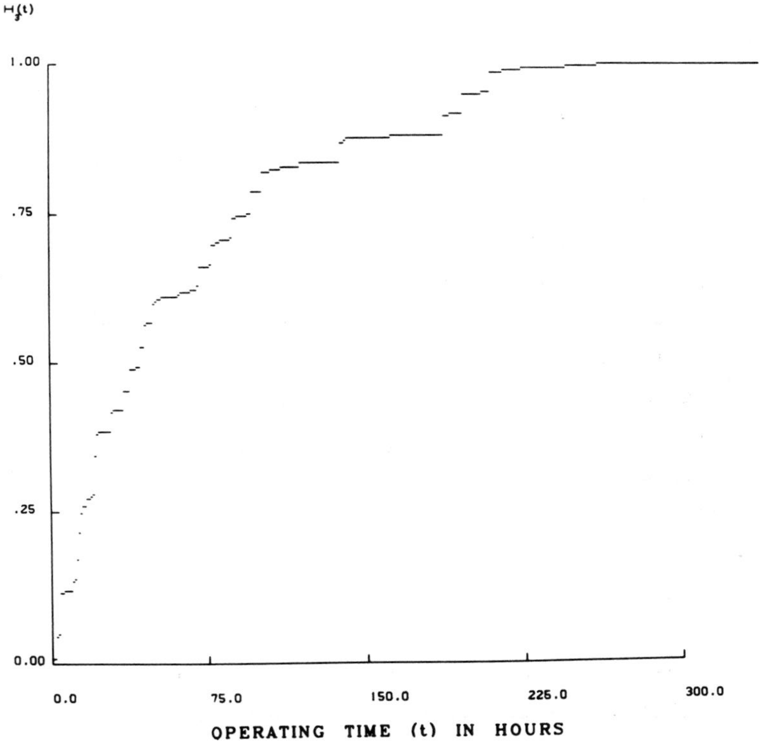

Figure 1c. Plots of $\hat{F}_3$ and H_3 continued

Acknowledgments. We are grateful to Richard Gurnee and Alan Zinsmeister for the programming that yielded Figures 1a, 1b, and 1c.

REFERENCES

[1] Aggarwal, O. P., "Some Minimax Invariant Procedures for Estimating a Cumulative Distribution Function", *Ann. Math. Statist. 26*, (1955), pp. 450-463.

[2] Antoniak, C. E., "Mixtures of Dirichlet Processes with Applications to Bayesian Nonparametric Problems", *Ann Statist. 2*, (1974), pp. 1152-1174.

[3] Berk, R. H. and Savage, I. R., "Dirichlet Processes Produce Discrete Measures: An Elementary Proof", Unpublished manuscript, (1975).

[4] Blackwell, D., "Discreteness of Ferguson Selections", *Ann. Statist. 1*, (1973), pp. 356-358.

[5] Blackwell, D. and Macqueen, J. B., "Ferguson Distributions Via Pólya Urn Schemes", *Ann. Statist. 1*, (1973), pp. 353-355.

[6] Doksum, K. A., "Decision Theory for Some Nonparametric Models", *Proc. Sixth Berkeley Symp. 1*, (1972), pp. 331-341.

[7] Dubins, L. E. and Freedman, D. A., "Random Distribution Functions", *Proc. Fifth Berkeley Symp. 2*, (1966), pp. 183-214.

[8] Dvoretzky, A., Kiefer, J. and Wolfowitz, J., "Asymptotic Minimax Character of the Sample Distribution Function and of the Classical Multinomial Estimator", *Ann. Math. Statist. 27*, (1956), pp. 642-669.

[9] Ferguson, T. S., "A Bayesian Analysis of Some Nonparametric Problems", *Ann. Statist. 1*, (1973), pp. 209-230.

[10] Ferguson, T. S., "Prior Distributions on Spaces of Probability Measures", *Ann. Statist. 2*, (1974), pp. 615-629.

[11] Goldstein, M., "A Note on Some Bayesian Nonparametric Estimates", *Ann. Statist. 3*, (1975), pp. 736-740.

[12] Grenander, U., "On the Theory of Mortality Measurement, Part II", *Skan. Aktuarietidskr* 39, (1956), pp. 125-153.

[13] Korwar, R. M. and Hollander, M., "Contributions to the Theory of Dirichlet Processes", *Ann. Prob.* 1, (1973), pp. 705-711.

[14] Korwar, R. M. and Hollander, M., "Empirical Bayes Estimation of a Distribution Function", *Ann. Statist.* 4, (1976), pp. 580-587.

[15] Kraft, C. H., "A Class of Distribution Function Processes Which Have Derivatives", *J. Appl. Probability* 1, (1964), pp. 385-388.

[16] Marshall, A. W. and Proschan, F., "Maximum Likelihood Estimation for Distributions with Monotone Failure Rate", *Ann. Math. Statist.* 36, (1965), pp. 69-77.

[17] Marshall, A. W. and Proschan, F., "Inconsistency of Maximum Likelihood Estimator of Distributions Having Increasing Failure Rate Average", *Ann. Statist.* (to appear).

[18] Phadia, E. G., "Minimax Estimation of a Cumulative Distribution Function", *Ann. Statist.* 1, (1973), pp. 1149-1157.

[19] Proschan, F., "Theoretical Explanation of Observed Decreasing Failure Rate", *Technometrics* 5, (1963), pp. 375-383.

[20] Robbins, H. E., "The Empirical Bayes Approach to Statistical Decision Problems", *Ann. Math. Statist.* 35, (1964), pp. 1-20.

A CLASS OF DISTRIBUTIONS USEFUL IN LIFE TESTING AND RELIABILITY WITH APPLICATIONS TO NONPARAMETRIC TESTING*

by Z. GOVINDARAJULU
University of Kentucky

Abstract. A new class of distributions having a finite range that are useful in life testing and reliability is proposed. Methods of estimating the unknown parameters are studied. Explicit expressions for lower moments of order statistics in random samples of any size are derived. A uniformly most powerful test is derived for the 'shape' parameter and it is illustrated by the data on times to failure of refrigerator motors. This class of distributions and their generalization constitute augmented Lehmann alternatives and locally most powerful rank tests with respect to these alternatives are derived and their Pitman efficiencies are evaluated. For a certain hypothesis - testing problem, an asymptotically distribution - free test is obtained.

1.0 *Introduction.* Many probability models for failure times are available in the literature. (See, for instance Hahn and Shapiro [2]). Hazard function suitable for many phenomena including human life is the U-shaped curve or "bath-tub" curve. For an initial period the hazard function is relatively large, but decreasing in value, because of "infant mortality" or early failures attributable to manufacturing defects. Subsequently the hazard function remains approximately constant for some time after which it increases because of wear-out failures. Thus, U-shaped candidates for hazard functions are quite useful in

* Research supported in part by the Navy under Office of Naval Research Contract No. N00014-75-1003 task order NR042-295.

in reliability theory.

NOTATION.

X	Random variable denoting time to failure
$F(x;\theta,\sigma,\lambda)$, $f(x;\ldots)$, $\overline{F}(x;\ldots)$	cdf, pdf and survivor function respectively of the time to failure
θ,σ,λ	location, scale and 'shape' parameter respectively
$h(x) = f(x;\ldots)/\overline{F}(x;\ldots)$	hazard function associated with F
$Z = (X-\theta)/\sigma$	standardized variable X
μ, σ_Z^2	EZ, Var Z respectively
$X_1,\ldots,X_n$	random sample of size n from F
X_{in}	ith smallest of $(X_1,\ldots,X_n)$, $i = 1,\ldots,n$
Z_{in}	$(X_{in}-\theta)/\sigma$, $i=1,\ldots,n$
$\mu_{in}^{(k)}$	$E(Z_{in}^k)$, $k=1,2$, $i=1,\ldots,n$
μ_{ijn}	$E(Z_{in}Z_{jn})$, $1\leq i,<j\leq n$
σ_{ijn}	$Cov(Z_{in},Z_{jn})$, $1\leq i$, $j\leq n$
ML	Maximum likelihood
χ_k^2	chi-square variable having k degrees of freedom
α	level of significance of the test
$\chi_{k,1-\alpha}^2$	$100(1-\alpha)$th percentile of the chi-square distribution with k degrees of freedom
$\overline{X},\overline{Z}$	sample mean of X's and Z's respectively
$Y_1\ldots,Y_n$	A random sample from G(y)
N	m + n
$W_1<\ldots<W_N$	The combined order $X_1\ldots X_m$ and $Y_1\ldots,Y_n$

$Z=(Z_1,\ldots Z_N)$	The rank order where $Z_i=1$ if W_i is a Y and equals zero if W_i is an X, $i=1,\ldots,N$
H_{L1}	Lehmann alternative such that $G(x) = [F(x)]^{\delta}$, $\delta>1$
LMP	Locally most powerful
$h(F;\theta)$	The nonparametric alternative
$U_{1N}<\ldots<U_{NN}$	The standard (0,1) uniform order statistics in a random sample of size N.
$U_1,\ldots U_N$	A random sample from the uniform distribution on (0,1)
$F_m(x)$	Empirical distribution function based on $(X_1,\ldots,X_m)$.

2.0 *A class of Distributions*. Consider the class of distributions for failure times given by their inverse functions:

$$x = \theta + \sigma\{(\lambda+1)u^{\lambda} - \lambda u^{\lambda+1}\}, \tag{2.1}$$

where $u = F(x;\theta,\sigma,\lambda)$ with $F(\theta;\ldots) = 0$, $F(\theta+\sigma;\ldots) = 1$ and $\lambda>0$. One easily obtains:

$$\mu = 2/(\lambda+2) \tag{2.2}$$

$$EZ^2 = \{4(2\lambda+1)\}^{-1} + 9\{4(2\lambda+3)\}^{-1} \tag{2.3}$$

and hence

$$\sigma_Z^2 = (5\lambda+3)(2\lambda+1)^{-1}(2\lambda+3)^{-1} - 4(\lambda+2)^{-2} \; . \tag{2.4}$$

By studying the derivative of σ_Z^2 one can surmize that there exists a $\lambda_0 (2.5 < \lambda_0 < 2.6)$ such that σ_Z^2 increases for $0 < \lambda \leq \lambda_0$ and decreases for $\lambda > \lambda_0$. Clearly μ decreases with λ.

Next, let us study the hazard function of Z, and hereafter we shall suppress the parameter λ in the distribution function of Z. Let

$$z = (\lambda+1)u^\lambda - \lambda u^{\lambda+1}, \quad u=F(z;\lambda)=F(z) \tag{2.5}$$

where $F(0) = 0$, $F(1) = 1$. Then

$$dz/du = \lambda(1+\lambda)u^{\lambda-1}(1-u) = 1/f(z), \tag{2.6}$$

and the hazard function is given by

$$h(z) = \{\lambda(\lambda+1)u^{\lambda-1}(1-u)^2\}^{-1}. \tag{2.7}$$

Thus

$$\lim_{z\to 0} h(z) = \lim_{u\to 0} h(z) = \begin{cases} 0 & \text{if } \lambda<1 \\ 1/2 & \text{if } \lambda=1 \\ \infty & \text{if } \lambda>1 \end{cases} \tag{2.8}$$

and

$$\lim_{z\to 1} h(z) = \lim_{u\to 1} h(z) = \infty. \tag{2.9}$$

Notice that larger the value of λ, faster the hazard function tends to infinitely at the initial period.

$$\left.\begin{array}{l} \text{When } u \text{ is close to zero, } u \simeq [z/(\lambda+1)]^{1/\lambda} \\ \text{and when } u \text{ is close to unity, } u \simeq z^{1/\lambda}. \end{array}\right\} \tag{2.10}$$

From (2.7) we have

$$h'(z) = \frac{\partial h}{\partial u}\frac{du}{dz}$$

and $$\frac{\partial h}{\partial u} = -\frac{[(\lambda-1)(1-u) - 2u]}{\lambda(\lambda+1)u^\lambda (1-u)^3}.$$

since $\frac{du}{dz} > 0$,

$$h'(z) \geq 0 \text{ according as } (\lambda-1)(1-u) \lesseqgtr 2u$$

$$\text{i.e. according as } u \gtreqless \frac{\lambda-1}{\lambda+1} = u_0 \text{ (say)}$$

The corresponding value of z_0 is given by

$$z_0 = (\lambda+1)\left(\frac{\lambda-1}{\lambda+1}\right)^{\lambda} - \lambda\left(\frac{\lambda-1}{\lambda+1}\right)^{\lambda+1}$$

$$= (\lambda-1)(3\lambda+1)/(\lambda+1)^{\lambda+1} \tag{2.11}$$

Hence, the distribution has IFR for $z > z_0$ and has DFR for $0 \leq z \leq z_0$. Notice that if $\lambda<1$, F has IFR for $0 \leq z \leq 1$.

Letting $R(z) = \int_0^t h(z)dz = \int_0 h(z)\frac{dz}{du}du$

$$= \int_0 [\lambda(1+\lambda)u^{\lambda-1}(1-u)^2]^{-1}\lambda(\lambda+1)u^{\lambda-1}(1-u)du$$

$$= \int_0^{u^*}(1-u)du = [1-(1-u^*)^2]/2 ,$$

where u^* is the solution of $t = (\lambda+1)u^{*\lambda} - \lambda u^{*\lambda+1}$.

Also $Eh(Z) = \int_0^1 h(z)f(z)dz = \int_0^1 \frac{f(z)}{1-u}du$

$$= [\lambda(1+\lambda)]^{-1}\int_0^1 u^{1-\lambda}(1-u)^{-2}du = \infty.$$

Further, it is of interest to note that the underlying density satisfies the differential equation:

$$\frac{f'(z)}{f(z)} = \frac{1}{1-F(z)} - \frac{1-\lambda}{F(z)} .$$

Solution is: $f(z) = \left(\frac{dz}{dF}\right)^{-1} = c^{-1}(1-F)^{-1}F^{\lambda-1}$,

$$z = c\int_0^u (1-v)v^{-1+\lambda}dv = (\lambda+1)u^{\lambda} - \lambda u^{\lambda+1},$$

Since $u = 1$ when $z = 1$.

3.0 *Estimation of Location and Scale Parameters.* We wish to estimate the location and scale parameters θ and σ on the basis of $X_1,\ldots,X_n$. We assume that λ is known. Then reasonable estimates of θ and σ are

$$\hat{\theta} = X_{1,n}, \ \hat{\sigma} = X_{n,n} - X_{1,n}. \tag{3.1}$$

If one is interested in the best unbiased estimates of θ and σ, they can be obtained using the generalized least squares theory. (For instance, see Sarhan and Greenberg [3].) In order to evaluate **these** estimates, we need the lower moments of order statistics in random samples drawn from the distribution of Z. In the following we shall obtain explicit expressions for these lower moments of order statistics. Notice that in our notation, $\mu_{in}^{(1)} = \mu_{in}$ and $\mu_{i,i,n} = \mu_{i,n}^{(2)}$.

Consider

$$\begin{aligned}\mu_{in} &= \frac{n!}{(i-1)!(n-i)!} \int_0^1 [(\lambda+1)u^{\lambda} - \lambda u^{\lambda+1}] u^{i-1}(1-u)^{n-i} du \\ &= \frac{n!}{(i-1)!} \left[(\lambda+1) \frac{(\lambda+i-1)!}{(n+\lambda)!} - \lambda \frac{(\lambda+i)!}{(n+\lambda+1)!}\right] \\ &= \frac{n!(\lambda+i-1)!}{(i-1)!(n+\lambda+1)!} [\lambda(n+2-i)+n+1]\end{aligned} \tag{3.2}$$

where, when λ is not an integer, the factorials should be replaced by the appropriate gamma functions. A similar computation yields

$$\begin{aligned}\mu_{i.n}^{(2)} = \frac{n!(2\lambda+i-1)!}{(i-1)!(n+2\lambda+2)!} &[(\lambda+1)^2(n+2\lambda+1)(n+2\lambda+2) \\ &+ \lambda^2(2\lambda+i+1)(2\lambda+i) \\ &- 2\lambda(\lambda+1)(2\lambda+i)(n+2\lambda+2)].\end{aligned} \tag{3.3}$$

Next, consider

$$\mu_{ijn} = \frac{n!}{(i-1)!(j-i-1)!(n-j)!} \iint_{0\le u\le v\le 1} [(\lambda+1)u^{\lambda} - \lambda u^{\lambda+1}]$$

$$\cdot [(\lambda+1)v^{\lambda} - \lambda v^{\lambda+1}] u^{i-1}(v-u)^{j-i-1}(1-v)^{n-j} du\, dv$$

$$= \frac{n!}{(i-1)!(j-i-1)!(n-j)!} \sum_{k=0}^{j-i-1} (-1)^k \binom{j-i-1}{k}$$

$$\iint_{u<v} [(\lambda+1)^2 u^{\lambda} v^{\lambda} - \lambda(\lambda+1)u^{\lambda} v^{\lambda+1} - \lambda(\lambda+1)u^{\lambda+1} v^{\lambda} + \lambda^2 u^{\lambda+1} v^{\lambda+1}]$$

$$\cdot\ u^{i+k-1} v^{j-i-1-k} (1-v)^{n-j} du\, dv$$

$$= \frac{n!}{(i-1)!} \sum_{k=0}^{j-i-1} \frac{(-1)^k}{k!(j-i-1-k)!}$$

$$\cdot\left[\frac{(\lambda+1)^2}{(i+\lambda+k)}\cdot\frac{(j+2\lambda-1)!}{(n+2\lambda)!}-\frac{\lambda(\lambda+1)}{(i+\lambda+k)}\cdot\frac{(j+2\lambda)!}{(n+2\lambda+1)!}\right.$$

$$\left.-\frac{\lambda(\lambda+1)}{(i+\lambda+k+1)}\cdot\frac{(j+2\lambda)!}{(n+2\lambda+1)!}+\frac{\lambda^2}{(i+\lambda+k+1)}\cdot\frac{(j+2\lambda+1)!}{(n+2\lambda+2)!}\right].$$

Algebraic simplification yields

$$\mu_{ijn}=\frac{n!}{(i-1)!}\left[(\lambda+1)\{\lambda(n+3-j)+n+1\}\frac{(j+2\lambda-1)!}{(n+2\lambda+1)!}\right.$$

$$\cdot\sum_{k=0}^{j-i-1}\frac{(-1)^k}{k!(j-i-1-k)!(i+\lambda+k)}-\lambda\{\lambda(n+3-j)+n+2\}\frac{(j+2\lambda)!}{(n+2\lambda+2)!}$$

$$\left.\cdot\sum_{k=0}^{j-i-1}\frac{(-1)^k}{k!(j-i-1-k)!(i+\lambda+k+1)}\right] \qquad (3.4)$$

From the above, one can easily compute

$$\sigma_{ijn}=\mu_{ijn}-\mu_{in}\mu_{jn},\quad 1\le i<j\le n,$$

and (3.5)

$$\sigma_{iin}=\mu_{in}^{(2)}-\mu_{in}^2,\quad 1\le i\le n.$$

Notice that θ and σ can be unbiasedly estimated by the generalized least squares method even if the random sample is censored or truncated.

4.0 *Estimation of* λ. In order to estimate θ and σ by the best unbiased procedure outlined in Sarhan and Greenberg [3] , one should know the value of λ. Thus it is of interest to estimate λ. In many practical problems, the value of θ can be roughly taken to be zero. Further, one has some feeling about the range of variation of X. Hence, without loss of generality set $\theta=0$ and $\sigma=1$. (If θ is different from zero and σ is not equal to 1, one can subtract known θ from each X_i and divide the difference $X_i-\theta$ by known σ.) Solving for λ from (2.2) we have

$$\lambda = 2(1-\mu)/\mu. \qquad (4.1)$$

Thus, a simple estimate of λ is

$$\hat{\lambda} = 2(1-\bar{Z})/\bar{Z} \qquad (4.2)$$

Note that $\hat{\lambda}$ is not unbiased for λ.

Remark 4.1. It should be noted that the method of ML is not applicable here since some of its regularity conditions are not satisfied.

Now, one might ask whether there exists an unbiased estimate for λ. Suppose there exists a function $g(Z)$ the expectation of which is λ. That is

$$Eg(Z) = \int_0^1 g(z)dF(z) \equiv \lambda. \tag{4.3}$$

That is,

$$\int_0^1 g[(\lambda+1)u^{\lambda} - \lambda u^{\lambda+1}]du \equiv \lambda. \tag{4.4}$$

One can easily verify that g cannot be a linear, a quadratic or a log function. Let us consider the case $\lambda=1$. Then

$$\int_0^1 g(2u - u^2)du = 1.$$

Letting $2u-u^2=w$, we obtain $1-u=(1-w)^{1/2}$. Hence

$$\int_0^1 g(w)(1-w)^{-1/2}dw = 2. \tag{4.5}$$

A class of possible solutions for g are

$$g(w) = 2(k+1)(1-w)^{k+(1/2)}, \quad k>-1. \tag{4.6}$$

Thus, there do exist unbiased estimates of λ for at least some values of λ. (For instance $\lambda=1/2$ is another possibility when one may be able to solve (4.4) explicitly for $g(w)$).

Remark 4.2. If all the parameters are unknown one can take

$$\hat{\theta} = X_{1n},$$

$$\hat{\sigma} = X_{nn}-X_{1n},$$

and

$$\hat{\lambda} = 2(\hat{\sigma}-\bar{X}+\hat{\theta})/(\bar{X}-\hat{\theta}). \tag{4.7}$$

Unbiased estimates for θ and σ for known value of λ.

Since $X_{1n} = \theta + \sigma Z_{in} \quad i = 1, \ldots, n$

one can easily show that

$$E\hat{\theta} = \theta + \sigma\mu_{1n} > \theta \tag{4.8}$$

and

$$E\hat{\sigma} = \sigma(\mu_{nn} - \mu_{1n}) = \sigma\left\{1 - \frac{\lambda(\lambda + 1)}{(n + \lambda)(n + \lambda + 1)} - \right.$$

$$- \frac{(n+1)!(\lambda+1)!}{(n+\lambda+1)!} \Bigg\} \ . \tag{4.9}$$

That is, θ is positively biased whereas σ is negatively biased. An unbiased estimate of σ would be:

$$\hat{\hat{\sigma}} = (X_{nn} - X_{1n})/(\mu_{nn} - \mu_{1n}) . \tag{4.10}$$

Then, an unbiased estimate of θ would be

$$\hat{\hat{\theta}} = X_{1n} - \hat{\hat{\sigma}}\mu_{1n} . \tag{4.11}$$

Using the lower moments of Z_{in} one can easily compute the variances of $\hat{\hat{\theta}}$ and $\hat{\hat{\sigma}}$ for given values of λ. Now we set

$$\hat{\hat{\lambda}} = 2(\hat{\hat{\sigma}} - \overline{X} + \hat{\hat{\sigma}})/(\overline{X} - \hat{\hat{\sigma}}) . \tag{4.12}$$

Next, using $\hat{\hat{\lambda}}$ obtain new estimates for θ and σ from (4.12) and (4.13) and iterate. Alternatively, using $\hat{\hat{\lambda}}$ and the lower moments of the Z_{in}, obtain the best linear unbiased estimates (BLUE) of θ and σ **and then iterate.**

Bias and Mean-square error of $\hat{\lambda}$ *for known* θ *and* σ

When θ and σ are known, $\hat{\lambda} = -2 + (2/\overline{Z})$.

Hence,

$$E\hat{\lambda} = -2 + 2E(1/\overline{Z}) \geq -2 + 2(1/E\overline{Z}) = -2 + \frac{2(\lambda + 2)}{2} = \lambda .$$

That is, $\hat{\lambda}$ is positively biased. Since $\overline{Z}$ converges to μ in probability, $\hat{\lambda}$ converges to λ in probability. That is, $\hat{\lambda}$ is consisent. Now, it is of interest to bound the bias and mean-square error of $\hat{\lambda}$. Towards this we have

Result 4.1:

$$\text{(i)} \quad E\hat{\lambda}-\lambda = \frac{2}{n\mu^3}\sigma_Z^2 - \frac{2}{\mu^4} E(\overline{Z} - \mu)^3 + \frac{2}{\mu^5} E(\overline{Z} - \mu)^4 - \ldots \tag{4.13}$$

and

$$\text{(ii)} \quad E(\hat{\lambda}-\lambda)^2 = 4\Big[\frac{\sigma_Z^2}{n\mu^2} - \frac{2}{\mu^3} E(\overline{Z} - \mu)^3 + \frac{3}{\mu^4} E(\overline{Z} - \mu)^4 - \ldots\Big] . \tag{4.14}$$

Proof:
$$\hat{\lambda} = -2 + (2/\overline{Z})$$
$$= -2 + (2/\mu)\left[1 + \frac{\overline{Z} - \mu}{\mu}\right]^{-1}$$
$$= -2 + (2/\mu) E \sum_{k=0}^{\infty} (-1)^k \left(\frac{\overline{Z} - \mu}{\mu}\right)^k . \qquad (4.15)$$

Hence,

$$E\hat{\lambda} = \lambda + \frac{2}{\mu^3}\frac{\sigma_Z^2}{n} - \frac{2}{\mu}\sum_{k=3}^{\infty} (-1)^k \, E\left(\frac{\overline{Z} - \mu}{\mu}\right)^k$$

from which we get (i). Further we have

$$(\hat{\lambda} - \lambda)^2 = \frac{4(\overline{Z} - \mu)^2}{\mu^2 \overline{Z}^2}$$

and

$$(\overline{Z})^{-2} = \mu^{-2}\left(1 - \frac{\overline{Z} - \mu}{\mu}\right)^{-2} = \mu^{-2}\sum_{k=0}^{\infty} (-1)^k (k+1)\left(\frac{\overline{Z} - \mu}{\mu}\right)^k .$$

Hence

$$E(\hat{\lambda}-\lambda)^2 = 4\sum_{k=0}^{\infty} (-1)^k (k+1) \frac{E(\overline{Z} - \mu)^{k+2}}{\mu^{k+2}}$$

from which (ii) follows.

Since $E(\overline{Z} - \mu)^3 = n^{-2}E(Z - \mu)^3 = 0(n^{-2})$

and $E(\overline{Z} - \mu)^4 = n^{-3}E(Z - \mu)^4 + \frac{6(n-1)}{n^3}(\sigma_Z^2)^2 = 0(n^{-2})$,

We have

Corollery 4.1.1. $E\hat{\lambda}-\lambda = \frac{2}{n\mu^3}\sigma_Z^2 + 0(n^{-2})$

and

$$\text{MSE } \hat{\lambda} = E(\hat{\lambda}-\lambda)^2 = \frac{4\sigma_Z^2}{n\mu^2} + 0(n^{-2}).$$

5.0 *Tests for Hypothesis for* λ. In this section we shall construct uniformly most powerful tests for λ, assuming that θ and σ are known. Suppose we are interested in testing

$$H_0: \lambda = \lambda_0 \text{ against } H_1: \lambda > \lambda_0 . \qquad (5.1)$$

Let us consider a simple alternative namely $H_1': \lambda=\lambda_1>\lambda_0$. Then according to Neyman-Pearson Lemma, the most powerful test of H_0 versus H_1' is:

Reject H_0 when

$$\prod_{i=1}^{n} \{f(X_i;\lambda_1)/f(X_i:\lambda_0)\} > K$$

ie, when

$$\prod_{i=1}^{n} \frac{\lambda_0(1+\lambda_0)U_i^{\lambda_0-1}(1-U_i)}{\lambda_1(1+\lambda_1)U_i^{\lambda_1-1}(1-U_i)} > K$$

ie, when

$$(\lambda_0-\lambda_1) \sum_{i=1}^{n} \log U_i > K'$$

or when

$$-2 \sum_{i=1}^{n} \log U_i > K'' \qquad (5.2)$$

where

$$Z_i = (\lambda_0+1)U_i^{\lambda_0} - \lambda_0 U_i^{1+\lambda_0} \quad i=1,\ldots,n, \qquad (5.3)$$

Determination of the Critical Point Under H_0, U_i given by (5.3) are distributed uniformly on (0,1). Hence $-2 \log U_i$ is distributed as χ_2^2. Thus, $K'' = \chi^2_{2n,1-\alpha}$. Now, since the test is free of the simple alternative H_1', the above test is uniformly most powerful for H_0 against H_1. Also, when $\lambda_0 = 1$,

$$U_i = 1-(1-Z_i)^{1/2}, \quad i=1,\ldots,n. \qquad (5.4)$$

Suppose we are interested in computing the power of the above test when $\lambda = \lambda^*$. Then U_i^* are distributed uniformly on

(0,1), where

$$Z_i = (1+\lambda^*)U_i^{*\lambda^*} - \lambda^* U_i^{*\lambda^*+1}, \ i=1,\ldots,n. \tag{5.5}$$

That is,

$$(1+\lambda_0)U_i^{\lambda_0} - \lambda_0 U_i^{1+\lambda_0} = (1+\lambda^*)U_i^{*\lambda^*} - \lambda^* U_i^{*1+\lambda^*}. \tag{5.6}$$

Then, the power of the test at λ^* is given by $\pi(\lambda^*)$ where

$$\pi(\lambda^*) = P(-2\sum_{i=1}^{n} \log U_i > \chi^2_{2n,1-\alpha}). \tag{5.7}$$

When $\lambda_0=1$,

$$U_i = 1-[1-(1+\lambda^*)U_i^{*\lambda^*} - \lambda^* U_i^{*1+\lambda^*}]^{1/2}. \tag{5.8}$$

It is nontrivial to obtain the exact distribution of $\log U_i$ when $\lambda = \lambda^*$. However, if the mean and variance of $\log U_i$ when $\lambda = \lambda^*$ are approximately known, then one can approximate the power $\pi(\lambda^*)$ by appealing to the central limit theorem.

When θ and σ are unknown one can obtain the BLUE of θ and σ using $\lambda = \lambda_0$. Consider the reduced data obtained by subtracting the BLUE of θ from each X_i and dividing it by the BLUE of σ. Then carry out the test procedure specified by (5.2) - (5.4). The resultant test will be asymptotically optimal.

6.0 *An Application*. In the hypothesis-testing problem (5.1) $\lambda_0=1$ plays an important role, since $\lambda_0=1$, implies that the hazard function is J-shaped. We shall test such a hypothesis for the data on life failures of a set of refrigerator motors which is taken from Chatfield [1]. Twenty refrigerator motors were run to destruction under advanced stress conditions and the times to failures (in hours) were recorded as follows:

104.3	158.7	193.7	201.3	206.2
227.8	249.1	307.8	311.5	329.6
358.5	364.3	370.4	380.5	394.6
426.2	434.1	552.6	594.0	691.5

Here, we will be interested in estimating λ and in testing the hypothesis: H_0: $\lambda = 1$ against H_1: $\lambda > 1$.

Let us take $\theta=100$ and $\sigma=600$. Then the transformed data $Z_i = (X_i-100)/600$, $i=1,\ldots 20$ is as follows:

.007	.098	.156	.169	.177
.213	.248	.346	.352	.383
.431	.441	.451	.468	.491
.544	.557	.754	.823	.985

Hence,

$$\bar{Z} = .405 \tag{6.1}$$

and using (4.2) we have

$$\hat{\lambda} = 2.94. \tag{6.2}$$

Letting $U_i = 1-\sqrt{1-Z_i}$, $i=1,2,\ldots,20$, we obtain the U sequence:

-5.521	-2.996	-2.513	-2.430	-2.375
-2.180	-2.017	-1.655	-1.635	-1.537
-1.402	-1.378	-1.351	-1.306	-1.248
-1.124	-1.097	-0.685	-0.546	-0.130

yielding

$$-2\Sigma \log U_i = 70.252.$$

Since $\chi^2_{40,.99} = 63.69$, we reject H_0: $\lambda=1$ at the 1% level of significance. Hence, most likely the hazard function is U-shaped.

Remark 6.1: In an analogous manner, one can study the class of distribution functions for failure times, given by

$$x = \theta + \{(\sigma+\lambda)u^{\lambda} - \lambda u^{\sigma+\lambda}\},\ u = F(x;\theta,\sigma,\lambda)$$

with $F(\theta;\ldots) = 0$, $F(\theta+\sigma;\ldots)=1$. Here notice, that σ is not only a scale parameter but also a part of the shape parameter. One can also obtain a two-parameter family as the solution of the differential equation:

$$\frac{f'}{f^2} = \frac{b-1}{1-F} - \frac{a-1}{F}, \quad a,b > 0$$

i.e. $\frac{dx}{du} = cu^{a-1}(1-u)^{b-1}$,

$$x = c\int_0^u v^{a-1}(1-v)^{b-1}dv. \tag{6.3}$$

$x = 1$ when $u = 1$ implies that $c = 1/B(a,b) = \Gamma(a+b)/\Gamma(a)\Gamma(b)$. When $b = 1$ and $a = \lambda$, we obtain (2.1) with $\theta = 0$, $\sigma = 1$.

7.0 *Applications to Nonparametric Testing Hypotheses.* In this section we shall employ the classes of distributions given in Sections 2 and 6 and their generalizations as nonparametric alternatives and derive LMP rank tests for the two-sample hypothesis-testing problems. If δ is a positive integer, then under HL1 [HL2] Y is distributed as the largest [smallest] in a random sample of size δ drawn from $F(x)$. Savage [9] has derived a LMP rank test for HL2. Then, we have the following results.

Result 7.1: Let $h(u;\theta) = 1-(1-u)^{\theta+1}(\theta u+1), 0 \leq u \leq 1, \theta > 0$. The LMP rank test of H_0 against H_L: $G = h(F;\theta)$ rejects H_1 when

$$T = \sum_{i=1}^{N} [2EU_{iN} + E\{\ln(1-U_{iN})\}]Z_i > K_\alpha \tag{7.1}$$

Where K_α is a constant determined by α.

Proof: $P(Z = z|HL) = m!n!\int\ldots\int_{-\infty< x_1<\ldots<x_N<\infty} \prod_{i=1}^{N} dG(x_i;\theta z_i)$

$$= m!n!\int\ldots\int_{0<u_1<\ldots<u_N<1} \prod_{0=1}^{N} h'(u_i;\theta z_i) \prod_{i=1}^{N} du_i$$

$$= m!n!\int_{0<u_1<\ldots<u_N<1} \prod_{i=1}^{N} (1-u_i)^{\theta z_i}\{\theta z_i(\theta z_i u_i+z_i u_i+1)+1\} \prod_{i=1}^{N} du_i.$$

Now, considering $\frac{\alpha}{\alpha\theta} P(Z=z|H_L)$ evaluated at $\theta = 0$, we obtain the test criterion (7.1) which is a combination of the Mann-Whitney-Wilcoxon test and the Savage's [9] test, because $-(\log 1-U_{iN})$ is distributed as the ith smallest standard negative exponential order statistic in a random sample of size N.

Remark 7.1: If $h(u;\theta) = u^{\theta+1}(\theta + 1 - \theta u)$, $0 \le u \le 1$, $\theta > 0$, then the LMP test rejects H_0 when

$$T = \sum_{i=1}^{N} [2E(1-U_{iN}) + E\ln U_{iN}]Z_i \ge K_\alpha. \tag{7.2}$$

Results 7.2: If $h(u;\theta) = 1 - (1-u)^{1+\theta}(1-\theta u)$, $0 < \theta \le 1$ then the LMP test rejects H_0 when

$$T = \sum_{i=1}^{N} [2E(1-U_{iN}) + E\{\ln(1-U_{iN})\}]Z_i \ge K_\alpha. \tag{7.3}$$

Remark 7.2: If $h(u;\theta) = u^{1+\theta}(\theta u+1-\theta)$, $0 < \theta \le 1$, then the LMP test rejects H_0 when

$$T = \sum_{i=1}^{N} [2EU_{iN} + E\ln U_{iN}]Z_i \ge K_\alpha. \tag{7.4}$$

Remark 7.3: In Result 7.2 and Remark 7.2, the alternative is a mixture of two Lehmann alternatives of the same kind.

Remark 7.4: If $h(u;\theta) = [(1-u)(1-\frac{\theta u}{1+\theta})]^{1+\theta}$, $\theta > 0$, then the LMP test criterion coincides with that in (7.3)

Results 7.3: Let $h(u;\theta) = \frac{\Gamma(2\theta+2)}{\Gamma^2(\theta+1)} \int_0^u [v(1-v)]^\theta dv$, $\theta > 0$ (which is obtained from (6.3) by setting $a = b = \theta + 1$). Then the LMP test rejects H_0 when

$$T = -\sum_{i=1}^{N} [E\{\ln(1-U_{iN}]\} + E\ln U_{iN}]Z_i \le K_\alpha \quad . \tag{7.5}$$

Proof: $P(Z = z|HL) = m!n!$

$$\cdot\left[\frac{\Gamma(2\theta+2)}{\Gamma^2(1+\theta)}\right]^n \int_{0\leq u_1<\cdots<u_N<1} \cdots \int \prod_{i=1}^{N} (u_i)(1-u_i)^{\theta z_i} \prod_{i=1}^{N} du_i \ .$$

Letting $L(\theta) = n[\ln(2\theta+1,+\ln\Gamma(2\theta+1)-2\ln\Gamma(\theta+1) + \sum_{i=1}^{N} \theta z_i \ln[u_i(1-u)]$.

We have $L'(0) = 2n + \sum_{i=1}^{N} z_i \ln\{u_i(1-u)\}$.

Thus, we reject H_0 for large value of

$$\sum_{i=1}^{N} [\, 2 + E\ln U_{iN} + E\{\ln(1-U_{iN})\}]Z_i .$$

8.0 *An Asymptotically Distribution-Free Test*. Consider the following problem. Let $X_1,\ldots,X_m$ denote the failure times of m items selected randomly from manufacturer I. Suppose we test nk items of manufacturer II in groups of **k** and record the minimum failure time in each group. Let these times be denoted by $Y_1,\ldots,Y_n$. If the X's have a common distribution F and the Y's have a common distribution G(x), then we wish to test H_0 : $G(x) = 1-[1-F(x)]^k$, $k > 1$ against the alternative H_1: $G(x) = 1-[1-F(x)^{k'}$, $1 \leq k' < k$. There does not exist an LMP distribution-free procedure for testing H_0 against H_1. In the following we shall obtain an asymptotically distribution-free test procedure.

If F is known, then by the Neyman-Pearson lemma, the uniformly most powerful test (uniformity is with respect to k') is to reject H_0 when

$$-\sum_{j=1}^{n} \ln(1-F(Y_j)) > \text{const.}$$

That is, when

$$T = -2k \sum_{j=1}^{n} \ln\{(1-F(Y_j)\} > K_\alpha \tag{8.1}$$

where K_α is determined by α. Under H_0 $=-2k\ln\{1-F(Y_j)\} =$ $-\ln\{1-G(Y_j)\}= -2\ln U_j = \chi^2_2$.

Hence, $K_\alpha = \chi^2_{2n,1-\alpha}$. However, since F is unknown, we use the following rule: reject H_0 when

$$-2k \sum_{j=1}^{n} \ln\{1-F_m(Y_j)\} > \chi^2_{2n,1-\alpha} \tag{8.2}$$

For sufficiently large m, the test will be close to the one given by (8.1).

Remark 8.1: In the practical application motivating the hypothesis-testing problem, k is a positive integer, which need not be in the general case. The alternative hypothesis of special interest is when $k' = 1$.

9.0 *Asymptotic Considerations*. Each statistic derived in Section 7 is a linear combination of statistics which belong to the Chernoff-Savage Class of Statistics the asymptotic normality of which is well known under all the hypotheses. (see, for instance, Govindarajulu, et al [3]). Thus, one can readily compute the Pitman efficiency in each case. In the following, let us do so for the statistics in (7.3) and (7.5) for location and scale alternatives respectively.

Test Procedure based on (7.3) with location alternatives.

For this statistics

$$J(u) = J_1(u) + J_2(u)$$

where $J_1(u) = 2(1-u)$ and $J_2(u) = \ln(1-u)$, $0 < u < 1$.

The Pitman efficiency of the Procedure (7.3) relative to the Student's t for location alternative is

$$e = \sigma^2[\int_{-\infty}^{\infty} J'(F)f(x)dF(x)]^2/I = \sigma^2A^2/I \text{ (say)} \tag{9.1}$$

where σ^2 denotes the variance of the distribution F,

$$J'(u) = -\{2 + (1-u)^{-1}\},$$

$$I = \int_0^1 J^2(u)du - [\int_0^1 J(u)du]^2 = 7/3 - 0 = 7/3. \tag{9.2}$$

and

$$A = [\int_{-\infty}^{\infty} J'(F)f(x)dF(x)]. \tag{9.3}$$

The following table gives the asumptotic relative efficiencies (AREA) of the Test (7.3) for some distributions of interest.

Table 9.1 ARE of the Test Procedure (7.3) for Location Alternatives

	NORMAL	LOGISTIC	DOUBLE EXPONENTIAL	UNIFORM $[-1/2,1/2]$
A	-1.467	-7/16	-1-1u2=1.693	$-\infty$
σ^2	1	$\pi^2/3$	2	1/12
ARE(7.3)	.922	$7\pi^2/36=1.919$	2.457	∞
ARE of Wilcoxon-test	$3/\pi=.955$	$\pi^2/9=1.097$	3/2	1

When f is the standard normal density

$$A = -2\int_{-\infty}^{\infty} f^2(x)dx - \int_{-\infty}^{\infty} \frac{f^2(x)}{1-F}dx = -\pi^{-1/2} - .903 = -1.467$$

(See Govindarajulu and Haller [2]. From Table 9.1 one concludes that, except for the normal case, the test (7.3) has higher efficiency than the Wilcoxon test.

Test Procedure (7.5) with scale alternatives. For this statistic

$$J(u) = J_1(u) + J_2(u)$$

where

$$J_1(u) = -\ln(1-u),\; J_2(u) = -\ln u, \quad 0<u<1.$$

The efficiency of (7.5) relative to the F-test with scale alternatives is

$$e = [\int_{-\infty}^{\infty} xJ'(F)f(x)dF(x)]^2 / I(2+\gamma_2) \tag{9.4}$$

where

$$J'(u) = (1-u)^{-1} - u^{-1},$$

$$I = \int_0^1 J^2(u)du - [\int_0^1 J(u)du]^2 = (8-\frac{\pi^2}{3}) - 4 = 4 - \frac{\pi^2}{3} .$$

and γ_2 denotes the Kurtosis of $F = -3 + \frac{\text{(Fourth Central Moment)}}{\text{(Second Central Moment)}^2}$.

Let

$$B = \int_{-\infty}^{\infty} x(\frac{1}{1-F} - \frac{1}{F})f(x)dF(x) = 2\int_{-\infty}^{\infty} \frac{xf(x)}{1-F}dF(x)$$

$$= 2\int_{-\infty}^{\infty} y(\frac{1}{1-F} - \frac{1}{F})f(y)dF(y). \tag{9.5}$$

The following table summarizes the values of β, γ_2 and ARE for some distributions of interest.

Table 9.2 ARE of the test procedure (7.5) for scale alternatives

	NORMAL	LOGISTIC	DOUBLE EXPONENTIAL	UNIFORM $[-1/2,1/2]$
γ_2	0	6/5	3	-6/5
B	.595	1/2	$1+1/2(\ln 2)^2 - \frac{\pi^2}{12} = .418$	$-\infty$
ARE of (7.5)	.997	$\frac{15}{16(12-\pi^2)} = .440$	.188	∞
ARE of Savage's [9] test	.175	5/64 = .078	.035	∞

The value for B when f is the standard normal density is taken from Govindarajulu and Haller [2]. The value of γ_2 for the logistic distribution is taken from Gupta and Shah [4]. Also, for the logistic

$$F(x) = (1 + e^{-x})^{-1}, f(x) = F(x) - F^2(x) \text{ and } x = \ln F - \ln(1-F)$$

and hence,

$$B = \int_0^1 u \ln u\, du - \int_0^1 u \ln(1-u)du = 2\int_0^1 u \ln u\, du - \int_0^1 \ln u\, du.$$

In the case of the double exponential, $x = -\ln\{2(1-F(x))\}$ for $x > 0$ and

$$B = 1 + (\ln 2)^2 + \int_{1/2}^{1} \frac{\ln(1-u)}{u} du = 1 + 1/2(\ln 2)^2 - \frac{\pi^2}{12} = .41777.$$

(See Ryshik and Gradstein [7]). Recall that Savage's [9] test is based on

$$\sum_1^n E[\mathrm{lu}(1-U_{iN})]Z_i \qquad \text{or} \qquad \sum_1^n E[\ln U_{iN}]Z_i \; .$$

From Table 9.2 we conclude that one can increase the ARE to 5 1/2 times by considering the procedure (7.5) instead of Savage's [9] procedure. The pleasant surprise is that the test (7.5) is almost as efficient as the F-test in the normal case.

Remark 9.1: For location alternatives with symmetry, the test (7.5) is iappropriate since its efficacy is zero (that is,

$\int_{-\infty}^{\infty} J'(F)f(x)dF = 0$ if $f(x)$ is symmetric about zero).

REFERENCES

[1] Chatfield, C., Statistics for Technology, Penguin Books, Harmondsworth, Middlesex, England, (1970), p. 16.

[2] Govindarajulu, Z. and Haller, S., "C-Sample Rank-Order Statistics." Chakrabarty Memorial Issue of the *Journal of Indian Statistical Association 10*, (1972), pp. 17-35.

[3] Govindarajulu, Z., LeCam, L. and Raghavachari, M., Generalizations of Theorems of Chernoff and Savage on the Asymptotic Normality of Test Statistics. *Proc. Fifth Berkeley Symposium Math. Statist. and Prob.*, Vol. 1, University of California Press, (1967), pp. 609-638.

[4] Gupta, S. S. and Shah, B. P., "Exact-Moments and Percentage Points of the Order Statistics and the Distribution of the Range From the Logistic Distribution," *Ann. Math. Statist.* *36*, (1965), pp. 907-920.

[5] Hahn, G. J., and Shapiro, S. S., Statistical Models In Engineering, New York: John Wiley, Section 3.5 (1965), pp. 103-118.

[6] Lehmann, E. L., The power of the rank tests. *Ann. Math. Statist.* *24*, (1953), pp. 23-42.

[7] Ryshik, I. M. and Gradstein, I. S., Tables of Series Products, and Integrals. Berlin: Veb Deutscher Verlag Der Wissenschaften, (1963), pp. 9 and 187.

[8] Sarhan, A. E. and Greenberg, B. G., Contributions To Order Statistics, New York: John Wiley, Part I: Chapter 3, (1962) pp. 20-27.

[9] Savage, I. R., "Contributions to the Theory of Rank Order Statistics: The Two-Sample Case". *Ann. Math. Statist.* *29* (1956), pp. 590-615.

A GENERALIZED WILCOXON MANN-WHITNEY STATISTIC WITH SOME APPLICATIONS IN RELIABILITY*

by A. P. BASU
University of Missouri, Columbia

Abstract. Let $X_1, X_2, \ldots X_m$ and $Y_1, Y_2, \ldots Y_n$ be two independent samples of sizes m and n $(m + n = N)$ from two populations with continuous cumulative distribution functions $F(x)$ and $G(y)$ where F and G belong to the same family $\mathcal{F}$ of distribution functions indexed by a parameter θ. To test the hypothesis

$$H_0:\ F = G$$

against the alternative that they are different, based on the first r ordered observations out of a combined sample of size N, we consider the following statistic:

$$V_r = \sum_{i=1}^{r} (nm_i - mn_i) + 1/2(N-r-1)(nm_r - mn_r),\quad (r \leq N)$$

where m_i and n_i are the number of x and y failures, respectively, among the first i ordered observations of the combined (x,y) sample.

Properties of V_r and its various k-sample extensions are reviewed. Suitable modification of V_r is used to classify an unknown population into one of two given populations. The classification rule is shown to be consistent and its asymptotic relative efficiency with respect to a classification rule for the normal distribution is computed. The problem of estimating the reliability function $P(y<x)$ is also discussed.

1.0 Introduction. Let $X_1, X_2, \ldots, X_m$ and $Y_1, Y_2, \ldots, Y_n$ be two independent samples of sizes m and n from two populations with

---

* Research supported by the Air Force Office of Scientific Research under Grant No. AFOSR-75-2795

continuous cumulative distribution functions (cdf) $F(x)$ and $G(y)$ where F and G belong to the same family $\mathcal{F}$ of distribution functions indexed by a parameter θ. Let all the $m+n = N$ observations be ordered in a sequence and suppose we want to test the hypothesis

$$H_0: \quad F = G \tag{1.1}$$

against the alternative that they are different based on the first r of the combined set of N observations. That is, we test the above hypothesis based on a (combined) right-censored sample of size at most r, where r is a given fixed number.

Such a censored sample occurs naturally in many physical situations. Thus in problems of life testing and reliability analysis we may be interested in comparing the mean life of two physical systems, or we may be interested in estimating the reliability function

$$P = P(Y < X)$$

but we can not afford to wait indefinitely to get information on all the sampling units put on test.

To test the hypothesis in (1.1) Sobel [31] considers the statistic V_r which we define below. Let m_i and n_i be the numbers of x and y failures respectively, among the first i ordered observations of the combined sample, so that

$$m_i + n_i = i, \qquad i = 1,2,\ldots,r. \tag{1.2}$$

The statistic V_r is given by

$$V_r = \sum_{i=1}^{r} (nm_i - mn_i) + \frac{1}{2}(N-r-1)(nm_r - mn_r) \tag{1.3}$$

$$= V_r^{(N)} + \frac{1}{2}(N-r-1)(nm_r - mn_r),$$

where

$$V_r^{(N)} = \sum_{i=1}^{r} (nm_i - mn_i) \quad . \tag{1.4}$$

The exact and asymptotic properties of V_r and $V_r^{(N)}$ have been considered by Sobel [30], [31] and Basu [2], [3]. It is also shown that V_r is related to a number of well known statistics. Define a new sequence $\{Z_i\}$ $(i=1,2,\ldots,N)$ derived from the combined ordered sample, always counting ordered observations from the left as follows:

$$Z_i = 1 \text{ if the ith ordered observation is an X} \qquad (1.5)$$
$$= 0 \quad \text{otherwise.}$$

Then it is shown [3] that

$$V_r = -N^2 G_r = -NW_r + \frac{mN(N+1)}{2}. \qquad (1.6)$$

Here

$$G_r = \sum_{i=1}^{r} \left(\frac{2i-N-1}{2N}\right) Z_i + \frac{r(m-m_r)}{2N} \qquad (1.7)$$

is asymptotically equivalent to the Gastwirth modification [13] of the Wilcoxon statistic, and

$$W_r = \sum_{i=1}^{r} iZ_i + \frac{(N+r+1)}{2} (m-m_r) \qquad (1.8)$$

is the locally most powerful rank test when the underlying distribution is given by

$$H(x,\theta) = (1-\theta)J(x)+\theta J^2(x) \qquad 0\leq\theta\leq 1, \qquad (1.9)$$

where $J(x)$ is a distribution function with density $j(x)$ (see [29]).

Similarly, let

$$s(u) = \begin{cases} +1, & u>0 \\ 0, & u=0 \\ -1, & u<0 \end{cases} \qquad (1.10)$$

and

$$c(u) = \frac{s(u)+1}{2} = \begin{cases} 1, & u>0 \\ 1/2, & u=0 \\ 0, & u<0 \end{cases} \qquad (1.11)$$

Then, it is well known that

$$W_N = \sum_{i=1}^{N} iZ_i = (\text{sum of ranks of X's})$$

$$= \frac{m(m+1)}{2} + \sum_{i=1}^{m} \sum_{j=1}^{n} c(X_i - Y_j) \qquad (1.12)$$

$$= \frac{m(N+1)}{2} + \frac{1}{2} \sum_{i=1}^{m} \sum_{j=1}^{n} S(X_i - Y\ ),$$

where W is the Wilcoxon's rank statistic. Thus,

$$W_r = \sum_{i=1}^{m_r} iZ_i + \frac{N+r+1}{2} (m-m_r)$$

$$= \frac{m_r(m_r+1)}{2} + \sum_{i=1}^{m_r} \sum_{j=1}^{n_r} c(X_i - Y_j) + \frac{N+r+1}{2} (m-m_r)$$

$$= U_G + \frac{m(m+1)}{2},$$

where

$$U_G = \sum_{i=1}^{m} \sum_{j=1}^{n} c(X_i - Y_j)$$

$$= \sum_{i=1}^{m_r} \sum_{j=1}^{n_r} c(X_i - Y_j) + \frac{1}{2}(m-m_r)(n+n_r)$$

is the Gehan-Gilbert [14], [15] statistic and is related to the statistic considered by Halperin [21]. Note that Gehan-Gilbert-Halperin considered r to be a random variable so that their test is conditional. Here we consider r to be fixed. Clearly for r = N, all the statistics described above are equivalent to the Wilcoxon statistic $W = \sum_{i=1}^{N} iZ_i$ [36] and the Mann-Whitney statistic [25] defined for a sequence of m x's and n y's as the number of y's preceding each x_i, summed from i = 1 to m.

In section 2 two k-sample extensions of W_r are given and in section 3 a classification rule based on censored data is considered and its asymptotic properties are studied. Finally, in section 4, the problem of estimating the reliability function is briefly described.

2.0 k-Sample Case The statistic V_r and W_r have been extended to the k-sample case by Basu [4]. Let $X_{ij}(j=1,2,\ldots,n_i;\ i=1,2,\ldots,k)$ be k independent samples of sizes $n_1,n_2,\ldots,n_k$ respectively from k populations with continuous cumulative distribution functions $F_1,F_2,\ldots,F_k$, respectively. We assume that the F_i's belong to a family $\mathcal{F}$ of distribution functions indexed by a parameter θ. Let all the $N = \sum_{i=1}^{k} n_i$ observations be put together and ordered to form a single sequence and suppose that only the first r ordered observations are available. That is, let us have a combined (right) censored sample of total size r. To test the null hypothesis

$$H_0:\ F_1(x) = F_2(x) = \ldots = F_k(x) \tag{2.1}$$

(or equivalently, H_o: $\theta_1=\theta_2=\ldots=\theta_k=0$ say,

under location alternatives)

against the alternative hypothesis

$$H_1:\ F_1(x) \neq \ldots \neq F_k(x)$$

Basu [4] has considered the statistic $B_r^{(N)}$ which is an extension of the V_r statistic as seen below.

$$B_r^{(N)} = G \sum_{i=1}^{k} \left(\frac{1}{n_i}\right)(S_i + rn_i/2N)^2. \tag{2.2}$$

Here

$$G = 12N^2(N-1)/r\,[(r^2 - 1) + 3N(N-4)] \tag{2.3}$$

$$S_i = \sum_{\alpha=1}^{r} [(2\alpha-N-r-1)/2N] Z_\alpha^{(i)} \qquad (i=1,2,\ldots,k) \qquad (2.4)$$

and

$$Z_\alpha^{(i)} = 1 \text{ if the th ordered observations, among the combined N observations is from the ith population.} \qquad (2.5)$$
$$= 0 \text{ otherwise, } (\alpha=1,2,\ldots,N;\ i-1,2,\ldots,k).$$

$B_r^{(N)}$ is a generalization of the Kruskal statistic [24] in the sense that when $r = N$, $B_N^{(N)}$ is the H-statistic proposed by Kruskal and is related to Terpstra's k-sample statistic [33] . It is also shown that, when $k = 2$,

$$G_r^{(N)} = \frac{-V_r}{N^2} = S_1 + n_1 r/2N$$

where $G_r^{(N)}$ and V_r were defined in Section 1 . Here n_1, n_2 are replaced by m and n respectively.

Similarly, to test the hypothesis (2.1) against the ordered alternative hypothesis

$$H:\ F_1(x) < F_2(x) < \ldots < F_k(x)$$

the following k sample extension of the V_r statistic has been considered by Basu [4] . Let $n_{i\alpha}$ be the cumulative number of observations from the ith population among the first α-ordered observations so that

$$\sum_{i=1}^{k} n_{i\alpha} = \alpha \qquad (\alpha=1,2,\ldots,r).$$

Defined V_{ij} by

$$V_{ij} = \sum_{\alpha=1}^{r} (n_j n_{i\alpha} - n_i n_{j\alpha}) + \frac{1}{2}(N-r-1)(n_j n_{ir} - n_i n_{jr})$$

$$(i,\ j=1,2,\ldots,k;\ i<j).$$

Then the statistics $V(N,r)$ is defined by

$$V(N,r) = \sum_{i<j} V_{ij}$$

where the summation is over all pairs (i,j) with $i<j$. For $r=N$, $V(N,r)$ reduces to the Jonckhere statistic [22] which is related to a statistic of Terpstra [34]. For $k=2$, $V(N,r)$ is the V_r statistic considered in section 1. Exact and asymptotic properties of $B_r^{(N)}$ and $V(N,r)$ have been considered by Basu [4]. In particular, the exact and the asymptotic distribution of both $B_r^{(N)}$ and $V(N,r)$ are obtained.

3.0 A Classification Rule. In this section we consider the following classification problem. Let Π_i, $i=0,1,2$ denote 3 populations with distribution functions $F_i(x)$. It is known that $\Pi_o = \Pi_i$ $(i=1,2)$ for exactly one i. Let X_{ij}, $j=1,2,\ldots,n_i$ denote a random sample of size n_i from $F_i(x)$ $(i=0,1,2)$. Let the $N=n_0+n_1+n_2$ observations be combined and only the first r ordered observations among the combined sample of size N are available. We wish to classify Π_0 to one of the two populations Π_1 and Π_2 based on the first r ordered observations, where r, n_0, n_1 and n_2 are assumed sufficiently large.

The problem of classification, both parametric and nonparametric, have been considered by several authors in the case $r=N$. See Anderson et al [1] and Das Gupta [11] for a bibliography. So far, as the author is aware, Basu and Gupta [5], [6] were the first to consider the classification problem based on censored samples in the special case when the underlying populations are exponential.

Consider the following classification rule R_1.

$$R_1: \text{ Classify } F_0 = F_1 \text{ if } V(T) > 0$$
$$F_0 = F_2 \text{ if } V(T) < 0, \tag{3.1}$$

where

$$V = V(T) = 2T_0 - T_1 - T_2, \tag{3.2}$$

and

$$n_i T_i = \sum_{\alpha=1}^{r} \frac{2\alpha-N-1}{2N} Z_\alpha^{(i)} + \frac{r(n_i - n_{ir})}{2N} \tag{3.3}$$

That is accept hypothesis H_1: $F_0 = F_1$ if $V(T) > 0$ and accept hypothesis H_2: $F_0 = F_2$ if $V(T) < 0$. Here $Z_\alpha^{(i)}$ and N_{ir} have definitions similar to those in section 2.

Statistics T_i's have been introduced by Basu [4] in connection with the k-sample problem and are related to the S_i's given in section 2 through the relation

$$n_i T_i = S_i + \frac{rn_i}{2N} . \tag{3.4}$$

Rule R_1 was considered by Kinderman [23] in the special case when $r = N$ and $n_0=n_1=n_2$. To study the asymptotic properties of R_1 we need the following results from Basu [4] .

We assume that N, n_i and r become infinitely large in such a way that

$$\lim_{N\to\infty} \frac{r}{N} = p>0, \lim_{N\to\infty} \frac{n_i}{N} = \lambda_i, \tag{3.5}$$

where $0<\lambda_0 \le \lambda_i \le 1-\lambda_0<1$ $(i=0,1,2)$, and λ_0 is a constant not greater than 1/3. Define

$$H(x) = \sum_{i=1}^{k} \lambda_i F_i(x) \tag{3.6}$$

and

$$H_N(x) = \sum_{i=1}^{k} \lambda_i F_i(x;n_i) \tag{3.7}$$

where $F_i(x;n_i)$ is the empirical cdf based on $x_{i1}, x_{i2}, \ldots, x_{in_i}$.

Then $T_{Ni} = T_i$ can be represented by

$$T_{Ni} = \int_{-\infty}^{\infty} J_N[H_N(x)]dF_i(x;n_i) \qquad (i=0,1,2), \tag{3.8}$$

where

$$\lim_{N\to\infty} J_N(u) = J(u) = u - \frac{1}{2}, \quad 0 \leq u \leq p \tag{3.9}$$

$$= p/2 \qquad u > p$$

The asymptotic distributions of the T_{Nj} is given by the following <u>Theorem 3.1</u>. Under the usual Cherroff-Savage regularity conditions as modified by Govindarajulu et al $\sqrt{N}\,(T_{Nj} - \mu_{Nj})$, $(j=0,1,2)$ has a limiting normal distribution with mean vector $(0,0,0)$ and covariance matrix $\Sigma = (\sigma_{ij})$ under either of the hypotheses H_1: $F_0 = F_1$ and H_2: $F_0 = F_2$. Here

$$\mu_{Nj} = \int_{-\infty}^{\infty} J\ H(x)\ dF_j(x),$$

$$\sigma_{jj} = \left((1-\lambda_j)/\lambda_j\right) A^2, \tag{3.10}$$

$$\sigma_{ij} = -A^2 \qquad ,$$

where

$$A^2 = \int_0^1 J^2(u)du - \left(\int_0^1 J(u)du\right)^2 = \frac{p}{12}(p^2-3p+3). \tag{3.11}$$

Note that $H(x)$ and hence μ_{Nj} depends on which of the hypotheses H_1 and H_2 is true.

From Theorem 3.1 it follows that the asymptotic distribution of $N^{\frac{1}{2}}(V(T) - \mu_N)$ is normal with mean 0 and variance C^2A^2 where

$$C^2 = \frac{4\lambda_1\lambda_2 + \lambda_0\lambda_1 + \lambda_0\lambda_2}{\lambda_0\lambda_1\lambda_2}, \tag{3.12}$$

and

$$\mu_N = 2\mu_{N0} - \mu_{N1} - \mu_{N2} = \begin{cases} \mu_{N1} - \mu_{N2} & \text{if } H_1 \\ -(\mu_{N1} - \mu_{N_2}) & \text{if } H_2. \end{cases} \tag{3.13}$$

It follows that

$$\lim_{N\to\infty} P(V_N > 0) = \lim_{N\to\infty} P\left(Z > -N^{\frac{1}{2}}\mu_N/(CA)\right) \tag{3.14}$$

where Z is the standard normal variable.

To compare R_1 with other classification procedures we need to compute the Pitman asymptotic relative efficiency of V with respect to other classification rules. To this end we assume that the $F_j(x)$ are close to each other in the sense of Pitman. In particular, let

$$F_2(x) = F_1(x + N^{-\frac{1}{2}}\theta) \tag{3.15}$$

in case F_1 and F_2 differ through a location parameter, and

$$F_2(x) = F_1\left(x(1 + N^{-\frac{1}{2}}\theta)\right) \tag{3.16}$$

when F_1 and F_2 differ through a scale parameter. We obtain the following theorem.

Theorem 3.2. Under the regularity assumptions of Theorem 3.1 if (3.15) holds then

$$\lim_{N\to\infty} N^{\frac{1}{2}}(\mu_{N1} - \mu_{N2}) = \theta\int_{-\infty}^{\infty} \frac{d}{dx} J\left(F_1(x)\right) dF_1(x) \tag{3.17}$$

and if (3.16) holds, then

$$\lim_{N\to\infty} N^{\frac{1}{2}} (\mu_{N1}-\mu_{N2}) = \theta \int_{-\infty}^{\infty} x \frac{d}{dx} J\left(F_1(x) \right) dF_1(x) \tag{3.18}$$

Proof. Consider the scale parameter case. Assume H_1 to be true so that

$$F_2(x) = F_1(x\tau) = F_0(x\tau)$$

where (3.19)

$$\tau = 1 + N^{-\frac{1}{2}} \theta$$

$$\mu_{N1} = \int_{-\infty}^{\infty} J[H(x)] \; dF_1(x)$$

$$= \int_{\infty}^{\infty} J[(\lambda_0 + \lambda_1)F_1(x) + \lambda_2 F_1(x\tau)] \, dF_1(x) \tag{3.20}$$

$$= g(\tau), \text{ say}$$

Expanding $g(\tau)$ in a Taylor series around $\tau=1$ and assuming $g''(\tau)$ exists and finite we have

$$g(\tau) = g(1) + (\tau-1)g'(1) + 0(\tau-1)^2$$

Hence

$$\mu_{N1} = \int_{-\infty}^{\infty} J\left(F_1(x)\right) dF_1(x) + (\tau-1)\lambda_2 \int_{-\infty}^{\infty} x \frac{d}{dx} J\left(F_1(x)\right) dF_1(x) + 0(\tau-1)^2 \tag{3.21}$$

Similarly,

$$\mu_{N2} = \int_{-\infty}^{\infty} J \left((\lambda_0 + \lambda_1)F_1 \left(\frac{x}{\tau}\right) + \lambda_2 F_1(x) \right) dF_1(x)$$

$$= \int_{-\infty}^{\infty} \left(J \quad F_1(x) \right) dF_1(x)$$

$$- \frac{(\tau-1)}{\tau^2} (\lambda_0 + \lambda_1) \int_{-\infty}^{\infty} x \frac{d}{dx} \; J\left(F_1(x)\right) dF_1(x) + 0(\tau-1)^2 \tag{3.22}$$

Thus, under H_1,

$$\mu_{N1}-\mu_{N2} = (\tau-1)\left[\lambda_2 + \frac{\lambda_0+\lambda_1}{\tau^2}\right]\int_{-\infty}^{\infty} x\,\frac{d}{dx}J\left(F_1(x)\right)dF_1(x) + 0(\tau-1)^2$$

Since $\tau = 1 + N^{-\frac{1}{2}}\theta$, we have

$$N^{\frac{1}{2}}(\mu_{N1}-\mu_{N2}) = \theta\left[\lambda_2 + \frac{\lambda_0+\lambda_1}{(1+N^{-\frac{1}{2}}\theta)^2}\right]\int_{-\infty}^{\infty} x\,\frac{d}{dx}J\left(F_1(x)\right)dF_1(x) + 0(N^{-\frac{1}{2}})$$

(3.23)

Hence (3.18) follows. The proof for H_2 is similar. (3.17) can be proven similarly.

From (3.14) and Theorem 3.2 follows the important corollary.

Corollary: Rule R_1 is consistent.

Since under H_1: $\mu_{N0}=\mu_{N1}$ and $\mu_N=\mu_{N1}-\mu_{N2}$, while under H_2: $\mu_{N0}=\mu_{N2}$ and $\mu_N=-(\mu_{N1}-\mu_{N2})$, we have, from (3.9) and Theorem 3.2, $N^{\frac{1}{2}}\mu_N$ converges to θa or $-\theta a$ depending on whether H_1 or H_2 obtains in the location parameter case. Similarly, in the scale parameter case $N^{\frac{1}{2}}\mu_N$ converges to θb or $-\theta b$ depending on whether H_1 or H_2 obtains.

Here,

$$a = \int_{-\infty}^{F^{-1}(p)} f_1^{\,2}(x)dx, \tag{3.24}$$

$$b = \int_{-\infty}^{F^{-1}(p)} xf_1^{\,2}(x)dx. \tag{3.25}$$

Hence under H_1, $P\left(V(T)\leq 0\right)$ converges to

$$\Phi(-\theta a/CA), \tag{3.26}$$

and $\Phi(-\theta b/CA)$ in the location and scale case respectively, where Φ is standard normal distribution. Kinderman [23] obtained similar result in the special case $r=N$ $(p=1)$ and $\lambda_0 = \lambda_1 = \lambda_2 = 1/3$.

Consider now the following classification rule R_2 for the normal distribution based on a combined sample of size M.

$$\text{Rule } R_2: \quad \begin{array}{ll} \text{Accept } H_1: \ F_0 = F_1, & \text{if } U>0 \\ \text{Accept } H_2: \ F_0 = F_2, & \text{if } U\leq 0 \end{array} \tag{3.27}$$

where $U = 2\overline{X}_{0M} - \overline{X}_{1M} - \overline{X}_{2M}$,

$\overline{X}_{jM}$ = jth sample mean, based on a random sample of size $M_j \sim N(\mu_j, \sigma^2/M_j)$.

$$\nu_j = \lim_{M\to\infty} M_j/M$$

Let

$$\mu_1-\mu_2 = M^{-\frac{1}{2}}\eta$$

Hence

$$E(U_M) = \begin{cases} \mu_1 - \mu_2 & \text{if } H_1 \\ -(\mu_1 - \mu_2) & \text{if } H_2 \end{cases}$$

and

$$\sqrt{M}\ (U_M - M^{-\frac{1}{2}}\eta)/d\sigma \xrightarrow{L} \Phi \text{ under } H_1, \tag{3.28}$$

and

$$\sqrt{M}\ (U_M + M^{-\frac{1}{2}}\eta)/d\sigma \xrightarrow{L} \phi \text{ under } H_2, \tag{3.29}$$

where

$$d^2 = \left(\frac{4}{\nu_0} + \frac{1}{\nu_1} + \frac{1}{\nu_2}\right).$$

Hence the limiting probability of error in this case is given by

$$\Phi(-\eta/d\sigma). \tag{3.30}$$

To compare rules V and U we assume sequences of N and M are chosen such that

$$N^{-\frac{1}{2}}\theta \sim M^{-\frac{1}{2}}\eta.$$

If the limiting probabilities are to be equal, from (3.26) and (3.30), we have

$$-\theta a/CA = -\eta/d\sigma.$$

Hence,

$$\frac{M}{N} \sim \frac{\eta^2}{\theta^2} = \frac{\sigma^2 a^2 d^2}{c^2 A^2} .$$

Thus, in the special case r=N,

$$\text{ARE}(V(T), U) = \frac{\sigma^2\left[\int_{\infty}^{\infty} f_1{}^2(x)dx\right]^2 d^2}{C^2A^2} . \tag{3.31}$$

Kinderman also obtained the same result when

$$\lambda_i = \nu_i = 1/3 \quad (i-0,1,2).$$

That is, when $d^2 = c^2$. It should be noted that

$$d^2 = c^2$$

any time the N_i's and the M_i's are in the same ration, that is, when

$$N_0: N_1: N_2 = M_0: M_1: M_2,$$

or (3.32)

$$N_0: N_1: N_2 = M_0: M_2: M_1.$$

(3.32) is specially important for design consideration as, depending on relative cost of sampling one may obtain more observations from F_1 or from F_2.

Similarly, for the scalar case, consider the following classification rule R_3.

$$\text{Rule } R_3: \quad \begin{array}{ll} \text{Accept } H_1: \; F_0 = F_1, & \text{if } W > 0 \\ \text{Accept } H_2: \; F_0 = F_2, & \text{if } W \leq 0 \end{array} \tag{3.33}$$

where

$$W = 2S_0^{\ 2} - S_1^{\ 2} - S_2^{\ 2} \ ,$$

$S_i^{\ 2}$ = ith sample variance based on a random sample of size M_i from a normal distribution with mean 0 and variance θ_i.

As before, let $\nu_j = \lim_{M\to\infty} \frac{M_j}{M}$ where $M = M_0 + M_1 + M_2$. Then, for large M, under H_i (i=1,2)

$$\sqrt{M} \quad (W - EW) \sim N(0, K_i^{\ 2}) \tag{3.34}$$

where

$$\begin{aligned} k_1^{\ 2} &= \frac{8}{\nu_0} + \frac{2}{\nu_1} + \frac{2}{\nu_2} \left(1 + M^{-\frac{1}{2}}\eta\right), \\ k_2^{\ 2} &= \frac{2}{\nu_1} + \left(\frac{8}{\nu_0} + \frac{2}{\nu_2}\right) \left(1 + M^{-\frac{1}{2}}\eta\right), \end{aligned} \tag{3.35}$$

$$\theta_1 = 1 \ ,$$

and

$$\theta_2 = 1 + M^{-\frac{1}{2}} \eta. \quad \text{Note } \lim_{M\to\infty} k_1^2 = \lim_{M\to\infty} k_2^2 = k^2 \ \text{(say)}.$$

As in the location parameter case, the ARE of V(T) with respect to W can then be shown to be

$$\text{ARE}(T,W) = \lim_{N\to\infty} \frac{M}{N} \sim \frac{\eta^2}{\theta^2} = \frac{b^2k^2}{C^2A^2} \ . \tag{3.36}$$

4.0 Concluding Remarks. The problem of estimating the reliability function

$$R = P \ (Y < X)$$

has been considered by Birnbaum [7] , Birnbaum and McCarty [8] ,

Church and Harris [10], Govindarajulu [16] , [17] and others. It is possible to estimate R in a similar manner based on the statistic V_r or the Gehan-Gilbert-Halperin version. However, as pointed out by Stedl and Saxena [32] , a simpler method based on binomial distribution seems to work better in the sense that it provides shorter confidence intervals. The method based on binomial distribution has the additional advantage that it does not require the assumption of independence.

Other two-sample and k-sample problems, as well as related classification problems corresponding to different modes of censoring or different rank statistics are being considered. Some of these will be reported elsewhere.

REFERENCES

[1] Anderson, T. W., Das Gupta, S. and Styan, G. P. H., A Bibliography of Multivariate Statistical Analysis, Oliver & Boyd, Edinburgh, (1972).

[2] Basu, A. P., On some two-sample and k-sample rank tests with applications to life testing, Tech. Report No. 77, Department of Statistics, University of Minnesota, (1966).

[3] Basu, A. P., On the large sample properties of a generalized Wilcoxon Mann-Whitney statistic, *Ann. Math. Statist.*, 38 (1967), pp. 905-915.

[4] Basu, A. P., On two k-sample rank tests for censored data, *Ann. Math Statist.* 38 (1967), pp. 1520-1535.

[5] Basu, A. P., and Gupta, A. K., Classification rules for exponential populations. *Proc. Conference on Reliability and Biometry*, SIAM Publ. (1974), pp. 637-650.

[6] Basu, A. P., and Gupta, A. K. Classification rules for exponential populations: two parameter case, Theory and Applications of Reliability, edited by I.N. Shimi and C.P. Tsokos.

[7] Birnbaum, Z. W., On a use of the Mann-Whitney Statistic, *Proceedings of the Third Berkeley Symp. on Math. Stat. and Prob.*, Univ. of Calif. Press, Vol. 1, (1956), pp. 13-17.

[8] Birnbaum, Z. W. and McCarty, R. C., "A distribution-free upper confidence bound for P_r $(Y<X)$ based on independent samples of X and Y, *Ann. Math. Statist. 29*, (1958), pp. 558-562.

[9] Chernoff, H. and Savage, I. R., Asymptotic normality and efficiency of certain non-parametric test statistics, *Ann. Math. Statist.*, *29* (1958), pp. 972-994.

[10] Church, J. D. and Harris, B., The estimation of reliability from stress-strength relationships, *Technometrics*, *12* (1970), pp. 49-54.

[11] Das Gupta, S., Theories and Methods in Classification: A Review, Discriminant Analysis and Applications, Academic Press, New York, (1973), pp. 77-137.

[12] Efron, B., The two sample problem with censored data, *Proc. Fifth Berkeley Symp. Math. Stat. and Prob.*, Vol. 4, (1967), pp. 831-853.

[13] Gastwirth, J. L., Asymptotically most powerful rank tests for the two-sample problem with censored data, *Ann. Math. Statist.*, *36* (1965), pp. 1243-1247.

[14] Gehan, E. A., A generalized Wilcoxon test for comparing arbitrarily singly-censored samples, *Biometrika*, *52* (1965), pp. 203-223.

[15] Gilbert, J. P., Random censorship, Ph. D. Thesis, University of Chicago, (1962).

[16] Govindarajulu, Z., Distribution-fre confidence bounds for $P(X<Y)$, *Ann. Inst. Statist. Math.* *20*(1968), pp. 229-238.

[17] Govindarajulu, Z., Fixed width confidence intervals for $P(X<Y)$, *Proc. Conference on Reliability and Biometry*, SIAM Publ. (1974), pp. 747-756.

[18] Govindarajulu, Z., Lecam, L. and Raghavachari, M., Generalizations of theorems of Chernoff and Savage on the asymptotic normality of test statistics, *Proc. Fifth Berkeley Symp. Math. Statist. and Prob.* Univ. of Calif. Press, Vol. 1, (1967), pp. 609-638.

[19] Govindarajulu, Z., and Gupta, A. K., Certain nonparametric classification rules: univariate case, Tech. Report No. 17, University of Michigan, Ann Arbor, (1972).

[20] Hájek, J., Asymptotically most powerful rank-order tests, *Ann. Math. Statist.*, *33* (1962), pp. 1124-1147.

[21] Halperin, M., Extension of the Wilcoxon Mann-Whitney test to samples censored at the same fixed point, *J. Amer. Statist. Assoc.*, *55* (1960), pp. 125-138.

[22] Jonckheere, A. R., A distribution-free k-sample test against ordered alternatives, *Biometrika*, *41* (1954), pp. 133-145.

[23] Kinderman, A. J., On some problems in classification: classifiability, asymptotic relative efficiency, and a complete class theorem, Tech. Report No. 178, University of Minnesota, (1972).

[24] Kruskal, W. H., A nonparametric test for the several sample problem, *Ann. Math. Statist.*, *23* (1952), pp. 525-540.

[25] Mann, H. B. and Whitney, D. R., On a test of whether one of the two random variables is stochastically greater than the other, *Ann. Math. Statist.*, *18* (1947), pp. 50-60.

[26] Patel, K. M., Hájek-Šidák approach to the asymptotic distribution of multivariate rank order statistics, *Jour. Multivariate Analysis*, *3* (1973), pp. 57-70.

[27] Pitman, E. J. G., Lecture notes on nonparametric statistical inference, Columbia University, (1949).

[28] Puri, M. L., Asymptotic efficiency of a class of c-sample tests, *Ann. Math. Statist.*, *35* (1964), pp. 102-121.

[29] Rao, U. V. R., Savage, I. R., and Sobel, M., Contributions to the theory of rank order statistics: the two-sample censored case, *Ann. Math. Statist.*, *31* (1960), pp. 415-426.

[30] Sobel, M. On a generalization of Wilcoxon's rank sum test for censored data, Tech. Report No. 69, University of Minnesota, (1965).

[31] Sobel, M. On a generalization of Wilcoxon's rank sum test for censored data, Tech. Report No. 69 (Revised), University of Minnesota, (1966).

[32] Stedl, J., and Saxena, K. M. L., A note on distribution free confidence intervals for P(Y<X), Unpublished manuscript, (1975).

[33] Terpstra, T. J., A non-parametric k-sample test and its connection with the H-test, Report S(92) (VP2) of the Statistical Department of the Mathematical Center, Amsterdam, (1952).

[34] Terpstra, T. J., The asymptotic normality and consistency of Kendall's test against trend, when ties are present in one ranking, *Nederl. Akad. Wetensch. Proc. A.*, 55 (1952) pp. 329-333.

[35] Vorlickova, D., Asymptotic properties of rank tests under discrete distributions, *Z. Wahrscheinlichkeitheorie Verw. Geb.*, 14 (1970) pp. 275-289.

[36] Wilcoxon, F., Individual comparisons by ranking methods, Biometrics, (1945), pp. 80-83.

THE ASYMPTOTIC THEORY OF EXTREME ORDER STATISTICS

by JANOS GALAMBOS
Temple University

Abstract. Let X_j denote the life length of the jth component of a machine. In reliability theory, one is interested in the life length Z_n of the machine where n signifies its number of components. Evidently, $Z_n = \min(X_j: 1 \leq j \leq n)$. Another important problem, which is extensively discussed in the literature, is the service time W_n of a machine with n components. If Y_j is the time period required for servicing the jth component, then $W_n = \max(Y_j: 1 \leq j \leq n)$. In the early investigations, it was usually assumed that the X's or Y's are stochastically independent and identically distributed random variables. If n is large, then asymptotic theory is used for describing Z_n or W_n. Classical theory thus gives that the (asymptotic) distribution of these extremes (Z_n or W_n) is of Weibull type. While the independence assumptions are practically never satisfied, data usually fits well the assumed Weibull distribution. This contradictory situation leads to the following mathematical problems: (i) What type of dependence property of the X's (or the Y's) will result in a Weibull distribution as the asymptotic law of Z_n (or W_n)? (ii) given the dependence structure of the X's (or Y's), what type of new asymptotic laws can be obtained for Z_n (or W_n)? The aim of the present paper is to analyze the recent development of the (mathematical) theory of the asymptotic distribution of extremes in the light of the questions (i) and (ii). Several dependence concepts will be introduced, each of which leads to a solution of (i). In regard to (ii), the following result holds: the class of limit laws of extremes for exchangeable variables is identical to the class of limit laws of extremes for arbitrary random variables. One can therefore limit attention to exchangeable variables. The basic references to this paper are the author's recent papers in Duke Math. J. 40(1973), 581-586, J. Appl. Probability 10(1973, 122-129 and 11(1974), 219-222 and Zeitschrift fur Wahrscheinlichkeitstheorie 32 (1975), 197-207. For multivariate extensions see H. A. David and the author, J. Appl. Probability 11(1974), 762-770 and the author's paper in J. Amer. Statist. Assoc. 70(1975), 674-680. Finally, we shall point out the difficulty of distinguishing between several distributions based on data. Hence, only a combination of theoretical results and experimentations can be used as conclusive evidence on the laws governing the behavior of extremes.

1.0 *Introduction.* Let X_j denote the life length of the j-th component of a machine. In reliability theory, one is interested in life length Z_n of the machine, where n signifies its number of components. Evidently, $Z_n = \min(X_j: 1 \leq j \leq n)$. Another important problem, which is extensively discussed in the literature, is the service time W_n of a machine with n components. If U_j is the time period required for servicing the j-th component and if service is possible concurrently on all components, then $W_n = \max(U_j, 1 \leq j \leq n)$. There are several other practical problems which lead to the investigation of the maximum or minimum of n random variables, where n is either fixed in advance or is itself a random variable. Our aim is to investigate the asymptotic behavior of W_n or Z_n as n increases indefinitely. Since any monotonically decreasing transformation of random variables transforms a minimum into a maximum, it is sufficient to analyze a theory for W_n, the maximum of random variables Y_j. For example, if we put $Y_j = -0X_j$ in our first example, then

$W_n = -Z_n$. In order to use a common language for all problems, we shall call the measurement Y_j of the j-th component its life length and W_n the life length of the equipment.

In the early investigations, it was usually assumed that the components operate stochastically independently and that each Y_j has the same distribution. Then the classical result of Gnedenko [8] yields that, if there are sequences a_n and b_n of real numbers such that, as $n \to +\infty$,

$$\lim P(W_n - a_n < b_n x) = L(x) \tag{1.1}$$

exists, then a_n and b_n can be chosen so that $L(x)$ is one of the following three functions:

$$L_1(x) = \exp(-\exp(-x));$$

$$L_2(x) = 0 \text{ for } x \leq 0 \text{ and } \exp(-x^{-a}) \text{ for } x > 0;$$

$$L_3(x) = \exp(-(-x)^a) \text{ for } x \leq 0 \text{ and } 1 \text{ for } x > 0;$$

where $a > 0$. The occurrence of $L_3(x)$, which is a transformed version of the Weibull distribution

$$F_a(x) = \begin{cases} 1 - \exp(-x^a) & \text{if } x > 0 \\ 0 & \text{otherwise,} \end{cases}$$

is one of the "justification" of practitioners to use the Weibull distribution for the description of life length of equipments with large number of components. An additional "justification" is the practical fact that data usually well fit one Weibull distribution $F_a(x/b)$, when one permits a scale parameter b. These two seemingly supporting factors do satisfy applied scientists when simple questions are to be answered. However, if deeper analysis of the data is required for which an accurate model for W_n is needed (e.g. to recognize characteristic properties by a transformation to the exponential distribution, using exact distributions), one immediately finds the error in the theoretical approximation: the independence of the components is practically never satisfied. But if models taking into account the dependence of the components lead to non-Weibull limiting laws for W_n, how is it possible that data never detected this possibility? The answer is simple: with varying a and b, the family $F_a(x/b)$ is so rich that one member of the family (with possibly a location transformation) would fit most data taken from a continuous population. We shall make this point more specific, through an example, after presenting the class of limit laws for W_n for dependent samples (section 5). Here, and throughout this paper, a limit law for W_n is meant in the extended sense of (1.1).

In regard to the possible limit laws for W_n, we investigate two questions: (i) what type of dependence concepts lead to $L_j(x)$, $1 \leq j \leq 3$, as limit laws for W_n and (ii) what are the possible limit laws for W_n for arbitrary (but not trivial) dependent systems? For (i), we present a general model in the next section. For (ii), a complete solution is given in section 4 for the class of non-trivial dependent systems (the term is to be specified accurately). These solutions were developed in a series of papers by the present author. It is hoped that by presenting this theory, which was developed on mathematical lines, to scientists with interest in applications, it will come closer to, and its future will be influenced by, such scientists.

2.0 *A Model Leading to $L_j(x)$ for W_n.* The following model, in a slightly more mathematical setting, was proposed in Galambos [3]. It covers the case of independence, m-dependence and several kinds of mixing concepts, recurring in the mathematical literature. Our aim here is to reformulate this model into a simpler language and to give an actual method of constructing the so called exceptional sets. The latter part appears here for the first time and thus the whole model becomes simpler and complete.

The Description of the Model. Let $Y_1, Y_2, \ldots, Y_n$ be random variables, where n is prescribed. We classify pairs (i, j) of indices as exceptional and non-exceptional (the model being unaffected by the actual rule of classification). If (i, j) is non-exceptional, then we require that, as $c \to \alpha$, the supremum of the natural bounds of the Y's,

$$P(Y_i \geq c, Y_j \geq c) = P(Y_i \geq c)P(Y_j \geq c)(1 + o(1)). \quad (2.1)$$

An arbitrary subset $Y_{i_1}, Y_{i_2}, \ldots, Y_{i_k}$ of the Y's will then be called non-exceptional if each pair (i_t, i_s) is non-exceptional. For these, we require that, as $c \to \alpha$,

$$P(Y_{i_1} \geq c, Y_{i_2} \geq c, \ldots, Y_{i_k} \geq c) \quad (2.1a)$$
$$\sim P(Y_{i_1} \geq c)P(Y_{i_2} \geq c) \ldots P(Y_{i_k} \geq c).$$

(This asymptotic expression has similar meaning to that of (2.1).) About exceptional subsets $Y_{i_1}, Y_{i_2}, \ldots, Y_{i_k}$, we make a very weak assumption for a very special case and no assumption in the remaining cases. If there is exactly one exceptional pair (i_t, i_u) among the subscripts $(i_1, i_2, \ldots, i_k)$ then we assume that there is a real number d_k such that, as $c \to \alpha$,

$$P(Y_{i_1} \geq c, Y_{i_2} \geq c, \ldots, Y_{i_k} \geq c) \quad (2.2)$$
$$\leq d_k\, P(Y_{i_t} \geq c, Y_{i_u} \geq c) \prod_k^{*} P(Y_{i_j} \geq c),$$

where $\prod_k^{*}$ signifies the product over j with $1 \leq j \leq k$, $j \neq t, u$. When there is more than one exceptional pair among $(i_1, i_2, \ldots, i_k)$, then no assumption is made on the interdependence of the Y's.

Notice that (2.1) and (2.1a) are weaker assumptions than the asymptotic independence of the Y's. As a matter of fact, we assume an asymptotic independence of the events $\{Y_j \geq c\}$ only (for non-exceptional sets of subscripts), where c is the same value for all j. This is why we selected the term "non-exceptional set" rather than asymptotic independent subsets.

The Conclusion for W_n. Assume that there are sequences a_n and b_n of real numbers such that, as $n \to +\infty$.

$$\lim \sum_{j=1}^{n} P(Y_j \geq b_n x + a_n) = w(x) \tag{2.3}$$

exists and

$$\lim \sum^{*} P(Y_i \geq b_n x + a_n,\ Y_j \geq b_n x + a_n) = 0, \tag{2.4}$$

where $\sum^{*}$ signifies summation over exceptional pairs (i, j). Furthermore, we assume that the number N(n) of exceptional pairs (i, j) satisfies the relation

$$\lim N(n)/n^2 = 0,\ n \to +\infty . \tag{2.5}$$

Then, as $n \to +\infty$.

$$\lim P(W_n < b_n x + a_n) = \exp\{-w(x)\}, \tag{2.6}$$

where, as before,

$$W_n = \max(Y_1, Y_2, \ldots, Y_n).$$

When the preceding conditions are satisfied then a_n and b_n can be chosen so that

$$L(x) = \exp\{-w(x)\}$$

is one of the functions $L_j(x)$, $1 \leq j \leq 3$.

A Method of Construction of Exceptional Pairs. The first natural step is to check if, for a particular pair (i, j), it is reasonable to assume the validity of (2.1). This may already lead to some exceptional pairs. We than proceed as follows. We take the observations $Y_1, Y_2, \ldots$ one after another, and look at the sets of subscripts to check whether (2.1a) or (2.2) hold. If neither is reasonable, then we include at least 2 pairs as exceptional in which case no condition is imposed on the interdependence of the Y's. We are completely free in this procedure to the extent that (2.5) should hold. That is, we now consider Y_1, Y_2, Y_3. If the first step already resulted in 2 exceptional pairs, then the interdependence of Y_1, Y_2, Y_3 does not come into account in our model. If the first step resulted in one exceptional pair (1,3) say, then we have to decide if (2.2) holds, that is, if

$$P(Y_1 \geq c,\ Y_2 \geq c,\ Y_3 \geq c) < d_3\ P(Y_1 \geq c,\ Y_3 \geq c)\ P(Y_2 \geq c)$$

with some d_3 not depending on c. If we do not feel this to be reasonable then we arbitrarily include one additional pair, (2,3) say, as exceptional, in which case again there is no condition to be satisfied. Finally, if the first step resulted in no exceptional pairs, then we first check (2.1). If decided to be unreasonable then we check (2.2) with one pair taken as exceptional. If this is also found to be unreasonable then we take 2 pairs, (1,2) and (2,3) say, as exceptional.

By this, we made a decision on the interdependence of (Y_1,Y_2,Y_3) and thus turn to (Y_1,Y_2,Y_3,Y_4). We go through all possibilities and then take one additional term, Y_5, and go on with the procedure. Actual problems in practice are of course simpler than the above procedure which applies to an arbitrary case. We show this through the examples below.

Example 1. Let the Y's be independent and identically distributed. Then the set of exceptional pairs is empty and (2.3) becomes the only assumption in the form

$$nP(Y_1 \geq b_n x + a_n) \to w(x). \tag{2.3a}$$

Our model thus reduces to Gnedenko's classical result.

Example 2. Observations are taken regularly from a machine. The observations can be considered independent and identically distributed, but a defect in the machine taking the observations mixes some observations which are considered dependent among each other. The defect is self-corrected and lasted for a short time only. Since the procedure is automatic, the error was not recognized. What is the effect on W_n?

If the observations are $Y_1, Y_2, \ldots,$ and Y_m is the first observation taken after the defect and Y_{m+t} is the first observation after the correction, then $Y_1, Y_2, \ldots, Y_{m-1}, Y_{m+t}, Y_{m+t+1}$ $\ldots, Y_n$ are independent and identically distributed, while the observations $Y_m, Y_{m+1}, \ldots, Y_{m+t-1}$ are independent of the others, but dependent on each other. Thus (i, j) is exceptional if, and only if, $m \leq i, j < m + t$. Thus $N(n) = \binom{t}{2}$ which satisfies (2.5) by the assumption of the defect having been of a short period. (2.2) holds with equality and with $d_k = 1$. Thus the only condition is (2.4). In order to see its implications, let us take in our example the common distribution of the Y's to be unit exponential with some form of bivariate exponential for exceptional pairs (i, j). Equation (2.3) again reduces to (2.3a) and thus $b_n = 1$, $a_n = \log n$ and $w(x) = e^{-x}$. Since $\sum^*$ consists of $N(n)$ terms, if even $t^2/n \to 0$ with $n \to +\infty$, then whatever the dependence of the exceptional Y's may be, the sum in (2.4) becomes smaller than

$$N(n)P(Y_m \geq x + \log n) = \binom{t}{2}n^{-1}e^{-x} \to 0.$$

Otherwise the form of

$$P(Y_i \geq c, Y_j \geq c)$$

does come in through (2.4), although not its exact form, only forms sufficient for estimates.

Example 3. In problems of life length or service time of equipment with a large number of components, one needs to know the interrelation of the components. The components can usually be grouped in such a way that within groups, the components are dependent (hence we take any pair within one group as exceptional), but components from different groups are independent. In this case, if we have k groups and if n_j components are collected in the j-th group then

$$N(n) = \sum_{j=1}^{k} \binom{n_j}{2}, \quad n = n_1 + n_2 + \dots + n_k$$

Hence (2.5) requires that each n_j should be a smaller and smaller proportion of the number n of all components as $n \to +\infty$ (n_j/n should tend uniformly to zero). Our assumptions imply again (2.1a) and (2.2) with equality ($d_k = 1$). The remaining assumption is again only (2.4), which requires only bivariate distributions within groups and only estimates even in this case.

Example 4. Rainfalls on consecutive days are clearly dependent but the dependence becomes weak as the days of observation draw apart from one another. Thus, if Y_j is the amount of rain on day j, (i, j) will be exceptional if $|i - j| \leq m$ (a number which can be estimated easily from data). With this choice, (2.1a) and (2.2) are evidently satisfied. $N(n) \leq nm$ and thus if m is fixed, (2.5) also holds. (2.4) follows from estimates by data and thus the conclusion (2.6) applies.

Example 4 exemplifies the so-called concept of m-dependence. This can easily be extended to various versions of mixing, the details of which are omitted.

3.0 *The General Form of Limit Laws for W_n.* Evidently, one is not interested in the whole class of dependent random variables. In fact, if we take a random variable Y repeatedly n times, then $W_n = Y$ and thus the class of limit laws for W_n would be the class of all univariate distributions. The following defintion is aimed at excluding the trivial cases.

Definition. Let $Y_{1,n} \leq Y_{2,n} \leq \dots \leq Y_{n,n}$ be the order statistics of $Y_1, Y_2, \dots, Y_n$. We say that the Y's form a non-trivial system of dependent observations if there are sequences a_n and b_n such that for at least one fixed t, as $n \to +\infty$

$$\lim P(Y_{n-t,n} < b_n x + a_n) = F_t(x)$$

if positive for $x > x_0$ and if for any function $t(n) \to +\infty$ **with n,**

$$\lim P(Y_{n-t(n),n} < b_n x + a_n) = 1, \ x > x_0.$$

Since the constants a_n and b_n determine the order of magnitude of $Y_{n-t,n}$, our requirement says that $Y_{n-t,n}$ should have different normalizing constants according as t is fixed or increases with n.

Now the following major result was obtained in Galambos [7].

Theorem. Let $Y_1, Y_2, \ldots, Y_n$ be random variables. Assume that there are sequences a_n and b_n such that, as $n \to +\infty$,

$$\lim P(W_n < b_n x + a_n) = F(x)$$

exists. Then, if the Y's form a non-trivial system,

$$F(x) = \int_0^{+\infty} e^{-a} \, d\, U(a, x), \tag{3.1}$$

where U(a, x) is a proper distribution function in a.

Note that if U(a, x) is degenerate at the point w(x), we get back the classical forms

$$F(x) = e^{-w(x)},$$

but many other distributions enter the theory in (3.1). In particular,

$$F(x) = (1 + e^{-x})^{-1}, \tag{3.2}$$

the logistic distribution, gets a central role. This was observed earlier by Berman [2] and Galambos [6] for different systems.

The formula (3.1) is mainly of theoretical value at present, but it should play an important part in future developments. Although U(a, x) is obtained in a constructive way if the multivariate distribution of the Y's is known, it is quite complicated.

The result expressed in the above theorem was achieved by recognizing in an earlier paper of mine, Galambos [4], that the distribution of W_n of an arbitrary system can be reduced to a problem on exchangeable events. One can then use a representation theorem of Kendall [10] to deduce (3.1).

It is interesting to remark that (3.1) also represents the possible limit laws for W_n when the Y's are independent and n is random, independently distributed of the Y's. With some additional restriction, such a case was earlier investigated by Thomas [13]. For independent Y's and random n, but not necessarily independent of the Y's, some results are available in Barndorff-Nielsen [2] and Mogyorodi [11]. Their result is extended in Galambos [5].

4.0 *Fitting Data with a Weibull Distribution.* As remarked earlier, one argument for justifying the application of classical theory is that data will fit one of the Weibull distributions. It is however a weak claim if not supported by theoretical results. As a matter of fact, a three parameter family $F_a((x-c)/b)$ of Weibull distributions would fit most data from a continuous population. The following example is presented here for showing the difficulty of using a goodness of fit test as a justification for the choice of a specific distribution.

Let $X_1, X_2, \ldots, X_n$ be unit exponential variates. Assume that their multivariate distribution is such that, with $y_n = \log n + y$, as $n \to +\infty$,

$$\lim \sum_k P(X_{i_1} \geq y_n, X_{i_2} \geq y_n, \ldots, X_{i_k} \geq y_n) = e^{-ky} \quad (4.1)$$

where $\sum_k$ is for summation over all choices $1 \leq i_1 < i_2 < \ldots < i_k \leq n$ of the subscripts i_j, $1 \leq j \leq k$. Then (see Galambos [6]), for

$$W_n = \max(X_1, X_2, \ldots, X_n),$$

$$\lim P(W_n < \log n + y) = 1/(1 + e^{-y}), \quad (n \to +\infty). \qquad (4.2)$$

Note that (4.1) and (4.2) are in complete contradiction of an assumption of independence of the X's. An approximation by a classical model is thus unjustified. Ignoring, however, this fact, a set of data on W_n with large n would lead the practitioner to a "justification" of using classical theory i.e. to the Weibull distribution as an approximation to W_n. This is due to a property of the family of Weibull distributions. Namely, the detailed analysis of the Weibull distributions by Plait [12] and properties of the logistic distribution (4.2) (see Johnson and Kotz [9], Vol. I, pp. 250-253 and Vol. II, p. 6) reveal that, by a suitable change in location and scale, the distributions

$$1/(1 + e^{-y}) \text{ and } 1 - \exp(-y^{3.25\ldots})$$

will differ from each other by less than 0.01. Therefore, if we assume the latter as the true distribution, a classical chi-squared test would accept the hypothesis at a very fine significance level. Such a conclusion definitely contradicts facts. Of course, if we were able to observe the X's, and thus to check that they are indeed identically distributed exponential variates, then the classical theory would result in $\exp(-\exp(-y))$ as asymptotic law for W_n. Theorists then could easily win against practitioners in the above case. But if we observe W_n only (as in cases of breakdowns), then the argument and discussion goes on.

5.0 *Summary*. The paper gives an analysis of recent developments in the asymptotic theory of extreme order statistics. The analysis is centered on reliability applications. This viewpoint gives a new light to available results as well as enabling one to formulate certain results in a somewhat extended form. The emphasis is on dropping the assumption of independence of the "components" and to present the class of possible limit laws for extremes we have shown that non-Weibull limit laws are possible as asymptotic laws for extremes and that this fact does not contradict the claim of practitioners that data will fit the class of Weibull distributions.

REFERENCES

[1] Barndorff-Nielsen, O., "On the Limit Distribution of the Maximum of a Random Number of Independent Random Variables", *Acta Math. Acad. Sci. Hungar. 15*, (1964), pp. 399-403.

[2] Berman, S. M., "Limiting Distribution of the Maximum Term in Sequences of Dependent Random Variables", *Ann. Math. Statist. 33*, (1962), pp. 894-908.

[3] Galambos, J., "On the Distribution of the Maximum of Random Variables", *Ann. Math. Statist., 43*, (1972), pp. 516-521.

[4] Galambos, J., "A General Poisson Limit Theorem of Probability Theory", *Duke Math. J., 40*, (1973a), pp. 581-586.

[5] Galambos, J., "The Distribution of the Maximum of a Random Number of Random Variables with Applications", *J. Appl. Probability, 10*, (1973b), pp. 122-129.

[6] Galambos, J., "A Limit Theorem with Applications in Order Statistics", *J. Appl. Probability, 11*, (1974), pp. 219-222.

[7] Galambos, J., "Limit Laws for Mixtures with Applications to Asymptotic Theory of Extremes", *Zeitschrift fur Wahrscheinlichkeit, 32*, (1975), pp. 197-207.

[8] Gnedenko, B. V., "Sur la Distribution Limite du Terme Maximum d'une Serie Aleatoire", *Ann. of Math., 44*, (1943), pp. 423-453.

[9] Johnson, N. L. and Kotz, S., Distributions in Statistics: Continuous Univariate Distributions I - II, John Wiley, New York, (1970).

[10] Kendall, D. G., "On Finite and Infinite Sequences of Exchangeable Events", *Studia Sci. Math. Hungar., 2*, (1967), pp. 319-327.

[11] Mogyorodi, J., "On the Limit Distribution of the Largest Term in the Order Statistics of a Sample of Random Size (in Hungarian)", *Magyar Tud. Akad. Mat. Fiz. Oszt. Kozl.*, *17*, (1967), pp. 75-83.

[12] Plait, A., "The Weibull Distribution with Tables", *Industrial Quality Control 19*, (1962), pp. 17-26.

[13] Thomas, D. L., "On Limiting Distributions of a Random Number of Dependent Random Variables", *Ann. Math. Statist.*, *43*, (1972), pp. 1719-1726.

ESTIMATING AND FORECASTING FAILURE-RATE PROCESSES BY MEANS OF THE KALMAN FILTER*

by H. F. MARTZ, JR., K. CAMPBELL
Los Alamos Scientific Laboratory
and H. T. DAVIS
University of New Mexico

Abstract. A new method is described for analyzing failure data. The approach used is to model the logarithm of the failure-rate process as a linear dynamic system with observations. This formulation permits the underlying failure-rate process to be corrupted by noise from various sources. In addition, the observations of the process are functions of simple non-parametric failure-rate estimates which are assumed to be noisy. The Kalman filter equations are used to provide the estimates and future forecasts. The procedure is non-parametric with regard to the underlying failure-time distribution. An example is provided.

1.0 *Introduction*. The failure-rate function, or hazard function, is of fundamental importance in both the theory and applications of reliability. Numerous parametric and non-parametric methods have been proposed for estimating the failure-rate function based on failure data. Parametric methods assume that the failure data arise from a specified distribution, but with unknown parameters which must be estimated from the data. A large portion of the book by Mann, Schafer, and Singpurwalla [14] is devoted to a discussion of such techniques. On the other hand, non-parametric methods do not require a distributional assumption. Barlow and Van Zwet [3] summarize and compare several non-parametric estimators for monotone failure-rate functions. Grenander [9] also discusses several non-parametric methods. Additional references may be found in [3].

---

*Work performed under the auspices of the Energy Research and Development Adm. and ONR Contract N00014-75-C-0832.

As Singpurwalla [25] points out, a basic disadvantage of both approaches is the inflexibility due to the assumed model and lack of a theory for forecasting. Further, we cannot account for contamination of the failure-rate estimates from such sources as periodicities due to inspection, data recording or reporting errors, or maintenance policy effects. In an effort to account for such contamination and to provide a theory for forecasting, Castellino and Singpurwalla [7] and Singpurwalla [25] have presented a new and novel approach for estimating and forecasting failure-rate functions. In their approach they think of the time-ordered sequence of certain non-parametric estimates of the failure-rate function as being generated by a time series process. The estimated failure-rate function is thus a stochastic process which they refer to as the *failure-rate process*. An appropriate Box-Jenkins time series model is then fitted to either the process itself [7], or a simple functional of the process [25]. The fitted model is then used to provide the required failure-rate estimates and forecasts. The approach is free of any assumptions regarding the failure distribution or the parametric form of its failure-rate function.

In this paper, we likewise consider the problem of estimating a failure-rate function, and then use Kalman filtering techniques to forecast its future values based on failure and withdrawal data up to some point in time. The approach used is to consider a simple functional of the true failure-rate function which satisfies a certain (specified) linear random differential equation, referred to as the *state equation*. The unknown value of the specified functional of the true failure-rate function at any time is referred to as the *state of the system* (or *system state* or *state*) at the time. Consequently, a general parametric form for the failure-rate function will be assumed in order to identify and fit the state equation. However, this equation does include a random error (noise) component to account for errors in identifying and fitting the state equation. Likewise, this error accounts for the obvious fact that any such mathematical model is at best an imperfect representation of reality. A second equation is adjoined to the state equation in order to relate the state of the system at any time to a simple non-parametric estimate of the state at that time. This equation is referred to as the *observation equation*. This equation also includes a random error component to account for the statistical error associated with the non-parametric estimate. The set of both equations is referred to as a *linear dynamic system with observations*. Once the system has been identified (Section 3), the unknown parameters are then estimated from the failure data (Section 4). The Kalman filter equations (Section 2) are then used to generate minimum mean square error estimates and forecasts of the system state, which are then transformed to the required failure-rate estimates and forecasts. An expository introduction to the use of the Kalman filter in reliability is given by Breipohl [5].

The idea of using the Kalman filter in time series forecasting is not new. McWhorter [16], and McWhorter et al [17], [18] have all considered the use of the Kalman filter for forecasting certain economic time series in which structural regression models with randomly varying time-dependent coefficients are used. Belsley and Kuh [4] and Rosenberg [23] also discuss some of the theoretical research relating to the use of the Kalman filter in time series forecasting. McWhorter [16] empirically compared the performance of the Kalman filter and the BEA macroeconometric forecasting models [10] for five quarterly economic time series. Narasimham et al [21] have also compared the predictive performance of the BEA model with certain Box-Jenkins models. Kamat and Cox [12] also discuss the use of the Kalman filter in time series forecasting. Also, Duncan and Horn [8] examine linear dynamic estimation from a regression viewpoint.

The manner in which the Kalman filter is used here is entirely different from its previous use in forecasting economic time series. Regression models with randomly varying time-dependent coefficients are not used in estimating and forecasting the failure-rate function. Rather, a completely different approach is taken, as discussed in detail in Section 3.

The Kalman filter method proposed here is a philosophically different alternative to the Box-Jenkins approach taken by Castellino and Singpurwalla. Let us consider some of the differences. One important difference is that it is not necessary to consider equispaced time points. In certain instances, the necessity for considering equispaced time points may make inefficient use of the failure data (see Section 5).

Another major difference is that there are two more or less distinct random error terms in the Kalman filter formulation. There is a random model identification error term which accompanies the mathematical model identified for the failure-rate process. The second random error term accounts for the statistical error associated with the data used to fit the model. There are several advantages in separating these two error sources. These two error terms are confounded in the Box-Jenkins approach. These errors will be further discussed in the next section.

The third major difference is that the entire procedure is less non-parametric. Although there is no assumption regarding the failure distribution, the random error components in the state and observation equations will be assumed to follow specified distributions. The justification for these will be considered in Section 3. The proposed procedure may be thought of as lying somewhere between a completely non-parametric and a completely parametric approach.

A brief introduction to linear dynamic estimation will be presented in the next section. Failure-rate estimation within the framework of linear dynamic estimation will be considered in Section 3. Section 4 discusses procedures for fitting the proposed model to the data. A real-data example application will be presented in Section 5.

2.0 *Linear Dynamic Estimation.* Linear dynamic estimation concerns the estimation of a physical process from observations of the process which may be corrupted by random "noise." The physical process is considered to be a random process which is linear in the state of the process.

As pioneers in this area, Wiener [27] and Kolmogorov [13] presented the basic theoretical solution to the problem of estimating the random process. The end result of their work was the specification of a weighting function for the optimal physically realizable estimator as the solution of a complicated integral equation. This estimator subsequently became known as the Wiener-Kolmogorov or Wiener filter. The details may be found in numerous modern textbooks on statistical control, communication, or information theory, such as Åström [2] and Morrison [20].

The practical problem of solving the integral equation of Wiener represented an additional degree of difficulty in applying the Wiener filter. Kalman [11] and Bucy and Kalman [6] recognized this shortcoming and proposed that the solution should be an algorithm which provides the numerical estimate from numerical observations with the aid of a digital computer. They converted the integral equation of Wiener into a non-linear differential equation which could be solved efficiently. Their basic method became known as the Kalman-Bucy or Kalman filter.

The mathematical statement of the general dynamic estimation problem will now be given. Consider the random linear differential equation (the *state equation*) given by

$$\frac{dx(t)}{dt} = A(t)\ x(t) + U(t) \tag{2.1}$$

where x(t) is an rxl vector which represents the *state of the system* at time t, A(t) is a specified time-varying rxr matrix of coefficients, and U(t) is a rxl vector-valued Wiener process representing the *state error* driving the system. From the Wiener process assumption it follows that $E[U(t)] = \phi$ and $E[U(t)U^T(t')] = K(t)\,\delta(t - t')$, where $K(t)$ is a specified $r \times r$ non-negative definite matrix whose elements are functions of t and $\delta(t - t')$ is the Dirac delta function. Here ϕ is used to represent a vector of "zeros" and the superscript "T" denotes matrix transposition.

It is shown in Morrison [20] that the function given by

$$x(t) = \Phi(t, t_{i-1})x(t_{i-1}) + \int_{t_{i-1}}^{t} \Phi(t, \lambda)U(\lambda)\, d\lambda \qquad (2.2)$$

satisfies (2.1) for initial condition $x(t_{i-1})$, where $\phi(t,t_{i-1})$ satisfies the differential equation

$$\frac{\partial\Phi(t,t_{i-1})}{\partial t} = A(t)\Phi(t, t_{i-1}) \qquad (2.3)$$

with initial condition $\Phi(t_{i-1}, t_{i-1}) = I$. Initial condition $x(t_{i-1})$ means that $x(t_{i-1})$ is precisely known. If we now define

$$u(t_{i-1}) = \int_{t_{i-1}}^{t_i} \Phi(t, \lambda)U(\lambda)\, d\lambda\ , \qquad (2.4)$$

then the discretized counterpart of (2.1) can be written as

$$x(t_i) = \Phi(t_i, t_{i-1})x(t_{i-1}) + u(t_{i-1}), \qquad (2.5)$$

where $\{t_i\}$ is a specified sequence of time points. The selection of this sequence will be explained later. From the Wiener process assumption on $U(t)$ it follows that $\{u(t_i)\}$ is a sequence of Gaussian random r-vectors with $E[u(t_i)] = \phi$ and $E[u(t_i)u^T(t_j)] = \delta_{ij}Q(t_i)$, where δ_{ij} is the Kronecker delta function. Thus, $\{u(t_i)\}$ is a sequence of independent Gaussian random vectors with mean ϕ and time-dependent covariance matrix

$$Q(t_i) \equiv \int_{t_{i-1}}^{t_i} \Phi(t, \lambda)K(\lambda)\Phi^T(t, \lambda)\, d\lambda. \qquad (2.6)$$

It is observed that the state equation in (2.5) is basically an autoregressive model with time-varying coefficients and a nonstationary independent Gaussian shock process with time-dependent covariance matrix. Such a process is sometimes referred to as a non-stationary Gaussian white noise shock process.

Now consider the discrete-time linear *observation equation* given by

$$y(t_i) = H(t_i)x(t_i) + v(t_i) \ , \tag{2.7}$$

where $y(t_i)$ is a pxl vector of observations, $H(t_i)$ is a specified pxr matrix relating $x(t_i)$ to $y(t_i)$, and $\{v(t_i)\}$ is a sequence of Gaussian random p-vectors with $E[v(t_i)] = \phi$ and $E[v(t_i)v^T(t_j)] = \delta_{ij}R(t_i)$. Thus $\{v(t_i)\}$ is also a non-stationary Gaussian white noise process known as the *observation error* within the system. Suppose we further assume that $u(t_i)$ is independent of $v(t_j)$ for all i and j. Equations (2.5) and (2.7) together are referred to as a *linear (discrete-time) dynamic system with observations*. The problem is to estimate $x(t_i)$ from the available sequence of observations $y(t_1), \ldots, y(t_i)$ and to forecast $x(t_m)$, where $t_m > t_i$.

Under the assumptions outlined above, and further if $\Phi(t_i, t_{i-1})$, $H(t_i)$, $Q(t_i)$, and $R(t_i)$ are known for time points $t_1, \ldots, t_i$, the Kalman filter equations are known to provide the minimum mean square error estimate of $x(t_i)$. The Kalman estimates are also known to be MVU estimates as well. The details may be found in most modern textbooks on control theory such as Åström [2]. The Kalman filter equations are given in [20] as

$$\hat{x}_i = \bar{x}_i + \bar{P}_i H_i^T \left(H_i \bar{P}_i H_i^T + R_i \right)^{-1} \left(y_i - H_i \bar{x}_i \right) \tag{2.8}$$

$$P_i = \bar{P}_i - \bar{P}_i H_i^T \left(H_i \bar{P}_i H_i^T + R_i \right)^{-1} H_i \bar{P}_i \tag{2.9}$$

$$\bar{x}_i = \Phi_{i,i-1} \hat{x}_{i-1} \tag{2.10}$$

$$\bar{P}_i = \Phi_{i,i-1} P_{i-1} \Phi_{i,i-1}^T + Q_{i-1} \ , \tag{2.11}$$

for i = 1, 2, ..., where for convenience in notation we have let $x_i \equiv x(t_i)$, $y_i \equiv y(t_i)$, $\Phi_{i,i-1} \equiv \Phi(t_i, t_{i-1})$, and so forth. Here $\hat{x}_i$ is the minimum mean square estimate of x_i and P_i is the covariance matrix of the estimation error $(\hat{x}_i - x_i)$. In a similar way, $\bar{x}_i$ is the minimum mean square estimate of x_i, given the observations $y_1, \ldots, y_{i-1}$, and $\bar{P}_i$ is its estimation error covariance matrix. The initial state estimate $\hat{x}_0$ and its error covariance matrix P_0 are also required to start the filtering process. Note that $\hat{x}_i$ is recursively computed and depends only upon $\hat{x}_{i-1}$ and y_i. Further note that P_i does not depend upon the observations, provided that Φ, H, Q, R, and P_0 are known. In practice, however, some or all of these quantities are unknown and must be estimated from the available data. This will be considered in Section 4. The matrix $\Phi_{i,i-1}$ is sometimes called the *state transition matrix* and the matrix $\bar{P}_i H_i^T (H_i \bar{P}_i H_i^T + R_i)^{-1}$ is often referred to as the *gain matrix (or gain)* of the Kalman filter.

3.0 *Failure-Rate Processes as Linear Dynamic Systems*. Let h(t), the true failure-rate at time t, represent the state of the system at time t. Consider a lifetest experiment in which n items are initially placed on test and in which $r \leq n$ failures are recorded as they occur at times $0 \equiv T_0 < T_1 < T_2 < \cdots < T_r$. Let Z_i denote the total time on test between the (i - 1)st and the ith failure. In the case of either censored or truncated testing and no progressive withdrawals, we have that $Z_1 = nT_1$, $Z_2 = (n - 1)(T_2 - T_1)$, ..., $Z_i = (n - i + 1)(T_i - T_{i-1})$, ..., $Z_r = (n - r + 1)(T_r - T_{r-1})$. In the case of progressive withdrawals, Z_i is calculated by appealing directly to the definition of total time on test. The MLE, $\hat{h}(t)$, is a step function, constant between observations, and is given by

$$\hat{h}(t) = Z_i^{-1}, \quad T_{i-1} < t \leq T_i, \quad i = 1, \ldots, r. \tag{3.1}$$

Although the MLE is defined for all time points over the range of observed failures, we shall use the MLE only at the observed failure times. Unfortunately, the asymptotic variance of $\hat{h}(t)$ depends upon $h(t)$ [3]. This violates the error variance condition of (2.7), since this error variance is not permitted to be a function of the system state. The logarithmic transformation is the appropriate variance stabilizing transformation, as illustrated later. Thus, we consider the use of $\ell n h(t)$ here.

Correspondingly, we now consider the state equation for the logarithm of the failure-rate function, $\ell n h(t)$. Suppose that we assume the state equation corresponding to a univariate version of (2.1) to be

$$\frac{d\ell n h(t)}{dt} = \left[\frac{abt^{b-1} + cd^t \ell n d + e/t}{at^b + cd^t + e\,\ell n t + f}\right] \ell n h(t) + U(t) \qquad (3.2)$$

Here $A(t)$ of (2.1) represents the expression in brackets, where a, b, c, d, e and f are parameters of the expression. Motivation for this choice of $A(t)$ is based on the following. Consider the deterministic counterpart of (3.2) obtained by ignoring $U(t)$, the random error process driving the system. A solution to this deterministic equation is given by

$$\ell n\, h(t) = at^b + cd^t + e\,\ell n t + f\,, \qquad (3.3)$$

from which

$$h(t) = \exp\,[at^b + cd^t + e\,\ell n t + f]. \qquad (3.4)$$

Equations (3.3) and (3.4) will be referred to as the *nominal system model* for the log failure-rate and failure-rate process, respectively.

Several important parametric failure-rate functions are special cases of (3.4). The constant failure-rate (exponential) model $h(t) = \lambda$ is obtained by letting $a \equiv \ell n\ \lambda$, $b=c=d=e=f \equiv 0$. The linearly increasing (Rayleigh) model $h(t) = \alpha t$, $\alpha > 0$, is obtained by setting $a \equiv \ell n\ \alpha$, $b=c=f \equiv 0$, and $e \equiv 1$. The polynomial (Weibull) and exponential (extreme value) failure-rate models are likewise easily shown to be special cases of (3.4). Also, the first-order autoregressive log failure-rate model for equally spaced time points, $\ln h(t+1) = \alpha \ln h(t)$, is obtained by letting $d \equiv \alpha$, $a=e=f \equiv 0$. It is noted here that higher-order autoregressive models can be obtained by considering a suitably dimensioned vector state variable consisting of an appropriate number of lags in the log failure-rate function. Each of the models and their combinations discussed above is, of course, assumed to be contaminated by Gaussian random noise input to the system as in (3.2). Thus, an actual log failure-rate process in practice is assumed to depart from the nominal system model in (3.3) depending upon the magnitude of the parameters in the assumed Wiener noise process contaminating (or corrupting) the system. The system state equation in (3.2) thus appears to be sufficiently flexible for use in many practical applications. It is also noted here that any assumed periodicities in the log failure-rate process can be accounted for in two ways. The first is to add appropriate periodic terms to the nominal system model (3.3). The second is to include higher order autoregressive terms in the state equation and consider a vector state variable of corresponding dimension. In the absence of these, it is impossible to forecast periodic behavior of the log failure-rate process when using the Kalman filter approach presented here. However, if there are periodicities in the MLE given in (3.1) as a result of periodically contaminated data, these will also likely be present to some extent in the Kalman filter estimates, even though (3.3) is used without adding periodic terms. The basis for this statement will be illustrated in Section 5.

The discrete-time version of (3.2) as given in (2.5) becomes

$$\ln h(t_i) = \left[\frac{a t_i^b + c d^{t_i} + e \ln t_i + f}{a t_{i-1}^b + c d^{t_{i-1}} + e \ln t_{i-1} + f} \right] \ln h(t_{i-1}) + u(t_{i-1}), \tag{3.5}$$

where $\Phi(t_i, t_{i-1})$ is obtained by solving (2.3) and is the expression given in brackets. Now $\Phi(t_i, t_{i-1})$ may be thought of as the time-dependent coefficient which maps the expected log failure-rate at time t_{i-1} into the expected log failure-rate at time t_i. The additive Gaussian error $u(t_{i-1})$ in (3.5) accounts for potential modeling errors as well as other contaminants which are likely to perturb the nominal log failure-rate process model. The assumption of additive Gaussian error in (3.5) is equivalent to an assumed multiplicative log-Gaussian noise component in the failure-rate system model corresponding to (3.5). It is argued that a positively skewed distribution, such as the log-Gaussian, is appropriate for a multiplicative error in which the mapped nominal system state is more often expected to underestimate the true system state. That is, underestimation of the true failure-rate is perhaps more frequent than overestimation based on the assumed system model.

Now let us consider the observation equation (2.7) in a form tentatively given by

$$\ln \hat{h}(t_i) = \ln h(t_i) + v(t_i) \ , \tag{3.6}$$

where $t_i \equiv T_i$, $i = 1, 2, \ldots, r$, and $\hat{h}(\cdot)$ is the MLE given in (3.1). For the case of testing without progressive withdrawals, let us determine the mean and variance of $\ln \hat{h}(t_i)$ in order to find the mean and variance of the observation error $v(t_i)$. It is well-known that, if x is an exponentially distributed random variable with mean μ, then $E[\ln x] = \ln\mu - \gamma$ and $V[\ln x] = \pi^2/6$, where γ is Euler's constant. Watson and Leadbetter [26] state that if $x_1, \ldots, x_n$ are iid failure time random variables then $(n - i + 1)[H(T_i) - H(T_{i-1})]$ are independent and exponentially distributed with mean 1, where $H(x)$ is the integrated failure-rate function corresponding to x. It then follows by use of the Mean Value Theorem that $H(T_i) - H(T_{i-1}) = h(\xi)[T_i - T_{i-1}]$ for some ξ such that $T_{i-1} \leqslant \xi \leqslant T_i$. From this it is easily shown that

$$E[\ln \hat{h}(t)] = \ln h(t) + \gamma + 0\left[\frac{h'(t)}{h(t)}\right] \tag{3.7}$$

and

$$V[\ln \hat{h}(t)] = \frac{\pi^2}{6} + 0\left[\frac{h'(t)}{h(t)}\right] . \tag{3.8}$$

Assuming the failure-rate function $h(t)$ to be reasonably smooth, the terms of $0[h'(t)/h(t)]$ can be neglected, thus yielding $E[\ln \hat{h}(t)] \doteq \ln h(t) + \gamma$ and $V[\ln \hat{h}(t)] \doteq \pi^2/6$. Thus, we redefine the observation equation (3.6) for use here as

$$\ln h^*(t_i) \equiv \ln \hat{h}(t_i) - \gamma = \ln h(t_i) + v^*(t_i) , \tag{3.9}$$

where now $v^*(t_i)$ is approximately normally distributed with mean 0 and variance $\pi^2/6$. Upon comparing (3.9) to (2.7), it is observed that $H(t_i) \equiv 1$ and $R(t_i) \equiv R = \pi^2/6$.

4.0 *Model Fitting and Parameter estimation.* We now discuss the procedures to be used in fitting the linear dynamic system with observations to a given set of failure data. We begin by considering the nominal system model given in (3.3) which is used in calculating $\Phi(t_i, t_{i-1})$ in (3.5). Since all of the parameters a, b, ..., f appearing in (3.3) will not likely be known *a priori*, a procedure for estimating these from the failure data is needed. We currently propose using nonlinear ordinary least squares (OLS) to estimate the unknown parameters in (3.3) according to the following scheme. Suppose that $N \leqslant r$ failures have been observed up to the present time, where N is large relative to the number of parameters a, b, ..., f to be estimated. These N observations will be used to identify the nominal system model in (3.3). Consider the sum of squares function given by

$$\begin{aligned} S(a, b, \ldots, f) &= \sum_{i=1}^{N} [\ln h^*(T_i) - (aT_i^b + cd^{T_i} + e \ln T_i + f)]^2 \\ &= \sum_{i=1}^{N} [- \ln Z_i - \gamma - (aT_i^b + cd^{T_i} + e \ln T_i + f)]^2 , \end{aligned} \tag{4.1}$$

where $\gamma = 0.5772157 \cdots$ is Euler's constant. Use the techniques of constrained nonlinear least squares estimation to find the values of the parameters a, b, ..., f which minimize S. These are the OLS estimates of a, b, ..., f and will be labeled $\hat{a}$, $\hat{b}$, ..., $\hat{f}$. These estimates are then used in computing $\hat{\Phi}(t_i, t_{i-1})$, where the OLS estimates replace the unknown parameters.

Several aspects regarding this procedure should be mentioned. First, in many practical applications some of the parameters a, b, ..., f are either likely to be known, or can be assumed to be known, *a priori*. For example, if the failure-rate process is justifiably believed to be basically a contaminated Weibull process, then b, c, and f can be set equal to zero and the general six-parameter nonlinear OLS problem reduces to a simple two-parameter linear OLS problem. Generally, it is unlikely that the full six-parameter model will be required. It has been included here for the sake of generality.

Secondly, the form of $\Phi(t_i, t_{i-1})$ in (3.5) is such that a high degree of accuracy for $\hat{a}$, $\hat{b}$, ..., $\hat{f}$ is unnecessary. That is, slight changes in $\hat{a}$, $\hat{b}$, ..., $\hat{f}$ have relatively small effect on $\hat{\Phi}(t_i, t_{i-1})$ and thus somewhat "rough" estimates will usually be sufficient. In fact, due to the flexibility of a six-parameter model, different lower order subsets of these six parameters can sometimes be used to calculate $\hat{\Phi}(t_i, t_{i-1})$ values which do not have a significant effect on the estimates generated by the Kalman filter equations. This will be illustrated in the next section.

Thirdly, each of the terms appearing in (3.3) may be loosely interpreted as follows. The term (at^b) in (3.3) may be loosely thought of as accounting for exponential tendencies in the failure-rate process. The terms (cd^t) and $(e \ln t)$ loosely account for autodependent and polynomial behavior, respectively, in the model. Finally the parameter f in (3.3) loosely acts as a scale parameter in the nominal failure-rate process model. Taken together, these terms can account for dissimilar process tendencies such as may occur during the infant mortality and wear-out regions of useful life of a device. This will also be illustrated in the next section.

Now consider estimation of the initial state $\ell n\ h(t_0)$, initial state error variance P_0, and state error variance Q_i. For convenience we shall assume that the state error variance Q_i is constant over time and consider estimating the common value $Q_i \equiv Q$. Thus, the additive error contaminating the log failure-rate state equation (3.5) is assumed to be a stationary white noise Gaussian process with mean 0 and variance Q. For convenience we shall also set P_0 equal to Q, since the long-range performance of the Kalman filter is known to be rather insensitive to initial starting conditions.

Numerous estimation procedures have been proposed for estimating the state error variance and initial state of a linear dynamic system with observations. Pearson [22] provides an extensive bibliography and gives an excellent survey of available methods. Shellenbarger [24] and Abramson [1] develop maximum likelihood estimators of both the state and observation error covariance matrices. Mehra [19] also develops estimators of both error covariance matrices based on the use of residuals.

We consider estimators for Q and $E[\ell n\ h(t_0)]$ based on the method of moments. By repeated use of (3.5), it is easily shown that the distribution of $\ell n\ h_i^*$, conditional on the initial state $\ell n\ h_0$, has mean and variance given by

$$E[\ell n\ h_i^* \mid \ell n\ h_0] = \Psi_{0i}\ \ell n\ h_0 \tag{4.2}$$

and

$$V[\ell n\ h_i^* \mid \ell n\ h_0] = \Theta_i Q + \frac{\pi^2}{6}, \tag{4.3}$$

respectively, where we have defined

$$\Psi_{ki} = \prod_{j=1}^{i-k} \Phi_{j+k,j+k-1},\ i = 1, 2, \ldots;\ k = 0, 1, \ldots, i-1, \tag{4.4}$$

$$\Theta_i = \sum_{k=1}^{i-1} \Psi_{ki}^2 + 1,\ i = 2, 3, \ldots, \tag{4.5}$$

and

$$\Theta_1 = 1. \tag{4.6}$$

Now, by use of the fact that $E(X) = E[E(X|Y)]$ and $V(X) = V[E(X|Y)] + E[V(X|Y)]$, we find that

$$\frac{1}{N}\sum_{i=1}^{N} E[\ell n\, h_i^*] = E[\ell n\, h_0] \sum_{i=1}^{N} \Psi_{0i}/N \tag{4.7}$$

and

$$\frac{1}{N}\sum_{i=1}^{N} V[\ell n\, h_i^*] = Q \sum_{i=1}^{N} (\Theta_i + \Psi_{0i}^2)/\ N + \pi^2/6. \tag{4.8}$$

Define the first two sample moments of the sequence $\{\ell n\, h_i^*, i=1, \ldots, N\}$ as

$$u_1 \equiv \frac{1}{N}\sum_{i=1}^{N} \ell n\, h_i^* \tag{4.9}$$

and

$$u_2 \equiv \frac{1}{N}\sum_{i=1}^{N} (\ell n\, h_i^* - u_1)^2 . \tag{4.10}$$

Equating (4.7) to (4.9) and (4.8) to (4.10) and solving for $E[\ell n\, h_0]$ and Q yields the moment estimators given by

$$\widehat{E[\ell n\, h_0]} = Nu_1/\sum_{i=1}^{N} \Psi_{0i} = \sum_{i=1}^{N} \ell n\, h_i^*/\sum_{i=1}^{N} \Psi_{0i} \tag{4.11}$$

and

$$\hat{Q} = N(u_2 - \pi^2/6)\ /\sum_{i=1}^{N} (\Theta_i + \Psi_{0i}^2) . \tag{4.12}$$

Two things should be pointed out here. First, since the nominal system model (3.3) is fitted to the data by means of OLS as in (4.1), $\widehat{E[\ell n\, h_0]}$ will be identically equal to the value of (3.3), in which the appropriate OLS estimates have been inserted, at the initial time t_0. Thus, the initial Kalman filter estimate required to start the filter is the initial nominal system estimate. If setting $t = t_0 \equiv 0$ in (3.3) yields a value of $-\infty$, i.e., if $e \neq 0$, then an initial time t_0 should be selected such that $0 < t_0 < T_1$. Secondly, it may happen that $\hat{Q} < 0$, in which case an arbitrary non-negative value must be selected for Q. It is noted here that the above estimates are considerably simpler to compute than corresponding maximum likelihood estimates which cannot be obtained in closed form.

Once $\hat{\Phi}(t_i, t_{i-1})$, $\widehat{E[\ell n\ h_o]}$, and $\hat{Q}$ have been obtained, the Kalman filter equations can be applied. The Kalman filter equations (2.8)-(2.11) now become

$$\ell n\ \tilde{h}_i = \ell n\ \bar{h}_i + 6\bar{P}_i(\ell n\ h_i^* - \ell n\ \bar{h}_i)/(6\bar{P}_i + \pi^2) \quad (4.13)$$

$$P_i = \bar{P}_i - 6\bar{P}_i^2\ /(6\bar{P}_i + \pi^2) \quad (4.14)$$

$$\ell n\ \bar{h}_i = \hat{\Phi}_{i,i-1}\ \ell n\ \tilde{h}_{i-1} \quad (4.15)$$

$$\bar{P}_i = \hat{\Phi}_{i,i-1}^2\ P_{i-1} + \hat{Q}, \quad i = 1, 2, \ldots, N \quad (4.16)$$

where $\ell n\ \tilde{h}_0 \equiv \widehat{E[\ell n\ h_0]}$ and $P_0 \equiv \hat{Q}$ are used as initial starting values. Recall that the subscript i denotes the ith observed failure time T_i. Kalman filter estimates of the log failure-rate function are thus calculated at each of the observed failure times. Future forecasts of the log failure-rate function at any time $t > t_N$ are calculated by means of

$$\ell n\ \bar{h}(t) = \hat{\Phi}(t, t_N)\ \ell n\ \tilde{h}_N, \quad (4.17)$$

and the variance associated with the forecast error is estimated to be

$$\bar{P}(t) = \hat{\Phi}^2(t, t_N)P_N + \hat{Q}. \quad (4.18)$$

5.0 *An Example Application.* We shall illustrate the Kalman filter procedure by the following example taken from NAILSC Report ILS 04-21-72. Castellino and Singpurwalla [7] used this same example. Singpurwalla [25] also considers this same example and Table 1 of that paper gives the failure and withdrawal times (in hours) for an A/C generator. A total of 55 failures were reported ranging from a minimum of 1.0 hour to a maximum of 1097.3 hours. However, only 53 generators failed at distinctly different times, since two generators failed at 3.0 hours and two generators failed at 252.8 hours.

Singpurwalla observed periodicity at lag 7 in the sequence of MLE's computed from (3.1) at 24-hour equispaced time points. This periodicity was the result of a weekly inspection policy wherein items soon expected to fail were withdrawn from the test. Castellino and Singpurwalla [7] fitted a Box-Jenkins ARIMA model of the form $(1, 0, 0) \times (2, 1, 0)_7 + \Theta_0$ directly to the MLE's of the failure-rate function. By use of this model, they were able to satisfactorily estimate and forecast the failure-rate function. Their estimates and forecasts preserved the periodicities in the data.

In order to fit the nominal model in (3.3) without the complexity of adding periodic terms, we shall deliberately ignore the periodic contaminants in the data and proceed directly to fit the general linear dynamic model presented in Sections 3 and 4. However, we are well aware that there is ample evidence that periodic-type terms should be included in the nominal system model (3.3). In this case, the Kalman model is not as compatible with the data as the Box-Jenkins model. However, it will still serve the purpose of illustration.

Upon examining the plot of $(T_i, \ln h_i^*)$, $i = 1, \ldots, 53$, it was tentatively decided that parameters e and b in the nominal system model (3.3) could be set equal to 0 and 1, respectively. The resulting OLS curves fitted to the data $\{(T_i, \ln h_i^*),$ $i = 1, 2, \ldots, 53\}$ tended to confirm this somewhat ad hoc choice. The remaining parameters a, c, d, and f were estimated by use of the software program Z05LSQS* using the method due to Marquardt [15]. The OLS estimates of these parameters were found to be $\hat{a} = 1.09113$, $\hat{c} = 2.61270$, $\hat{d} = 0.83065$ and $\hat{f} = -7.53429$. Thus, the nominal system model for the log failure-rate function was taken to be

$$\ln h(t) = 1.09113t + 2.61270(0.83065)^t - 7.53429. \quad (5.1)$$

---

*Internal nonlinear OLS program available at the Los Alamos Scientific Laboratory, Los Alamos, New Mexico.

Both the input observation data $\ell n\ \hat{h}(T_i)$ as well as the nominal system model (5.1) are plotted in Figure 1. It is observed that the apparent rapid decrease in the input data occurring during the brief infant mortality or break-in period is captured in the nominal model. It is further observed that the generators tend to begin wearing out sometime after the break-in period as evidenced by the increasing trend of the input data over time. The nominal model captures this increasing trend and is nearly linear during this region. Also, the chance-failure region is nominally estimated to be of fairly short duration. Finally, the input data marked with a dot (•) in Figure 1 represents the data used by Castellino and Singpurwalla [7] and Singpurwalla [25] when considering 24-hour equispaced time intervals. It is clearly apparent that this particular subset of input data is nonrepresentative of all the data. Consequently, it must be remembered that failure-rate estimates and forecasts based on this subset should be interpreted *only* at 24-hour intervals and *not* at arbitrary points in time. On the other hand, the Kalman filter procedure effectively uses all of the input data, thus providing a composite view of the entire failure-rate function.

Moment estimates of $E[\ell n\ h_0]$ and Q were computed to be -4.92 and -0.00055, respectively. Since $\hat{Q}$ is negative, we shall arbitrarily set $\hat{Q}$ equal to several non-negative values and observe the corresponding performance of the filter. The Kalman filter estimates given by (4.13) are plotted in Figure 1. For these estimates, Q was taken to be 0.02. It is observed that the Kalman estimates are significantly smoother than the input MLE's and effectively represent a compromise between the input data and nominal system model. The Kalman estimates are observed to account nicely for the log failure-rate function during the break-in as well as the wear-out regions. In Figure 1 we have also plotted the Kalman estimates forecasted ahead at each observed failure to the time of the next failure. These estimates are given in (4.15) and illustrate the short-range forecasting ability of the Kalman filter procedure. These short-range forecasts are observed to be in good agreement with the Kalman estimates themselves.

The same estimates are plotted in Figure 2 except that now $\hat{Q} = 2.0$. In this case, less smoothing occurs and the ragged nature of the input data is largely preserved in the Kalman estimates. This situation corresponds to somewhat imprecise knowledge of the nominal system model. It is interesting to observe that the Kalman estimates seem to be in phase with the input data. Periodicities in the input data are also likely to be preserved.

Figure 3 considers the case where $\hat{Q} = 0.001$. This corresponds to precise system knowledge. It is observed that the Kalman estimates nearly coincide with the nominal system model and are extremely smooth relative to the input data. In the case where $\hat{Q} \equiv P_0 = 0$, the Kalman estimates reduce to the nominal system estimates and no state noise is assumed to be driving the system.

In Figure 4 we have plotted the antilogs of corresponding estimates in Figure 1. This graph illustrates the Kalman filter's performance in estimating the failure-rate function. The performance appears to be satisfactory, and is analogous to Figure 1.

Figure 5 presents a plot of the Kalman estimates and forecasts of the failure-rate function for $\hat{Q} = 0.02$. The estimates are plotted up to time $T_{55} = 1097.3$ hours and the forecasts are given at 25-hour equispaced time intervals beginning with t = 1100 hours through t = 1700 hours. The forecasts were obtained by taking the antilog of (4.17). As pointed out in Section 3, periodicities cannot be forecast without the use of suitable periodic terms in the nominal system model. In Figure 5 we have also plotted approximate 95 percent probability limits for the underlying failure-rate function. At the observed failure times up to 1097.3 hours, the limits were computed according to $\exp[\ell n\, \tilde{h}_i \pm 1.96\sqrt{P_i}]$. The 95 percent limits on the failure-rate forecasts were computed from $\exp[\ell n\, \overline{h}(t) \pm 1.96\sqrt{\overline{P}(t)}]$, where

$\ell n\ \bar{h}(t)$ and $\bar{P}(t)$ are given in (4.17) and (4.18).

In Figure 5, we have also plotted the estimates and forecasts given by the Box-Jenkins technique of Castellino and Singpurwalla. These estimates and forecasts are exclusively given at 24-hour intervals. Since the Kalman filter results are based on nonequispaced input data, a direct comparison cannot be made.

The sensitivity of the resulting Kalman filter estimates to the choice of a nominal system model (3.3) was investigated by fitting a different nominal model to the input data. When the nominal model

$$\ell n\ h(t) = 0.03979t^{0.5} + 2.82400(0.85277)^{t} - 7.83338, \tag{5.2}$$

was used rather than (5.1), the resulting estimates were not noticeably different from those in Figure 1 based on the use of (5.1). This is explained by noticing that $\Phi(t_i, t_{i-1})$ in (3.5) is a ratio and that changes in the nominal model tend to "cancel out" to a large extent.

Now the entire procedure may be interpreted as a Bayesian technique for smoothing nonparametric failure-rate estimates. The degree of smoothing is governed by the magnitude of the state error variance estimate $\hat{Q}$, and the smoothing occurs in the direction of the nominal system model.

In conclusion, we have shown that the Kalman filter equations can be used to provide realistic failure-rate estimates and forecasts in the presence of imprecise knowledge of the actual form of the failure-rate function.

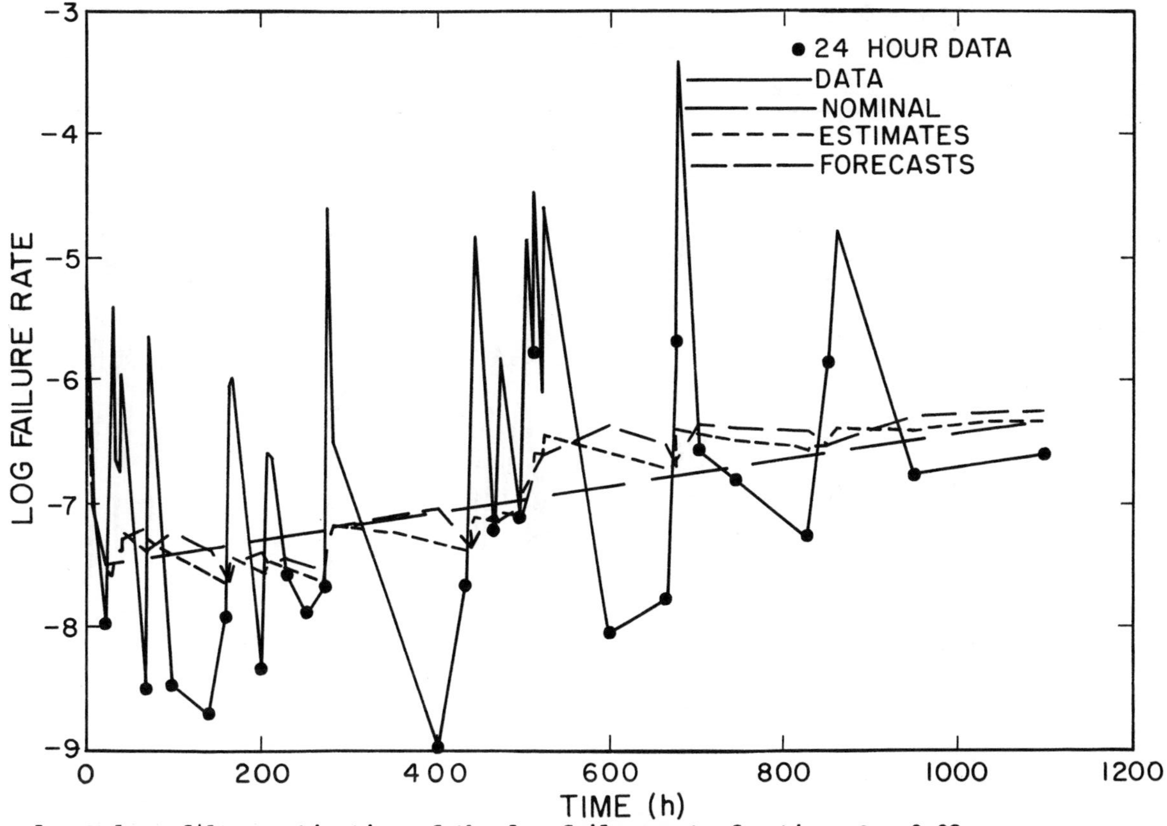

Figure 1. Kalman filter estimation of the log failure-rate function $Q = 0.02$.

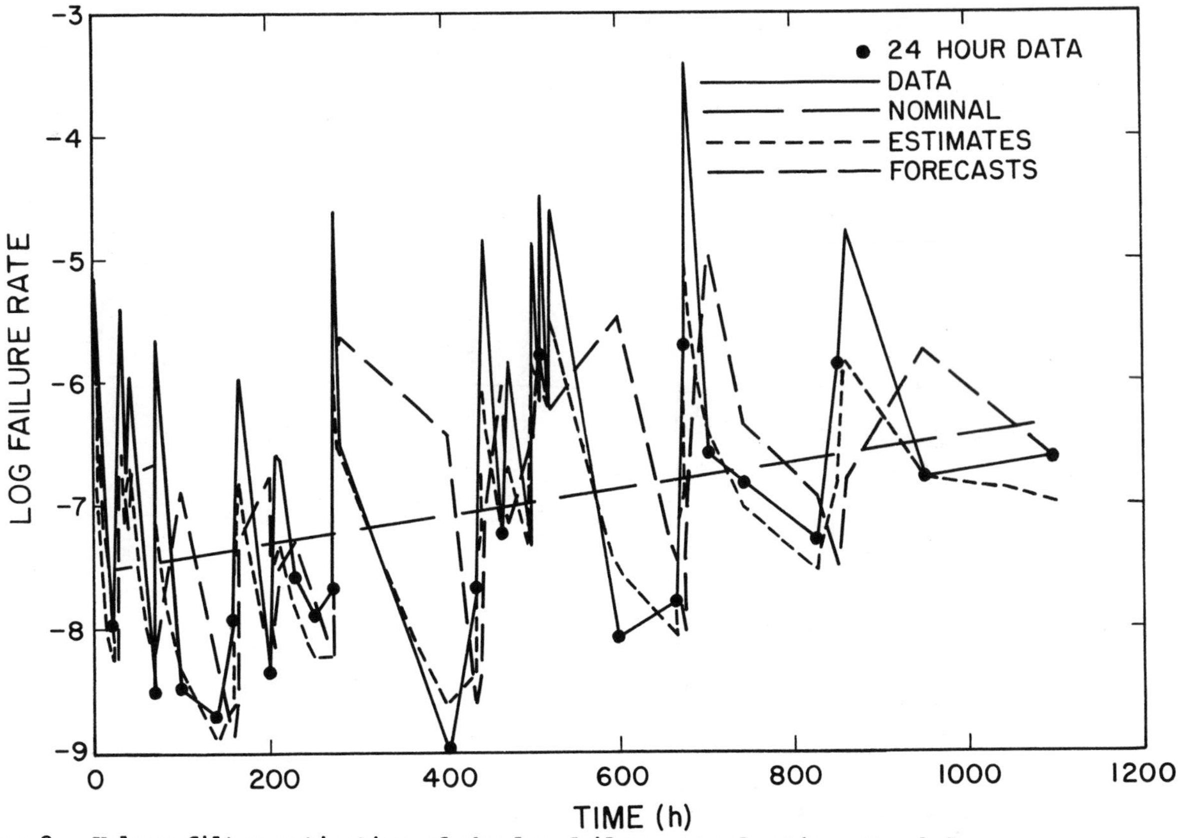

Figure 2. Kalman filter estimation of the log failure-rate function $Q = 2.0$.

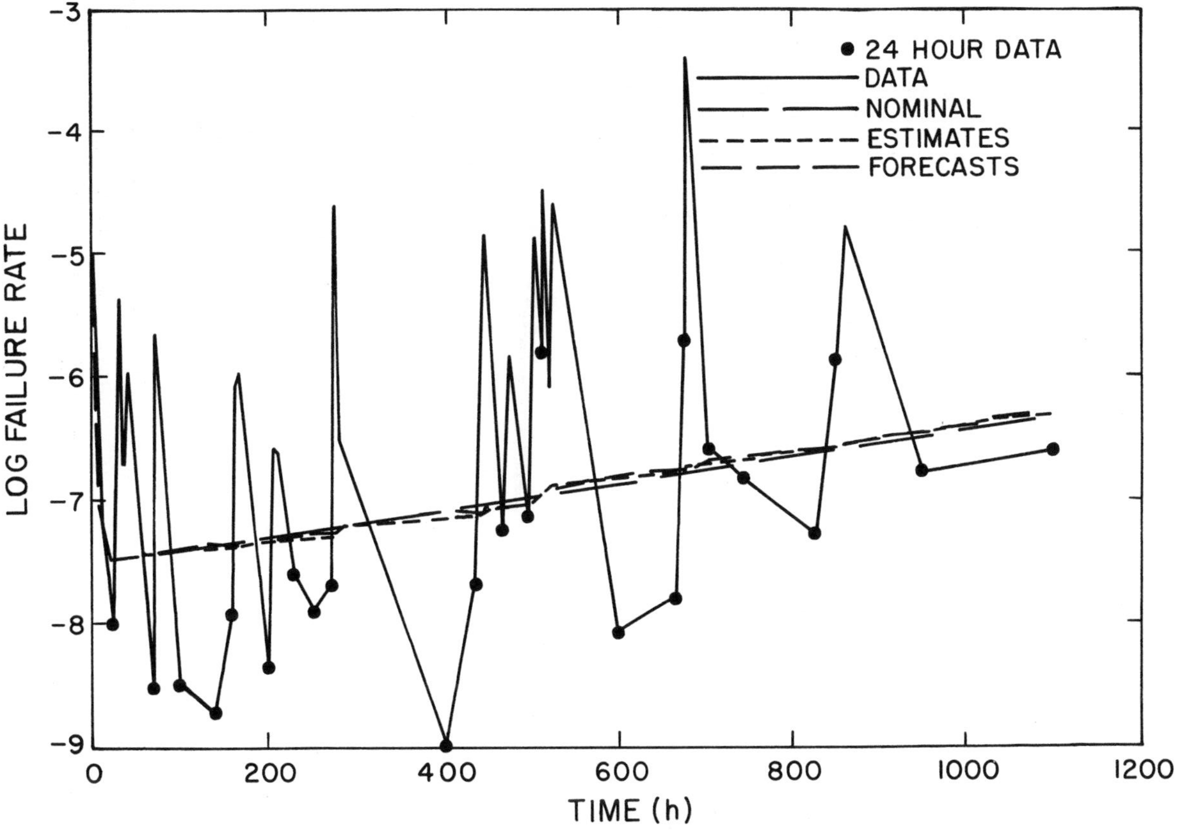

Figure 3. Kalman filter estimation of the log failure-rate function Q = 0.001.

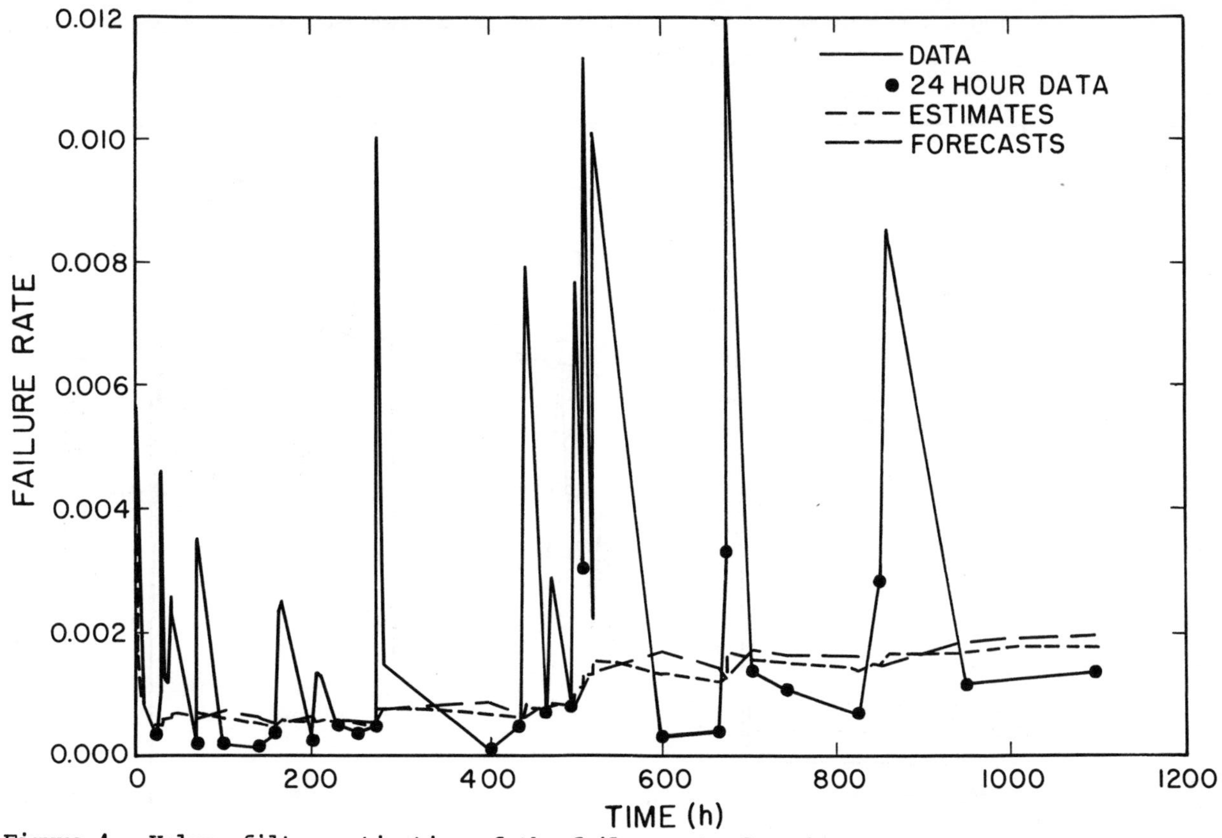

Figure 4. Kalman filter estimation of the failure-rate function Q = 0.02.

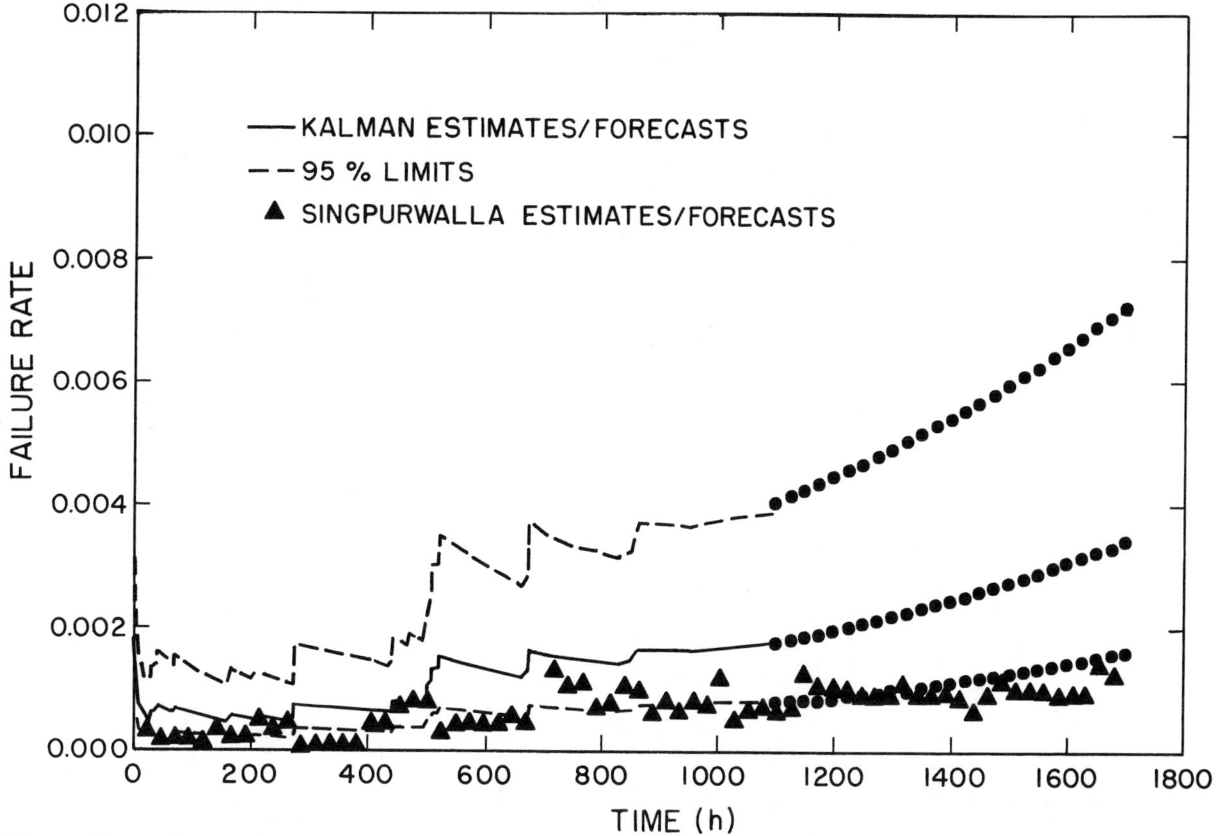

Figure 5. Kalman filter estimates, forecasts and probability limits of the failure-rate function $Q = 0.02$.

REFERENCES

[1] P. D. Abramson, Jr., "Simultaneous Estimation of the State and Noise Statistics in Linear Dynamical Systems," NASA report R-332, March, (1970).

[2] K. J. Åstrom, Introduction to Stochastic Control Theory Academic Press, New York, (1970).

[3] R. E. Barlow and W. L. VanZwet, "Comparison of Several Non-parametric Estimators of the Failure Rate Function," *Proc. of the N.A.T.O. Conference on Reliability at Turino*, Italy, (Israel Publishing House), June 30 - July 4, (1969).

[4] D. A. Belsley and E. Kuh, "Time-Varying Parameter Structures: An Overview," *Annals of Econ. and Social Meas.*, 2, October, (1973).

[5] A. M. Breipohl, "Kalman Filtering and Its Application to Reliability," *IEEE Trans. on Rel.*, R-18, August (1969), pp. 127-130.

[6] R. S. Bucy and R. E. Kalman, "New Results in Linear Filtering and Prediction Theory," *Trans. of ASME, Series D: Journal of Basic Eng.*, 83, (1961), pp. 95-108.

[7] V. F. Castellino and N. D. Singpurwalla, "On the Forecasting of Failure Rates--A Detailed Analysis," Technical Memorandum Serial TM-62197, Program in Logistics, The George Washington University, Washington, D.C. (1973).

[8] D. B. Duncan and S. D. Horn, "Linear Dynamic Recursive Estimation from the Viewpoint of Regression Analysis," *JASA*, 67, December (1972), pp. 815-821.

[9] V. Grenander, "On the Theory of Mortality Measurement, Parts I and II," *Skand. Aktuarietidskr.*, 39, (1956), pp. 125-153.

[10] A. A. Hirsch, M. Liebenberg, and G. Green, "The BEA Quarterly Model," BEA Staff Paper No. 22, U.S. Dept. of Commerce, Washington, D.C. (1973).

[11] R. E. Kalman, "A New Approach to Linear Filtering and Prediction Problems," *Basic Eng.*, 82, March, (1960), pp. 35-45.

[12] S. J. Kamat and L. D. Cox, "Use of the Kalman Filter in Time-Series Forecasting," presented at the II Interamerican Conference on Systems and Informatics, Mexico City, November, (1974).

[13] A. N. Kolmogorov, "Interpolation und Extrapolation von Stationaren Zufalligen Folgen," *Bulletin of Academic Sciences*, USSR, SER. MATH (1941).

[14] N. R. Mann, R. E. Schafer, and N. D. Singpurwalla, Methods for Statistical Analysis of Reliability and Life Data (John Wiley and Sons, New York), (1974).

[15] D. W. Marquardt, "An Algorithm for Least Squares Estimation of Nonlinear Parameters," *SIAM J.*, 11, (1963), pp. 431-441.

[16] A. McWhorter, Jr., "Time Series Forecasting Using the Kalman Filter: An Empirical Study," presented at the 1975 Annual Meeting of ASA, Atlanta, Georgia, August, (1975).

[17] A. McWhorter, Jr., W. A. Spivey, and W. J. Wrobleski, "Varying Parameter Econometric Models for Interrelated Time Series," *Proc. Bus. and Econ. Sect. of ASA*, St. Louis, August, (1974).

[18] A. McWhorter, W. A. Spivey, and W. J. Wrobleski, "Computer Simulation in Varying Parameter Regression Models," *Proc. Bus. and Econ. Sect. of ASA*, New York, December, (1973).

[19] R. K. Mehra, "On the Identification of Variances and Adaptive Kalman Filtering," *IEEE Trans. on Auto. Control*, AC-15, April, (1970), pp. 175-184.

[20] N. Morrison, Introduction to Sequential Smoothing and Prediction, (McGraw-Hill, New York), (1969).

[21] G. V. L. Narasimham, V. F. Castellino, and N. D. Singpurwalla, "On the Predictive Performance of the BEA Quarterly Economic Model and Box-Jenkins ARIMA Models," *Proc. Bus. and Econ. Sect. of ASA*, St. Louis, August, (1974).

[22] J. O. Pearson, "Estimation of Uncertain Systems," *Control and Dynamic Systems*, 10, (1973), pp. 255-343.

[23] B. Rosenberg, "A Survey of Stochastic Parameter Regression," *Annals of Econ. and Social Meas.*, 2 October, (1973).

[24] J. C. Shellenbarger, "Estimation of Covariance Parameters for an Adaptive Kalman Filter," *Proc. of the National Electronics Conf.*, Chicago, October 3-5, (1966), pp. 698-702.

[25] N. D. Singpurwalla, "Time Series Analysis and Forecasting of Failure-Rate Processes," in *Reliability and Fault Tree Analysis* (SIAM, Philadelphia), (1975), pp. 483-507.

[26] G. S. Watson and M. R. Leadbetter, "Hazard Analysis I," *Biometrika*, 1 & 2, (1964), pp. 175-184.

[27] N. Wiener, "The Extrapolation, Interpolation, and Smoothing of Stationary Time Series," OSRD 370, Report to the Services 19, Research Project DIC-6037, MIT, February, (1942).

SHOCK AND WEAR MODELS AND MARKOV ADDITIVE PROCESSES*

by ERHAN ÇINLAR
Northwestern University

Abstract. Our objective is to introduce a general shock and wear model which extends the available models in several directions. The model can be used to study fatigue loading due to random vibrations, wear in vacuum tubes due to hits by electrons whose energy levels vary stochastically, and quite generally, to study the damage process in situations where individual shocks do not cause any measurable damage, but there are very many shocks during even very small intervals.

The model views the cumulative deterioration process Z as the second component of a Markov additive process (X, Z). We obtain the failure time distribution under random threshold and multiplicative killing type failure mechanisms. In the latter case, when Z is further assumed to be a gamma process whose shape parameter varies as a function of a Brownian motion process, it is shown that the lifetime distribution is Weibull.

1.0 *Introduction*. Our overall aim is to introduce some shock and wear processes which depict situations where "shocks" occur very rapidly and the damage caused by a shock is very small. The deterioration suffered by a car due to shocks it receives from the road is an example. Another, more serious, example is the fatigue deterioration at a crucial spot on the wing of an airplane due to vibratory pressure variations caused by acoustically transmitted jet noise and/or boundary layer turbulence.

---

* Research supported by the Air Force Office of Scientific Research, Air Force Systems Command, USAF, under Grant No. AFOSR-74-2733. Reproduction in whole or in part is permitted for special purposes within the United States Government.

The model takes the cumulative deterioration process $Z = (Z_t)$ as a generalization of increasing Lévy processes (i.e. processes with stationary and independent increments). As such this model generalizes those of Gaver [11], Morey [13], Reynolds and Savage [15], A-Hameed and Proschan [1], and all the models in the remarkable paper by Esary, Marshall, and Proschan [7] except the Markovian ones on pages 636 and 641. We assume that the cumulative deterioration process (Z_t) has conditionally independent increments given the paths of an excitation process (X_t) which is Markovian. Such processes were introduced in Çinlar [3] under the name of Markov additive processes. We will give a simplified description of them in Section 2, with numerous examples indicating their scope of applicability in reliability theory.

We derive the distribution of the failure time under a random threshold type failure mechanism, and the distribution and the hazard rate function for the time of failure under a multiplicative killing type failure mechanism. In a particular case, we show in Example (3.24) that the lifetime distribution is Weibull if the deterioration process is a gamma process whose shape parameter is a function of a Brownian motion. This result might serve as a theoretical explanation of the frequent occurrence of the Weibull distribution in creep and fatigue failure situations.

A different formulation using Markov additive processes was successfully employed by Feldman [8], [9], [10]. There, starting with a Markov additive process (X, Y), the cumulative damage process Z is modeled by putting $Z_t = X_{s-}$ if s is such that $Y_{s-} \leq t < Y_s$ and $Z_t = X_s$ if $t = Y_s$ for some s. The resulting process Z is a generalized semimarkov process which admits infinitely many (possibly uncountable) "jumps" in very small intervals. Feldman's model, then, includes the semimarkov models discussed in Barlow and Proschan [2] and the models on pages 636 and 641 of Esary, Marshall, and Proschan [7].

The two models, the one here and Feldman's, emphasize different aspects of the same mathematical structure. We have so far been able to answer only the most immediate questions. It is hoped that this paper generates interest in models of this sort so that more important questions concerning reliability and the properties of the failure time distribution, etc. can also be answered.

2.0 *Markov Additive Processes*. Our aim is to describe Markov additive processes in an informal setting and to give various examples with reliability interpretations. For the precise definition we refer to Çinlar [3]; there are some measurability assumptions which we will make without explicitly stating them here.

We are working on a sample space Ω with a σ-algebra $\mathbf{M}$ on it. For each $\omega \in \Omega$, there are two functions

$$X(\omega): R_+ \to E, \qquad Z(\omega): \mathbf{R}_+ \to \mathbf{R}_+; \tag{2.1}$$

here $R_+ = [0, \infty)$ and E is some locally compact space with a countable base ;(in most applications E is either a countable set or is some subset of the real line). We let $X_t(\omega)$ and $Z_t(\omega)$ denote the values of $X(\omega)$ and $Z(\omega)$ at time t respectively. We assume that $t \to X_t(\omega)$ and $t \to Z_t(\omega)$ are both right continuous and have left hand limits everywhere, and that $t \to Z_t(\omega)$ is increasing.

For each $x \in E$ there is a probability measure P^x on $(\Omega, \mathbf{M})$ such that

$$P^x\{X_0 = x, Z_0 = 0\} = 1. \tag{2.2}$$

We define

$$Q_t(x, A, B) = P^x\{X_t \in A, Z_t \in B\} \tag{2.3}$$

for $t \in R_+$, $x \in E$, and Borel subsets $A \subset E$ and $B \subset R_+$.

The main hypothesis on the process (X, Z) is that

$$P^x\{X_{s+t} \in A,\ Z_{s+t} - Z_s \in B | H_s\} = Q_t(X_s, A, B) \tag{2.4}$$

for all $t, s \in R_+$, $x \in E$, $A \subset E$, $B \subset R_+$ (A and B Borel), where H_s is the history of the process (X, Z) until the time s, that is, H_s is the σ-algebra generated by $\{X_u, Z_u; u \leq s\}$.

Then, (X, Z) is a Markov additive process in the sense of Çinlar [3].

As a result of the defining property (2.4), the transition function (Q_t) defined by (2.3) satisfies

$$Q_{s+t}(x, A, B) = \int_{E \times R_+} Q_s(x,dy,dz)Q_t(y,A,B-z) \tag{2.5}$$

where $B - z = \{b - z \geq 0: b \in B\}$. Such (Q_t) are called semi-markov transition functions. Given a semimarkov transition function (Q_t), putting

$$P_t(x, A) = Q_t(x, A, R_+), \tag{2.6}$$

we obtain a Markov transition function.

The following is the main result on the structure of Markov additive processes. The first statement is immediate from (2.4) upon putting $B = R_+$; for the second, we refer to Çinlar [3], Theorem (2.22).

(2.7) Proposition. The process (X_t) is a Markov process with state space E and transition function (P_t). Given X, the conditional law of (Z_t) is that of an increasing additive process (usually non-stationary).

The structure of additive processes can be found in Doob [6], Chapter VIII, where they are called processes with independent increments. A detailed analysis of their sample paths can be found in Itô [12]. Using Proposition (2.7) together with such results, a detailed description of the structure of Markov additive processes (X, Z) was obtained in Çinlar [3]. Roughly, the probability law of Z during a small interval (t, t + dt) is that of an increasing Lévy process where the parameters depend on the state of X_t of the underlying process X.

We put below, for reasons of completeness, some of the essential facts about increasing Lévy processes, which are in fact special Markov additive processes where the Markov process X is trivial: E has only one state.

(2.8) Increasing Lévy processes. Such a process (Z_t) is an appropriate model for the cumulative deterioration if the deterioration during the interval (s, s + t] is independent of the past deterioration before s, and moreover, the probability law of the increment $Z_{s+t} - Z_s$ is the same as that of Z_t. It is well known that, then,

$$E[\exp(-\lambda Z_t)] = \exp[-t\lambda a - t \int_0^\infty \nu(dz)(1 - e^{-\lambda z})]. \qquad (2.9)$$

Here is a constant, called the drift rate; and ν is a measure on $(0, \infty)$, called the Lévy measure of Z, satisfying the integrability condition

$$\int_0^\infty \nu(dz)(z \wedge 1) < \infty. \qquad (2.10)$$

This implies that ν is σ-finite. If ν is infinite, that is, if $\nu(0, \infty) = +\infty$, then Z has infinitely many jumps in any open interval $(t, t + \varepsilon)$ with probability one, however small $\varepsilon > 0$ may be.

The following are some special cases.

(2.11) Compound Poisson processes. This is an increasing Lévy process with no drift and a finite Lévy measure. Then, $\nu(dz) = c\phi(dz)$ for some constant $c > 0$ and some distribution ϕ on R_+. The jumps occur according to a Poisson process with rate c, and the jump sizes are i.i.d. random variables with distribution ϕ.

(2.12) Increasing stable processes. Here either $a > 0$ and $\nu = 0$, or else $a = 0$ and the Lévy measure ν has the form

$$\nu(dz) = \frac{c}{z^{1+\alpha}}\, dz, \qquad z > 0,$$

for some $c > 0$ and some constant (called the shape index) $\alpha \in (0, 1)$.

(2.13) Gamma processes. This is an important class because of its tractability and apparent applicability. Such a process Z increases by jumps only, and there are infinitely many jumps in any open interval. In the canonical representation (2.9) we have $a = 0$ and

$$\nu(dz) = \frac{be^{-cz}}{z}\, dz, \qquad z > 0.$$

The parameters b and c are called the shape and scale parameters respectively. The name "gamma process" comes from the fact that the distribution of Z_t has the gamma density

$$\frac{ce^{-cz}(cz)^{bt-1}}{\Gamma(bt)}, \qquad z \geq 0,$$

which has mean bt/c and variance bt/c^2.

Next we pass on to give some examples of more genuine Markov additive processes.

(2.14) Compound Poisson shocks in random environments. For simplicity let (X_t) be a two state Markov process, say $E = \{r,w\}$. Interpret $X_t = w$ as saying that a device is in the working state at time t, and interpret r as the rest (or repair) state. Such a Markov process is very simple in structure: X stays in w an exponential time with some parameter $k(w)$, then jumps to state r, stays at r an exponential time with parameter $k(r)$, then jumps back to w, and so on. We now assume that shocks occur according to a Poisson process with rate c when X is in state w, each shock causing a random amount of damage with distribution ϕ independent of everything else. When X is in state r, there are no shocks. The picture for a possible realization ω of such a process is shown below in Figure 1.

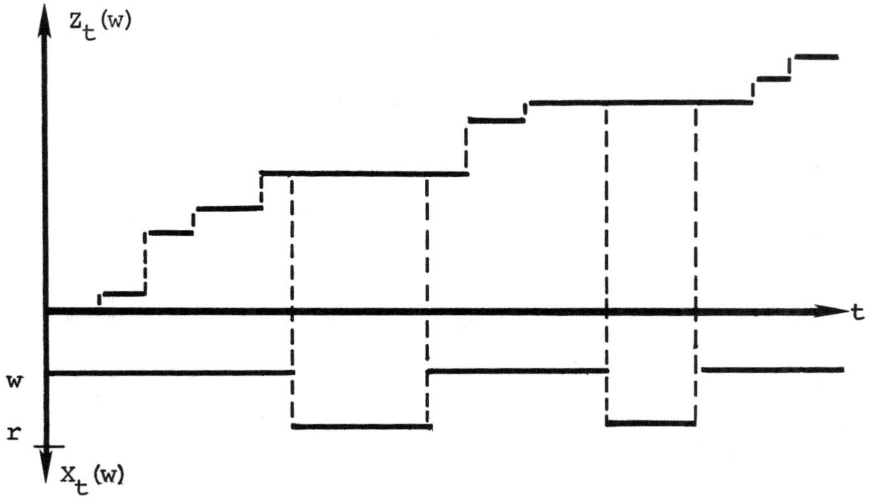

Figure 1

Shocks Occur Only When The Device Is Working; Each Shock Causes a Random Amount of Damage

The conditional probability law of Z given X is described by

$$E^x[\exp(-\lambda Z_t)\,|X] = \exp[-A_t \int_0^\infty c\phi(dz)(1 - e^{-\lambda z})] \tag{2.15}$$

where A_t is the amount of time spent working during (o, t].

(2.16) General Markov additive processes with E finite. Let (X_t) be a Markov process with a finite state space E. We think of X_t as the state of the environment at time t. Concerning the cumulative deterioration process Z, we assume that, while X is in state i, Z increases as a Lévy process (see (2.8) above) with drift rate a(i) and Levy measure $\nu(i, dz)$. Increases in deterioration over different environments are added up linearly.

In addition, every change of state from i to j is accompanied by a shock which causes an additional amount of damage with distribution F(i, j, ·). Figure 2 shows a possible realization for such a process (X, Z).

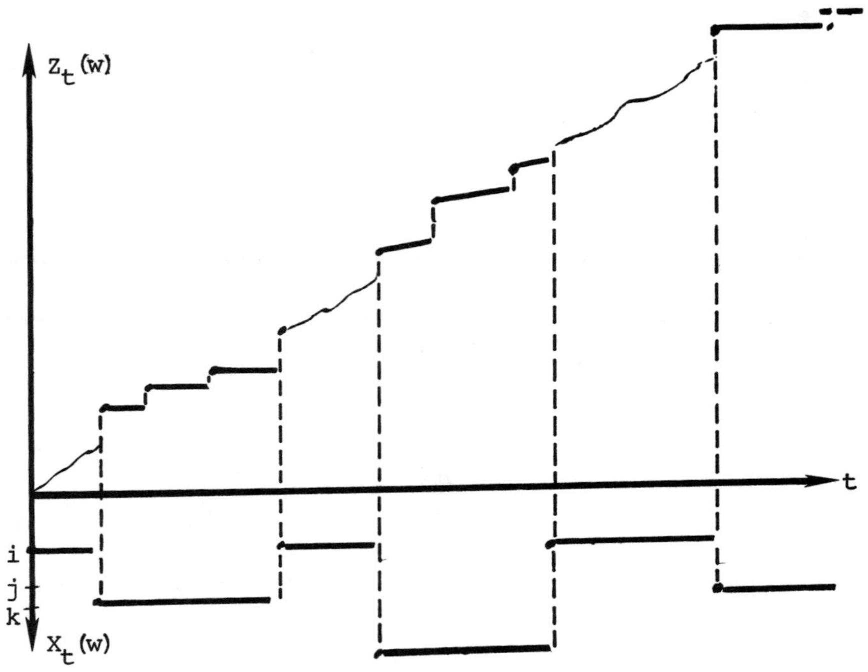

Figure 2.

When the Device is Working Under Environmental Condition i, Deterioration is a Gamma Process with Parameters $b(i)$ and $c(i)$; when it is Working Under Condition j, Deterioration is a Compound Poisson Process with Rate of Shocks $c(i)$ and Damage Distribution ϕ_i.

The conditional law of (Z_t) is described by

$$E^x[\exp(-\lambda Z_t)\,|X] = \exp[-\lambda\int_0^t a(X_s)ds - \int_0^t ds\int_0^\infty \nu(X_s,dz)(1-e^{-\lambda z})]\cdot\prod_{s\leq t} F^\lambda(X_{s-},X_s) \qquad (2.17)$$

where $F^\lambda(i, j) = \int_0^\infty F(i, j, dz)e^{-\lambda z}$ if $i \neq j$ and $F^\lambda(i,j) = 1$ if $i = j$.

The preceding example also holds in countable state case provided that X does not have instantaneous states. More generally, even when the space E is continuous, the situation is similar, and (2.17) holds, provided that X be a regular step process (that is, the jump times of X can be ordered as $\tau_1 < \tau_2 < \ldots$ and either $\tau_n \to \infty$ as $n \to \infty$, or else, if $\tau_n \to \tau$ as $n \to \infty$, then $X_t = \Delta$ for all $t \geq \tau$ for some distinguished absorbing state Δ.

(2.18) Failures in Markovian environments. Let T_0 be the time of failure of a metal device kept at a constant temperature x_0. We are interested in the failure time of the same device if it were subjected to temperature fluctuations. Let X_t denote the difference between the temperature at time t and the constant level x_0. For termorheologically simple materials the effect of temperature fluctuations can be compensated by a change of the time scale. Thus, Parkus and Bargman [14] define

$$Z_t = \int_0^t a(X_s)\, ds$$

as the reduced time, (the function a is to be obtained by experimental means,) and then the actual time of failure T is given by

$$T = \inf\{t: Z_t > T_0\}.$$

When X is a Markov process, Z is a continuous additive functional of X. Then (X, Z) is a special case of a Markov additive process.

Note that, in this example, we may think of Z as the deterioration process, and then the failure time T becomes the first time the process Z crosses the random threshold value T_0.

(2.19) Gamma processes in random environments. This seems to have wide applicability to modelling the creep and fatigue type failures; (we have some experimental evidence for this which will be reported elsewhere). Let Z_t be the creep level at time t, and let X_t be the environmental state prevailing at t. We suppose that X is a Markov process and that Z is locally a gamma process whose shape and scale parameters at t are $b(X_t)$ and $c(X_t)$ respectively. If b and c were both constants, then Z would be a gamma process as in example (2.13). Generally, we have

$$E^x[\exp(-\lambda Z_t)|X] = \exp\{-\int_0^t ds\, b(X_s) \int_0^\infty \frac{1-e^{-z}}{z} \exp[-c(X_s)z]dz\}$$

$$= \exp[-\int_0^t b(X_s)\log(1 + \frac{1}{c(X_s)})\, ds].$$

3.0 *Markov Additive Shock and Wear with Multiplicative Failure Rule.* Let (X, Z) be a Markov additive process with semimarkov transition function (Q_t) and with state space E for X. We think of X_t as the state of the excitation process at t, and of Z_t as the cumulative damage during (0, t]. For the failure mechanism we suppose that, given the knowledge of X and Z, the conditional probability that failure has not occurred at or before t is

$$M_t = \exp(-\lambda Z_t) \tag{3.1}$$

for some fixed $\lambda > 0$; in other words, if L is the lifetime of the device in question, we are assuming that

$$P^x\{L > t | X,Z\} = M_t = \exp(-\lambda Z_t). \tag{3.2}$$

Then (M_t) is a multiplicative functional of (X,Z), and hence our term "multiplicative failure rule." Note that

$$P^x\{L > t+u | X,Z;\ L > t\} = \frac{M_{t+u}}{M_t} = \exp[-\lambda(Z_{t+u} - Z_t)] \tag{3.3}$$

which depends on X only through $\{X_s;\ t \le s \le t + u\}$. Therefore, this failure mechanism is applicable in situations where the device is as good as new as long as it has not failed.

If the excitation process is trivial, that is, if E consists of a single state, then (Z_t) becomes an increasing Lévy process (see (2.8)). In that case, $Z_{t+u} - Z_t$ has the same distribution as Z_u and therefore (3.3) implies

$$P\{L > t+u | L > t\} = E(\exp(-\lambda Z_u)) = f(u) \tag{3.4}$$

independent of t; hence, $P\{L > t\} = e^{-ct}$ for some $c > 0$; (this fact seems to have escaped notice in some papers).

Consequently, the present failure mechanism is of interest only when X is non-trivial, that is, when E has at least two states. There are several more general looking models which reduce to the present case; the following are two such situations.

Suppose that failure can occur only when a shock occurs, that is, only at the jump times of Z. And suppose that, given that failure did not occur before, the probability of failure at t is a function of the jump size $Z_t - Z_{t-}$; that is,

$$P^x\{L > t | X,Z\} = \prod_{s \le t} q(Z_t - Z_{t-}) \tag{3.5}$$

for some "survival function" q. To see that this model is not too different from the original one, we define

$$\hat{Z}_t = - \sum_{s \le t} \log q(Z_t - Z_{t-}). \tag{3.6}$$

Then, $(X,\hat{Z})$ is still a Markov additive process, and (3.5) and (3.6) imply that

$$P^x\{L > t | X,Z\} = P^x\{L > t | X,\hat{Z}\} = \exp(-\hat{Z}_t) \tag{3.7}$$

which has the same form as (3.2).

Another case which reduces to the fundamental formulation (3.1), (3.2) is as follows. Suppose that the probability of instantaneous failure at t, given that the failure has not yet occurred, is a function of X_{t-}, X_t, Z_{t-}, Z_t; that is,

$$P^x\{L > t | X,Z\} = \prod_{s \le t} q(X_{s-}, X_s, Z_{s-}, Z_s) \tag{3.8}$$

for some survival function q. Now we define

$$\hat{X}_t = (X_t, Z_t), \quad \hat{Z}_t = - \sum_{s \le t} \log q(X_{s-}, X_s, Z_{s-}, Z_s). \tag{3.9}$$

Then, $\hat{X}$ is a Markov process with state space $\hat{E} = E \times R_+$, $(\hat{X}, \hat{Z})$ is again a Markov additive process, and

$$P^x\{L > t | X,Z\} = P^{x,0}\{L > t | \hat{X},\hat{Z}\} = \exp(-\hat{Z}_t), \tag{3.10}$$

which has the form (3.2) again.

In summarizing the preceding two paragraphs, we point out that the words "excitation," "damage," etc. do not have any absolute meanings; by choosing their relative meanings carefully, a researcher can reduce a number of different looking models to the present one.

From here on, we are working with the fundamental setup (3.1) - (3.2). First, note that (3.2) implies

$$P^x\{X_t \in A, Z_t \in B, L > t\} = \int_B Q_t(x, A, dz)\, e^{-\lambda z}, \tag{3.11}$$

and in particular,

$$\hat{P}_t(x, A) = P^x\{X_t \in A, L > t\} = \int_0^\infty Q_t(x, A, dz)\, e^{-\lambda z}. \tag{3.12}$$

The transition function $(\hat{P}_t)$ is a sub-Markovian transition function, that is, $\hat{P}_{t+s} = \hat{P}_t\hat{P}_s$ as usual but $\hat{P}_t(x,E) \leq 1$ instead of identically equal to 1.

Next we consider the computation of failure rate function

$$r_t(x) = \lim_{u\downarrow 0} \frac{1}{u} P^x\{L \leq t + u | L > t\}. \tag{3.13}$$

We will compute r_t separately for two important extreme cases; for the case where X is a regular step process with finitely many states, and for the case where X has continuous paths and Z does not have predictable jumps. More general results can be obtained by using our technique with the general results in Çinlar [4].

Suppose X is a finite Markov process with generator G, and recall the conditional law (2.18) for Z. With the numbers a(j), Lévy measures $\nu(j, \cdot)$, and the distributions $F(j, k, \cdot)$ described in the example (2.17), we define

$$r(j) = \lambda a(j) + \int_0^\infty \nu(j, dz)(1 - e^{-\lambda z})$$

$$+ \sum_{k\neq j} G(j, k) \int_0^\infty F(j, k, dz)(1 - e^{-\lambda z}). \tag{3.14}$$

(Recall that a generator G in this case is a matrix whose entries are all non-negative except on the diagonal, and G1 = 0.) The following is the main result for this case.

(3.15) Theorem. Suppose X is a Markov process with generator G and transition function (P_t) on a finite state space E. Then,

$$r_t(i) = \sum_{j\in E} P_t(i, j)r(j), \qquad i \in E, t \geq 0, \tag{3.16}$$

where r_t and r are as defined by (3.13) and (3.14).

Proof. Using first (3.3) and then (2.4) we have

$$P^i\{L > t + u \mid L > t\} = E^i[\exp[-\lambda(Z_{t+u} - Z_t)]]$$

$$= E^i\left[\int_0^\infty Q_u(X_t, E, dz)\, e^{-\lambda z}\right]$$

$$= \sum_j P_t(i, j) \int_0^\infty Q_u(j, E, dz)\, e^{-\lambda z}.$$

By an obvious definition for $f_u(j)$, this implies

$$P^i\{L \le t + u \mid L > t\} = \sum_j P_t(i, j)[1 - f_u(j)]. \tag{3.17}$$

By Theorem (3.7) in Çinlar [4], taking $f = 1$ and noting that $A1 = G1 = 0$ in that theorem, we obtain

$$\lim_{u \downarrow 0} \frac{1 - f_u(j)}{u} = r(j). \tag{3.18}$$

Putting (3.17) and (3.18) together,

$$r_t(i) = \lim_{u \downarrow 0} \frac{1}{u} \sum_j P_t(i, j)(1 - f_u(j))$$

$$= \sum_j P_t(i, j) \lim_{u \downarrow 0} \frac{1}{u}(1 - f_u(j)) = \sum_j P_t(i, j) r(j)$$

as desired.

(3.19) Example. Suppose X has only two states, and let the generator be

$$G = \begin{bmatrix} -0.2 & 0.2 \\ 0.3 & -0.3 \end{bmatrix}$$

Suppose further that

$$a(1) = 1,\ \nu(1,dx) = 2\phi(dx),\quad F(1,2,dx) = e^{-x}dx;$$

$$a(2) = 0,\ \nu(2,dx) = \frac{e^{-x}}{x}\,dx,\quad F(2,1,dx) = e^{-x}dx;\ x > 0,$$

where ϕ is a distribution function. In other words, in state 1, the drift rate of Z is 1, shocks occur according to a Poisson process with rate 2, each shock causes a random damage with distribution ϕ; a changeover shock occurs when X jumps from 1 to 2 causing an exponential amount of damage; in state 2, there is no drift, shock and wear process is gamma with both parameters equal to 1; and a changeover shock occurs when X jumps from 2 to 1 causing an exponential damage. Then,

$$r(1) = \lambda + 2(1 - \phi^{\lambda}) + 0.2(1 - \frac{1}{1 + \lambda}) = 1.5$$

$$r(2) = \log(1 + \lambda) + 0.3(1 - \frac{1}{1 + \lambda}) = \log 2 + 0.15$$

assuming that $\lambda = 1$ in the model and that the Laplace transform ϕ^{λ} of ϕ is such that $\phi^{1} = 0.8$.

The transition function corresponding to the given G is easy to compute; by standard textbook methods, we get

$$P_t = \begin{bmatrix} 0.6 & 0.4 \\ 0.6 & 0.4 \end{bmatrix} + e^{-0.5t} \begin{bmatrix} 0.4 & -0.4 \\ -0.6 & 0.6 \end{bmatrix}$$

Then,

$$r_t(1) = 0.96 + 0.4 \log 2 + (0.54 - 0.4 \log 2)\, e^{-0.5t}$$

which is evidently decreasing.

Next we consider the case where X is continuous, Z has no predictable jumps, and the Lévy system (a, L, H) as defined in Çinlar [4] is such that $H_t = t$ identically. Under these conditions, the shock and wear process behaves, locally at t, as an increasing Lévy process with drift rate $a(y)$ and Lévy measure $\nu(y, \cdot\,)$ if $X_t = y$. We define, similar to (3.14),

$$r(y) = \lambda a(y) + \int_0^{\infty} \nu(y, dz)(1 - e^{-\lambda z}). \tag{3.20}$$

Then, we have the following

(3.21) Proposition. Let (X, Z) be as described in the preceding paragraph. Then, supposing that r is right continuous,

$$r_t(x) = \int_E P_t(x, dy) r(y)$$

with r_t and r defined by (3.13) and (3.20).

Proof. By (3.3) and (2.4), as before in the proof of (3.15),

$$P^x\{L \leq t + u | L > t\} = \int_E P_t(x, dy) E^y[1 - e^{-\lambda Z_u}] \tag{3.22}$$

$$= 1 - \hat{P}_t(x, E)$$

where (P_t) is as defined in (3.12). We also have

$$E^y[1 - e^{-\lambda Z_u}|X] = 1 - \exp[-\lambda \int_0^u a(X_s)\, ds$$

$$- \int_0^u ds \, . \int_0^\infty \nu\ (X_s, dz)(1-e^{-\lambda z})]$$

$$= 1 - \exp[- \int_0^u r(X_s)\, ds].$$

Therefore, by the right continuity of r,

$$\lim_{u \downarrow 0} \frac{1}{u} E^y[1 - e^{-\lambda Z_u}] = \lim \frac{1}{u} E^y[1 - \exp[- \int_0^u r(X_s)\, ds]] \tag{3.23}$$

$$= E^y[\lim_{u \to 0} \frac{1}{u} [1 - \exp[- \int_0^u r(X_s)\, ds]]$$

$$= E^y[r(X_0)] = r(y).$$

Putting (3.22) and (3.23) together, applying the usual differentiability arguments to the transition function $(\hat{P}_t)$, we obtain the desired result.

(3.24) Example. Consider the example (2.20) again, that is, X is the Brownian motion, and Z is conditionally locally gamma. Suppose that when X is at y, the shape and scale parameters are

$$b(y) = b \cdot y^2, \qquad c(y) = c$$

where b and c are fixed constants. Then,

$$r(y) = by^2 \int_0^\infty \frac{e^{-cz}}{z} (1 - e^{-\lambda z})\, dz = \mu \cdot y^2$$

where the constant μ is equal to $b \log(1 + \lambda/c)$. Then, the preceding proposition gives, for the initial state x = 0,

$$r_t(0) = \int_{-\infty}^{\infty} \frac{1}{\sqrt{4\pi t}} \exp(-\frac{y^2}{4t}) r(y)\, dy = 2\mu t.$$

So, the lifetime of the device in question has linearly increasing failure rate, and in fact,

$$P^0\{L > t\} = \exp(-\mu t^2), \qquad t \geq 0.$$

Similar computations give for arbitrary x

$$r_t(x) = 2\mu t + \mu x^2, \qquad P^x\{L > t\} = \exp(-\mu t^2 - \mu x^2 t).$$

More generally, if for some positive integer β, and constants b and c,

$$b(y) = b \cdot y^{2\beta}, \qquad c(y) = c,$$

then

$$r_t(0) = \alpha' t^\beta$$

where α' is some constant depending on λ, β, b, c. So,

$$P^0\{L > t\} = \exp(-\alpha t^{\beta+1})$$

for some α; that is, L has the Weibull distribution with shape parameter β.

4.0 *Threshold Failure Models*. Throughout, (X, Z) is a Markov additive process with semimarkov transition function (Q_t) and with state space E for X. We think of (X_t) as the excitation process and of Z_t as the cumulative deterioration during $(0, t]$. We suppose that failure occurs as soon as the cumulative deterioration exceeds a random threshold T. Then, the lifetime is

$$L = \inf\{u: Z_u > T\}. \tag{4.1}$$

We assume that T is independent of (X, Z) and its distribution π is known. We are interested in the distribution of L. In addition, in determining the cost of failure, it is probably important to know the state of the excitation process and the level of deterioration at the time of failure. Hence, we will derive the joint distribution of the triple

$$(L, X_L, Z_L).$$

In view of the independence assumption regarding T and (X, Z),

$$P^x\{L \in A, X_L \in B, Z_L \in C\} = \int_0^\infty \pi(db) M_b(x, A, B, C) \tag{4.2}$$

where $M_b(x, A, B, C)$ is the same probability assuming that the threshold is fixed and is equal to b, that is,

$$M_b(x, A, B, C) = P^x\{L \in A, X_L \in B, Z_L \in C\} \quad \text{for } T = b. \tag{4.3}$$

A related problem was solved in Çinlar **[5]**: namely the joint distribution of X_{L-}, X_L, X_{L-}, Z_L was obtained. Unfortunately the result there does not include L in its formulation. We will derive the distribution desired directly from a general formula in Çinlar [4] in one case, and use a space-time argument to get it from a result in Çinlar [5] in another case.

For our present purpose we will assume that the Lévy system (a, **L**, H) obtained in Çinlar [4] is such that $H_t = t$ identically -- in fact, Proposition (2.11) of Çinlar [5] shows that there is no loss of generality in this; all that one needs is to redefine certain quantities.

The Lévy system (a, L) is made up of two things: a positive function a on E, and a transition kernel L from E into $E \times R_+$; the latter is defined by

$$L(x, A, B) = I(x, A)\nu(x, B) + \int_A K(x, dy)F(x, y, B) \tag{4.4}$$

where I is the identity kernel, $\nu(x, \cdot)$ is the Lévy measure of the Z process locally when X is at x, F(x, y, B) is the distribution of the amount of damage accompanying a change of state from x to y by X, and finally K(x, dy) is the rate (per unit time) at which the jumps of X from x to dy occur. If X is a regular step process (see example (2.18)), then K(x, dy) = G(x, dy) for $x \neq y$ and K(x, dy) = 0 for x = y, where G is the infinitestimal generator of X.

Note that the Lévy kernel L defines all the $\nu(x, A)$ and the K(x, A) and F(x, y, B). The following is the major portion of the result we are after.

(4.5) <u>Theorem</u>. Let M_b be defined by (4.3). Then,

$$M_b(x, A, B, C) = \int_A dt \int_{E \times [0,b]} Q_t(x, dy, dz)L(y, B, C - z) \tag{4.6}$$

for all $x \in E$, and Borel sets $A \subset R_+$ and $B \subset E$, and compact sets $C \subset (b, \infty)$.

<u>Proof</u>. By Theorem (2.29) of Çinlar [4] modified for the present case, we have

$$E^x[e^{-\alpha L}; X_T \in B, S_T \in C] = E^x \int_0^L e^{-\alpha t} L(X_t, B, C - Z_t)\, dt. \tag{4.7}$$

But, by the definition of L, $Z_s > b$ on $\{t > L\}$. Hence, we can write the right-hand side as

$$E^x \int_0^\infty e^{-\alpha t} 1_{[0,b]}(Z_t) L(X_t, B, C-Z_t)\, dt \tag{4.8}$$

$$= \int_0^\infty e^{-\alpha t} \int_{E\times[0,b]} Q_t(x,dy,dz) L(y,B,C-z)\, dt.$$

The left side of (4.7) is a Laplace transform as a function of α, and it is clear from the form of the right-hand side of (4.8) that inversion of that Laplace transform is

$$\int_A dt \int_{E\times[0,b]} Q_t(x,dy,dz) L(y,B,C-z).$$

This is the desired result (4.6).

The remaining problem is the possibility that $Z_L = b$ with a positive probability, that is, the threshold is hit without jumping. In general this is not probable unless there are some positive drift terms -- that is, $a(x) > 0$ for some $x \in E$. The following result is obtained by applying Proposition (1.7) in Çinlar [5] to the Markov additive process $(\hat{X}_t, Z_t)$ where $\hat{X}_t = (X_t, t)$.

First note that $Q_t(x, A, B) \le P_t(x, A)$ always; therefore the Radon-Nikodym derivative

$$G_t(x, y, B) = \frac{Q_t(x,dy,B)}{P_t(x,dy)}$$

exists. It is shown in Çinlar [3] that, in fact, it is possible to pick G_t such that it is jointly measurable in (x, y) and is a probability measure in B.

(4.9) <u>Proposition</u>. Suppose that the measure $G_t(x, y, \cdot)$ admits a density $z \to g_t(x, y, z)$ for every t, x, y. Then, for any $x \in E$, and Borel sets $A \subset R_+$ and $B \subset E$,

$$P^x\{L \in A, X_L \in B, Z_L = b\} = \int_A dt \int_B P_t(x,dy) a(y) g_t(x,y,b).$$

Proof is omitted. The distribution of (L, X_L, Z_L) for constant threshold value b follows from Theorem (4.5) and Proposition (4.9) above.

As can be seen from the results above, in order to have explicit results, the main function which needs to be derived is the transition function (Q_t). Presumably, the Lévy system (a, L) which describes the infinitesimal behavior of X and Z will be known or obtained experimentally. The problem of getting (Q_t) from (a, L) parallels the same problem in the theory of Markov processes, where starting with the infinitesimal generator G, the object is to compute the transition function (P_t). We refer to Çinlar [4] for some results pertaining to this problem.

REFERENCES

[1] M. S. A-Hameed and F. Proschan, "Nonstationary Shock Models", *Stochastic Processes Appl. 1*, (1973), pp. 383-404.

[2] R. E. Barlow and F. Proschan, Mathematical Theory of Reliability, Wiley, New York, (1965).

[3] E. Çinlar, "Markov Additive Processes, II", *Z. Wahrscheinlichkeitstheorie Verw Geb. 24*, (1972), pp. 94-121.

[4] E. Çinlar, "Levy Systems of Markov Additive Processes", *Z. Wahrscheinlichkeitstheorie Verw. Geb. 31*, (1975), pp. 175-185.

[5] E. Çinlar, "Entrance-exit Distributions for Markov Additive Processes", (1976), To appear.

[6] J. L. Doob, Stochastic Processes, Wiley, New York, (1953).

[7] J. D. Esary, A. W. Marshall, and F. Proschan, "Shock Models and Wear Processes", *Ann. Prob. 1*, (1973), pp. 627-649.

[8] R. M. Feldman, Optimal Replacement for Systems with Semi-Markovian Deterioration. Doctoral Dissertation, Northwestern University, Evanston, Illinois, (1975).

[9] R. M. Feldman, "Optimal Replacement for Systems Subject to Shock and Wear", *J. Appl. Prob.*, (1975), To appear.

[10] R. M. Feldman, "Optimal Replacement for Systems Governed by Markov Additive Shock Processes", (1975), To appear.

[11] D. P. Gaver, "Random Hazard in Reliability Problems", *Technometrics*, 5, (1963), pp. 211-226.

[12] K. Itô, Stochastic Processes, Lecture Notes Series No. 16, Matematisk Institut, Aarhus Universitet, (1969).

[13] R. C. Morey, "Some Stochastic Properties of a Compound Renewal Damage Model", *Oper. Res. 14*, (1966), pp. 902-908.

[14] H. Parkus and H. Bargman, "Note on the Behavior of Thermorheologically Simple Materials in Random Temperature Fields", *Acta Mechanica 9*, (1970), pp. 152-157.

[15] D. S. Reynolds and I. R. Savage, "Random Wear Models in Reliability Theory", *Adv. Appl. Prob. 3*, (1971), pp. 229-248.

THE MAINTENANCE OF SYSTEMS GOVERNED BY SEMI-MARKOV SHOCK MODELS*

by RICHARD M. FELDMAN
Texas A&M University

Abstract. Consider a system whose failure is due to damage caused from a sequence of randomly occurring shocks. The purpose of this paper is to consider the optimal replacement policy for the maintenance of such a system whose cumulative damage process is semi-Markovian. The semi-Markov property allows for both the time between shocks and the damage due to the next shock to be dependent on the present cumulative damage level.

When the system fails, it must be replaced immediately at a fixed cost. If the system is replaced before failure, a lower cost is incurred and that replacement cost can depend on the state of the system at the time of replacement. Only replacement policies within the class of control limit policies will be considered; namely, policies with which no action is taken if the damage is below a fixed level, and a replacement is made if the damage is above that.

A closed form expression is obtained for the optimal policy using a long run average cost criterion.

1.0 *Introduction*. Consider a system that is subject to a sequence of randomly occurring shocks; each shock causes some damage of random magnitude to the system. Any of the shocks might cause the system to fail, and the probability of such a failure is a function of the sum of the magnitudes of damage caused from all previous and present shocks.

Taylor [11] derives an optimal replacement rule for such a system when the cumulative damage process is a compound Poisson

*Research supported by the Air Force Office of Scientific Research through their Grant No. AFOSR-74-2733 under the guidance of Professor Erhan Çinlar.

process. In other words, Taylor assumes that the times at which the shocks occur form a Poisson process and the magnitudes of damage caused by each shock form a sequence of independent and identically distributed random variables.

The purpose of this paper is to derive the optimal replacement rule for a system whose cumulative damage process is a semi-Markov process. The semi-Markov property allows for both the time between shocks and the damage due to the next shock to be dependent on the present cumulative damage level. Optimality is based on the expected long run cost per unit time where the replacement cost of a system can depend on the cumulative damage level at replacement time. Feldman [9] presents the semi-Markov replacement problem except that the replacement cost is not a function of the cumulative damage at replacement time (assuming no failure).

Morey [10] derives some properties of a system for which the cumulative damage process is a compound renewal process similar to Cox ([5], p. 91); that is, the times that the shocks occur form a renewal process, and the magnitudes of damage caused by each shock form a sequence of independent identically distributed random variables. Morey considered the case where failure occurred at the time that the cumulative damage process first became greater than or equal to a certain threshold level. Taylor [11] included the threshold model as a special case.

Cherniavsky [2] introduces a shock process identical to the one presented in this paper. However, Cherniavsky only uses the process for modeling purposes and does not consider the maintainability problem.

Esary, Marshal, and Proschan [7] investigate the properties of a failure process for which the cumulative damage process is a Poisson process; that is, the time between shocks is exponential and independent of the past history, and the magnitude of damage equals one. A-Hameed and Proschan [1] generalize the above paper to a nonhomogeneous Poisson process. These two latter papers deal with the relationship between the raw data (i.e., the probability of failing after a given number of shocks) and the system reliability. This is important in determining whether a "control limit" type replacement policy is best or not; however, these relationships for the general problem are outside the scope of this paper.

We will always search for a policy that is optimal within the class of control limit policies. The term "control limit policy" refers to a policy in which no action is taken if the damage is less than a fixed number and a replacement is made if the damage is equal to or greater than the fixed number.

In Section 2, the cumulative damage process will be defined and the problem formulated; in Section 3, the cost equation will be minimized to obtain the optimal replacement policy; and in Section 4, two examples will be presented illustrating computational procedures.

The following will be standard notation used throughout the paper: $\mathbb{N} = \{0, 1, \ldots\}$, $\mathbb{N}_+ = \{1, 2, \ldots\}$, $\mathbb{R}_+ = [0, \infty)$. The space $(\Omega, \underline{\underline{F}}, P)$ denotes a fixed probability space that is big enough to include all random variables defined. For notational convenience, the random variable Z_t will be written as $Z(t)$.

The reader is referred to Çinlar [4], (ch. 10) for an excellent treatment of semi-Markov processes and Markov renewal processes. For arbitrary state spaces, see Çinlar [3].

2.0 *Problem Formulation*. Two processes will be defined: $\hat{Z}$ and Z. The process $\hat{Z}$ will represent the complete cumulative process, and Z will represent the cumulative damage process only until the time of system failure.

Let $\hat{Z} = (\hat{Z})_{t \geq 0}$ be a non-decreasing semi-Markov process with state space $(E, \underline{\underline{E}})$ and imbedded Markov renewal process $(\hat{X}, \hat{T}) = (\hat{X}_n, \hat{T}_n)_{n \in \mathbb{N}}$. Let $\hat{Q} = \{\hat{Q}(x, y, t): x, y \in E, t \geq 0\}$ be the semi-Markov kernel associated with (X, T). That is,

$$\hat{Q}(x, y, t) = P^x \{\hat{X}_1 \leq y, \hat{T}_1 \leq t\}. \tag{2.1}$$

Let $g(x)$ denote the probability that the system does not fail at time $\hat{T}_n$ given $\hat{X}_n = x$ and given that the system was working at $\hat{T}_{n-1}$. To illustrate, say at time t the system has not yet failed and its current accumulation of damage has a value of x and the next shock occurs at time $s > t$ causing an additional damage of y. In this case, $\hat{Z}(t) = \hat{Z}(s-) = x$, $\hat{Z}(s) = x + y$, and the probability of the system failing at time s is $1 - g(x + y)$. We shall call g the damage survival function.

Two conditions will be imposed on g.

Assumption. Let g be the survival function described in the above paragraph. Then (2.2)

a) g is non-increasing, and

b) there exist $y \in E$ and $n \in \mathbb{N}_+$ such that $g(y) < 1$ and $P^x\{\hat{X}_n \leq y\} < 1$ for all $x \leq y$.

Since the system can only fail at those times for which shocks occur, an integer valued random variable L is defined which denotes the n such that $\hat{T}_n$ is the failure time of the system. Its associated probability law is given by

$$P^x\{L > n \mid \hat{X}_0, \hat{X}_1, \ldots, \hat{X}_n\} = g(\hat{X}_1)\, g(\hat{X}_2) \ldots g(\hat{X}_n). \tag{2.3}$$

Note that by using Equation (2.3) we assume that the system cannot fail initially even if a shock is realized during installation.

The time of failure is defined by

$$\zeta(\omega) = \hat{T}_{L(\omega)}(\omega). \tag{2.4}$$

The killed process Z and its imbedded process (X, T) are defined by

$$Z_t(\omega) = \begin{cases} \hat{Z}_t(\omega) & \text{if } t < \zeta(\omega), \\ \Delta & \text{if } t \geq \zeta(\omega), \end{cases} \tag{2.5}$$

$$X_n(\omega) = \begin{cases} \hat{X}_n(\omega) & \text{if } n < L(\omega), \\ \Delta & \text{if } n \geq L(\omega), \end{cases} \tag{2.6}$$

$$T_n(\omega) = \begin{cases} \hat{T}_n(\omega) & \text{if } n \leq L(\omega), \\ +\infty & \text{if } n > L(\omega). \end{cases} \tag{2.7}$$

The process Z is a non-decreasing semi-Markov process with state space $(E, \underline{\underline{E}})$ and its imbedded Markov renewal process is (X, T).

Consider the semi-Markov kernel Q of (X, T). Now

$$Q(x, y, t) = P^x\{X_1 \leq y, T_1 \leq t\}$$

$$= P^x\{\hat{X}_1 \leq y, \hat{T}_1 \leq t, L > 1\}, \tag{2.8}$$

and thus

$$Q(x, y, t) = \int_{[0,y]} \hat{Q}(x, du, t)g(u) \quad \text{for } x, y \in E, t \geq 0. \tag{2.9}$$

Let $R = \{R(x, y, t);\ x, y \in E, t \geq 0\}$ denote the Markov renewal kernel corresponding to Q. That is,

$$R(x, y, t) = \sum_{n=0}^{\infty} Q^n(x, y, t) \quad \text{for } x,y \in E,\ t \geq 0. \tag{2.10}$$

where

$$Q^0(x, y, t) = 1_{[0,y]}(x) \cdot 1_{[0,\infty]}(t),$$

and

$$Q^{n+1}(x, y, t) = \int_{E\times[0,t]} Q(x, du, ds)Q^n(u, y, t-s).$$

Using the above processes, the optimal replacement problem can be stated rigorously. Replacement must occur at time ζ or before. The cost of replacement depends only on the state of the process at the time of replacement. Let f(x) denote the cost of replacing the system if the process is in state x at the time of replacement; thus, $f(\Delta)$ is the cost of replacing a system that has failed.

A control limit policy is one in which there is a fixed real value α such that the system is replaced as soon as the accumulated damage is greater than α or when a failure occurs, whichever comes first; otherwise the system is left alone.

For a fixed α, let τ_α be a random variable denoting the replacement time; that is,

$$\tau_\alpha(\omega) = \zeta(\omega) \wedge \inf\{t > 0: \hat{Z}_t(\omega) \geqslant \alpha\} \text{ for } \omega \in \Omega. \tag{2.11}$$

To determine the long term expected cost per unit time, consider the renewal process formed by repeated replacements of identical systems, each system having a lifetime given by the distribution of τ_α. Let the cost of the first, second, ... replaced system be denoted by C_1, C_2, ... where $\{C_n;\ n \in \mathbb{N}_+\}$ forms a sequence of independent identically distributed random variables.

If N_t is the number of renewals in $(0, t]$, then by using standard results in renewal theory, we have

$$\psi_\alpha = \lim_{t\to\infty} \frac{1}{t} E[C_1 + \cdots + C_{N_t}] = \frac{E[C_1]}{E[\tau_\alpha]}. \tag{2.12}$$

The distribution of the random variable representing the damage caused by the installation shock is given by μ; that is

$$\mu(x) = P\{X_0 \leq x\}. \tag{2.13}$$

Problem. Let (X, T) be the Markov renewal process as (2.14) defined by Equations (2.6) and (2.7). Let $N_\alpha = \sup\{n \in \mathbb{N}: T_n \leq \tau_\alpha\}$ and let f be a non-negative cost function. Let μ be the initial distribution. Find α^* that minimizes the function $\psi(\cdot)$, where $\psi(\alpha)$ is defined by

$$\psi(\alpha) = \frac{E^\mu[f(X_{N_\alpha})]}{E^\mu[\tau_\alpha]} \quad \text{for } \alpha \in E.$$ □

Remark. The notation $E^\mu[f(y)]$ denotes $\int E^x[f(y)]\mu(dx)$. □

3.0 *The Optimal Replacement Problem*. With the formulation of Problem (2.14), the optimal replacement policy can be stated as find α^* such that $\alpha \to \psi(\alpha)$ is minimized.

The notation of this section is the same as in Section 2. Namely, (X, T) is the Markov renewal process defined by (2.6) and (2.7); μ is the initial distribution given by (2.13); τ_α is the replacement time given by (2.11); N_α is the index of the shock time at which the replacement is made as given in (2.14); and

$\psi(\alpha)$ is the long run expected replacement cost per unit time as given in (2.14). Two assumptions will be needed to insure that the cost equation has at most one minimum.

Assumptions. The payoff function f is bounded, non-negative, (3.1) non-decreasing, and convex. The function $x \to P^x\{X_1 - x \geq y, T_1 < t\}$ is non-decreasing for $y \in E$ and $t \geq 0$.

Remarks. A convex function must be continuous in the interior of its domain; therefore, f is continuous in E but not necessarily a Δ. The condition that $x \to P^x\{X_1 - x \geq y, T_1 \leq t\}$ be increasing implies that as the initial state gets worse, the incremental damage stochastically increases while the shock times stochastically decrease.

This assumption applied to the function $x \to P^x\{X_1 \geq y, T_1 < t\}$ is analogous to the assumption used in the Markov chain replacement model of Derman [6], p. 123, Condition B) in which Derman assumes that the probability of making a transition to a "bad" set of states increases as the initial state gets worse.

Before obtaining the expressions for $E^\mu[f(X_{N_\alpha})]$ and $E^\mu[\tau_\alpha]$, it will be convenient to introduce some simplifying notation. Let m(x) denote the mean sojourn time in state x, let $P(\cdot,\cdot)$ denote the transition function of X, and, by an abuse of notation, let $R(\cdot,\cdot)$ denote the potential matrix of X. Thus

$$m(x) = \int_{t \geq 0} [1 - Q(x, E, t)]dt \quad \text{for} \quad x \in E, \tag{3.2}$$

$$P(x, y) = \lim_{t \to \infty} Q(x, y, t) \qquad \text{for} \quad x, y \in E, \tag{3.3}$$

$$R(x, y) = \sum_{n=0}^{\infty} P^n(x, y) = \lim_{t \to \infty} R(x, y, t) \text{ for } x, y \in E. \tag{3.4}$$

For any probability measure ν on $(\Omega, \underline{\underline{F}})$, let

$$R(\nu, y) = \int_{x \in E} R(x, y)\nu(dx). \tag{3.5}$$

For bounded f, the function Pf is defined by

$$Pf(x) = E^x[f(X_1)] = \int_{y\varepsilon E} P(x, dy)f(y) + f(\Delta)[1 - P(x, E)] \text{ for } x \varepsilon E. \tag{3.6}$$

where $P(x, E) = \int_{y\varepsilon E} P(x, dy)$.

<u>*Proposition*</u>. Let $\alpha \varepsilon E$ be fixed. Then (3.7)

$$E^{\mu}[\tau_{\alpha}] = \int_{y\varepsilon[0,\alpha)} R(\mu, dy)\ m(y).$$

Proof. From Feldman [9],(Proposition (2.10)), we have that

$$E^x[\tau_{\alpha}] = \int_{y\varepsilon[0,\alpha)} R(x, dy)\ m(y) \quad \text{for } x < \alpha.$$

Since $E^x[\tau_{\alpha}] = 0$ for $x \geq \alpha$, $E^{\mu}[\tau_{\alpha}] = \int_{x\varepsilon[0,\alpha)} \mu(dx) \int_{y\varepsilon[0,\alpha)} R(x,dy)m(y)$.

Using the fact that $R(x, y) = 0$ for $x > y$ and using Fubini's Theorem, the proof is complete. (The reason that the integral is over the half open interval $[0, \alpha)$ instead of the closed interval is due to the inclusion of equality in Equation (2.11).)

<u>*Proposition*</u>. Let $\alpha \varepsilon E$ be fixed. Let $f(\mu) = \int f\, d\mu$. (3.8)

Then

$$E^{\mu}[f(X_{N_{\alpha}})] = f(\mu) + \int_{y\varepsilon[0,\alpha)} R(\mu, dy)[Pf(y) - f(y)].$$

Proof. Using a renewal theoretic type argument, we have

$$\begin{aligned}
E^x[f(X_{N_{\alpha}})] &= E^x[f(X_{N_{\alpha}})|X_1 \geq \alpha] + E^x[f(X_{N_{\alpha}})|X_1 < \alpha] \\
&= 1_{[0,\alpha)}(x)E^x[f(X_1)] + \int_{y\varepsilon[0,\alpha)} E^y[f(X_{N_{\alpha}})]P(x,dy) \\
&= \int_{y\varepsilon[0,\alpha)} R(x, dy)E^y[f(X_1)] \\
&= \int_{y\varepsilon[0,\alpha)} R(x, dy)\{ \int_{u\varepsilon[\alpha,\infty)} P(y, du)f(u) \\
&\qquad + f(\Delta)[1 - P(y, E)]\}.
\end{aligned}$$

By adding and subtracting $\int_{u\varepsilon[y,\alpha)} P(y, du)f(u)$ and rearranging terms, the left-hand side of the equations of (3.9) and (3.10) are increasing so the function $\psi(\cdot)$ has at most one minimum.

If f is one of the simpler form, Corollary (3.10) reduces to the following

Corollary. Let (X, T) and μ be as in Corollary (3.10). (3.11)
Define the function f by

$$f(i) = \begin{cases} c_1 & \text{if } i \in \mathbb{N}, \\ c_1 + c_2 & \text{if } i = \Delta. \end{cases}$$

Then the optimal control limit is the minimal $\alpha \in \mathbb{N}$ such that

$$\frac{1}{m(\alpha)} \sum_{k=0}^{\alpha-1} r(\mu, k)\{m(k)[1-P(\alpha, \mathbb{N})] - m(\alpha)[1 - P(k, \mathbb{N})]\} \geq c_1/c_2.$$

Proof. Follows immediately from Corollary (3.10) and Equation (3.6).

4.0 *Examples.* If the state space of the Markov renewal process is $\mathbb{N}$, then the optimal solution to Problem (2.14) is not as hard as it might appear to be since the matrix $(r(i, j))$ is the inverse of an upper triangular matrix, where $p(i, j) = P(i, j) - P(i, j-1)$ and $r(i, j) = \sum_{n=0}^{\infty} p^n(i, j)$. For an unbounded state space, the payoff function must be of the form in Corollary (3.11) since a non-constant convex function is unbounded if its domain is unbounded. An example illustrating the computations involved for the unbounded state space can be found in Feldman [9], Section 4. An example of using a bounded state space is given below.

Example. Bounded State Space. Consider a batch queuing system with a capacity of three. The cost of processing a batch is $\$100 + k^2$ if the batch consists of k people. If the queue contains more than three people when a batch is to be serviced, then three people are processed and the remaining leave the system with an estimated lost service cost of \$400. Arrivals to the queue are in batches also. The probability an individual arrives by himself is .8; the probability of a group of size two

$$E^x[f(X_{N_\alpha})] = \int_{y \varepsilon [0,\alpha)} R(x, dy)Pf(y) - \int_{u \varepsilon [0,\alpha)} f(u) \int_{y \varepsilon [0,\alpha)} R(x, dy)P(y, du).$$

Using the fact that

$$RP(x, y) = R(x, y) - 1_{[0,y]}(x),$$

we now obtain

$$E^x[f(X_{N_\alpha})] = \int_{y \varepsilon [0,\alpha)} R(x, dy)Pf(y) - \int_{u \varepsilon [0,\alpha)} R(x,du)f(u) + f(x).$$

Then, since $E^\mu[f(X_{N_\alpha})] = \int_{x \varepsilon [0,\alpha)} \mu(dx)E^x[f(X_{N_\alpha})] + \int_{x \varepsilon [\alpha,\infty)} (dx)f(x)$,

and since $R(x, y) = 0$ for $x > y$, the proof is complete.

Theorem. Let (X, T) be a Markov renewal process as defined (3.9) by Equations (2.6) and (2.7) and let μ be its initial distribution. Assume $y \to R(x, y)$ is absolutely continuous for each $x \varepsilon E$. Then the optimal control limit for Problem (2.14) is $\alpha \varepsilon E$ such that

$$\frac{1}{m(\alpha)} \int_{y \varepsilon [0,\alpha)} R(\mu, dy)\{m(y)[Pf(\alpha) - f(\alpha)] - m(\alpha)\ [Pf(y) - f(y)]\} = f(\mu).$$

Proof. Follows from Propositions (3.7) and (3.8) by taking derivatives with respect to α.

Corollary. Let (X, T) be a Markov renewal process defined (3.10) by Equations (2.6) and (2.7) with state space $\mathbb{N}$. Let μ be the initial distribution. Then the optimal control limit for Problem (2.14) is the minimal $\alpha \varepsilon \mathbb{N}$ such that

$$\frac{1}{m(\alpha)} \sum_{k=0}^{\alpha-1} r(\mu, k)\{m(k)[Pf(\alpha) - f(\alpha)] - m(\alpha)[Pf(k) - f(k)]\} \geq f(\mu),$$

where $r(\mu, k) = R(\mu, k) - R(\mu, k-1)$.

Proof. Same as Theorem (3.9) except differences are used.

Remark. By Assumption (3.1), the function $m(\cdot)$ is decreasing. It can be shown that $x \to Pf(x) - f(x)$ is increasing by using Assumption (3.1) (see Feldman [8], Equation 4.7). Therefore,

arrives is .1; and the probability of a group of size three arrives is .1. The attractiveness of the system increases as the queue increases so the mean time between batch arrivals is $1/(k+1)$ if there are k in the queue. (Note that from (3.10) it is seen that only the mean of the interarrival times is needed.) The problem is to determine when to process the queue.

The above description leads to the following:

$$E = \{0,1,2,3,4\},$$

$$f(k) = \begin{cases} 100 + k^2 & \text{for } k \in E, \\ 509 & \text{for } k = \Delta, \end{cases}$$

$$g(k) = \begin{cases} 1 & \text{for } k = 0,1,2,3, \\ 0 & \text{for } k = 4, \end{cases}$$

$$m(k) = 1/(k+1) \text{ for } k \quad E,$$

$$P = \begin{bmatrix} 0 & .8 & .1 & .1 & 0 \\ & 0 & .8 & .1 & 0 \\ & & 0 & .8 & 0 \\ & & & 0 & 0 \\ & & & & 0 \end{bmatrix},$$

$$\mu(\{0\}) = 1,$$

$$r(0,0) = 1,$$

$$r(0,1) = .8,$$

$$Pf(0) - f(0) = 80.8 + 10.4 + 10.9 - 100 = 2.1,$$

$$Pf(1) - f(1) = 83.2 + 10.9 + 50.9 - 101 = 44,$$

$$Pf(2) - f(2) = 87.2 + 101.8 - 104 = 85.$$

Let the left-hand side of the equation in Corollary (3.10) be denoted by $h(\alpha)$. Then

$$h(1) = 2\{44 - (1/2)(2.1)\} = 85.9$$

$$h(2) = 3\{[85 - (1/3(2.1)] + .8[(1/2)(85) - (1/3)(44)]\}$$
$$= 319.7.$$

Since the first time $h(\alpha) \geq 100$ is for $\alpha = 2$, the optimal policy is to process the queue as soon as there are 2 or more in the quence.

REFERENCES

[1] A-Hameed, M. S. and Proschan, F. Nonstationary shock models. *Stochastic Processes Appl.* 1, (1973), pp. 383-404.

[2] Cherniavsky, E. A., Some contributions to failure models: the shock and continuous danger processes. Ph.D. thesis, Cornell University, Ithaca, New York, (1973).

[3] Cinlar, E., On semi-Markov processes on arbitrary spaces. *Proc. Cambridge Philos. Soc.* 66, (1969), pp. 381-392.

[4] Cinlar, E., Introduction to Stochastic Processes. Prentice-Hall, Englewood Cliffs, N.J., (1975).

[5] Cox, D. R., Renewal Theory. Methuen & Company, London, (1962).

[6] Derman, C., Finite State Markov Decision Processes. Academic Press, New York, (1970).

[7] Esary, J. D., Marshall, A. W., and Proschan, F., Shock models and wear processes. *Ann. Probability* 1, (1973), pp. 627-649.

[8] Feldman, R. M., Optimal replacement with semi-Markov shock models using discounted costs. Submitted to *Mathematics of Operations Res.*, (1975).

[9] Feldman, R. M., Optimal replacement with semi-Markov shock models. *J. Appl. Probability*, 13, (1976), pp. 108-117.

[10] More, R. C., Some stochastic properties of a compound-renewal damage model. *Operations Res.* 14, (1966), pp. 902-908.

[11] Taylor, H. M., Optimal replacement under additive damage and other failure models. *Naval Res. Logist. Quart.* 22:1-18, (1975).

BOUNDS FOR THE DISTRIBUTIONS AND HAZARD GRADIENTS OF MULTIVARIATE RANDOM MINIMUMS*

by MOSHE SHAKED
The University of New Mexico

Abstract. Let $\underline{Y} = (Y_1, \ldots, Y_m)$ be a random vector whose distribution can be represented as the joint distribution of $\min(X_{1,i}, X_{2,i}, \ldots, X_{N_i,i})$, $i = 1, 2, \ldots, m$, where $\underline{X}_j = (X_{j,1}, \ldots, X_{j,m})$, $j = 1, 2, \ldots$ are i.i.d. non-negative random vectors and $\underline{N} = (N_1, \ldots, N_m)$ is a positive, integer valued random vector which is independent of the $\underline{X}_j$'s. We obtain upper and lower bounds for the survival function and for the hazard gradient of $\underline{Y}$; the bounds are determined by the distribution of $\underline{X}$, $E\,\underline{N}$ and $\underline{k} = (k_1, \ldots, k_m)$ where $k_i = \min\{n: P\{N_i=n\} > 0\}$. A condition under which the bounds are valid is that the probability generating function (p.g.f.) of $\underline{N}$ is TP_2 in each pair of variables. Examples and situations in which such p.g.f.'s arise are discussed.

1.0 *Introduction.* Let $\underline{X}_j = (X_{j,1}, \ldots, X_{j,m})$, $j = 1, 2, \ldots$ be non-negative i.i.d. random vectors and let $\underline{N} = (N_1, \ldots, N_m)$ be a positive, integer valued random vector which is independent of the $\underline{X}_j$'s. Define

$$Y_i = \min(X_{1,i}, X_{2,i}, \ldots, X_{N_i,i}), \quad i = 1, 2, \ldots, m. \tag{1.1}$$

*Part of this work was done while the author was in the University of Rochester and was supported by NSF Grant MPS74-15239. The paper was also partially supported by NSF Grant MPS 7509450-000 while the author was a visitor at Stanford University.

In this paper we discuss some probabilistic properties of $\underline{Y} = (Y_1, \ldots, Y_m)$, motivated by reliability considerations.

Random variables that admit the representation (1.1) (with m = 1) have been discussed by Cohen [2] and Shaked [10] [see additional references there]. Cohen suggested to use Formula (1.1), for a fixed i, as a description of a lifelength of a system (e.g. an electronic equipment) consisting of many components, a random number of which are weak components. In this case each of the $X_{j,i}$'s (of (1.1)) is a random lifelength of one of the weak components. The joint distribution of the lifelengths of m such systems is the distribution of $Y_1, \ldots, Y_m$ of (1.1). Note that if the N_i's are not independent then the Y_i's are not independent. Such dependent random lifelengths arise in practice when the m systems come, say, from the same production center or from the same production line, since, it is reasonable to assume in such circumstances that the number of weak components in one system is not independent of the number of weak components in another system.

Cohen suggested also that, for a fixed i, (1.1) can be used to describe a (random) strength limit of a material which contains a (random) number of structural defects. The strength of the material is then equal to the strength of the weakest defect. The joint distribution of m such strengths is, thus, the joint distribution of the m variables defined in (1.1).

In some practical situations the joint distribution of $N_1, \ldots, N_m$ (of (1.1)) may not be known. We may know, however, their expectations and we may be led to believe that $N_1, \ldots, N_m$ are positively dependent (in a sense to be formulated in Section 3). Under these assumptions we obtain, in Sections 3 and 4, bounds on the survival function of $\underline{Y}$ (for a non-negative random vector $(T_1, \ldots, T_m)$ with distribution function F the survival function is defined by $\overline{F}(\underline{t}) = P\{T_1 > t_1, \ldots, T_m > t_m\}$) and on other quantities of interest which are associated with the distribution of $\underline{Y}$.

A set of such quantities of interest are the components of the hazard gradient. For a survival function $\overline{F}(\underline{t})$ let $R(\underline{t}) = -\log \overline{F}(\underline{t})$, (R is defined on $\{\underline{t}: \overline{F}(\underline{t}) > 0\}$). The hazard gradient is

$$\underline{r}(\underline{t}) \equiv \nabla R(\underline{t}) \equiv (r_1(\underline{t}), \ldots, r_m(\underline{t})) \qquad (1.2)$$
$$\equiv \left(\frac{\partial}{\partial t_1} R(\underline{t}), \ldots, \frac{\partial}{\partial t_m} R(\underline{t})\right),$$

provided the derivatives exist. The hazard gradient has been discussed by, e.g., Johnson and Kotz [7] and Marshall [9] (and see additional references there).

If $\underline{r}(\underline{t})$ is the hazard gradient of $\underline{T}$ then, as noted by Johnson and Kotz (1975), $r_i(\underline{t})$ is the hazard rate function of T_i given that $T_{i'} > t_{i'}$ for all $i \neq i$. The (unconditional) hazard rate T_i can be obtained from $r_i(t)$ by letting $t_{i'} \to 0$, $i' \neq i$.

In Section 3 we obtain upper and lower bounds on the components of the hazard gradient of random vectors which can be represented as in (1.1). The bounds are determined by the components of the hazard gradient of $\underline{X}_1$ (of (1.1)) and EN_i ($\underline{N}$ of (1.1)).

The bounds which are obtained in Sections 3 and 4 hold if the joint distribution of $N_1, \ldots, N_m$ (of (1.1)) is positively dependent (see Theorem 3.2). In Section 5 we give some examples of such positively dependent integer valued random vectors.

2.0 *The Models*. Let $\underline{X}_j = (X_{j,1}, \ldots, X_{j,m})$, $j = 1, 2, \ldots$ be i.i.d. non-negative random vectors and let $\underline{N} = (N_1, \ldots, N_m)$ be a random vector of positive integers independent of the $\underline{X}$'s. Define

$$Y_i = \min(X_{1,i}, \ldots, X_{N_i,i}), \quad i = 1, 2, \ldots, m. \qquad (2.1)$$

Random vectors $\underline{Y}$ that have the representation (2.1) will be discussed. The analysis will be restricted to the following models:

Model A. The N_i's of (2.1) are identical, that is

$N_1 \equiv N_2 \equiv \ldots \equiv N_m$.

Denoting the (univariate) p.g.f. of N_1 by $\tilde{\psi}(u) = \sum_{n=1}^{\infty} P\{N_1=n\}u^n$ it is easy to see that under the assumption of Model A the survival function of $\underline{Y}$ (of (2.1)) is

$$\overline{G}(\underline{y}) = P\{Y_1 > y_1, \ldots, Y_m > y_m\} = \tilde{\psi}(\overline{F}(\underline{y})) \tag{2.2}$$

where $\overline{F}$ is the survival function of $\underline{X}_1$.

Model B. The components of $\underline{X}_1$ (and hence of each $\underline{X}_j$) are independent.

Let $\overline{F}_i(x_i) = P\{X_{1,i} > x_i\}$ be the marginal survival function of $X_{1,i}$, $i = 1, \ldots, m$, (then $\overline{F}(\underline{x}) = \prod_{i=1}^{m} \overline{F}_i(x_i)$). Denoting the joint p.g.f. of $\underline{N}$ by $\psi(\underline{u}) = \sum_{n_1=1}^{\infty} \cdots \sum_{n_m=1}^{\infty} P\{\underline{N}=\underline{n}\} \prod_{i=1}^{m} u_i^{n_i}$ it is easy to verify that under the assumption of Model B the survival function of $\underline{Y}$ (of (2.1)) is

$$\overline{G}(\underline{y}) = \psi(\overline{F}_1(x_1), \ldots, \overline{F}_m(x_m)). \tag{2.3}$$

3.0 *Bounds on the Components of the Hazard Gradient.* In this section we find lower and upper bounds on the components of the hazard gradient (see (1.2)) of random vectors of Model A and of Model B. The bounds are useful when the p.g.f. of $\underline{N} = (N_1, \ldots, N_m)$ [that is $\tilde{\psi}$ of (2.2) or ψ of (2.3)] is not known but EN_i (for determination of the upper bounds) and

$$k_i = \min\{n: P(N_i=n) > 0\}, \quad i = 1, 2, \ldots, m \tag{3.1}$$

(for determination of the lower bounds) are known. The bounds are shown to be sharp (see discussion after Theorem 3.2).

The first theorem deals with random vectors of Model A; it is an extension of a result of Shaked [10]. Denote by ρ_i i^{th} component of the hazard gradient of $\overline{F}$ (of (2.2)).

Theorem 3.1. If $\underline{Y}$ is a random vector of Model A (that is, the survival function of $\underline{Y}$ has the representation (2.2)), then $r_i(\underline{y})$ - the ith component of the hazard gradient of $\underline{Y}$ - has the bounds

$$k_1\rho_i(\underline{y}) \le r_i(\underline{y}) \le (EN_1)\rho_i(\underline{y}), \quad \underline{y} \in \{\underline{y}': \overline{F}(y') > 0\}. \tag{3.2}$$

The proof of the theorem is given at the end of this section.

The next theorem deals with random vectors of Model B. We will need the following definition. Let U_1 and U_2 be subsets of the real line. A kernel L: $U_1 \times U_2 \to (0, \infty)$ is said to be totally positive of order 2 (denoted by TP_2) in $u_1 \in U_1$ and $u_2 \in U_2$ if $L(u_1', u_2')\, L(u_1, u_2) \ge L(u_1', u_2)\, L(u_1, u_2')$ whenever $u_i \le u_i' \in U_i$, i = 1, 2. (For extensive study of total positivity see Karlin [8].

The next theorem deals with random vectors of Model B. Under the assumption of this model all the X's of (2.1) are independent. Denote by η_i the hazard rate function of $X_{1,i}$(of 2.1)), i = 1, ..., m.

Theorem 3.2. If $\underline{Y}$ is a random vector of Model B (that is, the survival function of $\underline{Y}$ has the representation (2.3)), and if $\psi(\underline{u})$ [of 2.3] is TP_2 in each pair of variables when the other variables are held fixed (the domain of each of the u's is [0, 1]) then for i = 1, ..., m, $r_i(\underline{y})$, the ith component of the hazard gradient of $\underline{Y}$, has the bounds

$$k_i\eta_i(y_i) \le r_i(\underline{y}) \le (EN_i)\eta_i(y_i), \quad \underline{y} \in \{\underline{y}': \overline{G}(\underline{y}') > 0\}. \tag{3.3}$$

Before providing the theorems we make some remarks.

The bounds of Theorem 3.1 are sharp in the following sense. Given a positive integer k and real numbers C > k and ε > 0 there exists a positive, integer valued random variable N_1 such that $k = \min\{n: P(N_1=n)\varepsilon > 0\}$, $EN_1 = C$ and for any choice of ρ_i in Theorem 3.1 it holds that

$$r_i(\underline{y}) - k\rho_i(\underline{y}) < \varepsilon \quad , \underline{y} \in \{\underline{y}': \overline{F}(\underline{y}') > 0\}. \tag{3.4}$$

This result, for m=1, is proved in Shaked [10]. The proof for $m \geq 2$ goes along the same lines. Inequality (3.4) shows the sharpness of the left hand side inequality of (3.2). The right hand side inequality is also sharp. Given a positive integer C there exists a positive, integer valued random variable N_1 such that $EN_1 = C$ and $r_i(\underline{y}) = (EN_1)\rho_i(\underline{y})$, $\underline{y} \in \{\underline{y}': \overline{F}(\underline{y}') > 0\}$, where r_i, ρ_i and $\overline{F}$ are defined in Theorem 3.1. The random variable N_1 in this case is degenerate ($P\{N_1=C\}= 1$). The bounds of Theorem 3.2 are sharp in a similar sense.

The condition that $\psi(\underline{u})$ of Theorem 3.2 is TP_2 in each pair of variables may seem strong, however, it turns out that it is satisfied by many well known p.g.f.'s. If N_i, $i = 1, \ldots, m$ (of Theorem 3.2) are independent, then this condition is satisfied. Some additional examples are given in Section 5. This condition is also satisfied when the N_i's are positively dependent in the sense of Propositions 5.6' and 5.7'. The assumption of positive dependence is reasonable in some applications. For example if $Y_1, \ldots, Y_m$ (of Theorem 3.2) are lifelengths of m systems where the ith system have N_i weak components, $i = 1, \ldots, m$ (see Section 1) and if the m systems come from the same production line, then it is reasonable to assume that the N_i's are positively dependent.

Recall that the hazard rate function of Y_i is equal to $\lim_{y_{i'} \to 0,\ i' \neq i} r_i(\underline{y})$, hence by letting $y_{i'} \to 0$, in (3.2) and in

(3.3) bounds are obtained for the hazard rate function of Y_i. These bounds have been already found in Shaked [10].

In the proof of Theorems 3.1 and 3.2 the following lemma will be used.

Lemma 3.3. Let $a(u) = \int_0^\infty u^z \, dH(z)$, $0 \le u \le 1$, where H is a distribution function on $[0, \infty)$ then $ua'(u)/a(u)$ is non-decreasing on $[0,1]$.

Proof. See Shaked [10]. ||

Proof of Theorem 3.1. It can be easily verified from (2.2) that

$$r_i(\underline{y}) = \left. \frac{u\tilde{\psi}'(u)}{\tilde{\psi}(u)} \right|_{u = \bar{F}(\underline{y})} \cdot \rho_i(\underline{y}).$$

Thus, by Lemma 3.3,

$$\rho_i(\underline{y}) \lim_{u \downarrow 0} \frac{u\tilde{\psi}'(u)}{\tilde{\psi}(u)} \le r_i(\underline{y}) \le \rho_i(\underline{y}) \lim_{u \uparrow 1} \frac{u\tilde{\psi}'(u)}{\tilde{\psi}(u)},$$

but $\lim_{u \uparrow 1} \frac{u\tilde{\psi}'(u)}{\tilde{\psi}(u)} = EN_1$ and by l'Hopital's rule $\lim_{u \downarrow 1} \frac{u\tilde{\psi}'(u)}{\tilde{\psi}(u)} = k_1$ and the proof is complete. ||

Proof of Theorem 3.2. It can be easily verified from (2.3) that for every $1 \le i \le m$

$$r_i(\underline{y}) = \left. \frac{u_i \frac{\partial}{\partial u_i} \psi(u_1, u_2, \ldots, u_m)}{\psi(u_1, u_2, \ldots, u_m)} \right|_{u_\ell = \bar{F}_\ell(y_\ell),\ 1 \le \ell \le m} \times \eta_i(y_i). \tag{3.5}$$

It will be shown now that

$$\chi_i(\underline{u}) \equiv \frac{u_i \frac{\partial}{\partial u_i} \psi(u_1, u_2, \ldots, u_m)}{\psi(u_1, u_2, \ldots, u_m)} \uparrow \underline{u} \tag{3.6}$$

(by writing $\uparrow \underline{u}$ we mean monotonicity in each u_ℓ, $\ell = 1, \ldots, m$

when the other u's are held fixed). By Lemma 3.3 $\chi_i(\underline{u}) \uparrow u_i$, hence we have to show that $\chi_i(\underline{u}) \uparrow u_\ell$ for $\ell \neq i$. By assumption $\psi(u_1, \ldots, u_m)$ is TP_2 in u_i and in u_ℓ when the other variables are held fixed, hence by Theorem 1.5, Karlin [8] $\frac{\partial}{\partial u_i}\psi(\underline{u})/\psi(\underline{u}) \uparrow u_\ell$, that is $\chi_i(u) \uparrow u_\ell$ and the proof of (3.6) is complete.

From (3.5) and (3.6) one obtains

$$\eta_i(y_i) \lim_{u_\ell \downarrow 0, 1 \leq \ell \leq m} \chi_i(\underline{u}) \leq r_i(\underline{y}) \leq \eta_i(y_i) \lim_{u_\ell \uparrow 1, 1 \leq \ell \leq m} \chi_i(\underline{u}) \tag{3.7}$$

but

$$\lim_{u_\ell \uparrow 1, 1 \leq \ell \leq m} \chi_i(\underline{u}) = EN_i \text{ and } \lim_{u_\ell \downarrow 0, 1 \leq \ell \leq m} \chi_i(\underline{u})$$

$$= \lim_{u_\ell \downarrow 0, \ell \neq i} [\lim_{u_i \downarrow 0} \chi_i(\underline{u})].$$

Denote $P_{n_1, n_2, \ldots, n_m} = P\{\underline{N}=\underline{n}\}$ and write

$$\chi_i(\underline{u}) = \frac{\sum_{n_i} n_i [\sum \ldots \sum_{n_\ell, \ell \neq i} P_{n_1, \ldots, n_m} \prod_{\ell \neq i} u_\ell^{n_\ell}] u_i^{n_i}}{\sum_{n_i} [\sum \ldots \sum_{n_\ell\, \ell \neq i} P_{n_1, \ldots, n_m} \prod_{\ell \neq i} u_\ell^{n_\ell}] u_i^{n_i}}$$

Note that for $u_\ell > 0, \ell \neq i$,

$$\sum_{n_\ell, \ell \neq i} \ldots \sum P_{n_1, \ldots, n_m} \prod_{\ell \neq i} u_\ell^{n_\ell} \neq 0$$

iff

$$\sum_{n_\ell, \ell \neq i} \ldots \sum P_{n_i, \ldots, n_m} = P(N_i = n_i) \neq 0.$$

Thus, $\lim_{u_i\ 0} \chi_i(\underline{u}) = k_i$ and hence $\lim_{u_\ell \downarrow 0, 1 \leq \ell \leq m} \chi_i(\underline{u}) = k_i$, that is, (3.7) is equivalent to (3.3). This fact completes the proof.

4.0 *Bounds on the Survival Functions*. Marshall [9] has observed that if $\underline{r}(\underline{y})$ is the hazard gradient of the survival function $\overline{G}(\underline{y})$ and if $\overline{G}(\underline{0}) = 1$ then $\overline{G}$ can be retrieved from $\underline{r}$ by the relation

$$\bar{G}(\underline{y}) = \exp\{-\int_C \underline{r}(\underline{z})\, d\underline{z}\}$$

where C is any sufficiently smooth path beginning at $\underline{0}$ and terminating at $\underline{y}$. Because of path independence the relation can be written without ambiguity as

$$\bar{G}(\underline{y}) = \exp\{-\int_{\underline{0}}^{\underline{y}} \underline{r}(\underline{z})\, d\underline{z}\}. \tag{4.1}$$

Using (4.1) and Theorems 3.1 and 3.2 the following bounds for random vectors of Models A and B are obtained.

Corollary 4.1. If $\underline{Y}$ is a positive random vector of Model A then its survival function $\bar{G}$ has the bounds

$$[\bar{F}(\underline{y})]^{EN_1} \leq \bar{G}(\underline{y}) \leq [\bar{F}(\underline{y})]^{k_1}, \tag{4.2}$$

where N_1, F and k_1 are defined respectively in (2.1), (2.2) and (3.1).

Proof. From (3.2) and (4.1) it is clear that

$$\exp\{-EN_1 \int_{\underline{0}}^{\underline{y}} \underline{\rho}(\underline{z})\, d\underline{z}\} \leq \bar{G}(\underline{y}) \leq \exp\{-k_1 \int_{\underline{0}}^{\underline{y}} \underline{\rho}(\underline{z})\, d\underline{z}\}, \tag{4.3}$$

but by replacing $\bar{G}$ and $\underline{r}$ in (4.1) by $\bar{F}$ and $\underline{\rho}$ it is proved that the left hand side of (4.3) is $[\bar{F}(\underline{y})]^{EN_1}$ and the right hand side of (4.3) is $[\bar{F}(\underline{y})]^{k_1}$. ||

Corollary 4.2. If $\underline{Y}$ is a positive random vector of Model B and if ψ (of (2.3) is TP_2 in each pair of variables when the other variables are held fixed then the survival function $\bar{G}$ of $\underline{Y}$ has the bounds

$$\prod_{i=1}^{m} [\bar{F}_i(y_i)]^{EN_i} \leq \bar{G}(\underline{y}) \leq \prod_{i=1}^{m} [\bar{F}_i(y_i)]^{k_i} \tag{4.4}$$

where $\underline{N}$, F_i and k_i are defined respectively in (2.1), (2.3) and (3.1).

The proof of this corollary is similar to the proof of the previous corollary.

Note that the inequalities (4.2) can be derived also directly from (2.2) [the left hand side inequality can be obtained using Jensen's Inequality]. For m=1 the inequalities (4.2) and (4.4) reduce to a result of Shaked [10].

5.0 *Examples*. In this section we give examples of positive random vectors of integers that have p.g.f.'s which are TP_2 in each pair of variables when the other variables are held fixed. Such random vectors satisfy the conditions of Theorem 3.2 and Corollary 4.2.

Example 5.1. (Bivariate geometric). Downton [3] discussed bivariate distributions with geometric marginals having the following p.g.f.

$$\psi(u_1,u_2) = \frac{p_0 u_1 u_2}{1-p_1u_1-p_2u_2-p_3u_1u_2} \quad 0 \le u_i \le 1,\ i = 1, 2 \tag{5.1}$$

where $p_i \ge 0$, $i = 0, \ldots, 3$ and $\sum_{i=0}^{3} p_i=1$. It is easy to verify that $\psi(u_1, u_2)$ of (5.1) is TP_2 in $0 \le u_1, u_2 \le 1$. If (N_1, N_2) has the p.g.f. ψ of (5.1) then $EN_1 = (1 - p_2)/p_0$, $EN_2 = (1-p_1)/p_0$ and $k_i = 1$, $i = 1, 2$ where the k_i's are defined in (3.1).

Example 5.2. (Positive multivariate negative binomial). The function

$$\psi(u_1,u_2,\ldots,u_m) = \left(\frac{p_0}{1 - \sum_{i=1}^{m} p_iu_i}\right)^{b} \prod_{i=1}^{m} u_i,\ 0 \le u_i \le 1 \tag{5.2}$$

where $p_i \ge 0$, $i = 0, 1, \ldots, m$, $\sum_{i=0}^{m} p_i = 1$ and $b > 0$, is a p.g.f. of a positive random vector of integers $\underline{N} = (N_1, \ldots, N_m)$. The joint distribution N_i-1, $i = 1, \ldots, m$ is called a multivariate negative binomial distribution (Johnson and Kotz [6]). It is easy to verify that ψ of (5.2) is TP_2 in each pair of variables when the other variables are held fixed. Here $EN_i = 1+bp_i/p_0$ and $k_i = 1$, $i = 1, 2, \ldots, m$.

The next proposition shows that some well known integer valued random vectors have p.g.f.'s that are TP_2 in each pair of variables.

Proposition 5.3. Let $M_1, \ldots, M_m$ and L be independent integer valued random variables. Then the p.g.f. of $\underline{N} = (N_1,\ldots,N_m)$ $(M_1 + L, \ldots, M_m + L)$ is TP_2 in each pair of variables when the other variables are held fixed.

Proof. Let g_i be the p.g.f. of M_i, $i = 1, \ldots, m$ and let h be the p.g.f. of L. Then the p.g.f. of $\underline{N}$ is

$$\psi(u_1, \ldots, u_m) = h\left(\prod_{i=1}^{m} u_i\right) \prod_{i=1}^{m} g_i(u_i). \tag{5.3}$$

The function $h(\prod_{i=1}^{m} u_i)$ is TP_2 in each pair of variables as can be seen from Theorem 5.1, of Karlin [8]. From this fact and Theorem 1.1, of Karlin [8] it follows that ψ is TP_2 in each pair of variables. ||

If the N's of Proposition 5.3 are positive then their joint p.g.f. satisfies the conditions of Theorem 3.2 and Corollary 4.2. Here are some examples:

Example 5.4 (i). (Bivariate positive-Poisson). Let M_1, M_2 and L be independent Poisson random variables with means μ_1, μ_2 and λ respectively. Then the joint p.g.f. of (N_1,N_2) $\equiv(M_1 + L + 1, M_2 + L + 1)$ is TP_2 in $0 \leq u_1, u_2 \leq 1$, and N_1 and N_2 are positive. The joint distribution of $N_1 - 1$ and $N_2 - 1$ is called a bivariate Poisson distribution (Holgate [4]). Here

$EN_i = 1 + \mu_i + \lambda$ and $k_i = 1$, $i = 1, 2$.

Example 5.4 (ii). (another bivariate positive-Poisson). Let M_1 and M_2 be independent Poisson random variables with means μ_1 and μ_2 and let L be a random variable independent of M_1 and M_2 such that $P(L=n) = (e^{-\lambda}\lambda^n)/[(1-e^{-\lambda})n!]$, $n = 1, 2, \ldots$ (L is said to have a positive-Poisson distribution, Smith and Kimelderf [12]). Then the joint distribution p.g.f. of $(N_1, N_2) = (M_1 + L, M_2 + L)$ is TP_2 in $0 \leq u_1, u_2 \leq 1$, and N_1 and N_2 are positive. Here $EN_i = \mu_i + \lambda/(1-e^{-\lambda})$ and $k_i = 1$, $i = 1, 2$.

Other bivariate positive-Poisson distributions with p.g.f.'s that are TP_2 in $0 \leq u_1, u_2 \leq 1$ can be obtained in a similar fashion (e.g. when M_1 and M_2 have positive-Poisson distributions and when L has a Poisson distribution or when zeroes are censored from the bivariate Poisson distribution of Holgate [4]).

Example 5.5. (Positive Neyman Type A). If (N_1-1, N_2-1) is distributed according to any of the three types of the bivariate Neymann Type A distributions introduced by Holgate [5] then the p.g.f. of (N_1, N_2) is TP_2 on $[0, 1] \times [0, 1]$. For the first two types this fact can be verified directly from the expressions of their p.g.f.'s. Proposition 5.3 provides a proof that the p.g.f. of the third type of bivariate Neyman Type A distribution is TP_2 on $[0, 1] \times [0, 1]$ (see Johnson and Kotz [6]).

In the following propositions it is shown that some 'positive dependent' random vectors have TP_2 p.g.f.'s. Consider a random vector of integers (N_1, N_2) with the probability function $P(N_1 = n_1, N_2 = n_2) = P_{n_1,n_2}$, $n_i \in \mathbb{Z}$, $i = 1, 2$ where $\mathbb{Z} = \{1, 2, \ldots\}$. Denote its distribution function $K(w_1, w_2) = P(N_1 \leq w_1, N_2 \leq w_2)$, $w_i \in R$, $i = 1, 2$. The vector (N_1, N_2) or its distribution function is said to be left corner set decreasing (denoted LCSD) if $P[N_1 \le w_1', N_2 \le w_2' | N_1 \le w_1, N_2 \le w_2]$ is non-increasing in w_1 and w_2 for every choice of w_1' and w_2' (Brindley and Thompson [1]).

Proposition 5.6. If $P(N_1 = n_1, N_2 = n_2) = P_{n_1,n_2}$ is TP_2 in $n_1 \in Z$ and $n_2 \in Z$ then the p.g.f. of (N_1, N_2) is TP_2 on $[0, 1] \times [0, 1]$.

Proposition 5.7. If (N_1, N_2) is LCSD then the p.g.f. of (N_1, N_2) is TP_2 on $[0, 1] \times [0, 1]$.

Proofs of Propositions 5.6 and 5.7. First it is shown that if p_{n_1,n_2} is TP_2 then (N_1, N_2) is LCSD, and then it is shown that if (N_1, N_2) is LCSD then it has a TP_2 p.g.f.

Assume p_{n_1,n_2} is TP_2. Define $H(w, n) = 0$ if $n > w$ and $H(w, n) = 1$ if $n \le w$. Then

$$K(w_1, w_2) = \sum_{n_1=1}^{\infty} \sum_{n_2=1}^{\infty} H(w_1, n_1)H(w_2, n_2)p_{n_1,n_2} \tag{5.4}$$

$$= \sum_{n_1=1}^{\infty} H(w_1, n_1) \Big[\sum_{n_2=1}^{\infty} H(w_2, n_2)p_{n_1,n_2} \Big].$$

By assumption p_{n_1,n_2} is TP_2 on $Z \times Z$ and by Karlin [8], $H(w_2,n_2)$ is TP_2 in $w_2 \in [0, \infty]$ and $n_2 \in Z$, hence by the Basic Conposition Formula (Karlin [8]) the Expression [ ] of (5.4) is TP_2 in $n_1 \in Z$ and $w_2 \in [0, \infty)$. Using the fact that $H(w_1, n_1)$ is TP_2 in $n_1 \in Z$ and $w_1 \in [0, \infty)$ and invoking the Basic Composition Formula once more one obtains that $K(w_1, w_2)$ is TP_2 in $w_i \in [0, \infty)$. However, the condition that $K(w_1, w_2)$ is TP_2 is equivalent to the condition that K is LCSD (see similar results in Shaked [10]).

Assume now that $K(w_1, w_2)$ is TP_2 in $w_i \in [0, \infty)$, $i = 1, 2$. We will show that

$$\rho(t_1, t_2) = \int_0^\infty \int_0^\infty e^{t_1 w_1 + t_2 w_2} \, dK(w_1, w_2) \text{ is } TP_2 \text{ in} \tag{5.5}$$
$$t_i \in (-\infty, 0], \; i = 1, 2,$$

and then by noting that the p.g.f. of (N_1, N_2) is $\psi(u_1, u_2) = \rho(\log u_1, \log u_2)$, $0 < u_i \leq 1$ and invoking Theorem 1.1 of Karlin [8] one sees that (5.5) implies that $\psi(u_1, u_2)$ is TP_2 on $[0, 1] \times [0, 1]$ (in fact (5.5) implies that $\psi(u_1, u_2)$ is TP_2 on $(o, 1]$, but this observation together with the continuity of ψ implies that $\psi(u_1, u_2)$ is TP_2 on $[0, 1] \times [0, 1]$).

To prove (5.5) note that

$$\rho(t_1, t_2) = t_1 t_2 \int_0^\infty \int_0^\infty \int_{w_1}^\infty \int_{w_2}^\infty e^{t_1 w_1' + t_2 w_2'} \, dw_2' \, dw_1' \, dK(w_1, w_2)$$

$$= t_1 t_2 \int_0^\infty e^{t_1 w_1'} \left[\int_0^\infty e^{t_2 w_2'} K(w_1', w_2') \, dw_2'\right] dw_1'.$$

By hypothesis $K(w_1, w_2)$ is TP_2 and by Karlin [8], the kernels $e^{t_1 w_1}$ and $e^{t_2 w_2}$ are TP_2, hence by invoking the Basic Composition Formula twice one obtains that $\rho(t_1, t_2)$ is TP_2 in $t_i \in (-\infty, 0]$, and the proofs of Proposition 5.6 and 5.7 are complete. ||

The following examples can be verified directly or by using similar results of Shaked [11].

Example 5.8. If M_1 and M_2 are positive integer valued random variables then $p_{n_1, n_2} = \{P \min(M_1, M_2) = n_1, \max(M_1, M_2) = n_2\}$

is TP_2 on $Z \times Z$. By Proposition 5.6 the associated p.g.f. is TP_2 on $[0, 1] \times [0, 1]$.

Example 5.9. If M_1, M_2 and M_3 are positive integer valued random variables then $P\{\max(M_1, M_3) \leq w_1, \max(M_2, M_3) \leq w_2\}$ is TP_2 in $w_i \in [0, \infty)$, $i = 1, 2$ (i.e. it is LCSD (see Brindley and Thompson [1]), and by Proposition 5.7 the associated p.g.f. is TP_2 on $[0, 1] \times [0, 1]$.

Example 5.10. If M_i, $i = 1, 2, 3$ are i.i.d. positive integer valued random variables then $P(\min(M_1, M_3) \leq w_1, \min(M_2, M_3) \leq w_2)$ is TP_2 in $w_i \in [o, \infty)$, $i = 1, 2$ and by Proposition 5.7 the associated p.g.f. if TP_2 on $[0, 1] \times [0, 1]$.

We state, in passing, generalizations of Proposition 5.6 and 5.7. Their proofs are similar to the proofs in Propositions 5.6 and 5.7.

Proposition 5.6'. Let $\underline{N} = (N_1, N_2, \ldots, N_m)$ be a random vector of positive integers. If $P(N_1 = n_1, \ldots, N_m = n_m)$ is TP_2 is each pair of variables when the other variables are held fixed, then the p.g.f. of $\underline{N}$ is TP_2 in each pair of variables when the other variables are held fixed.

Proposition 5.7'. Let $\underline{N}$ be as in Proposition 5.6'. If $P(N_1 \leq w_1, \ldots, N_m \leq w_m)$ is TP_2 in each pair of variables when the other variables are held fixed, then the p.g.f. of $\underline{N}$ is TP_2 in each pair of variables when the other variables are held fixed.

Acknowledgement. I thank A. W. Marshall for many fruitful conversations.

REFERENCES

[1] Brindley, E. C. and Thompson, W. A., "Dependence and Aging Aspects of Multivariate Survival", *Jour. Amer.*

Statist. Assoc., *67*, (1972), pp. 822-830.

[2] Cohen, J. W., "Some Ideas and Models in Reliability Theory", *Statistica Neerlandica*, *26*, (1974), pp. 1-10.

[3] Downton, F., "Bivariate Exponential Distributions in Reliability Theory", *J. R. Statist. Soc. B*, *32*, (1970), pp. 63-73.

[4] Holgate, P., "Estimation for the Bivariate Poisson Distribution", *Biometrika*, *51*, (1964), pp. 241-245.

[5] Holgate, P., "Bivariate Generalization of Neyman's Type A Distribution", *Biometrika*, *53*, (1966), pp. 241-244.

[6] Johnson, N. L. and Kotz, S., Distributions in Statistics: Discrete Distributions, Wiley, New York, (1969).

[7] Johnson, N. L. and Kotz, S., "A Vector Valued Multivariate Hazard Rate", *Jour. Multivariate Analysis*, *5*, (1975), pp. 53-66.

[8] Karlin, S., Total Positivity, Stanford University Press, Stanford, California, (1968).

[9] Marshall, A. W., "Some Comments on the Hazard Gradient", *Stochastic Processes and Their Applications*, *3*, (1975), pp. 293-300.

[10] Shaked, M., "On the Distribution of the Minimum and of the Maximum of a Random Number of i.i.d. Random Variables", *Statistical Distributions in Scientific Work, Vol. I.*, (ed. S. Kotz and G. P. Patil), D. Reidel, Boston, (1975a), pp. 363-380.

[11] Shaked, M., A Family of Concepts of Dependence for Bivariate Distributions. Technical Report, Department of Mathematics and Statistics, University of New Mexico, Albuquerque, N.M., (1975b).

[12] Smith, F. H. and Kimeldorf, G., "Discrete Sequential Search for One of Many Objects", *Ann. Statist.*, *3*, (1975), pp. 906-915.

THE ROLE OF THE POISSON DISTRIBUTION IN APPROXIMATING SYSTEM RELIABILITY OF k-OUT-OF-n STRUCTURES*

by R. J. SERFLING
The Florida State University

Abstract. The determination of the exact reliability of a complex system of "components" (or subsystems) is typically a formidable theoretical and computational problem. Often one resorts merely to bounds on system reliability. The present paper delineates, for *k-out-of-n* structures, a simple and easily implemented approach based on the *Poisson* distribution. The components are allowed to be statistically dependent and to have unequal reliabilities. Special attention is given to the cases of independent and Markov dependent components, and some numerical illustration is provided. Comparisons are made to other reliability bounds in the literature. Reliability bounds for k-out-of-n structures have wide application, including "series" systems, "parallel" systems, monitoring systems, and systems with "spares". The results are also applicable in the context of availability.

1.0 *Introduction.* The determination of the exact reliability of a complex system of "components" (or subsystems) is typically a formidable theoretical and computational problem. Often one resorts merely to *bounds* on system reliability. Broadly applicable lower and upper bounds are provided by Barlow and Proschan [1]. However, the general formulas for these bounds are somewhat cumbersome.

---

* Research supported by the Army, Navy and Air Force under Office of Naval Research Contract No. N00014-75-C-0551. Reproduction in whole or in part is permitted for special purposes within the United States Government.

The present paper delineates, for *k-out-of-n* structures, a simple and easily implemented approach based on the *Poisson* distribution. For such structures, where the n-component system functions if and only if *at least* k components function, the system reliability R may be represented in the form

$$R = P[\sum_{i=1}^{n} Y_i \leq n - k], \tag{1.1}$$

where $Y_i = 1$ or 0 according as the i-th component has failed or is still functioning. The random variable $\sum_{i=1}^{n} Y_i$ denotes the total number of failed components, i.e., the number of occurrences among the collection of "rare" (hopefully) events $\{Y_1 = 1\}, \ldots, \{Y_n = 1\}$, and thus its probability distribution is subject to approximation by a suitably chosen Poisson distribution. In this fashion one obtains a Poisson approximation for the system reliability R, as well as related lower and upper bounds for R.

A general theorem for Poisson approximation of system reliability of k-out-of-n structures is presented in Section 2. The components of the system are allowed to be statistically *dependent* and to have *unequal* reliabilities. Also, the special case of *independent* components is examined.

In Section 3 we consider k-out-of-n structures with *Markov-dependent* components. For convenience, attention is confined here to the case of equal component reliabilities. For the 2-out-of-3 structure, an exact analysis is given in order to illustrate quantitatively the consequences of erroneously assuming independence in computing system reliability. For the general k-out-of-n structure, the theorem of Section 2 is utilized to obtain approximations and bounds explicitly involving parameters of the Markov dependence. Numerical illustration for the 2-out-of-3 case is provided.

Comparisons between the bounds of the present paper and those of Barlow and Proschan [1], specialized to k-out-of-n structures, are made in Section 4. For example, in the case of independent components with unequal reliabilities, the bounds based on Poisson approximation are considerably easier to compute. On the other hand, in the case of dependent *associated* components, the bounds of the present paper require explicit computation of dependence parameters, whereas the Barlow and Proschan bounds do not entail such parameters. However, such computational advantage is offset by a loss of sharpness. Various examples and numerical illustrations are presented.

Our treatment is restricted to k-out-of-n structures for a technical reason, to accomodate the utilization of the Poisson distribution as the basis for a convenient approximation to system reliability. Nevertheless the results have wide practical application. Familiar special cases are "series" (n-out-of-n) and "parallel" (1-out-of-n) structures. The role of k-out-of-n structure in designing monitoring systems is discussed by Barlow and Proschan [1], p. 49. A k-series system with a "spares pool" of size m may be regarded as a k-out-of-(k+m) structure. A maintenance policy of calling a repairman or ordering a batch of replacements as soon as n - k - 1 units have been exhausted corresponds to a k-out-of-n system. Further, the results of the paper may be interpreted in the context of *availability*. Shaw and Shooman [3] study the availability of systems having k-out-of-n structure with respect to n subsystems with known availabilities. Our results provide an alternative approach for such studies. Finally, we note that the results of the paper could be extended to systems which are j-out-of-m *compositions* of k-out-of-n structures.

A useful feature of the Poisson approximation for system reliability is that, rather than directly estimating reliability of a given system, it may alternatively be used to *design* a k-out-of-n system to achieve desired reliability specifications.

2.0 *A Poisson Approximation for System Reliability of k-out-of-n Structures.* Our approximations to the reliability R defined by (1.1) will be expressed in terms of Poisson distributions. Let Q_λ denote the *Poisson* distribution with mean λ (where $\lambda > 0$), and let $Q_\lambda(m)$ denote the probability assigned by Q_λ to the set $\{0, 1, \ldots, m\}$. Thus

$$Q_\lambda(m) = \sum_{j=0}^{m} \frac{e^{-\lambda}\lambda^j}{j!}, \qquad m = 0, 1, \ldots \quad .$$

Random variables taking only the values 0 and 1 will be called *Bernoulli* variables. Thus the reliability R is expressed by (1.1) in terms of the distribution of a *sum* of Bernoulli variables. A general theorem on approximation of such distributions by Poisson distributions is given in Serfling [2]. An approximation to R follows as a direct corollary, which is now presented.

Specifically, consider a k-out-of-n structure with components having states indicated by the Bernoulli variables $Y_1, \ldots, Y_n$ defined in Section 1.0. Put

$$\theta_1 = P[Y_1 = 1]$$

and, for $2 \leq i \leq n$,

$$\theta_i(y_1, \ldots, y_{i-1}) = P[Y_i = 1 | Y_1 = y_1, \ldots, Y_{i-1} = y_{i-1}],$$

for $y_1, \ldots, y_{i-1}$ taking values 0 or 1. The θ_i denote (conditional) failure probabilities of the components. Restating (1.1), the system reliability is given by

$$R = P[\sum_{i=1}^{n} Y_i \leq n - k].$$

Theorem 2.1. Consider a k-out-of-n structure with conditional component failure probabilities $\theta_1, \ldots, \theta_n$ *and system reliability* R *as given above. Let* $\lambda_1, \ldots, \lambda_n$ *be any set of values such that* $0 < \lambda_1, \ldots, \lambda_n < 1$ *and put*

$$\lambda = \sum_{i=1}^{n} \lambda_i,$$

$$L = \frac{1}{2} \sum_{i=1}^{n} \lambda_i^2,$$

$$D = \sum_{i=1}^{n} E|\Theta_i - \lambda_i|,$$

$$\gamma_i = -\log(1-\lambda_i), \quad 1 \leq i \leq n,$$

$$\gamma = \sum_{i=1}^{n} \gamma_i,$$

and

$$G = \frac{1}{2} \sum_{i=1}^{n} \gamma_i^2.$$

Then

$$Q_\lambda (n - k) - D - L \leq R \leq Q_\lambda (n - k) + D + L \tag{2.1a}$$

and

$$Q_\gamma (n - k) - D \leq R \leq Q_\gamma (n - k) + D + G. \tag{2.1b}$$

Remarks. (i) Here $E|\Theta_i - \lambda_i|$ denotes the quantity

$$E|\Theta_i(Y_1, \ldots, Y_{i-1}) - \lambda_i|.$$

For i = 1, this is simply $|\Theta_1 - \lambda_1|$.

(ii) It is important to note that the values λ_i in this theorem may be selected arbitrarily. For example, a natural and convenient choice for λ_i is the *mean* $E(\Theta_i)$. On the other hand, if it is desired to minimize the quantity $E|\Theta_i - \lambda_i|$, the appropriate choice is λ_i equal to a *median* of the distribution of $\Theta_i(Y_1, \ldots, Y_{i-1})$.

(iii) For the case of *independent* components, the quantities Θ_i are no longer random, so that the choice $\lambda_i = \Theta_i$, $1 \leq i \leq n$, is feasible. This completely eliminates the term D. See Corollary 2.1 below.

(iv) Conditions (2.1a) and (2.1b) provide alternate sets of bounds for R. For small λ_i's, (2.1b) tends to give a tighter interval for R. See Sections 3 and 4 for numerical illustration. For theoretical inquiry in connection with the bounds given in Theorem 2.1 or Corollaries 2.1 and 3.1, see Serfling [2] and references cited therein.

The following result focuses upon the case of independent components and expresses the conditions in terms of the parameters

$$p_i = 1 - \Theta_i = P[\text{i-th component is functioning}], \quad 1 \leq i \leq n,$$

which are more conventional than the Θ_i's.

Corollary 2.1. For a k-out-of-n structure with independent components having reliabilities p_i, $1 \leq i \leq n$, *the system reliability* R *satisfies*

$$Q_\lambda(n - k) - \frac{1}{2}\sum_{i=1}^{n}(1 - p_i)^2 \leq R \leq Q_\lambda(n - k) + \frac{1}{2}\sum_{i=1}^{n}(1 - p_i)^2 \tag{2.2a}$$

and

$$Q_\gamma(n - k) \leq R \leq Q_\gamma(n - k) + \frac{1}{2}\sum_{i=1}^{n}\log^2(p_i), \tag{2.2b}$$

where

$$\lambda = \sum_{i=1}^{n}(1 - p_i) \text{ and } \gamma = -\sum_{i=1}^{n}\log p_i.$$

Another special case, that of Markov-dependent components, is treated in Section 3. However, there attention is confined for convenience to the case of equal component reliabilities $p_i \equiv p$.

3.0 *Reliability of k-out-of-n Structures with Identical Markov-dependent Components*. Let us indicate the states of the components by $X_1, \ldots, X_n$, where $X_i = 1$ or 0 according as the i-th component is still functioning or has failed. Suppose that the sequence of Bernoulli variables $X_1, \ldots, X_n$ is *Markov-dependent*, with transition probabilities

$$\alpha = P[X_i = 1 | X_{i-1} = 1]$$

and

$$\beta = P[X_i = 1 | X_{i-1} = 0]$$

for $2 \leq i \leq n$. Assume that $0 < \alpha < 1$ and $0 < \beta < 1$ and put

$$\delta = \alpha - \beta, \qquad p = \frac{\beta}{1 - \delta}.$$

Assume for convenience that the components have equal reliabilities, in which case it is found that

$$P[X_i = 1] = p, \qquad 1 \leq i \leq n.$$

Suppose also that $\alpha > \beta$ (or $1 - \beta > 1 - \alpha$), reflecting the assumption that failure of the (i - 1)-th component only enhances the chance of failure of the i-th component, $2 \leq i \leq n$. Thus $\delta > 0$ and $\alpha > p > \beta$.

The assumption that $\delta \geq 0$ makes $X_1, \ldots, X_n$ *conditionally increasing in sequence* in the sense of Barlow and Proschan [1] and hence, by their Theorem 4.7, p. 146, *associated*. This fact, namely that Markov-dependent Bernoulli variables $X_1, \ldots, X_n$ with $\delta \geq 0$ are associated, is apropos to the comparisons made in Section 4.

Let us further define

$$d = \frac{1 - \beta}{1 - \alpha}.$$

The conditions $\delta > 0$ and $d > 1$ are equivalent assertions, but the quantities δ and d measure the dependence in different ways.

The quantity d measures the relative severity of the influence of failure of the (i - 1)-th component upon the chance of failure of the i-th component. In considering examples, it is intuitively appealing to specify values for p and d and then solve for δ, α and β via

$$\delta = \frac{(1 - p)(d - 1)}{1 + (1 - p)(d - 1)} .$$

In order to see how Markov dependence alters the system reliability from its value in the case of independence, let us examine the *exact* reliability function in the simple case of 2-out-of-3 structure.

Example: 2-out-of-3 structure. Let X_1, X_2 and X_3 be Markov dependent as described above. Considering the system reliability as a function of the component reliability p, the *reliability function* is found to be

$$\begin{aligned} R(p) &= P[X_1 + X_2 + X_3 \geq 2] \\ &= P[X_1 = X_2 = X_3 = 1] + P[X_1 = X_2 = 1, X_3 = 0] \\ &\quad + P[X_1 = X_3 = 1, X_2 = 0] + P[X_1 = 0, X_2 = X_3 = 1] \\ &= P[X_3 = 1 | X_2 = 1]\ P[X_2 = 1 | X_1 = 1]\ P[X_1 = 1] + \ldots \\ &= \alpha^2 p + (1 - \alpha)\alpha p + \beta(1 - \alpha)p + \alpha\beta(1 - p) \\ &= \alpha p + \beta p + \alpha\beta - 2\alpha\beta p. \end{aligned}$$

In terms of simply the parameters p and δ, we have

$$\begin{aligned} R(p) = {} & 3p^2 - 2p^3 - 2p(1 - p)(2p - 1)\delta \\ & + (3p^2 - 2p^3 - p)\delta^2. \end{aligned}$$

The special case of independence ($\delta = 0$) is given by

$$R_0(p) = 3p^2 - 2p^3.$$

The disparity between R(p) and $R_0(p)$ is illustrated in Table 3.1.

Table 3.1. The reliability function R(p) for selected values of p and d.

p	$R_0(p)$	d = 2	d = 3
.8	.896	.871	.850
.9	.972	.966	.950
.95	.993	.989	.986
.99	.9997	.9995	.9993

The change in R(.99) in passing from independence (d = 1) to the dependence cases d = 2, d = 3 is rather substantial, if loss is measured in terms of the *unreliability* 1 - R.

Turning now to the general k-out-of-n structure, we consider the implications of Theorem 2.1 for Markov-dependent components. For the random variables $Y_i = 1 - X_i$, we have $\Theta_1 = P[Y_1 = 1] = 1 - p$ and, for $i \geq 2$,

$$\Theta_i(Y_1, \ldots, Y_{i-1}) = \begin{cases} 1 - \alpha \text{ if } Y_{i-1} = 0 \\ 1 - \beta \text{ if } Y_{i-1} = 1. \end{cases}$$

If follows, for $i \geq 2$, that Θ_i has *mean* 1 - p and (since $p > \frac{1}{2}$) *median* $1 - \alpha$. We find, for $i \geq 2$, that $E|\Theta_i - (1 - p)| = 2\delta p(1 - p)$, whereas $E|\Theta_i - (1 - \alpha)| = (1 - p)\delta$, and thus that $E|\Theta_i - \lambda_i|$ is not only minimized at $\lambda_i = 1 - \alpha$ but is actually about half the value corresponding to $\lambda_i = 1 - p$. Moreover, since $1 - \alpha < 1 - p$, the term $L = \frac{1}{2}\sum_{i=1}^{n} \lambda_i^2$ is smaller for $\lambda_i = 1 - \alpha$ than for $\lambda_i = 1 - p$, $i \geq 2$. Consequently, in order to minimize the contribution $D = \sum_{i=1}^{n} E|\Theta_i - \lambda_i|$ in the bounds (2.1), we choose $\lambda_1 = 1 - p$ and $\lambda_i = 1 - \alpha$, $i \geq 2$. We obtain

Corollary 3.1. Consider a k-out-of-n system of Markov-dependent components with parameters p *and* δ. *Put*

$$\lambda = (1 - p) + (n - 1)(1 - \alpha),$$

$$L = \frac{1}{2}[(1 - p)^2 + (n - 1)(1 - \alpha)^2],$$

$$D = (n - 1)(1 - p)\delta,$$

$$\gamma = -\log p - (n - 1)\log \alpha,$$

and

$$G = \frac{1}{2}[\log^2 p + (n - 1)\log^2 \alpha].$$

Then

$$Q_\lambda(n - k) - D - L \leq R \leq Q_\lambda(n - k) + D + L \tag{3.1a}$$

and

$$Q_\gamma(n - k) - D \leq R \leq Q_\gamma(n - k) + D + G. \tag{3.1b}$$

As an illustration, we apply Corollary 3.1 to the 2-out-of-3 structure with d = 2. Table 3.2 gives, for selected values of p, the exact reliability R(p) for d = 2, the exact reliability $R_0(p)$ for the case of independence, the bounds on R(p) provided by (3.1a) and (3.1b) for d = 2, and the bounds provided by Barlow and Proschan [1] for associated components. (We further discuss the latter bounds in Section 4.)

Table 3.2. Exact values and associated bounds for R(p) in the 2-out-of-3 structure, for selected values of p and d.

			(3.1a), d=2		(3.1b), d=2		Barlow & Proschan	
p	$R_0(p)$	R(p), d=2	Lower	Upper	Lower	Upper	Lower	Upper
.8	.896	.871	.789	1	.814	1	.64	.96
.9	.972	.966	.936	.998	.945	.991	.81	.99
.95	.993	.989	.981	.998	.985	.998	.903	.9975
.99	.9997	.9995	.9992	.9999	.99936	.9999	.980	.9999

4.0 *Comparisons with Other Reliability Bounds*. Barlow and Proschan [1], p. 37, provide bounds on system reliability for an arbitrary coherent structure. From these general bounds they derive more explicit bounds for the case of *associated* components (this form of dependence includes independence as a special case and also includes the Markov dependence treated in Section 3). Specialized to the case of a k-out-of-n structure with components having reliabilities $p_1, \ldots, p_n$ and with system reliability R, the latter bounds are

$$\max_{1 \le i_1 < \cdots < i_k \le n} \prod_{j=1}^{k} p_{i_j} \le R \tag{4.1}$$

$$\le \min_{1 \le i_1 < \cdots < i_{n-k+1} \boxed{\le n}} \coprod_{j=1}^{n-k+1} p_{i_j} ,$$

where $\coprod_{i=a}^{b} y_i = 1 - \prod_{i=a}^{b} (1 - y_i)$. For the case of *equal* component reliabilities $p_i \equiv p$, (4.1) reduces to

$$p^k \le R \le 1 - (1 - p)^{n-k+1} \tag{4.2}$$

We have utilized (4.2) in Table 3.2.

Barlow and Proschan [1], p. 34, also provide another set of bounds for the case of associated components. Specializing to the case of k-out-of-n structure with *independent* components having reliabilities $p_1, \ldots, p_n$, these bounds are:

$$\prod_{1 \le i_1 < \cdots < i_{n-k+1} \le n} \coprod_{j=1}^{n-k+1} p_{i_j} \le R \tag{4.3}$$

$$\le \coprod_{i \le i_1 \le \cdots < i_k < n} \prod_{j=1}^{k} p_{i_j} .$$

For the case of *equal* component reliabilities $p_i \equiv p$, (4.3) reduces to

$$[1 - (1 - p)^{n-k+1}]^{\binom{n}{n-k+1}} \le R \le 1 - [1 - p^k]^{\binom{n}{k}} . \tag{4.4}$$

Let us now focus upon the case of *independent* components with *equal* reliabilities $p_i \equiv p$. Corollary 2.1 yields

$$Q_{n(1-p)}(n - k) - \tfrac{1}{2}n(1 - p)^2 \le R \le Q_{n(1-p)}(n - k) + \tfrac{1}{2}n(1 - p)^2 \quad (4.5)$$

and

$$Q_{-n \log p}(n - k) \le R \le Q_{-n \log p}(n - k) + \tfrac{1}{2}n \log^2(p). \quad (4.6)$$

We may compare (4.2), (4.4), (4.5) and (4.6). As regards computational ease, these are all relatively simple. For a numerical comparison, Table 4.1 provides evaluations of the bounds for the 2-out-of-3 structure. It is seen that collectively the bounds (4.2), (4.4), (4.5) and (4.6) yield more information than any one of them individually. The composite information thus supplied by Table 4.1 is presented in Table 4.2. It is thus seen that the Barlow and Proschan bounds (4.2) and (4.4) and the Poisson approximation bounds (4.5) and (4.6) are competitive in the case of independent components with equal reliabilities.

Table 4.1. Various bounds on R(p) for the 2-out-of-3 structure with independent components, for selected values of p.

		(4.2)		(4.4)		(4.5)		(4.6)	
p	R(p)	Lower	Upper	Lower	Upper	Lower	Upper	Lower	Upper
.8	.896	.64	.96	.885	.953	.818	.938	.854	.929
.9	.972	.81	.99	.970	.99	.948	.978	.959	.976
.95	.993	.903	.9975	.9925	.9991	.986	.994	.989	.993
.99	.9997	.980	.9999	.9997	.9999	.9994	.9997	.99956	.9997

Table 4.2. Composite bounds on R(p), from Table 4.1.

p	R(p)	Bounds Lower	Bounds Upper
.8	.896	.885	.929
.9	.972	.970	.976
.95	.993	.9925	.993
.99	.9997	.9997	.9997

For the case of independent components with *unequal* reliabilities, the bounds given by (4.1) and (4.3) may be compared with those of Corollary 2.1. It is evident that (4.1) and (4.3) are considerably more cumbersome computationally than (4.2) and (4.4), whereas the general bounds of Corollary 2.1 are only slightly more cumbersome than (4.5) and (4.6). This difference in computational ease becomes increasingly pronounced for larger values of n and n - k. Consider the following example.

Example. k-out-of-20 structure. Let us consider *independent* components with reliabilities $p_1 = \ldots = p_5 = .99$, $p_6 = \ldots = p_{10} = .98$, $p_{11} = \ldots = p_{15} = .97$, and $p_{16} = \ldots = p_{20} = .96$. We shall not bother to compute (4.1) and (4.3), which are tedious. In order to utilize Corollary 2.1, we readily compute

$$\lambda = 5(.01 + .02 + .03 + .04) = .5,$$

$$\gamma = -5(\log .99 + \log .98 + \log .97 + \log .96) = .508,$$

$$L = \frac{1}{2} \cdot 5 \cdot [(.01)^2 + \ldots] = .0075,$$

and

$$G = \frac{1}{2} \cdot 5 \cdot [\log^2(.99) + \ldots] = .0078.$$

Using a table of the Poisson distribution to obtain $Q_{.5}(20 - k)$ and $Q_{.508}(20 - k)$, and then evaluating the bounds (2.2a) and

(2.2b) with the use of λ, γ, L and G, we find that in the present example the bounds (2.2b) are more effective than (2.2a). On the other hand, the quantity $Q_{.5}(20 - k)$ involved in (2.2a) may be interpreted as an *approximation* to R, whereas $Q_{.508}(20 - k)$ arises as a lower bound rather than as an approximation. The relevant values are given in Table 4.3.

Table 4.3. Approximation and bounds for R, for k-out-of-20 structure.

		(2.2a)		(2.2b)	
k	$Q_{.5}(20-k)$	Lower	Upper	Lower	Upper
20	.60653	.59903	.62153	.60180	.60960
19	.90980	.90230	.91730	.90732	.91512
18	.98561	.97811	.99311	.98495	.99275
17	.99825	.99075	1	.99813	1
16	.99983	.99233	1	.99981	1

Note that the second and third columns in Table 4.3 may be eliminated, being improved by the fourth and fifth columns, respectively.

Finally, we discuss the case of *dependent* components. Whereas the bounds given by Theorem 2.1 require explicit computation of dependence parameters, the Barlow and Proschan bounds (4.1) for *associated* components do not entail explicit dependence parameters. Whether or not (4.1) is actually easier to compute than (2.1a, b) depends on the nature of the actual dependence and the feasibility of computing the quantity D. In Section 3 we have seen that for Markov-dependent components the computation of the bounds (2.1) is easy and yields substantial sharpening of the bounds (4.3). Indeed, this illustrates the potential gains from utilizing when possible any knowledge of the specific dependence structure. Of course, the bounds (2.1) are applicable also to other varieties of dependence.

Acknowledgments. I am indebted to Frank Proschan for correcting a gross blunder and to Chris Ashley for useful remarks. The support of the U.S. Office of Naval Research, under Contract No. N00014-75-C-0551, is also acknowledged.

REFERENCES

[1] Barlow, Richard E. and Proschan, Frank, Statistical Theory of Reliability and Life Testing. Holt, Rinehart and Winston, Inc., (1975).

[2] Serfling, R. J., "Some Elementary Results on Poisson Approximation in a Sequence of Bernoulli Trials", FSU Statistics Report M374, Florida State University, (1976).

[3] Shaw, L. and Shooman, M., "Confidence Bounds and Propagation of Uncertainties in System Availability and Reliability Computations", Report No. EER-114, Department of Electrical Engineering and Electrophysics, Polytechnic Institute of New York, (1975).

CONVERTING DEPENDENT MODELS INTO INDEPENDENT ONES, WITH APPLICATIONS IN RELIABILITY*

by N. LANGBERG, F. PROSCHAN
and A. J. QUINZI
Florida State University

Abstract. Esary and Marshall show in the *Ann. Statist.* (1974) *2* that given a random vector $(T_1, \ldots, T_n)$ with exponential minima, there exists a random vector $(U_1, \ldots, U_n)$ with the Marshall-Olkin multivariate exponential (MVE) distribution such that $\min(T_{i_1}, \ldots, T_{i_k})$ and $\min(U_{i_1}, \ldots, U_{i_k})$ have the same distribution for every subset $\{i_1, \ldots, i_k\}$ of $\{1, \ldots, n\}$. In the present paper, we develop a simpler proof of this result, determine explicitly the parameters of the MVE, and, using our simpler approach, show how to replace each of a variety of reliability models involving dependent random variables by essentially equivalent models involving independent random variables.

1.0 *Introduction and Summary*. Many theorems in probability and statistics, particularly in reliability analysis, contain assumptions involving independent random variables. In actual practice, however, the random variables of interest are often mutually dependent. It is important, therefore, to have ways of converting distributions involving dependent variables into distributions involving independent ones. In this paper such results are obtained by first simplifying the proof of a known result.

---

* Research sponsored by the Air Force Office of Scientific esearch, AFSC, USAF, under Grant AFOSR 74-2581B. Reproduction in whole or in part is permitted for special purposes within the United States Government.

An important feature of any multivariate distribution, especially in reliability theory, is its relevance to realistic physical models. The multivariate exponential (MVE) distribution of Marshall and Olkin (1967) is significant in that respect. Because the MVE may be applied in a number of different situations, it is useful to know when a multivariate distribution of interest may be replaced by an MVE distribution which is equivalent (in some sense) to it. A recent result by Esary and Marshall [3] provides one answer.

The (multivariate distribution of the) random vector $\underline{T} = (T_1, \ldots, T_n)$ is said to have <u>exponential minima</u> if, for every nonempty subset I of $\{1, \ldots, n\}$, $\min_{i \in I} T_i$ is exponentially distributed. The (distributions of the) two random vectors $\underline{T} = (T_1, \ldots, T_n)$ and $\underline{U} = (U_1, \ldots, U_n)$ are <u>marginally equivalent in minima</u> ($\underline{T} \overset{m}{=} \underline{U}$, in symbols) if for every nonempty subset I of $\{1, \ldots, n\}$, $\min_{i \in I} T_i$ and $\min_{i \in I} U_i$ have the same distribution.

The main result of Esary and Marshall [3] is:

Theorem 1.1. (Esary and Marshall). Suppose that a random vector $\underline{T} = (T_1, \ldots, T_n)$ has exponential minima. Then there exists a random vector $\underline{U} = (U_1, \ldots, U_n)$ with the MVE distribution such that $\underline{T} \overset{m}{=} \underline{U}$.

Thus, when one is interested in the distribution of the nonnegative random vector $\underline{T}$ only through the distributions of $\min_{i \in I} T_i$, as is the case, for example, when considering coherent systems in reliability theory, then one might as well use the well-known and highly structured MVE distribution.

In Section 2 we obtain an alternative proof of Theorem 1.1. Our approach has at least three advantages over that of Esary and Marshall. First, our method of proof is simple, appealing only to elementary probability considerations. Second, our proof is constructive in the sense that we specify explicitly the MVE distribution in question. Third, we obtain new ways of converting distributions involving dependent variables into distributions involving independent variables. Section 3 is devoted to applications.

2.0 *From Dependence to Independence: The Exponential Minima Model*. Throughout this paper we use the following notation. Let $\mathcal{I}$ denote the class of all nonempty subsets of $\{1, \ldots, n\}$, $n \geq 2$. For $\mathcal{J} \subseteq \mathcal{I}$, let $\cup \mathcal{J} = \{i: i \in J \text{ for some } J \in \mathcal{J}\}$. Let $\mathcal{A} = \{\mathcal{J}: \mathcal{J} \subseteq \mathcal{I} \text{ and } \cup \mathcal{J} = \{1, \ldots, n\}\}$. For random variables X and Y, $X \overset{st}{=} Y$ indicates that X and Y have the same distribution. Unless otherwise stated, all random vectors are assumed to be n-dimensional.

For our purposes, the most important characterization of the MVE is given by the following generalization of Theorem 3.2 of Marshall and Olkin [4]:

Theorem 2.1. A random vector $\underline{U}$ has the (n-dimensional) MVE distribution if and only if there exists a collection $\{S_J, J \in \mathcal{J}\}$, $\mathcal{J} \in \mathcal{A}$, of independent exponential random variables such that

$$U_i = \min_{J \in \mathcal{J}: i \in J} S_J, \qquad i = 1, \ldots, n.$$

In view of Theorem 2.1, we can state Theorem 1.1 in the following equivalent form:

Theorem 2.2 (Esary and Marshall). Suppose that a random vector $\underline{T}$ has exponential minima. Then there exists a collection $\{S_J, J \in \mathcal{J}\}$, $\mathcal{J} \in \mathcal{A}$, of independent exponential random variables such that for each $I \in \mathcal{I}$,

$$\min_{i \in I} T_i \overset{st}{=} \min_{J \in \mathcal{J}: J \cap I \neq \emptyset} S_J.$$

Thus, whenever a random vector $\underline{T}$ has exponential minima, any distribution involving $\min_{i \in I} T_i$, where $T_1, \ldots, T_n$ are dependent, may be replaced by a distribution involving

$\min_{J \in \mathcal{J}: J \cap I \neq \emptyset} S_J$, where the S_J's are <u>independent</u>.

Lemma 2.3. Suppose that the random vector $\underline{T}$ has exponential minima with (positive) parameters $\{\mu_I, I \in \mathcal{I}\}$; i.e., for each $I \in \mathcal{I}$, $P(\min_{i \in I} T_i > t) = \exp(-\mu_I t)$, $t \geq 0$, $\mu_I > 0$. Then there exists a collection $\{S_J, J \in \mathcal{J}\}$, $\mathcal{J} \in A$, of independent exponential random variables such that for each $I \in \mathcal{I}$, $\min_{i \in I} T_i \overset{st}{=} \min_{J \in \mathcal{J}: J \cap I \neq \emptyset} S_J$ provided the following two conditions hold:

The system of equations

$$\mu_I = \sum_{J \in \mathcal{I}: J \cap I \neq \emptyset} \lambda_J, \qquad I \in \mathcal{I}, \tag{2.1}$$

has a set of nonnegative solutions $\{\lambda_I, I \in \mathcal{I}\}$; and

$$\text{If } \mathcal{J} = \{J \in \mathcal{I}: \lambda_J > 0\}, \text{ then } \mathcal{J} \in A . \tag{2.2}$$

Proof. Suppose that (2.1) and (2.2) hold. For each $J \in \mathcal{J}$, let S_J be distributed exponentially with parameter λ_J and let the S_J, $J \in \mathcal{J}$, be independent. Then for each $I \in \mathcal{I}$ and $t \geq 0$,

$$P(\min_{J \in \mathcal{J}: J \cap I \neq \emptyset} S_J > t) = \exp(-\sum_{J \in \mathcal{J}: J \cap I \neq \emptyset} \lambda_J t)$$
$$= \exp(-\sum_{J \in \mathcal{I}: J \cap I \neq \emptyset} \lambda_J t) = \exp(-\mu_I t) = P(\min_{i \in I} T_i > t). \quad ||$$

Remark 2.4. Note that the system of equations in (2.1) may be expressed as $\underline{\mu} = X\underline{\lambda}$, where $\underline{\mu} = (\mu_1, \mu_2, \ldots, \mu_{1\ldots n})'$, $\underline{\lambda} = (\lambda_1, \lambda_2, \ldots, \lambda_{1\ldots n})'$, X, a square matrix of order $2^n - 1$ consisting of 0's and 1's, is the matrix of coefficients of the λ's, and the subscripts on the components of $\underline{\mu}$ and $\underline{\lambda}$ are ordered lexicographically.

Remark 2.5. It can be shown that the system of equations in (2.1) has a unique solution (see Remark 2.8 below) no matter what the values of the μ_I's. In our particular situation, the question is whether or not there is always a <u>nonnegative</u>

solution when the μ_I's represent certain parameters. If $n = 3$, for example, the system of equations in (2.1) may be represented pictorially as in Figure 2.1:

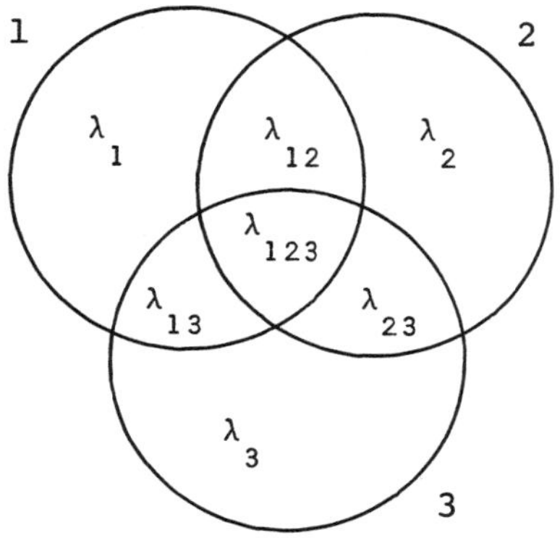

Figure 2.1.

For each $I \in \mathcal{I}$, μ_I is represented in Figure 2.1 by the union of the interiors of those circles whose labels belong to subset I. For example, μ_1 is represented by the interior of circle 1, μ_{12} is represented by the union of the interiors of circles 1 and 2, and so on. The λ_I's are represented by the partial areas pictured in Figure 2.1. It is easy to verify from the figure that $\lambda_{12} = -\mu_3 + \mu_{13} + \mu_{23} - \mu_{123}$. We will show that the λ_I's are nonnegative by showing that they may be defined via certain probabilities. First, we need to establish the following result of elementary probability theory:

Lemma 2.6. For every collection $\{A_1, \ldots, A_n\}$, $n \geq 2$, of events in a probability space $(\Omega, \mathcal{E}, P)$, where $\overline{A}_i$ denotes the event complementary to the event A_i, $i = 1, \ldots, n$, the following holds:

$$P(A_1 \ldots A_r \bar{A}_{r+1} \ldots \bar{A}_n) = (-1)^{r-1}[P(\bigcup_{\nu=1}^{n} A_\nu) - \sum_{i=1}^{r} P(\bigcup_{\substack{\nu=1 \\ \nu \neq i}}^{n} A_\nu) + \sum_{\substack{i_1,i_2=1 \\ i_1<i_2}}^{r} P(\bigcup_{\substack{\nu=1 \\ \nu \neq i_1,i_2}}^{n} A_\nu) - \ldots \quad (2.3)$$

$$+ (-1)^r P(\bigcup_{\nu=r+1}^{n} A_\nu)],$$

for every $r = 1, \ldots, n$.

Proof. By a remarkable theorem of Rényi [5], we have only to verify (2.3) for those 2^n special cases when each of the A_ν is either Ω or $\emptyset$. If $A_\nu = \Omega$ for some ν such that $r + 1 \leq \nu \leq n$, then (2.3) holds, since then $(-1)^{r-1}[1 - \binom{r}{1} + \binom{r}{2} - \ldots + (-1)^r]$ $= 0$. Thus, we may assume that $A_\nu = \emptyset$ for each ν such that $r + 1 \leq \nu \leq n$. The proof of the lemma will be complete if we show that

$$P(A_1 \ldots A_r) = (-1)^{r-1}[P(\bigcup_{\substack{\nu=1 \\ \nu \neq 1}}^{r} A_\nu) - \sum_{i=1}^{r} P(\bigcup_{\substack{\nu=1 \\ \nu \neq i}}^{r} A_\nu) \quad (2.4)$$

$$+ \sum_{\substack{i_1,i_2=1 \\ i_1<i_2}}^{r} P(\bigcup_{\substack{\nu=1 \\ \nu \neq i_1,i_2}}^{r} A_\nu) - \ldots + (-1)^{r-1} \sum_{\nu=1}^{r} P(A_\nu)].$$

But (2.4) is equivalent to the well-known inclusion-exclusion identity [5]:

$$P(\bigcup_{i=1}^{r} A_i) = \sum_{i=1}^{r} P(A_i) - \sum_{\substack{i_1,i_2=1 \\ i_1<i_2}}^{r} P(A_{i_1} A_{i_2}) + \ldots$$

$$+ (-1)^r P(A_1 \ldots A_r).$$

To see this, replace each A_ν by $\bar{A}_\nu$ in (2.4) and use the fact that $P(\bigcup_\nu \bar{A}_\nu) = 1 - P(\bigcap_\nu A_\nu)$. ||

Remark 2.7. In (2.3), the probability $P(A_1 \ldots A_r \bar{A}_{r+1} \ldots \bar{A}_n)$ is expressed as a linear combination of the probabilities $P(\bigcup_{i\varepsilon I} A_i)$, $I \varepsilon \mathcal{I}$. By suitably relabelling the sets $A_1, \ldots, A_n$, Lemma 2.6 allows every probability of the form $P(A_{i_1} \ldots A_{i_r} \bar{A}_{i_{r+1}} \ldots \bar{A}_{i_n})$ to be expressed as an analogous linear combination.

Proof of Theorem 2.2. Suppose that

$$P(\min_{i\varepsilon I} T_i > t) = \exp(-\mu_I t), \qquad t \geq 0, \tag{2.5}$$

for every $I \varepsilon \mathcal{I}$ and for some collection $\{\mu_I, I \varepsilon \mathcal{I}\}$ of positive constants. By Lemma 2.3 and Remark 2.4, the proof will be complete if the following two conditions are met:

(i) The system of equations

$$\underline{\mu} = X\underline{\lambda}, \tag{2.6}$$

where $\underline{\mu}$, $\underline{\lambda}$, and X are defined in Lemma 2.3, has a nonnegative solution; and

(ii) If $\mathcal{J} = \{J \varepsilon \mathcal{I}: \lambda_J > 0\}$, then $\mathcal{J} \varepsilon \mathcal{A}$.

For each $I \varepsilon \mathcal{I}$,

$$\bigcup_{i\varepsilon I} (T_i \leq t) = \bigcup_{j_1<\ldots<j_r} (T_{j_1} \leq t, \ldots, T_{j_r} \leq t, T_{j_{r+1}} > t, \ldots, T_{j_n} > t),$$

where the second union is over all $\{j_1, \ldots, j_r\}$ belonging to $\{J \varepsilon \mathcal{J}: J \cap I \neq \emptyset\}$, $r = 1, \ldots, n$. This is a consequence of the fact that any event may be expressed uniquely as the union of its mutually exclusive "basic functions" [5]. It follows that

$$P[\bigcup_{i\varepsilon I} (T_i \leq t)] = \sum_{j_1<\ldots<j_r} P(T_{j_1} \leq t, \ldots, T_{j_r} \leq t, T_{j_{r+1}} > t, \ldots, T_{j_n} > t)$$

for each $I \in \mathcal{I}$. More concisely,

$$\underline{a}(t) = X\underline{b}(t),$$

where $\underline{a}(t)$ is the vector of probabilities $P[\bigcup_{i \in I} (T_i \le t)]$, $I \in \mathcal{I}$, $\underline{b}(t)$ is the vector of probabilities $P[\bigcap_{i \in I} (T_i \le t) \bigcap_{i \notin I} (T_i > t)]$, $I \in \mathcal{I}$, X, the matrix of coefficients, is precisely that of the original system (2.6), and the components of the vectors $\underline{a}(t)$ and $\underline{b}(t)$ have subscripts which are ordered as in (2.6). By Lemma 2.6, each probability $P(T_{j_1} \le t, \ldots, T_{j_r} \le t, T_{j_{r+1}} > t, \ldots, T_{j_n} > t)$ may be expressed solely as a linear combination of the probabilities $P[\bigcup_{i \in I} (T_i \le t)]$, $I \in \mathcal{I}$. It follows that the matrix X is nonsingular. From (2.5),

$$\lim_{t \to 0^+} t^{-1} P[\bigcup_{i \in I} (T_i \le t)] = \mu_I \quad \text{for each } I \in \mathcal{I}.$$

Consequently,

$$\lim_{t \to 0^+} t^{-1} \underline{b}(t) = X^{-1} \lim_{t \to 0^+} t^{-1} \underline{a}(t)$$

$$= X^{-1} \lim_{t \to 0^+} (t^{-1} P(T_1 \le t), \ldots, t^{-1} P[\bigcup_{i=1}^{n} (T_i \le t)])'$$

$$= X^{-1} \underline{\mu}.$$

Let $\underline{\lambda} = \lim_{t \to 0^+} t^{-1} \underline{b}(t)$. The components of $\underline{\lambda}$, being defined in terms of limits of probabilities, are nonnegative and satisfy (2.6). Thus, condition (i) above holds.

Next, suppose, on the contrary, that condition (ii) fails. Then for some i_0, $1 \le i_0 \le n$, $i_0 \notin I$ whenever $\lambda_I > 0$. Suppose that $i_0 = 1$. It follows from Lemma 2.5 after some straightforward calculation (see Remark 2.8 below) that $\mu_1 = 0$, a contradiction. ||

Remark 2.8. Assume the hypothesis and notation of Theorem 2.2 and let $I = \{i_1, \ldots, i_r\} \in \mathcal{I}$. From the proof of Theorem 2.2,

$$\lambda_I = \lim_{t \to 0^+} t^{-1} P(T_{i_1} \le t, \ldots, T_{i_r} \le t, T_{i_{r+1}} > t, \ldots, T_{i_n} > t).$$

By Lemma 2.6,

$$P(A_{i_1} \cdots A_{i_r} \bar{A}_{i_{r+1}} \cdots \bar{A}_{i_n}) = (-1)^{r-1}[P(\bigcup_{\nu=1}^{n} A_{i_\nu}) - \sum_{j=1}^{r} P(\bigcup_{\substack{\nu=1 \\ \nu \ne j}}^{n} A_{i_\nu}) + \sum_{\substack{j_1, j_2 = 1 \\ j_1 < j_2}}^{r} P(\bigcup_{\substack{\nu=1 \\ \nu \ne j_1, j_2}}^{n} A_{i_\nu}) - \cdots + (-1)^r P(\bigcup_{\nu=r+1}^{n} A_{i_\nu})]. \tag{2.7}$$

Let A_{i_ν} in (2.7) represent the event $(T_{i_\nu} \le t)$, $\nu = 1, \ldots, n$. Note that for every subset $\{j_1, \ldots, j_s\}$ of $\{1, \ldots, n\}$,

$$\lim_{t \to 0^+} t^{-1} P(\bigcup_{\substack{\nu=1 \\ \nu \ne j_1, \ldots, j_s}}^{n} A_{i_\nu}) = \lim_{t \to 0^+} t^{-1} P[\bigcup_{\substack{\nu=1 \\ \nu \ne j_1, \ldots, j_s}}^{n} (T_{i_\nu} \le t)]$$

$$= \lim_{t \to 0^+} t^{-1} P[\min_{\substack{1 \le \nu \le n \\ \nu \ne j_1, \ldots, j_s}} T_{i_\nu} \le t]$$

$$= \mu_{\overline{\{j_1, \ldots, j_s\}}}.$$

By dividing both sides of (2.7) by t and taking the limit as $t \to 0^+$, it follows that

$$\lambda_I = (-1)^{\operatorname{card}(I)-1}(\mu_{1\ldots n} - \sum_{i \in I} \mu_{\overline{\{i\}}} + \sum_{\substack{i_1, i_2 \in I \\ i_1 < i_2}} \mu_{\overline{\{i_1, i_2\}}} - \cdots + (-1)^{\operatorname{card}(I)} \mu_{\bar{I}}), \tag{2.8}$$

where card (I) is the cardinality of I. Formula (2.8) provides an <u>explicit</u> solution for the parameters λ_I in terms of the known constants μ_I. Note that in the case n = 3, the formula yields $\lambda_{12} = -\mu_3 + \mu_{13} + \mu_{23} - \mu_{123}$ as may be verified from Figure 2.1. In general, formula (2.8) represents the unique solution to any system of linear equations which is equivalent to the system of equations (2.6).

Formula (2.8) also provides a way to test the validity of the exponential minima assumption of Theorem 2.2 in certain situations. For example, suppose it is known _a priori_ that due to the particular structure of the system, it is impossible for the components in certain subsets to fail simultaneously, say subsets $I_1, \ldots, I_m$. This is tantamount to knowing that $\lambda_{I_i} = 0$, $i = 1, \ldots, m$. By formula (2.8), the corresponding m linear combinations of the μ's must be zero. In such a situation, if the m linear combinations of the (known) constants $\{\mu_I\}$ do _not_ all vanish, then the assumption that the distribution of $\underline{T}$ has exponential minima must be wrong. Similarly, if some linear combination of the constants $\{\mu_I\}$ in (2.8) is negative, then the assumption of exponential minima is likewise incorrect.

In Section 3 we use the techniques appearing in the proof of Theorem 2.2 to establish new results concerning multivariate distributions with specified marginals. We also obtain a direct generalization of Theorem 2.2. Our technique of proof can also be used to simplify results arising from _different_ multivariate shock models.

The Cumulative Damage Model. Esary and Marshall [2] characterize the following discrete multivariate distribution in terms analogous to those of Theorem 2.1 for the MVE. The positive integer-valued random variables $K_1, \ldots, K_n$ have the _multivariate geometric distribution in the narrow sense (MVG-N)_ if and only if there exists a collection $\{M_J, J \in \mathcal{J}\}$, $\mathcal{J} \in A$, of independent geometric random variables such that

$$K_i = \min_{J \varepsilon \mathcal{J}: i \varepsilon J} M_J, \qquad i = 1, \ldots, n.$$

Consider n components (devices) for which exposure to failure occurs in discrete cycles. In each cycle there is a shock from exactly one source, say I, $I \varepsilon \mathcal{I}$. A shock from source I causes the same amount of damage to each component indexed by $i \varepsilon I$ and has no effect on components not belonging to subset I. Assume that the damages are observations on a nonnegative random variable W, are mutually independent, and accumulate additively. When the accumulated damage of the ith component exceeds Y_i, its breaking threshold, the component fails, $i = 1, \ldots, n$. An interesting consequence of this model is:

Theorem 2.9. (Esary and Marshall). Suppose

(i) $K_i = \min\{k: W_1 + \ldots + W_k \geq Y_i\}$, $i = 1, \ldots, n$, where W_1, W_2, ... are i.i.d. as a nonnegative random variable W, and $\underline{Y} = (Y_1, \ldots, Y_n)$ has the MVE distribution;

(ii) $P(W > 0) > 0$;

(iii) $\underline{Y}$ is independent of $\{W_1, W_2, \ldots\}$; and

(iv) W is infinitely divisible.

Then $\underline{K} = (K_1, \ldots, K_n)$ has the MVG-N distribution.

Esary and Marshall [2] present a relatively tedious proof of this theorem. The following proof employs the elementary techniques used in proving Theorem 2.2.

Proof of Theorem 2.9. By Theorems 4.1 and 5.1 of Esary and Marshall [2], it suffices to find a nonnegative solution $\{\lambda_I, I \varepsilon \mathcal{I}\}$ to the system of equations:

$$m_I = \sum_{J \varepsilon \mathcal{I}: J \cap I \neq \emptyset} \lambda_J, \qquad I \varepsilon \mathcal{I},$$

where $m_I = -\log P(\min_{i \varepsilon I} K_i > 1)$. By the infinite divisibility of W, for each integer r, there esist r i.i.d. random variables

$W_1^{(r)}, \ldots, W_r^{(r)}$ such that W and $\sum_{i=1}^{r} W_i^{(r)}$ have the same distribution. Let $W^{(r)}$ have the same distribution as $W_1^{(r)}$. Let $K_i^{(r)}$ denote the number of cycles up to and including failure of the ith component, $i = 1, \ldots, n$, where damage is an observation on the random variable $W^{(r)}$. Suppose that $P(\min_{i\varepsilon I} Y_i > t) = \exp(-\mu_I t)$, $I \varepsilon \mathcal{I}$. It follows that for each positive integer r,

$$P(\min_{i\varepsilon I} K_i^{(r)} > 1) = \exp(-m_I/r), \tag{2.9}$$

where $m_I = -\log P(\min_{i\varepsilon I} K_i > 1)$. To verify (2.8), note that

$$P(\min_{i\varepsilon I} K_i^{(r)} > 1) = P(W^{(r)} < \min_{i\varepsilon I} Y_i)$$

$$= E[\exp(-\mu_I W^{(r)})] = \{E[\exp(-\mu_I W)]\}^{r^{-1}} = \exp(-m_I/r),$$

where $m_I = -\log E[\exp(-\mu_I W)] = -\log P(\min_{i\varepsilon I} K_i > 1)$. It follows from (2.9) that

$$\lim_{r\to\infty} r^{-1} P[\bigcup_{i\varepsilon I} (K_i^{(r)} \leq 1)] = m_I.$$

In the proof of Theorem 2.2, replace the event $(T_i \leq t)$ by the event $(K_i^{(r)} \leq 1)$. The remainder of the present proof is similar to the proof of Theorem 2.2. ||

3.0 *Applications*. In this section we show how our simplified proof of Theorem 2.2 may be applied to a variety of reliability models. In each case a model involving dependent random variables is replaced by an essentially equivalent model involving independent random variables. The reader will observe that the conclusion of each theorem in this section asserts the existence of a specific collection of random variables. The explicit distributions of the random variables in each collection are easily obtained by solving a system of equations equivalent to (2.6) via formula (2.8) (cf. Remark 2.8), although we do not explicitly display these distributions.

Every (univariate) survival probability $\overline{F}(\cdot)$ ($\equiv 1 - F(\cdot)$) may be expressed as $\overline{F}(t) = \exp[-R(t)]$, where $R(t)$ is a hazard function, i.e., a nonnegative, increasing function satisfying $R(0) = 0$ and $R(\infty) = \infty$. A very popular univariate survival probability in reliability whose hazard function has a particularly simple form is the Weibull survival probability:

$$\overline{F}(t) = \exp[-(\lambda t)^{\alpha}], \qquad t \geq 0,$$

where λ, $\alpha > 0$. When $\alpha = 1$, the survival probability is exponential. Note that the ratio of two Weibull hazard functions with fixed $\alpha > 0$ is constant.

The distribution of a nonnegative random vector $\underline{T}$ is said to have proportional hazard minima if for each $I \in \mathcal{I}$,

$$P(\min_{i \in I} T_i > t) = \exp[-\mu_I R(t)],$$

where $\mu_I > 0$, $I \in \mathcal{I}$, and $R(\cdot)$ is a (fixed) hazard function. Thus, a random vector $\underline{T}$ whose minima have the Weibull distribution with fixed $\alpha > 0$ has a distribution with proportional hazard minima. The following theorem generalizes Theorem 2.2 of Esary and Marshall:

Theorem 3.1. Suppose that the distribution of a random vector $\underline{T}$ has proportional hazard minima with hazard function $R(.)$. Suppose further that $R(\cdot)$ is continuous at $t_0 =$ $\sup\{t: R(t) = 0\}$. Then there exists a collection $\{S_J, J \in \mathcal{J}\}$, $\mathcal{J} \in A$, of independent random variables with hazard functions proportional to $R(\cdot)$ such that for each $I \in \mathcal{I}$,

$$\min_{i \in I} T_i \overset{st}{=} \min_{J \in \mathcal{J} : J \cap I \neq \emptyset} S_J .$$

Proof. Note that in the special case $R(t) = t$, the conclusion holds by Theorem 2.2. Suppose that for each $I \in \mathcal{I}$,

$$P(\min_{i \in I} T_i > t) = \exp[-\mu_I R(t)], \qquad t \geq 0, \tag{3.1}$$

where $\mu_I > 0$, $I \in \mathcal{I}$. It suffices to find a nonnegative solution $\{\lambda_I, I \in \mathcal{I}\}$ to the system of equations

$$\mu_I R(t) = \sum_{J \in \mathcal{I}: J \cap I \neq \emptyset} \lambda_J R(t), \qquad I \in \mathcal{I},$$

i.e.,

$$\mu_I = \sum_{J \in \mathcal{I}: J \cap I \neq \emptyset} \lambda_J, \qquad I \in \mathcal{I}.$$

It follows from (3.1) and the continuity of $R(t)$ at $t = t_0$ that

$$\lim_{t \to t_0} [R(t)]^{-1} P[\bigcup_{i \in I} (T_i \leq t)] = \mu_I.$$

The remainder of the present proof follows the proof of Theorem 2.2. ||

Consider now the following discrete model:

Model 3.2. Suppose that during a fixed interval of time [0, t] an n-component system is exposed to shocks which are not necessarily fatal. For each $I \in \mathcal{I}$, a shock of type I simultaneously affects all components in subset I and no other components. For example, the shock pattern for a two-component system might be exhibited as in Figure 3.1 below:

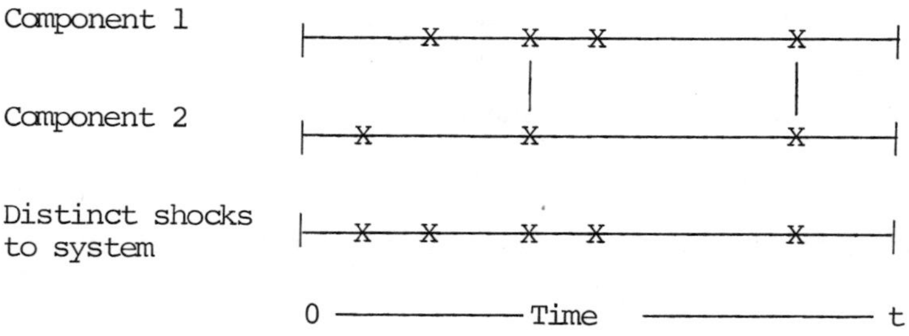

Figure 3.1

Figure 3.1 indicates that a total of 5 distinct shocks occurred in [0, t]: 2 shocks affected component 1 alone, 1 shock affected component 2 alone, and 2 shocks affected both components simultaneously. Let N_I denote the number of shocks in [0, t] simultaneously received by the components in subset I and by no other components. Let K_I denote the number of distinct shocks in [0, t] received by the components in subset I. Then

$$K_I = \sum_{J \cap I \neq \emptyset} N_J.$$

From Figure 3.1, $N_1 = 2$, $N_2 = 1$, $N_{12} = 2$, $K_1 = 4$, $K_2 = 3$, and $K_{12} = 5$. Note that $K_1 = N_1 + N_{12}$, $K_2 = N_2 + N_{12}$, and $K_{12} = N_1 + N_2 + N_{12}$. Thus, the random variables K_I, $I \varepsilon \mathcal{I}$, are, in general, dependent. Using the same type of proof as for Theorem 2.2, we prove

Theorem 3.3. Assume that Model 3.2 holds. Let K_I be a Poisson random variable with mean $\alpha_I > 0$, $I \varepsilon \mathcal{I}$. Then there exists a collection $\{N_J^*, J \varepsilon \mathcal{J}\}$ of independent Poisson random variables such that for every $I \varepsilon \mathcal{I}$,

$$K_I \overset{st}{=} \sum_{J \varepsilon \mathcal{J} : J \cap I \neq \emptyset} N_J^*.$$

Proof. It suffices to find a nonnegative solution to the system of equations

$$\alpha_I = \sum_{J \varepsilon \mathcal{I} : J \cap I \neq \emptyset} \beta_J, \qquad I \varepsilon \mathcal{I}, \tag{3.2}$$

since then we can define the collection $\{N_J^*, J \varepsilon \mathcal{J}\}$ to be independent Poisson random variables with respective means $\{\beta_J, J \varepsilon \mathcal{J}\}$, where $\mathcal{J} = \{J \varepsilon \mathcal{I} : \beta_J > 0\}$.

For each $I \varepsilon \mathcal{I}$, define β_I to be the mean number of shocks simultaneously received by the components in I. Then $\beta_I \geq 0$, $I \varepsilon \mathcal{I}$, and $\{\beta_I, I \varepsilon \mathcal{I}\}$ is a solution of (3.2). The matrix of coefficients of the β's in (3.2) is identical to the matrix X in (2.6). Since X is invertible, the solution to (3.2), given in Remark (2.8), is unique.||

Model 3.2 may be placed in a more general setting as follows:

Theorem 3.4. Let $\{T_I, I \varepsilon \mathcal{I}\}$ be a collection of random variables with the following properties:

(i) For each $I \varepsilon \mathcal{I}$, the distribution of T_I belongs to a family of distributions $\mathcal{F} = \{F_\theta, \theta \varepsilon \Theta\}$ such that $E(T_I) = \theta_I$; and

(ii) For each $\theta_1, \theta_2 \varepsilon \Theta$, $F_{\theta_1} * F_{\theta_2} = F_{\theta_1+\theta_2}$, where $*$ denotes convolution. Then there exists a collection $\{S_J, J \varepsilon \mathcal{J}\}$ of independent random variables such that the distribution of S_J belongs to $\mathcal{F}$, $J \varepsilon \mathcal{J}$, and for each $I \varepsilon \mathcal{I}$,

$$T_I \overset{st}{=} \sum_{J \varepsilon \mathcal{J}: J \cap I \neq \emptyset} S_J.$$

The following (continuous) reliability model yields an application of Theorem 3.4:

Model 3.5. Suppose that an n-component system is subject to repair (replacement). If any component fails, the system is down, and the repair (replacement) time for the component is recorded. Likewise, if any subset of components fails simultaneously, the system is down, and the time it takes to repair (replace) all of the components in that subset is recorded. For example, the operating record of a two-component system might be exhibited as in Figure 3.2 below:

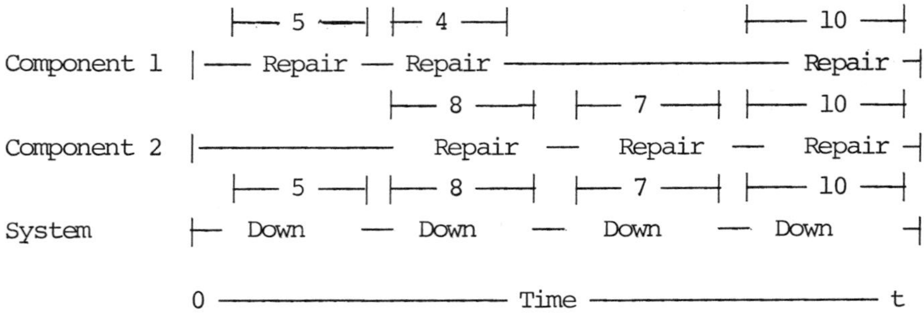

Figure 3.2

Figure 3.2 indicates that a total of 4 repair (replacement) periods occurred during [0, t]: 1 period was due to the failure of component 1 alone, 1 period was due to the failure of component 2 alone, and 2 periods were due to the simultaneous failure of components 1 and 2. Let D_I denote the system down time due to simultaneous failures of all the components in subset I. After each distinct failure of components in subset I, there is a corresponding system down time. Let T_I be the total of system down times corresponding to the distinct failures of one or more components in subset I. Then

$$T_I = \sum_{J \cap I \neq \emptyset} D_J.$$

From Figure 3.2, $D_1 = 5$, $D_2 = 7$, $D_{12} = 18$, $T_1 = 23$, $T_2 = 25$, and $T_{12} = 30$. First note that although the second failure of component 1 required only 4 units of repair (replacement) time, the system down time corresponding to that failure was 8 units of time since the system was down until both components were repaired (replaced). Next, note that $T_1 = D_1 + D_{12}$, $T_2 = D_2 + D_{12}$, and $T_{12} = D_1 + D_2 + D_{12}$. Thus, the random variables T_I, $I \in \mathcal{I}$, are, in general, dependent.

An immediate consequence of Theorem 3.4 is:

Corollary 3.6. Assume that Model 3.5 holds. Suppose

$$P(T_I > t) = \int_t^{\infty} \frac{x^{\alpha_I - 1}}{\Gamma(\alpha_I)} e^{-x}\, dx, \qquad t \geq 0, \quad I \in \mathcal{I},$$

where $\alpha_I > 0$, i.e., T_I has the gamma distribution with mean α_I and unit scale parameter. Then there exists a collection $\{D_J^*, J \in \mathcal{J}\}$ of independent gamma random variables such that for each $I \in \mathcal{I}$,

$$T_I \overset{st}{=} \sum_{J \in \mathcal{J} : J \cap I \neq \emptyset} D_J^* \quad .$$

REFERENCES

[1] Barlow, R. E. and Proschan, F., Statistical Theory of Reliability and Life Testing: Probability Models. Holt, Rinehart, and Winston, New York, (1975).

[2] Esary, J. D. and Marshall, A. W., Multivariate Geometric Distributions Generated by a Cumulative Damage Process. Naval Postgraduate School Technical Report, (1973).

[3] Esary, J. D. and Marshall, A. W., "Multivariate Distributions with Exponential Minimums", *Ann. Statist.*, *2*, (1974), pp. 84-98.

[4] Marshall, A. W. and Olkin, I., "A Multivariate Exponential Distribution", *J. Amer. Statist. Assoc.*, *62*, (1967), pp. 30-44.

[5] Rényi, A., Foundations of Probability, Holden-Day, New York, (1970).

STRUCTURAL INFERENCE ON RELIABILITY IN A LOGNORMAL MODEL

by DANNY DYER
University of Texas at Arlington

Abstract. The theory of structural inference, as developed by Fraser [10], is based on a group-theoretic approach using invariant Haar measures to Fisher's fiducial theory. Structural inference theory constructs a unique distribution, conditional on the given sample information only, for the parameters of a measurement model. Based on the structural density and distribution function for the reliability function are derived. Consequently, expressions for structural point and interval estimates of the reliability function are developed. Approximations for large sample sizes and/or moderately reliable components are also discussed. An example based on lognormal data is given to illustrate the theory.

1.0 *Introduction*. In life testing, the two-parameter lognormal density

$$f_T(t;\mu,\sigma)=\left[(2\pi)^{1/2}\sigma t\right]^{-1}\exp\left\{-\left[(\ln t-\mu)/\sigma\right]^2\Big/2\right\},\quad t,\ \sigma>0,\ -\infty<\mu<+\infty \tag{1.1}$$

has been widely used to represent the distribution of certain failure-times. The reliability function is

$$R(\tau;\mu,\sigma) = 1 - \Phi\left[(\ln \tau - \mu)/\sigma\right], \tag{1.2}$$

where τ is the "mission time", $\Phi(z) = \int_{-\infty}^{z}\phi(x)dx$, and $\phi(x) = (2\pi)^{-1/2}\exp(-x^2/2)$. Discussions of the lognormal distribution in reliability theory may be found in Epstein [6] - [7], Feinlieb [9], Goldthwaite [12], and Nowick and Berry [19]. It is felt, at times, the lognormal is a more realistic failure-time distribution than the normal which assigns positive probability to negative failure-times. Furthermore, even if a normal distribution seems appropriate, it might be closely approximated by a lognormal distribution. Kotz [16] and Klimko, et. al., [15] have derived tests for normality vs. lognormality. The lognormal and the (two-parameter) Weibull are often competitors when a skewed distribution for a non-negative random variable is needed. For certain components which become "work-hardened" *only* after a certain period of operating time, the lognormal might be a more appropriate failure-time distribution than the Weibull. This is due to the fact that the lognormal hazard function initially increases with time to a maximum then decreases with increasing time. The Weibull hazard function (for a given set of parameters) can only increase, decrease, or remain constant with time. Dumonceaux and Antle [5] provide a test for selecting between a lognormal model or a (two-parameter) Weibull model.

It is of considerable interest to construct point and interval estimates of (1.2) based on lognormal failure data. In most situations, by making a logarithmic transformation on the data, lognormal inferential problems may be reduced to corresponding normal inferential problems. Under the *classical* approach toward estimating (1.2) using transformed data from a complete sample of size n, the results are well known.

(a.) Point estimates of $R(\tau;\mu,\sigma)$. Because of the invariance property of maximum likelihood estimates, the maximum likelihood estimate of $R(\tau;\mu,\sigma)$ is given by

$$\hat{R}_{ML}(\tau;\mu,\sigma)$$

$$= 1 - \Phi[(\ln\tau - \hat{\mu})/\hat{\sigma}], \text{ where } \hat{\mu} = n^{-1}\sum_{i=1}^{n}\ln t_i \text{ and}$$

$$\hat{\sigma} = \left[n^{-1}\sum_{i=1}^{n}(\ln t_i - \hat{\mu})^2\right]^{1/2}.$$ Lieberman and Resnikoff [18] give a point estimate of $R(\tau;\mu,\sigma)$ which has the minimum variance unbiasedness property:

$$\hat{R}_{UMVU}(\tau;\mu,\sigma) = \begin{cases} 1 & , \text{ if } z(\hat{\mu},\hat{\sigma}) < 0 \\ 1 - I_{z(\hat{\mu},\hat{\sigma})}\left(\frac{n}{2} - 1, \frac{n}{2} - 1\right), & \text{if } 0 \leq z(\hat{\mu},\hat{\sigma}) \leq 1 \\ 0 & , \text{ if } z(\hat{\mu},\hat{\sigma}) > 1 , \end{cases} \quad (1.3)$$

where $\hat{\mu}$ and $\hat{\sigma}$ are the ML estimates, $I_z(p,q)$ is the incomplete beta function ratio, and

$$z(\hat{\mu},\hat{\sigma}) = (1/2)\ [1 + (\ln \tau - \hat{\mu})/(n-1)^{1/2}\hat{\sigma}] .$$

Zacks and Milton [23] have compared the mean square errors of $\hat{R}_{ML}(\tau;\mu,\sigma)$ and $\hat{R}_{UMVU}(\tau;\mu,\sigma)$ and found $\hat{R}_{UMVU}(\tau;\mu,\sigma)$ to be more efficient than $\hat{R}_{ML}(\tau;\mu,\sigma)$ only if $0.3 < R(\tau;\mu,\sigma) \leq 0.7$.

(b.) Lower confidence bound (LCB) for $R(\tau;\mu,\sigma)$. The problem of determining a LCB for $R(\tau;\mu,\sigma)$ may be solved by finding a lower tolerance limit on the normal distribution function. It can be shown (Zacks [22]), that a LCB for $R(\tau;\mu,\sigma)$ with confidence level $1 - \alpha$ is

$$\underline{R} = 1 - \Phi(-\lambda^*/\sqrt{n}) , \quad (1.4)$$

where λ^* is the value of the noncentrality parameter which satisfies the equation $\sqrt{n-1}\ \mathbf{a} = t'_{1-\alpha;n-1,\lambda}$,

where $\alpha = (\mu - \ln \tau)/\sigma$, and $t'_{1-\alpha;n-1,\lambda}$ is the $100(1-\alpha)$th percentile of a noncentral t-distribution with $n - 1$ degrees of freedom and parameter of noncentrality λ. For t', n, α given, the value of λ may be determined from tables in Owen [21].

In this paper we shall consider the structural approach to statistical inference on the reliability function $R(\tau;\mu,\sigma)$. The essentials of the structural model will be given in the next section. Based on the structural density for the two-parameter lognormal distribution, the structural density and distribution function for the reliability function are derived. Consequently, expressions for structural point and interval estimates of the reliability function are developed. Graphical displays of the expressions are given for a few special cases of interest. Approximations for large sample sizes and/or moderately reliable components are also presented. A numerical example based on lognormal data is given to illustrate the theory.

2.0 *The Structural Model.* The theory of structural inference, as developed by Fraser [10], is based on a group-theoretic approach using invariant Haar measures to Fisher's fiducial theory. Basically, the structural model may be described as follows. The model has two parts: an *error variable* **E** with a known distribution $f(E)$ on the sample space S (an open set in Euclidean space R^N) of the response X; and a *structural equation* $X = [\theta]E$ in which an observed response X in S is obtained by a transformation $[\theta]$ applied to a realized but unknown value from the error variable. Suppose the transformations $[\theta]$ are indexed by a parameter $[\theta]$ with values in a parameter space Ω and that the transformations $[\theta]$ are precisely the transformations of a group $G = \{g\}$ that is unitary in its application to the space S, i.e., if $g_1x = g_2x$ for any x, then $g_1 = g_2$.

The transformations g in G carry a point X into the orbit of X: $GX = \{gX : g \varepsilon G\}$. Suppose that a reference point is chosen on each orbit and let $[X]$ represent the unique transformation in G that carries the reference point on the orbit GX into the point X. The reference point for the orbit through X can then be designated by $D(X) = [X]^{-1}X$. An orbit is indexed by $D(X)$, while $[X]$ gives position on an orbit (see Fraser [10]).

The observed response X together with the structural equation $X = [\theta]E$ and the above assumptions imply that E must lie in the orbit of X, but its position $[E]$ is not known. However, the conditional distribution of $[E]$, given $D(X)$, (called the error probability distribution) may be found. The structural distribution is obtained from the error probability distribution by the transformation $[\theta] = [X][E]^{-1}$.

The structural probability element for θ (we now write θ instead of $[\theta]$ since the group G is isomorphic to the parameter space Ω) on the space G is (Fraser [10], p. 64)

$$g(\theta|X)d\theta = k([X]^{-1}X)f(\theta^{-1}X)J_N(\theta^{-1}X)\ \frac{J_L^*(\theta^{-1}[X])}{J_L(\theta^{-1}[X])}\ \frac{d\theta}{J_L(\theta)}, \quad (2.1)$$

where k is a normalizing constant, $J_N(\theta^{-1}X) = J_N(\theta^{-1}:X)J_N(X)$ $= J_N(\theta^{-1}:X)J_N([X]:D)$ is the Jacobian $\left|\frac{\partial\theta^{-1}X}{\partial X}\right|\cdot\left|\frac{\partial\ X D}{\partial D}\right| = \left|\frac{\partial\theta^{-1}[X]D}{\partial D}\right|$, $J_L(\theta^{-1}[X]) = J_L(\theta^{-1}[X]:i)$ is the Jacobian $\left|\frac{\partial\theta^{-1}[X]\ i}{\partial i}\right|$ associated with a left invariant Haar measure and i is the identity in G, and $J_L^*(\theta^{-1}[X])$ is the Jacobian $\frac{\partial i\theta^{-1}[X]}{\partial i}$ associated with a right invariant Haar measure.

Consider the group of affine transformations on R^n, $G = \{[\mu,\sigma]: -\infty<\mu<+\infty, 0<\sigma<+\infty\}$, where $[\mu,\sigma]\underline{x} = [\mu,\sigma](x_1,\ldots,x_n) = (\mu+\sigma x_1,\ldots,\mu+\sigma x_n)$, i.e., the group of location-scale transformations. The product of two affine transformations is $[A,C][a,c] = [A + Ca, Cc]$, the identity is $[0,1]$, and the inverse is $[a,c]^{-1} = [-c^{-1}a, c^{-1}]$. Under the observed response $X = \underline{x} = (x_1,x_2,\ldots,x_n)$ and $[X] = [\bar{x},s_x]$, where $\bar{x} = n^{-1}\sum_{i=1}^{n} x_i$

and $s_x = \left[n^{-1} \sum_{i=1}^{n} (x_i - \bar{x})^2\right]^{1/2}$, we find

$$J_N(\theta^{-1}X) = \left|\frac{\partial [\mu,\sigma]^{-1}[\bar{x},s_x]\underline{d}}{\partial \underline{d}}\right| = \left|\frac{\partial(-\frac{\mu}{\sigma} + \frac{\bar{x}}{\sigma} + \frac{s_x}{\sigma}\underline{d})}{\partial \underline{d}}\right| = \left(\frac{s_x}{\sigma}\right)^n ,$$

$$J_L{*}(\theta^{-1}[X]) = \left|\frac{\partial [a,c][\mu,\sigma]^{-1}\bar{x},s_x}{\partial [a,c]}\right| = \left|\frac{\partial[a+c\left(-\frac{\mu}{\sigma} + \frac{\bar{x}}{\sigma}\right),c\frac{s_x}{\sigma}}{\partial [a,c]}\right| = \frac{s_x}{\sigma} ,$$

$$J_L(\theta^{-1}[X]) = \left|\frac{\partial [\mu,\sigma]^{-1}[x,s_x][a,c]}{\partial [a,c]}\right| = \left|\frac{\partial[-\frac{\mu}{\sigma} + \frac{\bar{x}}{\sigma} + a\frac{s_x}{\sigma}, c\frac{s_x}{\sigma}]}{\partial [a,c]}\right| = \left(\frac{s_x}{\sigma}\right)^2 ,$$

$$J_L(\theta) = \left|\frac{\partial [\mu,\sigma][a,c]}{\partial [a,c]}\right| = \left|\frac{\partial [\mu + a\sigma, c\sigma]}{\partial [a,c]}\right| = \sigma^2,$$

where $\underline{d} = [\bar{x},s_x]^{-1}\underline{x} = \left(\frac{x_1 - \bar{x}}{s_x}, \ldots, \frac{x_n - \bar{x}}{s_x}\right)$.

We therefore find the structural probability element for μ and σ, given the observed value of $\underline{x}$, to be

$$g(\mu,\sigma|\underline{x})d\mu d\sigma = k(\underline{d}) \prod_{i=1}^{n} f\left(\frac{x_i - \mu}{\sigma}\right)\left(\frac{s_x}{\sigma}\right)^n \frac{\frac{s_x}{\sigma}}{\left(\frac{s_x}{\sigma}\right)^2} \frac{d\mu d\sigma}{\sigma^2} \tag{2.2}$$

$$= K \prod_{i=1}^{n} \left[\frac{1}{\sigma} f\left(\frac{x_i - \mu}{\sigma}\right)\right]\frac{d\mu d\sigma}{\sigma} , \tag{2.3}$$

where K is a normalizing constant.

Equation (2.3) will be the basis for the derivations in this paper. In essence, under the structural approach, the parameters of location-scale families of distributions are treated as *random variables*, their joint distribution being determined by information from the sample. Consequently, a classical model and a structural model are *fundamentally different* with regard to mathematical structure. In applications, they function differently, and different interpretations are given their results. Nevertheless, it is well known (Hora and Buehler [13]) that within a certain class (the invariantly estimable functions) of interval estimation problems for families of distributions exhibiting location-scale structure, there is a "numerical equivalency" between the structural solution and classical (confidence interval) solution.

It should also be pointed out that a Bayesian interpretation of (2.3) may be given. The structural density of μ and σ is precisely the Bayes *posterior* distribution when the prior measure is given by a right invariant Haar measure of $[\theta] = [\mu, \sigma]$, i.e., $\pi(\mu,\sigma) \propto 1/\sigma$. Similarities and dissimilarities of the classical, Bayes, and structural approaches to statistical inference are discussed in Fraser [11]. Applications of the theory of structural inference to other life-testing distributions may be found in Bury and Bernholtz (one-parameter exponential model [4]) and Bury (two-parameter Weibull model [3]).

3.0 *Structural Distribution of the Reliability Function.* The lognormal density (1.1) does not have location-scale structure. However, the density of the logarithmic transformation of the measurement variable T (which is, of course, a normal density) does. Consequently, the structural joint density of the location and scale parameters μ and σ is

$$f(\mu,\sigma|a(\underline{x})) = c \prod_{i=1}^{n} [(2\pi)^{1/2}\sigma]^{-1} \exp\{-[(x_i-\mu)/\sigma]^2/2\}\cdot\sigma^{-1}, \quad \begin{matrix} -\infty<\mu<+\infty \\ \sigma>0 \end{matrix} \qquad (3.1)$$

where $a(\underline{x})$ represents the sample information on which (3.1) is conditional, $x_i = \ln t_i$, and c is a normalizing constant. It should be noted that in the above density, μ and σ are random variables while the observations $\{x_i\}$ are given constants. The structural joint density of μ and σ, conditioned by $a(\underline{t})$, is

$$f(\mu,\sigma|a(\underline{t})) = c\sigma^{-n-1}(2\pi)^{-n/2}\exp\left[-\sum_{i=1}^{n}(\ln t_i-\mu)^2/2\sigma^2\right], \quad \begin{matrix}-\infty<\mu<+\infty\\ \sigma>0\end{matrix} \qquad (3.2)$$

where

$$c^{-1} = \int_0^\infty \int_{-\infty}^{\infty} f(\mu,\sigma|a(\underline{t}))\,d\mu d\sigma$$

$$= (2\pi)^{-n/2}\int_0^\infty\left(\int_{-\infty}^{\infty}\exp\left[-n(\mu-n^{-1}\sum_{i=1}^{n}\ln t_i)^2/2\sigma^2\right]d\mu\right)\sigma^{-n-1}\exp(-ns^2/2\sigma^2)d\sigma$$

$$= \left[(2\pi)^{(n-1)/2}n^{1/2}\right]^{-1}\int_0^\infty \sigma^{-n}\exp(-ns^2/2\sigma^2)d\sigma$$

$$= \Gamma\left(\frac{n-1}{2}\right)\left[2n^{1/2}(\pi ns^2)^{(n-1)/2}\right]^{-1},$$

and $$ns^2 = \sum_{i=1}^{n}(\ln t_i - \hat{\mu})^2 .$$

Upon comparing (3.1) and (3.2), it might appear that we are not actually dealing with a lognormal distribution but, in fact, a normal distribution. In a general setting this is true. However, since this paper is concerned with applications in reliability theory, there is the problem of the normal distribution assigning positive probabilities to negative lifetimes. In a general setting, the range of μ in (3.1) is the real line. However in a reliability theory setting, it is *not* clear what range should be given μ (it would seem that the range of μ should be some subset of the positive real line). We therefore shall only use (3.1) to reach (3.2), and all results will be based on (3.2). Of course, *in a general setting*, the results obtained by using (3.2) can be carried over to a normal distribution upon replacing $\ln t_i$ by x_i.

Since μ and σ are considered as random variables, the reliability function (1.2) is also a random variable.

Theorem 3.1. The structural density function of the reliability function $R(\tau;\mu,\sigma) = 1 - \Phi[(\ln\tau - \mu)/\sigma]$ is, for $n \geq 2$, (parabolic cylinder function representation)

$$f_R(r \mid \tau, a(\underline{t})) = \frac{\sqrt{n}\,\pi\,\Gamma(n-1)}{2^{(n-5)/2}\Gamma\left(\frac{n-1}{2}\right)(1+a^2)^{(n-1)/2}} \tag{3.3}$$

$$\times\ \phi\left(\sqrt{(1-1/n)(1+1/a^2)}\,\omega\right)\phi\left(\omega/\sqrt{-2}\right)D_{-n+1}(\omega),\quad 0<r<1$$

or

(infinite series representation)

$$f_R(r|\tau, a(\underline{t})) = \frac{\sqrt{n}\,\sqrt{2\pi}}{\Gamma\left(\frac{n-1}{2}\right)(1+a^2)^{(n-1)/2}} \tag{3.4}$$

$$\times\ \phi\left(\sqrt{(1-1/n)(1+1/a^2)}\,\omega\right)\sum_{j=0}^{\infty}\frac{(-1)^j\,\Gamma\left(\frac{j+n-1}{2}\right)}{j!}(\sqrt{2}\omega)^j,\quad 0<r<1$$

where $\omega = \sqrt{n}\Phi^{-1}(1-r)a/\sqrt{1+a^2}$, $D_{-n+1}(\omega)$ is the parabolic cylinder function, and $a = (\hat{\mu} - \ln\tau)/\hat{\sigma}$.

Proof: The structural density of R is a derived distribution from (3.2). We have

$$f_R(r|\tau, a(\underline{t})) = \int_0^{\infty} f(\ln\tau - \sigma\,\Phi^{-1}(1-r), \sigma|a(\underline{t}))\cdot\left|\frac{d(\ln\tau - \sigma\,\Phi^{-1}(1-r))}{dr}\right| d\sigma$$

$$= \frac{\sqrt{n}\,\sqrt{2\pi}\,\phi(\sqrt{(1-1/n)(1+1/a^2)}\,\omega)}{2^{(n-3)/2}\Gamma\left(\frac{n-1}{2}\right)(1+a^2)^{(n-1)/2}}$$

$$\times\int_0^{\infty} y^{n-2}\exp(-y^2/2 - \omega y)\,dy\ , \tag{3.5}$$

where ω and a are as defined in the theorem. We have also used the result that $(d/dr)\Phi^{-1}(1-r) = -\{\phi[\Phi^{-1}(1-r)]\}^{-1}$.

The integral on the right-hand side of (3.5) may be written as $\sqrt{2\pi}\,\Gamma(n-1)\,\phi(\omega/\sqrt{-2})\,D_{-n+1}(\omega)$, for $n \geq 2$, where $D_\gamma(x)$ is the parabolic cylinder function (Erdélyi, et. al., II [8].

Lebedev [17] has shown, through the use of Hermite functions, that a parabolic cylinder function may be represented by the series

$$D_{-n+1}(x) = \frac{2^{(n-3)/2}\exp(-x^2/4)}{\Gamma(n-1)} \sum_{j=0}^{\infty} \frac{(-1)^j\,\Gamma\left(\frac{j+n-1}{2}\right)}{j!} (\sqrt{2}x)^j$$

thus completing the proof.

To find the structural distribution function of R, we shall use the series representation of the structural density function of R. When $0 \leq r \leq 1/2$,

$$F_R(r|\tau,a(\underline{t})) = \frac{\sqrt{n}}{\Gamma\left(\frac{n-1}{2}\right)(1+a^2)^{(n-1)/2}} \sum_{j=0}^{\infty} \frac{\Gamma\left(\frac{j+n-1}{2}\right)}{j!} \left(\frac{-\sqrt{2n}\,a}{\sqrt{1+a^2}}\right)^j \xi_j(0,r),$$

where

$$\xi_j(a,b) = \int_a^b [\Phi^{-1}(1-t)]^j \exp\{-(n-1)[\Phi^{-1}(1-t)]^2/2\}dt.$$

Making the change of variable $y = n[\Phi^{-1}(1-t)]^2$, we have

$$\xi_j(0,r) = \frac{2^{(j-2)/2}\Gamma\left(\frac{j+1}{2}\right)}{\sqrt{\pi}\,n^{(j+1)/2}} \{1 - \chi^2_{j+1}(n[\Phi^{-1}(1-r)]^2)\},$$

where $\chi^2_k(x)$ is the chi-square distribution function with k degrees of freedom.

Upon using the duplication formula

$$\Gamma(2x) = (2\pi)^{-1/2}\,2^{2x-1/2}\,\Gamma(x)\,\Gamma(x+1/2),$$

we have

$$F_R(r|\tau,a(\underline{t})) = \frac{1}{2\Gamma\left(\frac{n-1}{2}\right)(1+a^2)^{(n-1)/2}} \times \tag{3.6}$$

$$\sum_{j=0}^{\infty} \frac{\Gamma\left(\frac{j+n-1}{2}\right)}{\Gamma\left(\frac{j}{2}+1\right)} \left(\frac{-a}{\sqrt{1+a^2}}\right)^j \{1 - \chi^2_{j+1}\ (n[\Phi^{-1}(1-r)]^2)\ \}.$$

When $1/2 \le r \le 1$,

$$F_R(r|\tau,a(\underline{t})) = F_R(1/2|\tau,a(\underline{t}))$$

$$+ \frac{\sqrt{n}}{\Gamma\left(\frac{n-1}{2}\right)(1+a^2)^{(n-1)/2}} \sum_{j=0}^{\infty} \frac{\Gamma\left(\frac{j+n-1}{2}\right)}{j!} \left(\frac{-\sqrt{2n}\ a}{\sqrt{1+a^2}}\right)^j \xi_j(1/2,r).$$

Making the change of variable $y = n[\Phi^{-1}(1-t)]^2$, we find

$$\xi_j(1/2,r) = \frac{(-1)^j\ 2^{j/2}\ \Gamma\left(\frac{j+1}{2}\right)}{\sqrt{\pi}\ n^{(j+1)/2}} \chi^2_{j+1}\left(n[\Phi^{-1}(1-r)]^2\right),$$

so that

$$F_R(r|\tau,a(\underline{t})) = F_R(1/2|\tau,a(\underline{t}))$$

$$+ \frac{1}{2\Gamma\left(\frac{n-1}{2}\right)(1+a^2)^{(n-1)/2}} \sum_{j=0}^{\infty} \frac{\Gamma\left(\frac{j+n-1}{2}\right)}{\Gamma\left(\frac{j}{2}+1\right)} \left(\frac{a}{\sqrt{1+a^2}}\right)^j \chi^2_{j+1}(n[\phi^{-1}(1-r)]^2). \tag{3.7}$$

The value of the distribution function at $r = 1/2$ will be of special interest. From (3.6),

$$F_R(1/2|\tau,a(\underline{t})) = \frac{1}{2\Gamma\left(\frac{n-1}{2}\right)(1+a^2)^{(n-1)/2}} \sum_{j=0}^{\infty} \frac{\Gamma\left(\frac{j+n-1}{2}\right)}{\Gamma\left(\frac{j}{2}+1\right)} \left(\frac{-a}{\sqrt{1+a^2}}\right)^j$$

$$= \frac{1}{2\Gamma\left(\frac{n-1}{2}\right)(1+a^2)^{(n-1)/2}} \left\{ \sum_{j=0}^{\infty} \frac{\Gamma\left(j+\frac{n-1}{2}\right)}{\Gamma(j+1)} \left(\frac{a^2}{1+a^2}\right)^j \right.$$

$$- \frac{a}{\sqrt{1 + a^2}} \sum_{j=0}^{\infty} \frac{\Gamma\left(j + \frac{n}{2}\right)}{\Gamma\left(j + \frac{3}{2}\right)} \left(\frac{a^2}{1 + a^2}\right)^j . \tag{3.8}$$

Each series on the right-hand side of (3.8) may be written in terms of a hypergeometric function

$${}_2F_1(\alpha, \beta; \gamma; z) = \sum_{k=0}^{\infty} \frac{(\alpha)_k (\beta)_k}{(\gamma)_k\, k!} z^k , \quad |z| < 1$$

where $(\alpha)_k = \Gamma(\alpha + k)/\Gamma(\alpha)$. The first series is (Erdélyi, et. al., I, [8].

$$\sum_{j=0}^{\infty} \frac{\Gamma\left(j + \frac{n-1}{2}\right)}{\Gamma(j + 1)} \left(\frac{a^2}{1 + a^2}\right)^j = \Gamma\left(\frac{n - 1}{2}\right) {}_2F_1\left((n-1)/2, \beta; \beta;\ a^2/(1+a^2)\right) \tag{3.9}$$

$$= \Gamma\left(\frac{n - 1}{2}\right)[1 - a^2/(1 + a^2)]^{-(n-1)/2} = \Gamma\left(\frac{n - 1}{2}\right)(1 + a^2)^{(n-1)/2}.$$

In order to evaluate the second series, we first establish a relationship between the hypergeometric function and an F-distribution function. It is well known that

$$1 - F_{\gamma_1, \gamma_2}(y) = I_x(\gamma_2/2, \gamma_1/2) , \tag{3.10}$$

where $F_{\gamma_1, \gamma_2}(y)$ is the F-distribution function with γ_1 and γ_2 degrees of freedom, $x = \gamma_2/(\gamma_2 + \gamma_1 y)$, and $I_x(\alpha, \beta)$ is the incomplete beta function ratio. However

$$I_x(\gamma_2/2, \gamma_1/2) = \frac{(\gamma_2/2)^{-1} x^{\gamma_2/2}}{B(\gamma_1/2, \gamma_2/2)} {}_2F_1(\gamma_2/2,\ 1 - \gamma_1/2; 1 + \gamma_2/2; x)$$

$$= \frac{(\gamma_2/2)^{-1} x^{\gamma_2/2} (1-x)^{\gamma_1/2}}{B(\gamma_1/2, \gamma_2/2)} {}_2F_1(1, (\gamma_1+\gamma_2)/2; 1+\gamma_2/2; x) \tag{3.11}$$

by a linear transformation of the hypergeometric function (Erdélyi, et. al., I [8]. Combining (3.10) and (3.11) yields the desired result

$$
{}_2F_1(1,(\gamma_1+\gamma_2)/2;1+\gamma_2/2;x) = \frac{B(\gamma_1/2,\gamma_2/2)\,(\gamma_2/2)}{x^{\gamma_2/2}(1-x)^{\gamma_1/2}}\,[1 - F_{\gamma_1,\gamma_2}(y)], \tag{3.12}
$$

where $y = \gamma_2(1-x)/\gamma_1 x$, $x \neq 0$. The second series on the right-hand side of (3.8) is now

$$
\sum_{j=0}^{\infty} \frac{\Gamma\left(j+\frac{n}{2}\right)}{\Gamma\left(j+\frac{3}{2}\right)}\left(\frac{a^2}{1+a^2}\right)^j = \frac{\Gamma\left(\frac{n}{2}\right)}{\Gamma\left(\frac{3}{2}\right)}\,{}_2F_1(1,n/2;3/2;\ a^2/(1+a^2))
$$

$$
= \begin{cases} \dfrac{\Gamma\left(\frac{n-1}{2}\right)(1+a^2)^{n/2}}{|a|}\,F_{1,n-1}[(n-1)a^2], & a \neq 0 \\[2ex] \Gamma(n/2)/\Gamma(3/2), & a = 0. \end{cases} \tag{3.13}
$$

Finally, upon substituting (3.9) and (3.13) into (3.8), we have the value of the distribution function at $r = 1/2$,

$$
F_R(1/2\,|\,\tau,a(\underline{t})) = (1/2)\{\,1 - \mathrm{sgn}(a)\ F_{1,n-1}[(n-1)a^2]\}, \tag{3.14}
$$

where $\mathrm{sgn}(a) = 1$, $a > 0$; $= -1$, $a < 0$; $= 0$, $a = 0$.

By combining (3.6), (3.7), and (3.14), we may now write the following

Theorem 3.2. The structural distribution function of the reliability function $R(\tau;\mu,\sigma)$ is, for $0 \le r \le 1$,

$$
F_R(r\,|\,\tau,a(\underline{t})) = (1/2)\ \{1 - \mathrm{sgn}(a)F_{1,n-1}[(n-1)a^2]\}
$$

$$
+ \frac{\mathrm{sgn}(r-1/2)}{(1+a^2)^{(n-1)/2}} \sum_{j=0}^{\infty} \frac{\Gamma\left(\frac{j+n-1}{2}\right)}{\Gamma\left(\frac{j}{2}+1\right)\Gamma\left(\frac{n-1}{2}\right)}\left[\frac{\mathrm{sgn}(r-1/2)a}{\sqrt{1+a^2}}\right]^j \chi^2_{j+1}\{n[\Phi^{-1}(1-r)]^2\}. \tag{3.15}
$$

Since the value of r in (3.15) is unknown, the following lemma will be useful. The proof is straightforward.

Lemma. Let a^* be the solution for a in the equation $F_R(1/2|\tau, a(\underline{t})) = \alpha (\alpha \leq 1/2)$. For fixed α and n, $\text{sgn}(r - 1/2) = 1$, i.e., $1/2 < r \leq 1$, iff $a > a^* = [(n-1)^{-1} F^{-1}_{1,n-1}(1 - 2\alpha)]^{1/2}$.

4.0 Structural Inference on the Reliability Function. We shall now use the distributions of Section 3 to obtain point and interval estimates of the reliability function $R(\tau;\mu,\sigma)$.

(i.) *Point estimates.* Measures of central tendency of the structural density would seem natural choices for point estimates. Furthermore, because of the Bayesian interpretation discussed in Section 2, the mean, median, and mode of the structural density correspond to the Bayesian point estimate of $R(\tau;\mu,\sigma)$ using a squared error, absolute error, and simple loss function, respectively, and a prior measure given by $\pi(\mu,\sigma) \propto 1/\sigma$ (a right invariant Haar measure).

Theorem 4.1. Based on the structural distribution of the reliability function $R(\tau;\mu,\sigma)$, the *mean* of the distribution of $R(\tau;\mu,\sigma)$ is

$$E[R(\tau;\mu,\sigma)] = (1/2)\left\{ 1 + \frac{a}{(1+a^2)^{n/2}} \times \sum_{j=0}^{\infty} \frac{\Gamma(j + n/2)}{\Gamma(j+3/2)\,\Gamma\left(\frac{n-1}{2}\right)} \left(\frac{a^2}{1+a^2}\right)^j F_{1,2(j+1)}[2(j+1)/n] \right\}; \tag{4.1}$$

the *median* of the distribution of $R(\tau;\mu,\sigma)$ is the value of r which satisfies the equation

$$r = 1 - \Phi\left[\frac{-\,\text{sgn}(a)}{\sqrt{n}}\,\Phi^{-1}\left(\frac{1+A}{2}\right)\right], \tag{4.2}$$

where

$$A = (1 + a^2)^{(n-1)/2} F_{1,n-1}[(n-1)a^2]$$

$$- \sum_{j=1}^{\infty} \frac{\Gamma\left(\frac{j+n-1}{2}\right)}{\Gamma(\frac{j}{2}+1)\Gamma\left(\frac{n-1}{2}\right)} \frac{\text{sgn}(a)a^j}{\sqrt{1+a^2}} \chi^2_{j+1}(n[\Phi^{-1}(1-r)]^2) ;$$

the *mode* of the distribution of $R(\tau;\mu,\sigma)$, r_{mode}, is given by

$$r_{mode} = 1 - \Phi\left(\frac{\sqrt{1+a^2}}{\sqrt{n}\,a}\,\xi\right), \tag{4.3}$$

where ξ is the solution to the equation

$$D_{-n+2}(\xi) = [n(1+a^2)^{-1} - 1]\; D_{-n}(\xi) . \tag{4.4}$$

Proof. Since $R(\tau;\mu,\sigma)$ is a positive random variable,

$$E[R(\tau;\mu,\sigma)] = \int_0^1 [1 - F_R(r|\tau,a(\underline{t}))]\; dr$$

$$= (1/2)\;\{1 + \text{sgn}(a)\, F_{1,n-1}\;[(n-1)a^2]\;\}$$

$$- \frac{a}{(1+a^2)^{n/2}} \sum_{j=0}^{\infty} \frac{\Gamma(j+n/2)}{\Gamma(j+3/2)\;\Gamma\left(\frac{n-1}{2}\right)} \left(\frac{a^2}{1+a^2}\right)^j \int_0^{1/2} \chi^2_{2j+2}(n[\Phi^{-1}(1-r)]^2)dr.$$

Using the relationship between the chi-squared distribution function and the incomplete gamma function $\Gamma(\alpha,x) = \int_x^{\infty} t^{\alpha-1}e^{-t}dt$, i.e., $\chi^2_{2j+2}(y) = 1 - \Gamma\{j+1,\, n[\phi^{-1}(1-r)]^2/2\}/\Gamma(j+1)$, we have

$$\int_0^{1/2} \chi^2_{2j+2}(n[\Phi^{-1}(1-r)]^2)\; dr$$

$$= 1/2 - \frac{1}{\Gamma(j+1)} \int_0^{1/2} \Gamma\{j+1,\; n[\Phi^{-1}(1-r)]^2/2\}\, dr$$

$$= 1/2 - \frac{1}{2\sqrt{n\pi}\,\Gamma(j+1)} \int_0^\infty y^{-1/2} \exp(-y/n)\,\Gamma(j+1, y)\,dy$$

$$= 1/2 - \frac{\Gamma(j+3/2)}{\sqrt{n\pi}\,\Gamma(j+1)(1+1/n)^{j+3/2}}\ {}_2F_1\left(1, j+3/2; 3/2; 1/(n+1)\right)$$

$$= 1/2 \left\{ 1 - F_{1,2(j+1)} \left[2(j+1)/n \right] \right\}$$

using (3.12). The result follows from (3.13).

The median of the distribution of $R(\tau;\mu,\sigma)$ is the value of r which satisfies the equation $(F_R\, r|\tau, a(\underline{t})) = 1/2$. The proof of (4.2) will be given in a slightly more general version toward the end of this section.

The mode of the distribution of $R(\tau;\mu,\sigma)$ is the value of r, say r_{mode}, which maximizes the structural density. We first differentiate (3.3) with respect to r using the following expression for the parabolic cylinder function from Erdelyi, et. al., II [8], p. 119

$$(d/dx)\ [\exp(x^2/4) D_\gamma(x)] = \gamma \exp(x^2/4) D_{\gamma-1}(x).$$

Equating the first derivative to zero yields

$$\Phi^{-1}(1-r)\, D_{-n+1}\left(\sqrt{n}\, \Phi^{-1}(1-r) a/\sqrt{1+a^2} \right)$$
$$+ (\sqrt{n}\, a/\sqrt{1+a^2})\, D_{-n}\left(\sqrt{n}\, \Phi^{-1}(1-r) a/\sqrt{1+a^2} \right) = 0$$

or equivalently,

$$\omega D_{-n+1}(\omega) + [na^2/(1+a^2)]\, D_{-n}(\omega) = 0,$$

where $\omega = \sqrt{n}\, \Phi^{-1}(1-r) a/\sqrt{1+a^2}$. By using the recurrence relation

$$D_{\gamma+1}(\omega) - \omega D_\gamma(\omega) + \gamma D_{\gamma-1}(\omega) = 0,$$

the proof is completed.

A few remarks concerning Theorem 4.1 are in order. In most cases, only a few terms are needed to evaluate the sum in equation (4.1). For the convenience of the reader, Figure 1 graphically displays na vs. $E[R(\tau;\mu,\sigma)]$ for $n = 5(5)25$. Equation (4.2) can be solved iteratively by using the value of $E[R(\tau;\mu,\sigma)]$ as a starting value. Again, in most instances, only a few terms are needed to evaluate the sum. Equation (4.4) can be readily solved using tables of the parabolic cylinder function given in Abramowitz and Stegun [1], pp. 702-711.

(ii.) *Interval estimates*. Since the reliability function $R(\tau;\mu,\sigma)$ is a random variable we wish to determine $r(1 - \alpha)$ where $P[\underline{r}(1-a) < R(\tau;\mu,\sigma)] = 1 - \alpha$, i.e., $F_R(\underline{r}(1-\alpha)|\tau,a(\underline{t})) = \alpha$, for some preassigned α. We shall call $\underline{r}(1 - \alpha)$ the $(1 - \alpha)$ - *structural lower bound* for $R(\tau;\mu,\sigma)$. It is, in this case, "numerically equivalent" to the $(1 - \alpha)$-Bayes lower confidence bound under the prior right invariant Haar measure and to the $(1 - \alpha)$-classical (in the sense of Neyman) lower confidence bound for $R(\tau;\mu,\sigma)$ given by (1.4).

Theorem 4.2. Based on the structural distribution of the reliability function $R(\tau;\mu,\sigma)$, the $(1 - \alpha)$-structural lower bound for $R(\tau;\mu,\sigma)$ is the solution to the equation

$$r = 1 - \Phi\left\{-\text{sgn}(r - 1/2)\Phi^{-1}\left[(1+ A^*)/2\right]/\sqrt{n}\right\}, \tag{4.5}$$

where

$$A^* = (1+a^2)^{(n-1)/2}\text{sgn}(r-1/2)\left\{\text{sgn}(a)F_{1,n-1}[(n-1)a^2] - (1-2\alpha)\right\}$$

$$- \sum_{j=1}^{\infty} \frac{\Gamma\left(\frac{j+n-1}{2}\right)}{\Gamma\left(\frac{j}{2}+1\right)\Gamma\left(\frac{n-1}{2}\right)}\left(\frac{\text{sgn}(r-1/2)a}{\sqrt{1+a^2}}\right)^j \chi^2_{j+1}\left(n[\Phi^{-1}(1-r)]^2\right) \tag{4.6}$$

Proof. The equation $F_R(r|\tau,a(\underline{t})) = \alpha$ may be written

$$\text{sgn}(a)F_{1,n-1}[(n-1)a^2] - (1-2\alpha)$$

$$= \frac{\operatorname{sgn}(r - 1/2)}{(1+a^2)^{(n-1)/2}} \sum_{j=0}^{\infty} \frac{\Gamma\left(\frac{j+n-1}{2}\right)}{\Gamma\left(\frac{j}{2}+1\right)\Gamma\left(\frac{n-1}{2}\right)} \left(\frac{\operatorname{sgn}(r-1/2)\ a}{\sqrt{1+a^2}}\right)^j \chi_{j+1}^2\left(n\left[\Phi^{-1}(1-r)\right]^2\right)$$

or equivalently,

$$\chi_1^2\left(n\left[\Phi^{-1}(1 - r)\right]^2\right) = A^* > 0, \qquad (4.7)$$

where A^* is given by (4.6). But it can be shown that if $\chi_1^2(y) = A^*$, then $\sqrt{y} = \Phi^{-1}[(1 + A^*)/2]$. Consequently, (4.7) becomes $|\Phi^{-1}(1 - r)| = \Phi^{-1}[(1 + A^*)/2]/\sqrt{n}$ so that $\Phi^{-1}(1 - r) = -\operatorname{sgn}(r - 1/2)\Phi^{-1}[(1 + A^*)/2]\sqrt{n}$. Equation (4.5) now follows.

Given the data, i.e., the values of n and a, the lemma of Section 3 indicates the value of $\operatorname{sgn}(r - 1/2)$ for specified α. Equation (4.5) can then be solved iteratively using any one of the three point estimates of Theorem 4.1 as a starting value. In most cases, only the first few terms in the series in (4.6) need be computed. Figures 2 and 3 graphically display a **vs.** $\underline{r}(1 - \alpha)$ **for** $\alpha = 0.10, 0.05$ and $n = 5(5)25$.

<u>5.0 *Approximations*</u>. The expressions for the structural point and interval estimates given in the previous section may be evaluated with the aid of existing tables and/or a computer. It would, however, be desirable to find approximations to these estimates which are both accurate and easier to calculate.

The starting point will be the approximation of the structural density of $R(\tau;\mu,\sigma)$ given by (3.3) and, in particular, the parabolic cylinder function $D_{-n+1}(\omega)$, where $\omega = \sqrt{n}\,\Phi^{-1}(1 - r)a/\sqrt{1 + a^2}$. For the moment, we shall focus our attention on a large-sample approximation. For convenience, we shall use the alternate (and more contemporary) expression

$$U(n - 3/2,\ \omega) = D_{-n+1}(\omega). \qquad (5.1)$$

Although there exist several asymptotic expansions for $U(n - 3/2, \omega)$, most do not allow for the doubly asymptotic nature of the parabolic cylinder function being considered (unbounded n and ω). Fortunately, a general theory for the asymptotic solution of second-order linear differential equations has been developed by Olver [20] which is applicable in this case. Specifically, for large n,

$$U(n - 3/2, \sqrt{2(2n-3)}\, t) \approx \bar{g}(n)(t^2 + 1)^{-1/4} \exp[-(n-3/2)\bar{\xi}(t)], \qquad (5.2)$$

provided $-1 + \delta < t < 1 - \delta$, where $\bar{g}(n)$ is a function of n and $\bar{\xi}(t) = t(t^2 + 1)^{1/2} + \ln[t + (t^2 + 1)^{1/2}]$. However,

$$t(t^2 + 1)^{1/2} = t + \frac{1}{2} t^3 - \frac{1 \cdot 1}{2 \cdot 4} t^5 + \cdots, \quad |t| < 1$$

and

$$\ln[t + (t^2 + 1)^{1/2}] = t - \frac{1}{2 \cdot 3} t^3 + \frac{1 \cdot 3}{2 \cdot 4 \cdot 5} t^5 - \ldots, \quad |t| < 1$$

so that $\bar{\xi}(t) \approx 2t$ provided $-1 + \delta < t < 1 - \delta$.

Recall that in (5.1), $\omega = \sqrt{n}\, \Phi^{-1}(1 - r)a/\sqrt{1 + a^2}$. Consequently, in (5.2) $t = (1/2)[2n/(2n-3)]^{1/2}\Phi^{-1}(1 - r)a/\sqrt{1 + a^2}$. It follows that $-1 + \delta < t < 1 - \delta$ for large n, all values of a, and all but extreme values (i.e., $r < 0.01$ or $r > 0.99$) of r.

We shall also use the expansion

$$(1 + t^2)^{-1/4} = 1 - \frac{1}{4} t^2 + \frac{1 \cdot 5}{4 \cdot 4} \frac{t^4}{2} - \cdots, \quad |t| < 1 .$$

Therefore, the large sample structural density of $R(\tau;\mu,\sigma)$ is taken to be

$$\tilde{f}_R(r|\tau, a(\underline{t})) = k\{1 - (1/16)[2n/(2n-3)]\,[\Phi^{-1}(1-r)]^2\, a^2/(1 + a^2)\}$$

$$\times \exp\{-(1/2)\left[\frac{n(2+a^2)}{2(1+a^2)} - 1\right]\left[\Phi^{-1}(1-r)\right]^2 \qquad (5.3)$$

$$- (1/2)[2n(2n-3)]^{1/2}\Phi^{-1}(1-r)a/\sqrt{1+a^2}\}, \quad 0 < r < 1$$

where k is a normalizing constant.

The following lemma will be needed in evaluating k as well as the distribution function of (5.3).

Lemma.

$$\int_0^{c_3} [\Phi^{-1}(1-t)]^{2j} \exp(-(1/2)\{c_1[\Phi^{-1}(1-t)]^2+2c_2\Phi^{-1}(1-t)\})\,dt$$

$$= (1+c_1)^{-1/2} \exp[c_2{}^2/2(1+c_1)] \times \tag{5.4}$$

$$\{1-\Phi[(1+c_1)^{1/2}\Phi^{-1}(1-c_3)+c_2(1+c_1)^{-1/2}]\}\mu'_{2j}$$

for $j = 0, 1, 2, \ldots$ and $c_1, c_3 > 0$,

where μ'_{2j} is the $2j^{th}$ moment of a right-truncated normal distribution with mean $c_2/(1+c_1)$ and variance $1/(1+c_1)$ with upper truncation point $-\Phi^{-1}(1-c_3)$.

Proof. In the integral on the left-hand side of (5.4), we make the transformation $y = -\Phi^{-1}(1-t)$, $dt = \phi(-y)dy$. Then the integral may be written

$$(2\pi)^{-1/2} \int_{-\infty}^{-\Phi^{-1}(1-c_3)} y^{2j} \exp\{-(1/2)[(c_1+1)y^2-2c_2y]\}\,dy$$

$$= (1+c_1)^{-1/2} \exp[c_2{}^2/2(1+c_1)] \times$$

$$\int_{-\infty}^{-\Phi^{-1}(1-c_3)} y^{2j} \frac{1}{(1+c_1)^{-1/2}}\phi\left[\frac{y-c_2(1+c_1)^{-1}}{(1+c_1)^{-1/2}}\right] dy$$

$$= (1+c_1)^{-1/2} \exp[c_2{}^2/2(1+c_1)]\Phi\left[\frac{-\Phi^{-1}(1-c_3)-c_2(1+c_1)^{-1}}{(1+c_1)^{-1/2}}\right] \times$$

$$\int_{-\infty}^{-\Phi^{-1}(1-c_3)} y^{2j} \frac{\dfrac{1}{(1+c_1)^{-1/2}}\,\phi\left[\dfrac{y - c_2(1+c_1)^{-1}}{(1+c_1)^{-1/2}}\right]}{\Phi\left[\dfrac{-\Phi^{-1}(1-c_3) - c_2(1+c_1)^{-1}}{(1+c_1)^{-1/2}}\right]}\,dy$$

The results follows by using the relationship $\Phi(-x) = 1 - \Phi(x)$.

The value of the normalizing constant k is now easily found by using (5.4) with $c_1 = [n(2 + a^2)/2(1 + a^2)] - 1$, $c_2 = (1/2)\sqrt{2n(2n-3)}\ a/\sqrt{1 + a^2}$, $c_3 = 1$, and noting that μ'_{2j} is the $2j^{th}$ moment of an (untruncated) normal distribution with mean $\lambda_1 = \sqrt{2(2n-3)}\ a\sqrt{1 + a^2}/\sqrt{n}\,(2 + a^2)$ and standard deviation $\lambda_2 = \sqrt{2(1 + a^2)/n(2 + a^2)}$. Upon simplification,

$$k^{-1} = (1+c_1)^{-1/2} \exp[c_2{}^2/2(1+c_1)] \left[1 - \frac{n\,a^2(\lambda_1{}^2+\lambda_2{}^2)}{8(2n-3)(1+a^2)}\right]. \tag{5.5}$$

The distribution function corresponding to (5.3) is , with the aid of (5.4),

$$\tilde{F}_R(r|\tau,a(\underline{t})) = \int_0^r \tilde{f}_R(t|\tau,a(\underline{t}))\,dt$$

$$= k(1+c_1)^{-1/2} \exp[c_2{}^2/2(1+c_1)] \left[1 - \frac{n\,a^2}{8(2n-3)(1+a^2)}\,\mu_2{}^{*\prime}\right]$$

$$\times\left\{1 - \Phi\left[\frac{\Phi^{-1}(1-r) + \lambda_1}{\lambda_2}\right]\right\}, \tag{5.6}$$

where $\mu_2{}^{*\prime}$ is the second moment of a right-truncated normal distribution with mean $\lambda_1 = \sqrt{2(2n-3)}\ a\sqrt{1 + a^2}/\sqrt{n}\ (2 + a^2)$ and standard deviation $\lambda_2 = \sqrt{2(1 + a^2)/n(2 + a^2)}$ and upper truncation point $-\Phi^{-1}(1 - r)$. From Johnson and Kotz [14], p. 83,

$$\mu_2{}^{*\prime} = \lambda_1{}^2 + \lambda_2{}^2 + \left[\frac{\Phi^{-1}(1 - r) - \lambda_1}{\lambda_2}\right] \lambda_2{}^2 \; \frac{\phi(u)}{1 - \Phi(u)} , \qquad (5.7)$$

where $u = [\Phi^{-1}(1 - r) + \lambda_1]/\lambda_2$.

By combining (5.5), (5.6), and (5.7) we obtain the following

Theorem 5.1. The large-sample structural distribution function of $R(\tau;\mu,\sigma)$ is

$$\tilde{F}_R(r|\tau,a(\underline{t})) = 1-\Phi(u) - \frac{(u-2\lambda_1/\lambda_2)\;\lambda_2{}^2\;\phi(u)}{8[(2n-3)(1+a^2)/na^2]-(\lambda_1{}^2+\lambda_2{}^2)} , \qquad (5.8)$$

where

$$u = [\Phi^{-1}(1 - r) + \lambda_1]/\;\lambda_2 ,$$

$$\lambda_1 = \sqrt{2(2n - 3)}\; a\sqrt{1 + a^2}/\sqrt{n}(2 + a^2) ,$$

$$\lambda_2 = \sqrt{2(1 + a^2)/n(2 + a^2)} .$$

Theorem 5.2. (i.) A large-sample point estimate of $R(\tau;\mu,\sigma)$ is

$$\tilde{E}[R(\tau;\mu,\sigma)] = \Phi\left(\frac{\lambda 1}{\sqrt{1+\lambda_2{}^2}}\right) - \theta\,\frac{\lambda 1(2+\lambda 2^2)}{\lambda_2(1+\lambda_2{}^2)}\;\phi\left(\frac{\lambda_1}{\sqrt{1+\lambda_2{}^2}}\right) , \qquad (5.9)$$

where $\theta = \dfrac{\lambda_2{}^2}{[8(2n - 3)(1+ a^2)/na^2] - (\lambda_1{}^2 + \lambda_2{}^2)}$;

(ii.) A large-sample $(1 - \alpha)$-structural lower bound for $R(\tau;\mu,\sigma)$ is given by

$$\underline{\tilde{r}}(1 - \alpha) = 1 - \Phi(\lambda_2\tilde{\Delta} - \lambda_1) , \qquad (5.10)$$

where $\tilde{\Delta}$ is the solution to the equation

$$\Delta = \Phi^{-1}\,[1 - \alpha - \theta(\Delta - 2\lambda_1/\lambda_2)\;\phi(\Delta)]\quad .$$

Although (5.3) was derived under the assumption of a large sample size n, it may also be used for *any* n provided $|a|$ is sufficiently small, say $|a| \leq 1/2$. To see this is the case, we first note that $1 >> [\Phi^{-1}(1 - r)]^2\, a^2/(1 + a^2)$ for small $|a|$. Consequently, by an expansion of the parabolic cylinder function given in Abramowitz and Stegun [2] p. 689, we have

$$U(n - 3/2,\omega) \approx \bar{h}(n)\,\exp(-\sqrt{n - 3/2}\;\omega)\quad ,$$

where $\omega = \sqrt{n}\,\Phi^{-1}(1 - r)\, a/\sqrt{1 + a^2}$ and $\bar{h}(n)$ is a function of n. Based on this result, an approximation to the structural density of $R(\tau;\mu,\sigma)$ is given by (5.3) except that the term $(1/16)[2n/(2n - 3)][\Phi^{-1}(1 - r)]^2 a^2/(1 + a^2)$ is deleted. It follows that

<u>*Corollary*</u> For arbitrary samples size and $|a| \leq 1/2$, an approximation to the structural distribution function of $R(\tau;\mu,\sigma)$ is

$$\tilde{\tilde{F}}_R(r|\tau,a(\underline{t})) = 1 - \Phi(\eta), \tag{5.11}$$

where $\eta = \sqrt{\dfrac{n(2 + a^2)}{2(1 + a^2)}}\;\Phi^{-1}(1 - r) + \sqrt{\dfrac{2n - 3}{2 + a^2}}\; a\;.$

Furthermore,

(i.) an approximate point estimate is

$$\tilde{\tilde{E}}[R(\tau;\mu,\sigma)] = \Phi\left\{\sqrt{\frac{2(2n-3)(1+a^2)}{(2+a^2)[2(1+a^2)+n(2+a^2)]}}\; a\right\}, \tag{5.12}$$

(ii.) and an approximate $(1 - \alpha)$-structural lower bound is

$$\tilde{\tilde{\underline{r}}}(1-\alpha) = 1 - \Phi\left\{\sqrt{\frac{2(1+a^2)}{n(2+a^2)}}\left[\Phi^{-1}(1-\alpha) - \sqrt{\frac{2n-3}{2+a^2}}\; a\right]\right\}. \tag{5.13}$$

6.0 Numerical Example. To illustrate numerically some of the results of the previous sections, we consider the following data, given by Antle [2], which have been generated from a lognormal distribution (with $\mu = 4.0$ and $\sigma = 0.5$):

26.96,	43.02,	57.99,	78.50,
28.06,	53.02,	60.39,	96.28,
39.52,	53.22,	65.92,	119.05,
41.01,	53.41,	66.99,	120.18,
41.41,	57.14,	69.73,	175.37.

Suppose the "mission time" is $\tau = 50.0$. Then based on the logarithmically transformed data,

$$\hat{\mu} = (1/n) \sum_{i=1}^{n} \ln t_i = 4.09688$$

$$\hat{\sigma} = \left[(1/n) \sum_{i=1}^{n} (\ln t_i - \quad)^2 \right]^{1/2} = 0.46225$$

and

$$a = 0.39991 \quad .$$

(i.) *Point estimates* of $R(50;\mu,\sigma)$.

Using Theorem 4.1, we find the mean, median, and mode of the structural density of $R(50;\mu,\sigma)$, which are the point estimates,

$$E\,[R(50;\mu,\sigma)] = 0.646 \,,$$

$$r_{median} = 0.650 \,,$$

$$r_{mode} = 0.657 \,.$$

An approximate point estimate based on equation (5.12) is

$$\tilde{\tilde{E}}[R(50;\mu,\sigma)] = \Phi(.374) = 0.646 \,.$$

By way of comparison, the maximum likelihood estimate is

$$\hat{R}_{ML}(50;\mu,\sigma) = 1 - \Phi(-.400) = 0.655 \,;$$

the minimum variance unbiased estimate is

$$\hat{R}_{UMVU}(50;\mu,\sigma) = 1 - I_{.454}(9,9) = 0.650.$$

(ii.) *Interval estimate* of $R(50;\mu,\sigma)$.

The 95% structural lower bound for $R(50;\mu,\sigma)$ is the iterative solution to the defining equation in Theorem 4.2. This solution,

$$\underline{r}(.95) = 0.501 ,$$

is also "numerically equivalent" to the 95% Bayes lower confidence bound under the improper prior $\pi(\mu,\sigma) \propto 1/\sigma$ and to the 95% classical lower confidence bound. An approximate 95% structural lower bound based on equation (5.13) is

$$\tilde{\tilde{\underline{r}}}(.95) = 1 - \Phi(-.002) = 0.501 .$$

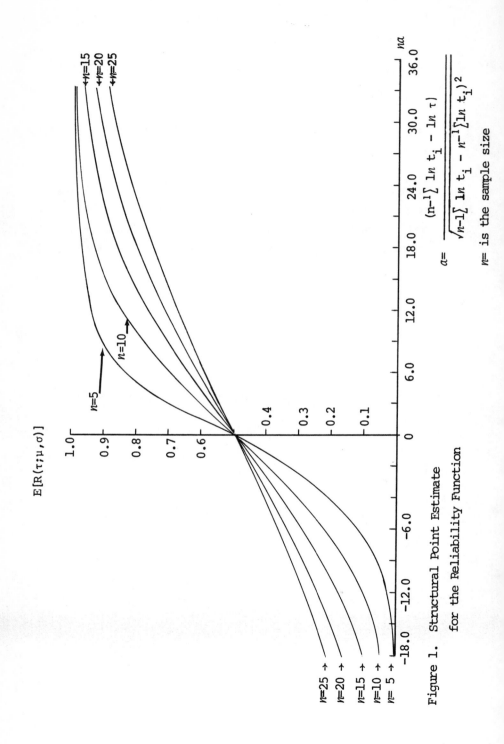

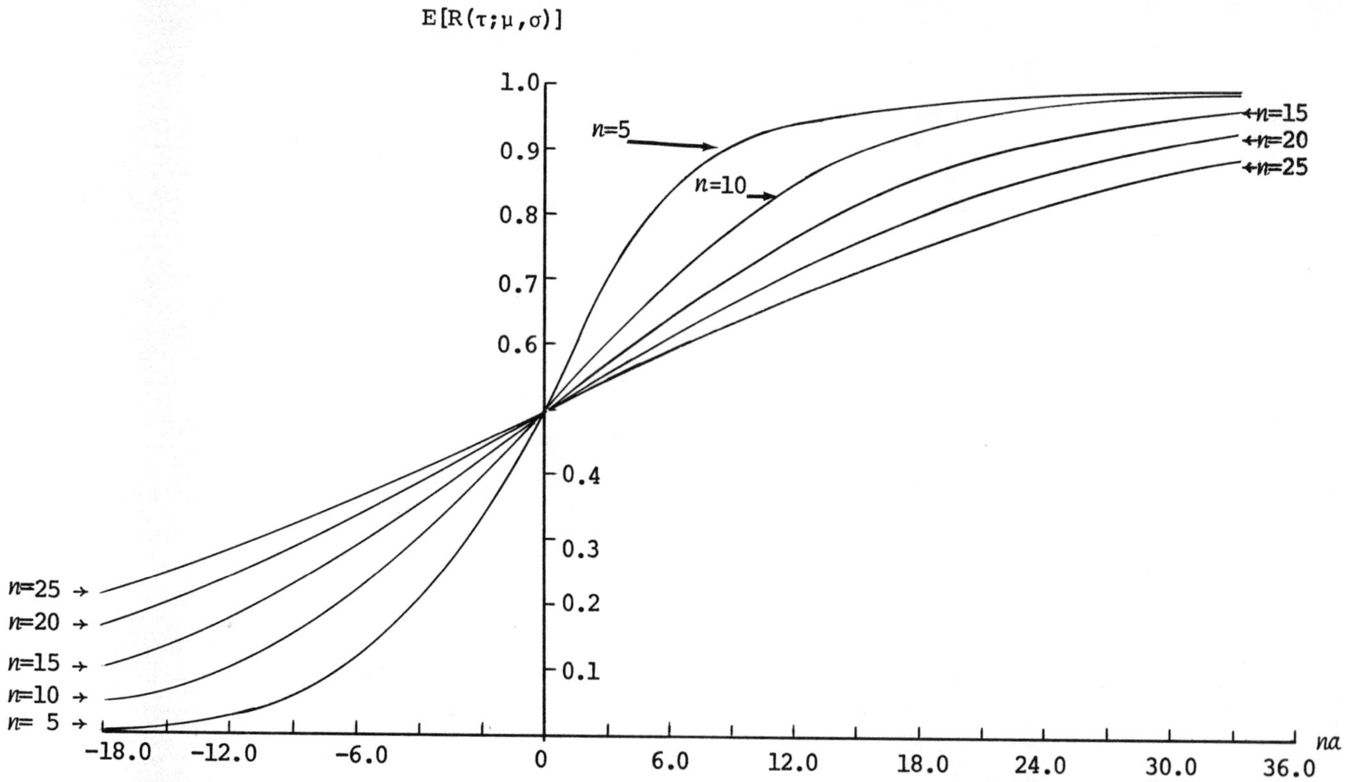

Figure 1. Structural Point Estimate for the Reliability Function

$$a = \frac{(n^{-1}\sum \ln t_i - \ln \tau)}{\sqrt{n-1\sum \ln t_i - n^{-1}\sum \ln t_i)^2}}$$

n= is the sample size

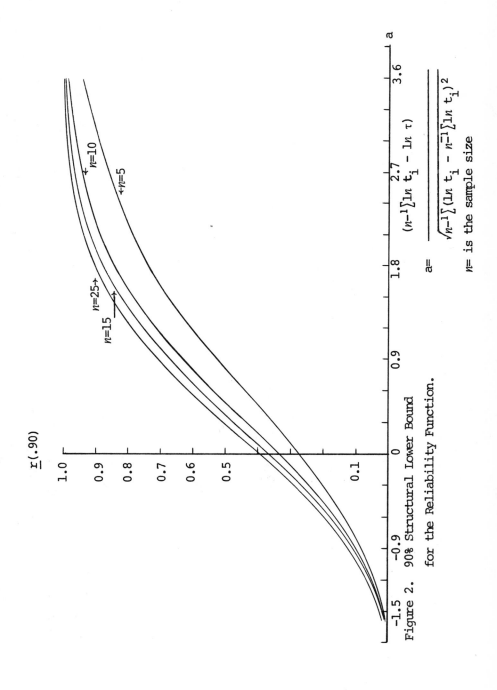

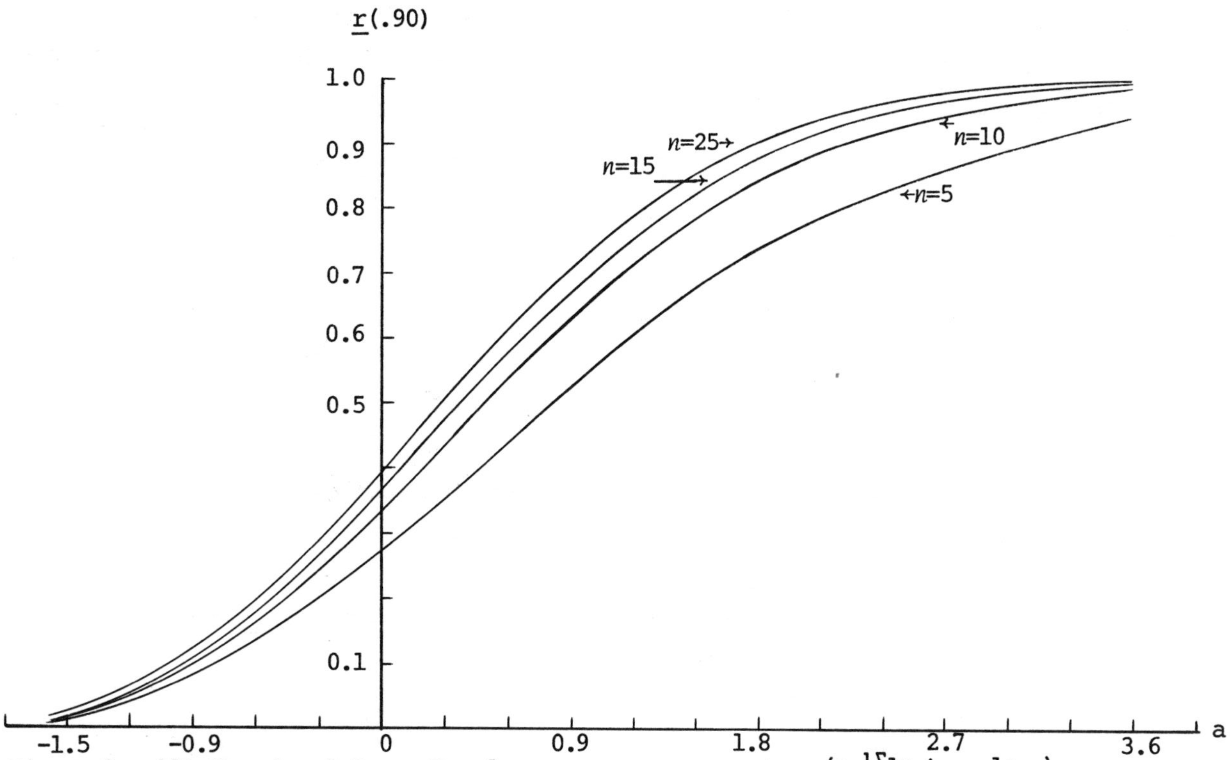

Figure 2. 90% Structural Lower Bound for the Reliability Function.

$$a = \frac{(n^{-1}\sum \ln t_i - \ln \tau)}{\sqrt{n^{-1}\sum(\ln t_i - n^{-1}\sum \ln t_i)^2}}$$

n= is the sample size

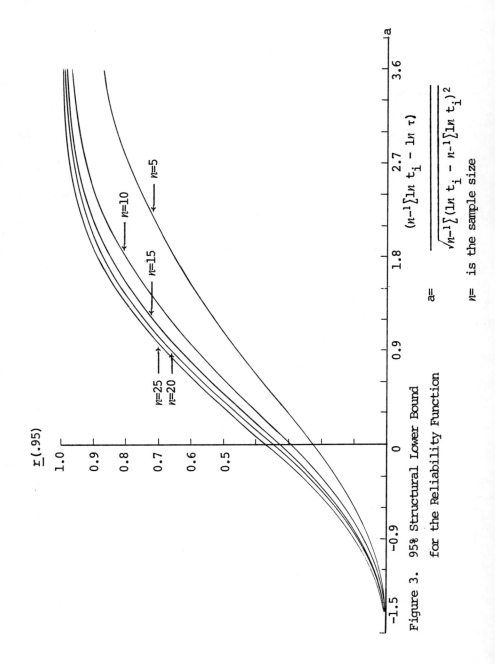

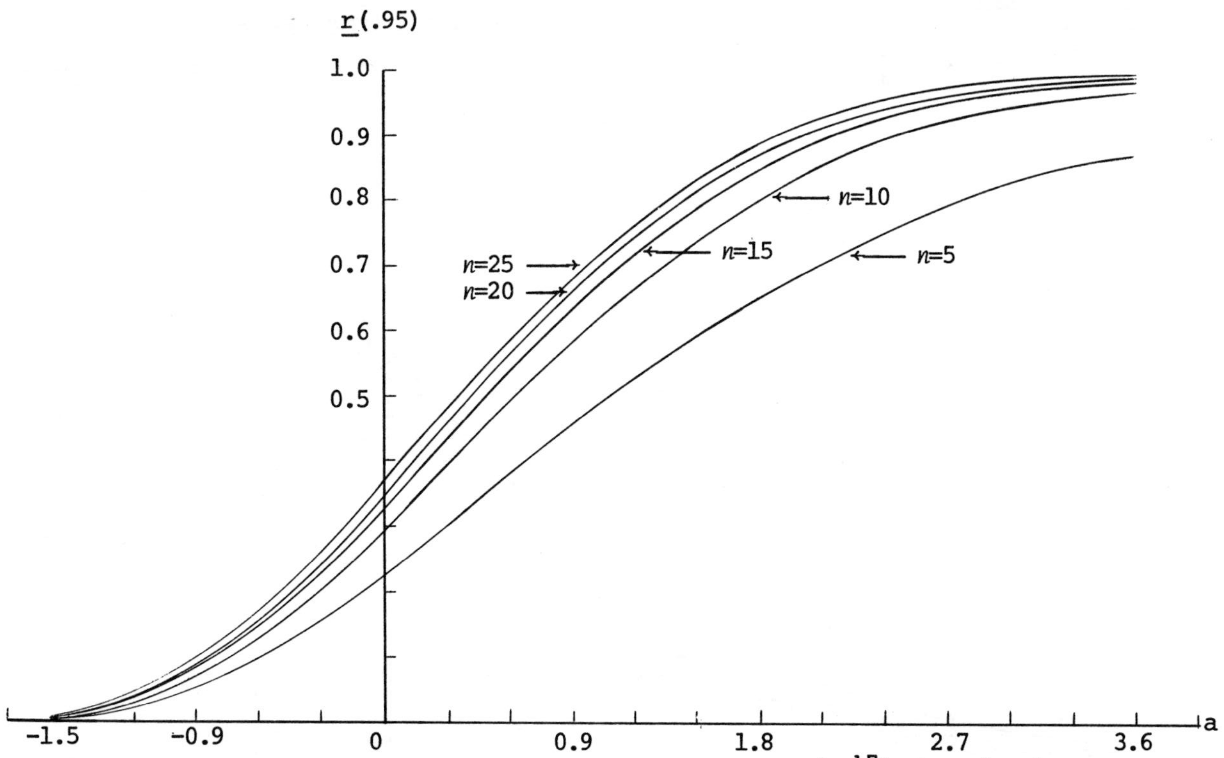

Figure 3. 95% Structural Lower Bound for the Reliability Function

$$a = \frac{(n^{-1}\sum \ln t_i - \ln \tau)}{\sqrt{n^{-1}\sum(\ln t_i - n^{-1}\sum \ln t_i)^2}}$$

n= is the sample size

REFERENCES

[1] Abramowitz, M. and Stegun, I. A., *Handbook of Mathematical Functions*, National Bureau of Standards, Applied Mathematics Series No. 55, U. S. Government Printing Office, Washington, D. C., (1965).

[2] Antle, C. E., Choice of Model for Reliability Studies and Related Topics, *Aerospace Research Laboratories TR 72-0108*, Wright-Patterson Air Force Base, Ohio, (1972).

[3] Bury, K. V., On the Reliability Analysis of a Two-Parameter Weibull Process, *INFOR, Vol. 10*, (1972), pp. 129-139.

[4] Bury, K. V. and Bernholtz, B., (1972), Life Testing: Structural Inference on the Exponential Model, *INFOR, Vol. 9*, (1971), pp. 148-160.

[5] Dumonceaux, R. and Antle, C. E., Discrimination Between the Log-Normal and the Weibull Distributions, *Technometrics, Vol. 15*, (1973), pp. 923-926.

[6] Epstein, B., The Mathematical Description of Certain Breakage Mechanisms Leading to the Logarithmico-Normal Distribution, *Journal of the Franklin Institute, Vol. 244*, (1947), pp. 471-477.

[7] Epstein, B., Statistical Aspects of Fracture Problems, *Journal of Applied Physics, Vol. 19*, (1948), pp. 140-147.

[8] Erdelyi, A., Magnus, W., Oberhettinger, F., and Tricomi, F. G., Higher Transcendental Functions (Bateman Manuscript Project), Vols. I and II, McGraw-Hill, New York, (1953).

[9] Feinlieb, M., A Method of Analyzing Log-Normally Distributed Survival Data with Incomplete Follow-Up, *Journal of the American Statistical Association, Vol. 55*, (1960), pp. 534-545.

[10] Fraser, D. A. S., The Structure of Inference, John Wiley and Sons, New York, (1968).

[11] Fraser, D. A. S., Bayes, Likelihood, or Structural, *Annals of Mathematical Statistics, Vol. 43*, (1972), pp. 777-790.

[12] Goldthwaite, L. R., Failure Rate Study for the Lognormal Lifetime Model, *Proceedings of the 7th National Symposium on Reliability and Quality Control*, (1961), pp. 208-213.

[13] Hora, R. B. and Buehler, R. J., Fiducial Theory and Invariant Estimation, *Annals of Mathematical Statistics, Vol. 37*, (1966), pp. 643-656.

[14] Johnson, N. L. and Kotz, S., Continuous Univariate Distributions-1, John Wiley and Sons, New York, (1970).

[15] Klimko, L. A., Antle, C. E., and Rademaker, A., Tests for Normality Versus Lognormality, *Communications in Statistics, Vol. 4*, (1975), pp. 1009-1019.

[16] Kotz, S., Normality vs. Lognormality with Applications, *Communications in Statistics, Vol. 1*, (1973), pp. 113-132.

[17] Lebedev, N. N., Special Functions and Their Applications, Prentice-Hall, Englewood Cliffs, New Jersey, (1965).

[18] Lieberman, G. J. and Resnikoff, G. J., Sampling Plans for Inspection by Variables, *Journal of the American Statistical Association, Vol. 50*, (1955), pp. 457-516.

[19] Nowick, A. S. and Berry, B. S., Lognormal Distribution Function for Describing Anelastic and Other Relaxation Processes, *IBM Journal of Research and Development, Vol. 5*, pp. 297-311, Ibid., (1961), pp. 312-230.

[20] Olver, F. W. J., Uniform Asymptotic Expansions for Weber Parabolic Cylinder Functions of Large Orders, *Journal of Research of the National Bureau of Standards-B, Vol. 63*, (1959), pp. 131-169.

[21] Owen, D. B., Handbook of Statistical Tables, Addison-Wesley, Reading, Massachusetts, (1962).

[22] Zacks, S., The Theory of Statistical Inference, John Wiley and Sons, New York, (1971).

[23] Zacks, S. and Milton, R. C., Mean Square Errors of the Best Unbiased and Maximum Likelihood Estimators of Tail Probability in Normal Distributions, *Journal of the American Statistical Association, Vol. 66*, (1971), pp. 590-593.

UNIFORMLY MOST POWERFUL UNBIASED TESTS FOR THE PARAMETERS OF THE GAMMA DISTRIBUTION

by MAX ENGELHARDT and
LEE J. BAIN
University of Missouri-Rolla

Abstract. Uniformly most powerful unbiased tests for the parameters of the gamma distribution are constructed for each parameter with the other parameter considered as an unknown nuisance parameter. Convenient approximations for the sampling distributions of the respective test statistics are provided.

1.0 *Introduction.* The two-parameter gamma distribution, as defined by the density function

$$f(x) = \lambda(\lambda x)^{\alpha-1}\exp(-\lambda x)/\Gamma(\alpha),\ x > 0,\ \alpha > 0,\ \lambda > 0,$$

provides a population model which is useful in many areas of statistics, including the areas of life-testing and reliability. An important property of the gamma distribution is that the hazard rate function, $h(x) = f(x)/[1 - F(x)]$, is increasing (decreasing) if $\alpha > 1(\alpha < 1)$ and it approaches a positive constant, namely $h(x) \to \lambda$, as $x \to \infty$. If $\alpha = 1$, then $h(x) = \lambda$ for all $x > 0$, and the distribution is exponential. Although the Weibull model is frequently used when an increasing or decreasing hazard rate function is desired the resulting hazard rate function approaches either zero or infinity as $x \to \infty$, except in the exponential case.

For an example where the gamma model might be useful, consider the results of Barlow and Proschan [2] which involve a system whose components are replaced as they fail. Under certain conditions the limiting distribution of times between system failures is exponential as time and the number of components tend to infinity. This suggests that for complex systems the times between system failures should be approximately exponential after an initial period of time. A similar result, based upon examination of failure data, is considered by Davis [6]. Specifically, Davis discusses bus motors, which after the second and subsequent overhauls consist of components in a "scattered state of wear". Although individual components may each follow some special failure-time distribution (lognormal, Weibull, etc.), "with components in random stages of wear, a motor has an equally likely chance of failing during any period of operation". In other words, a conditional uniform distribution in a time interval of a certain length given that a failure occurs in that interval. This rationale was used to justify a constant hazard rate function and subsequent analysis with the exponential model. Since, in an actual system, the time between failures may only be approximately exponential, a better analysis might result by using a more flexible model admitting hazard rate functions which are capable of increasing or decreasing, but approaching a positive constant asymptotically with increasing time. The bus motor failure data provided by Davis [6] appear to support this conjecture. The gamma distribution provides a model which extends the exponential model and possesses the desired flexibility.

In the present paper, both α and λ will be unknown parameters, and uniformly most powerful unbiased (UMPU) tests will be constructed for each parameter with the other parameter considered as an unknown nuisance parameter.

2.0 *UMPU Tests for α with λ Unknown.* Let $X_1, X_2, \ldots, X_n$ denote a random sample of size n from a gamma distributed population with both α and λ unknown. The gamma distribution belongs to the exponential family as defined by Lehmann [9]. Consider the joint density function of the random sample:

$$[\lambda^{\alpha}/\Gamma(\alpha)]^n \exp[\alpha \ln(\prod_{i=1}^{n} x_i) + \lambda(-\sum_{i=1}^{n} x_i)] \; (\prod_{i=1}^{n} x_i)^{-1}$$

for $\prod_{i=1}^{n} x_i > 0$, zero otherwise. It follows that $U = \ln(\prod_{i=1}^{n} X_i)$ and $T = -\sum_{i=1}^{n} X_i$ are joint complete sufficient statistics for α and λ. It is sometimes convenient to express these in terms of the arithmetic and geometric means, respectively

$$\overline{X} = \sum_{i=1}^{n} X_i/n = -T/n \text{ and } \tilde{X} = (\prod_{i=1}^{n} X_i)^{1/n} = \exp(U/n).$$

It will be convenient to base our UMPU tests for α on the the statistic $V = \ln(\tilde{X}/\overline{X})$. Since for any fixed α, T is a complete sufficient statistic for λ and since V is distributed independently of λ, it follows from the results of Basu [4] that V and T are independent statistics. Furthermore, $V = h(U, T)$ where $h(u, t) = (1/n)u - \ln(-t/n)$. It follows from Lehmann [9] that UMPU tests for α with λ unknown can be derived. Suppose for example, a test is desired for the hypothesis H_0: $\alpha \leq \alpha_0$ against the alternative H_1: $\alpha > \alpha_0$ for some specified $\alpha_0 > 0$. A UMPU level γ test for this hypothesis would be to reject H_0 if $V > C_0$ where $P_{\alpha_0}[V > C_0] = \gamma$. One problem with this test is the determination of a convenient method for finding critical values. The exact density function of $L = \tilde{X}/\overline{X} = \exp(V)$ was derived by Nair [10]. A series solution whose complexity increases with sample size, was obtained. The major difficulty with this method is the numerical evaluation of the sampling distribution of L. Other results which lead to approximate sampling distributions are given by Bartlett [3], Hartley [8], and Box [5]. These results pertain to Bartlett's test for homogeneity of normal population variances which, in the case of equal sample sizes and with proper identification of the parameters, has the same sampling distribution as L. These approximations are all fairly accurate for large α, but are not adequate for small α. More recently, an approximation which holds for all $\alpha > 0$ has been proposed by Bain and Engelhardt [1]. The approximation is based upon a two-moment chi-square fit of the variable $2n\alpha S$ where $S = \ln(\bar{X}/X) = -V$. The cumulant generating function of S is given by

$$K_S(t) = -t \ln n + n \ln \Gamma(\alpha - t/n) - \ln \Gamma(n\alpha - t) - n \ln \Gamma(\alpha) + \ln \Gamma(n\alpha).$$

Hence, the cumulant generating function of $2n\alpha S$ is given by $K_{2n\alpha S}(t) = K_S(2n\alpha t)$. Bartlett [3] verified that $K_{2n\alpha S}(t) \to -[(n-1)/2] \ln(1-2t)$ as $\alpha \to \infty$ which suggested the approximation $2n\alpha S \sim \chi^2(n-1)$ for large α. Bain and Engelhardt [1] noted that also $K_{2n\alpha S}(t) \to -(n-1) \ln(1-2t)$ as $\alpha \to 0$ which suggests the approximation $2n\alpha S \sim \chi^2[2(n-1)]$ for small α. The proposed approximation, which is quite accurate for all $\alpha > 0$, is $2n\alpha c(\alpha, n)S \sim \chi^2[\nu(\alpha, n)]$ where $c(\alpha, n)$ and $\nu(\alpha, n)$ are constants such that the mean and variance of $2n\alpha c(\alpha, n)S$ agree with those of a chi-square variate with $\nu(\alpha, n)$ degrees of freedom. It is shown by Bain and Engelhardt [1] that $c(\alpha, n) = [n\phi_1(\alpha) - \phi_1(n\alpha)]/[n\phi_2(\alpha) - \phi_2(n\alpha)]$ and $\nu(\alpha, n) = [n\phi_1(\alpha) - \phi_1(n\alpha)]c(\alpha, n)$ where $\phi_1(z) = 2z[\ln z - \psi(z)]$, $\phi_2(z) = 2z[z\psi'(z) - 1]$, and $\psi(z) = \Gamma'(z)/\Gamma(z)$ denotes the digamma function, and $c(\alpha, n)$ and $\nu(\alpha, n)/(n-1)$ are tabulated for a wide range of α and n. It is also noted that for fixed n, $c(\alpha, n) \to 1$ and $\nu(\alpha, n)/(n-1) \to 1$ as $\alpha \to \infty$. However, as $\alpha \to 0$, $c(\alpha, n) \to 1$, while $\nu(\alpha, n)/(n-1) \to 2$. Hence, this approximation is consistent with the chi-square approximations which were discussed previously for large or small α. The simple approximations $\phi_1(\alpha) \doteq 3-1/(1+\alpha)+\alpha/6(1+\alpha)^2-2\alpha \ln(1+1/\alpha)$, $\phi_2(\alpha) \doteq 1+(2+\alpha)/3(1+\alpha)^2+1/3(1+\alpha)^3$ may also be useful for computer implementation. The resulting approximate values of

$c(\alpha, n)$ and $\nu(\alpha, n)/(n-1)$ are in error by less than .001 and .002 respectively. It may also be useful to have simpler, though less accurate, approximations: $c(\alpha, n) \doteq A_1(\alpha)/A_2(\alpha)$ and $\nu(\alpha, n)/(n-1) \doteq A_3(\alpha)$, where $A_1(\alpha) = 1 + 1/(1 + 6\alpha)$, $A_2(\alpha) = 1 + 1/(1 + 2.5\alpha)$ for $0 < \alpha < 2$ and $1 + 1/3\alpha$ for $2 \leq \alpha < \infty$, and $A_3(\alpha) = 1 + 1/(1 + 4.3\alpha)^2$. The latter approximations are possible since the dependence of $c(\alpha, n)$ and $\nu(\alpha, n)/(n-1)$ on n is very slight. They also agree with the exact values asymptotically as $\alpha \to 0$ or $\alpha \to \infty$.

A UMPU level γ test for H_0: $\alpha \leq \alpha_0$ against the alternative H_1: $\alpha > \alpha_0$ would reject H_0 if $S < s_\gamma$ where $P_{\alpha_0}[S < s_\gamma] = \gamma$. A convenient approximate test for any $\alpha_0 > 0$ would result from $s_\gamma \doteq \chi_\gamma^2[\nu(\alpha_0,n)]/2n\alpha_0 c(\alpha_0, n)$ where $P[\chi^2(\nu) < \chi_\gamma^2(\nu)] = \gamma$. UMPU tests for other hypotheses concerning α can also be constructed from the results of Lehmann [9] although they involve some additional side conditions. The chi-square approximation is rather convenient in this situation. For example, a UMPU level γ test for H_0: $\alpha = \alpha_0$ against the alternative H_1: $\alpha \neq \alpha_0$ would reject H_0 if $S < k_1$ or $S > k_2$ where k_1 and k_2 are determined by $E_{\alpha_0}[I_A(S)] = P_{\alpha_0}[k_1 \leq S \leq k_2] = 1 - \gamma$ and $E_{\alpha_0}[SI_A(S)] = (1 - \gamma)E_{\alpha_0}(S)$ where $A = [k_1, k_2]$. Since, approximately $2n\alpha_0 c(\alpha_0, n)S \sim \chi^2[\nu(\alpha_0, n)]$ the problem is solved by following the procedure outlined by Lehmann [9] in deriving a two-sided UMPU test for a normal population variance. This test would accept H_0 if $C_1 \leq 2n\alpha_0 c(\alpha_0, n)S \leq C_2$ where $C_1 = 2n\alpha_0 c(\alpha_0, n)k_1$ and $C_2 = 2n\alpha_0 c(\alpha_0, n)k_2$. The values C_1 and C_2 are solutions of the equation $\nu(\alpha_0, n) = (C_1 - C_2)/\ln(C_1/C_2)$.

3.0 *UMPU Tests for λ with α Unknown*. Consider the joint density function of the random sample from Section 2.0. To be consistent with the notation of Lehmann [9], we will now denote $U = -\sum_{i=1}^{n} X_i = -n\overline{X}$ and $T = \ln(\prod_{i=1}^{n} X_i) = n\ln(\tilde{X})$. The approach of Section 2.0 will not work here, since no function $h(U, T)$ which is independent of T is available. However, comparable tests for λ with α unknown are possible using conditional tests. Since the gamma distribution belongs to the exponential family, it follows that the conditional density function of U given T = t has the exponential form and is independent of α. It also follows from Lehmann [9] that conditional tests which are UMPU for λ with α unknown can be derived. Suppose, for example, a test is desired for the hypothesis H_0: $\lambda \leq \lambda_0$ against the alternative H_1: $\lambda > \lambda_0$. A UMPU level γ test for this hypothesis would be to reject H_0 if $U > C_0(t)$ where $P_{\lambda_0}[U > C_0(t) | T = t] = \gamma$. If we denote $W = \overline{X}/\tilde{X}$ and $G_0 = \lambda_0\tilde{X}$, then it would be equivalent to reject H_0 if $W < K_0(g)$ where $P[W < K_0(g) | G_0 = g] = \gamma$. The latter modification allows us to consider the sampling distribution independent of parameters. It is possible to derive an expression for the exact conditional distribution of W given $G_0 = g$, although it is quite difficult to numerically evaluate this expression. A large-sample approximate distribution, based upon asymptotic normal results, was derived by Engelhardt and Bain [7]. The proposed approximation was $K_0(g) \doteq E(W|g) + z_\gamma\sqrt{\text{Var}(W|g)}$, for large n, where $z_\gamma = \phi^{-1}(\gamma)$ is the γ-quantile of the standard normal distribution, $E(W|g) = \psi^{-1}(\ln g)/g$ is an asymptotic approximation for the conditional mean and

$$\text{Var}(W|g) = [\psi^{-1}(\ln g) - 1/\psi'(\psi^{-1}(\ln g))]/g^2 n$$

is an asymptotic approximation for the conditional variance of W given $G_0 = g$. The asymptotic quantities $E(W|g)$ and $ng^2 Var(W|g)$ have been tabulated by Engelhardt and Bain [7] for a wide range of g. The simple rational approximations $E(W|g) = 1 + 1/2g$ and $ng^2 Var(W|g) = 1/2 - 1/4(1 + g)$ are fairly accurate for large g, say $g \geq 2$.

This asymptotic approximation was obtained by considering the variables $G = \lambda\tilde{X}$ and $Y = \sqrt{n}[W - \psi^{-1}(\ln G)/G]$. Let $F_n(y) = P(Y \leq y)$ and $F_n(y|g) = P(Y \leq y|g)$ denote distribution function of Y and the conditional distribution function of Y given G = g respectively. It can be shown that $F_n(y) \rightarrow F(y) = \phi(y\exp[\psi(\alpha)]/\sqrt{\alpha - 1/\psi'(\alpha)})$ as $n \rightarrow \infty$. Furthermore, an important property relating $F_n(y)$ and $F_n(y|g)$ is that $E[F_n(y|G)] = F_n(y)$. In fact, by completeness the function $F_n(y|g)$ is determined by this relationship. It can be shown that if we define $F(y|g) = \phi(yg/\sqrt{\psi^{-1}(\ln g) - 1/\psi'(\psi^{-1}(\ln g))})$ then asymptotically $E[F(y|G)] = F(y)$ as $n \rightarrow \infty$. In this sense, an approximate large sample distribution of W given G = g would be normal with mean $E(W|g)$ and variance $Var(W|g)$ as previously defined.

REFERENCES

[1] Bain, L. J. and Engelhardt, M. E., "A Two-Moment Chi-Square Approximation for the Statistic Log $(\overline{X}/\tilde{X})$", *Journal of the American Statistical Association*, *70*, (1975), pp. 948-950.

[2] Barlow, R. E. and Proschan, F., Mathematical Theory of Reliability, New York: John Wiley and Sons, Inc., (1965).

[3] Bartlett, M. S., "Properties of Sufficiency and Statistical Tests", *Proceedings, Royal Society of London*, *160*, *Ser. A.*, (1937), pp. 268-282.

[4] Basu, D., "On Statistics Independent of a Complete Sufficient Statistic", *Sankya, 15*, (1955), pp. 377-380.

[5] Box, G. E. P., "A General Distribution Theory for a Class of Likelihood Criteria", *Biometrika, 36*, (1949), pp. 317-346.

[6] Davis, D. J., "An Analysis of Some Failure Data", *Journal of the American Statistical Association, 47*, (1952), pp. 113-150.

[7] Engelhardt, M. E. and Bain, L. J., "Asymptotic Results for Uniformly Most Powerful Unbiased Tests on the Scale Parameter of a Gamma Distribution with a Nuisance Shape Parameter", Submitted for publication, (1975).

[8] Hartley, H. O., "Testing the Homogeneity of a Set of Variances", *Biometrika, 31*, (1940), pp. 249-255.

[9] Lehmann, E. L., Testing Statistical Hypotheses, New York: John Wiley and Sons, Inc., (1959).

[10] Nair, U. S., "The Application of the Moment Function in the Study of Distribution Laws in Statistics", *Biometrika, 30*, pp. 274-294.

UNBALANCED RANDOM-EFFECT MODEL WITH UNEQUAL GROUP VARIANCES: ESTIMATION OF DESIGN PARAMETERS ALLOWING FOR SCREENED BATCHES*

by K. W. FERTIG
and N. R. MANN
Rocketdyne Rockwell International Corporation

Abstract. A random effects model is considered in which the within group variance is allowed to vary from group to group. The model is complicated by the presence of a testing program which will be used to screen out those groups which have estimated lower percentiles less than a prescribed value. The aggregate distribution of the acceptable groups will then be skewed to the right. This type of screening thus tends to increase the value of any specific percentile. A method to estimate, prior to testing, the percentiles of the skewed distribution that would result from a given testing program is presented. Also, a method for constructing tolerance intervals for a particular group given only that it meets the acceptance criterion of the testing program is presented. The model is applied to a materials problem in which it is desired to increase the minimum strength of the material through the use of a screening program.

1.0 *Introduction*. The work in this paper is an extension of the work of Fertig and Mann [1974c]. As in that paper, we assume a model of the following form:

(a) $E\ (X_{jk} | k) = b_k$

(b) $\text{var}\ (X_{jk} | k) = \sigma_k^2$

---

*This research was initiated under the Interdivisional Technology Program of Rockwell International. It was completed with the support of the Air Force Office of Scientific Research under Contract #F44620-71-C-0029.

$$\begin{aligned}
&(c) \quad \underset{i \neq j}{\text{cov}} (X_{jk}, X_{ik} | k) = 0 \\
&(d) \quad \text{cov} (X_{jk}, X_{ik}) = \text{var} (b_k) = \sigma_b^2 \qquad (1.1) \\
&(e) \quad E (b_k) = \mu \text{ and } E (\sigma_k^2) = \sigma_e^2 \\
&(f) \quad \underset{k \neq \ell}{\text{cov}} (b_k, b_\ell) = \underset{k \neq \ell}{\text{cov}} (\sigma_k^2, \sigma_\ell^2) = \underset{k \neq \ell}{\text{cov}} (b_k, \sigma_\ell^2) = 0 \\
&\quad i,j = 1, \ldots, n_k \quad \ell,k = 1, \ldots, K
\end{aligned}$$

Further distributional assumptions will be discussed in Section 2. Model (1.1) is applicable to situations in which the random variable, X_{jk}, is formed in two stages. The first stage (process stage or k-stage) defines the random variables b_k and σ_k^2. The second stage (within stage or j-stage) defines the X_{jk} from a conditional distribution given the b_k and σ_k^2.

This model is described extensively in the literature. (See Harville [4] for a critical review of much of this literature.)

In this paper, the model described by (1.1) is complicated by the presence of a screening process that is to be used in the future. To be more specific, we describe the actual problem encountered by the materials personnel at Rocketdyne that led to this research. A certain material was produced in batches or heats. The observed strengths, X_{jk}, were recorded. It was noted that σ_k^2 as well as b_k probably varied from heat to heat. For this case, if we let $\sigma^2 = \sigma_e^2 + \sigma_b^2$, then $\hat{\mu} - z_\gamma \hat{\sigma}^2$ is a maximum-likelihood estimate of a percentile of the underlying process distribution if $\hat{\mu}$ and $\hat{\sigma}^2$ are maximum likelihood estimates of μ and σ^2, z_γ is the <u>100γ</u> th percentile of a standard normal distribution, and if the distributional assumption that the X_{jk} are multivariate normal is made. Fertig and Mann [1] give the actual data for this example. The design engineers were concerned that the estimate of the fifth percentile (when $z_\gamma = 1.645$) indicated the material was only marginally acceptable, if at all.

Nevertheless, they were required to use this type of material. It was therefore decided to accept the material in batches only if the batch met certain specifications. The question then to be answered is what would be the estimate of the fifth percentile of the accepted material after screening.

The distribution of screened material would be expected to be skewed to the right if the acceptance plan was of the form: accept the batch if $\bar{x} - As > L$, where $\bar{x}$ and s represent the sample mean and sample standard depiction for a given batch. This type of screening would then tend to increase the value of the minimum strength (defined as the first or fifth percentile). The question just posed is to be answered at the design stage in building hardware. The number requested (minimum strength estimate) is termed a *design allowable*. Typically one obtains a great deal of data at the design stage in order to evaluate the material. In this paper, a method is presented which would use these data to estimate minimum strengths of future materials given that the future material is in the form of heats which belong to the same population as the design data but are also subject to meeting given acceptance criteria.

Specifically, this paper provides means for (I) estimating any percentile of the distribution of strength of material in general, given that the material belongs to a heat that has passed inspection and (II) constructing confidence intervals on the percentile of the strength distribution of a heat of material given that heat has met the inspection criteria. The specific model is defined in Section 2. Also given there are certain density functions pertiment to the model. Percentile estimation is discussed in Section 3 and the method of obtaining confidence intervals on within-batch percentiles is discussed in Section 4. A numerical example is given in Section 5.

2.0 *Model Definition*. As mentioned in Section 1, we consider the two-stage process defined by (1.1). The distributional assumptions we make on the b_k and σ_k^2 as well as the residuals $e_{jk|k} = x_{jk} - b_k$ when conditioned on k are discussed at some length in Fertig and Mann [3]. We do not wish to reiterate the reasons for these assumptions here, but just state them without further comment. First, as is often done, the conditioned residuals, $e_{jk|k}$, are assumed to be normal with mean zero and variance σ_k^2. The joint distribution of b_k and $h_k (=1/\sigma_k^2)$ is assumed to belong to the family of natural conjugate prior distributions to the normal, i.e. the normal-gamma with parameters μ', n', v', r' (see Raiffa and Schlaifer [5]). This implies that the unconditional distribution of X_{jk} will be generalized Student t and not normal as first considered by Fertig and Mann [1]. We now list the specific densities for b_k, h_k and X_{jk}, using notation of Raiffa and Schlaifer [5]. The unconditional joint distribution of b_k and h_k (with subscripts dropped) is:

$$f_{N\gamma_2}(b,h|\mu', n', v', r') = f_N(b|\mu', n'h) . f_{\gamma_2}(h|v', r')$$

$$= \frac{1}{\sqrt{2\pi}} (n'h)^{1/2} \exp\left[-\frac{n'h}{2}(b-\mu')^2\right] \quad (2.1)$$

$$\frac{(r'v'/2)^{r'/2}}{\Gamma(r'/2)} h^{r'/2-1} \exp[-r'v'h/2] ,$$

$$-\infty < b < \infty \; h, v', n', r' > 0$$

The parameters μ', n', v', and r' are related to the moments of b_k and σ_k^2 as follows:

$$\mu = E(b_k) = \mu' \quad (2.2)$$

$$\text{var}(b_k|\sigma_k^2) = \sigma_k^2/n' \quad (2.3)$$

$$\sigma_b^2 = \text{var}(b_k) = \sigma_e^2/n' \quad (2.4)$$

$$\xi = \text{var}(\sigma_k^2) = 2\sigma_e^4/(r'-4) \qquad r' > 4 \quad (2.5)$$

$$\sigma_e^2 = E(\sigma_k^2) - r'v'/(r'-2) \qquad r' > 2 \quad (2.6)$$

The conditional distribution of X_{jk} given b_k and h_k (sampling distribution from a given heat) is normal and is given by

$$f_N(X_{jk}|b_k, h_k) = \frac{1}{\sqrt{2\pi}} h_k^{1/2} \exp\left[-\frac{h_k}{2}(X_{jk}-b_k)^2\right] \qquad (2.7)$$

The unconditional distribution of X_{jk} is a generalized Student t-distribution.

$$f_X(x|\mu', n', v', r') = f_S\left(x|\mu', \frac{n'}{v'(n'+1)}, r'\right) \qquad (2.8)$$

where f_S is given by

$$f_S(z|m,H,\nu) = \frac{\nu^{\nu/2}}{B(1/2,\nu/2)} [\nu+H(z-m)^2]^{-(\nu+1)/2} \sqrt{H_j} \quad \nu,H>0 \qquad (2.9)$$

The expression $B(a_1,a_2)$ is the complete Beta function with parameters a_1 and a_2.

If one defines $\bar{X}_k$ and S_k^2 as the sample mean and variance to be calculated from the k'th batch, i.e. $\bar{X}_k = \sum_j X_{jk}/n_k$ and $S_k^2 = \sum_j (X_{jk}-\bar{X}_k)^2/(n_k-1)$, it can be shown (see Fertig and Mann [3]) that the joint conditional distribution of $\bar{X}_k$ and S_k^2 is

$$f(\bar{x}, s^2|b,h) = \frac{1}{\sqrt{2\pi}}(n\,h)^{1/2} \exp[-nh/2(\bar{x}-b)^2]$$

$$\frac{((n-1)h/2)^{((n-1)/2)}}{\Gamma((n-1)/2)} s^{2((n-3)/2)} \exp[-(n-1)h\,s^2/2] \qquad (2.10)$$

and that the unconditional distribution of $\bar{X}$ and S^2 is

$$f_{Si\beta_2}(\bar{x}_k, s_k^2|\mu', n', v', r') \qquad (2.11)$$

$$= f_S\left(\bar{x}_k \middle| \mu', \frac{n'n_k}{n'+n_k}\; \frac{r'+n_k-1}{r'v'+(n_k-1)s_k^2}, r'+n_k-1\right)$$

$$f_{i\beta_2}\left(s_k^2 \middle| (n_k-1)/2, r'/2, r'v'/(n_k-1)\right)$$

where f_S is given by (2.9) and $f_{i\beta_2}$ is the inverted Beta-2 given for example by Raiffa and Schlaifer [5] as

$$f_{i\beta_2}(y|p,q,w) = \frac{1}{B(p,q)} \frac{y^{p-1}w^q}{(y+w)^{p+q}} \quad y > 0,\ p > q,\ w > 0 \tag{2.12}$$

Finally, it can be shown that the joint posterior density of b and h given $\bar{x}$ and s^2 is normal-gamma. That is,

$$f_{N\gamma_2}(b,h|\bar{x}, s^2,\mu',n',v',r') \tag{2.13}$$
$$= f_N(b|\mu'',n''h)\cdot f_{\gamma_2}(h|v'',r'')$$

where

$$\mu'' = (n'\mu'+n\bar{x})/(n'+n)$$
$$n'' = n'+n$$
$$v'' = (r'\nu'+(n-1)s^2+\frac{nn}{n+n'}(\bar{x}-\mu')^2)/(r'+n')$$
$$r'' = r'+n'$$

We now assume that the acceptance criteria will be of the form, accept if $\frac{\bar{x}-L}{s} > A$ where L and A are specified. In order to answer the first question (I) posed in the introduction, we require the density of the strength, X, given it comes from a heat that has passed inspection. Let $\bar{x}$ and s be observed from a future heat and let X be one additional observable value from that same heat. Specifically, what is needed is the distribution of X given $\theta = \frac{\bar{x}-L}{s} > A$ and given the process parameters μ', n', v', and r'. Let b and h be the conditional parameters for the heat from which x, $\bar{x}$, s^2 are samples values.

Since, conditioned on b and h, X is assumed independent of $\bar{X}$ and S^2, we have that

$$f(x|\bar{x},s^2;\mu',n',v',r') = \iint_{bh} f_N(x|b,h)\,f(b,h|\bar{x},s^2,\ \mu',n',v',r')\,dbdh$$

$$= f_S\left(x|\mu'',\ \frac{n'+n}{v''(n'+n+1)},\ r''\right) \tag{2.14}$$

where μ'', v'', and r'' are as in (2.13). Since it is only hypothesized that $\theta = \frac{\bar{x}-L}{s} > A$, in order to derive a density useful in answering question (I) we need to integrate (2.14) times the density of $\bar{X}$ and S^2 given $\theta > A$ over the allowable ranges of $\bar{X}$ and S^2. This is effectively done in Section 3 to derive the expression $P\{x < q_\alpha | \theta > A; \mu', n', v', r'\}$ where $P(E_1|E_2)$ stands for the probability of E_1 given E_2. Setting this expression equal to α implies q_α is the α'th percentile of the strength of the material given it comes from a heat that has been accepted. This percentile can be approximated at design time if we replace μ', n', v', and r' by estimates of these parameters based on data taken at design time. Such an estimation procedure was discussed in Fertig and Mann [1]. In that paper, a maximum likelihood scheme was developed. A computer program to implement the method was written and is documented in Fertig and Mann [2].

In the next section, we develop the estimation procedure for percentiles of the strength distribution of X, given the process parameters in the presence of a screening program.

3.0 *Percentile Estimation in the Presence of a Screening Program.*

We seek to answer the following question at design time. Given estimates of the process parameters μ', n', v', and r', what is an estimate of any 100α th percentile of the strength distribution of heats of material produced by that process that meet a given acceptance criterion. A heat will be assumed to meet the acceptance criterion if a sample of size n drawn from that heat is such that its mean and standard deviation fall in the acceptance region given by

$$\bar{x} - As > L \tag{3.1}$$

where A (usually positive) and L are parameters specified by the quality control personnel. Of course, this question is only hypothetical at the design stage in that the actual sample data $\bar{x}$ and s^2 would not be available until the production stage. The question still has meaning though in probabilistic terms. If we let $\bar{x}$ and s^2 be the hypothetical sample mean and standard deviation based on n samples from a specific heat and if we consider X to be one additional observable variate from that heat, we seek the number q_α such that

$$P\,\{X < q_\alpha \,|\, \bar{x}\text{-}As > L;\ \mu',n',v',r'\} = \alpha \tag{3.2}$$

Thus q_α represents the point at which the probability is $1-\alpha$ that material passing inspection will have a strength measurement at least equal to q_α. We note that the larger L and A are, the more skewed to the right the density function implied by (3.2) will be.

An estimate of q_α, $\hat{q}_\alpha$, will be given by substituting the estimates $\hat{\mu}',\hat{n}',\hat{v}',\hat{r}'$ into (3.2) for μ',n',v',r' and solving for q_α. The expression given by (3.2) can be computed according to the following method. First note that

$$P\left\{X < q_\alpha \,\middle|\, \frac{\bar{x}-L}{s} > A;\mu',\ n',\ v',\ r'\right\} \tag{3.3}$$

$$= \iint_{b,h} P\,\{X < q_\alpha \,|\, \theta > A,b,h,\mu',n',v',r'\} \cdot f(b,h \,|\, \theta > A,\mu',n',v',r')\ db\,dh$$

where b and 1/h are the heat mean and variance. The first term is given by

$$P\,\{X < q_\alpha \,|\, \theta > A,b,h,\mu',n',v',r'\} = \Phi[h^{1/2}(q_\alpha - b)] \tag{3.4}$$

where $\Phi(\cdot)$ is the cumulative standard normal probability distribution

$$\Phi(z) = \int_{-\infty}^{z} \frac{1}{\sqrt{2\pi}} \exp(-t^2/2)\ d\,t \tag{3.5}$$

The second term is given by

$$f(b,h \,|\, \theta > A,\mu',n',v',r') \tag{3.6}$$

$$= \frac{f(b,h \,|\, \mu',n',v',r')}{P\{\theta > A \,|\, \mu',n',v',r'\}} \cdot P\ (\theta > A \,|\, b,h,\mu',n',v',r')$$

We note that the distribution of $\theta = \frac{\bar{X}-L}{S}$ given b and h is a scaler multiple of the noncentral t with n-1 degrees of freedom and non-centrality parameter $(b-L)\ (nh)^{1/2}$. Letting $T\ (t|\nu,\delta)$ denote the cumulative non-central t with ν degrees of freedom and non-centrality parameter δ, we have that

$$P\ \{\theta > A|b,h,\mu',n',v',r'\} = 1 - T\ (\sqrt{n}A|n-1,\ (nh)^{1/2}\ (b-L)) \tag{3.7}$$

Thus

$$P\ \{X < q_\alpha\ |\theta > A,\mu',n',v',r'\} \tag{3.8}$$

$$= \left\{ \iint\limits_{b,h} \Phi(h^{1/2}(q_\alpha-b))\ (1-T(\sqrt{n}\ A|n-1,(nh)^{1/2}(b-L))) \cdot f\ (b,h|\mu',n',v',r')\ dbdh \right\} \Big/ P\{\theta>A|\mu',n',v',r'\}$$

The denominator, which is just a normalizing constant, is given by

$$P\ \{\theta>A|\mu',n',v',r'\} \tag{3.9}$$

$$= \iint\limits_{b,h} [1-T\ (\sqrt{n}\ A\ |n-1,\ (nh)^{1/2}(b-L))] \cdot f\ (b,h|\mu',n',v',r')\ dbdh$$

The density $f(b,h|\mu',n',v',r')$ is given by equation (2.1)

Equations (3.8) and (3.9) can be integrated numerically. A computer program which solves equation (3.2) for q_α by integrating (3.8) (and (3.9)) numerically has been written and is available in Fertig and Mann [2]. The integration is done by successively using a 32 point Gauss-Laguerre quadrature formula for the inner "h integral" as a function of b and a 64 point Gauss-Hermite quadrature formula for the outer "b integral." An example output from this program is given in Section 5.

4.0 *Confidence Bounds on Percentiles of Given Future Heat.* In the previous section, the quantity q_α was defined as the 100α th percentile of the strength distribution of future accepted material if that acceptance process continues indefinitely. An estimate of q_α is given by estimating μ',n',v', and r'. In this section, we derive a confidence bound on a given percentile of a fixed future heat that undergoes acceptance testing. The confidence bound derived is Bayesian in nature. It does not represent a confidence statement concerning q_α, the 100α th percentile of accepted heats in general, but is restricted to fixed accepted heats. A confidence statement in the former case would read, for example, *one is* 100(1-β)% *confident that the 100 α th percentile of the strength distribution of heats produced by manufacturer XYZ that pass the prescribed acceptance testing is greater than* $q^*_{\alpha,\beta}$ (α is percentile level and 1-β is confidence level). The confidence statement in the latter case would read, *one is 100(1-β)% confident that the 100 α th percentile of a given heat that is accepted is greater than* $W_{\alpha,\beta}$. In this latter situation, one is concerned with within-heat variability. Since it is assumed that the strength distribution of a specified heat is normal, the 100α th percentile of that heat is just $b+z_\alpha c$ where $c^2 = 1/h$ is the within heat variance, b is the heat mean and z_α is the 100α th percentile of a standard normal distribution. Given μ', n', v', and r', one can make probability statements concerning the quantity $b+z_\alpha c$. If these probability statements are made taking into account that the heat has a sample which passes inspection, then a Bayesian confidence bound is formed relevant to the percentiles of the screened heats. Since the confidence statement is to be made at design time, all one wishes to specify is the fact that $\theta=(\bar{X}-L)/s$ is greater than A. This part is the same as in Section 3. Specifically, then the quantity $W_{\alpha,\beta}$ is defined through the equation

$$1-\beta = P\,\{b+z_\alpha c > W_{\alpha,\beta} \,|\, \theta>A;\mu',n',v',r'\} \qquad (4.1)$$

where the probability distribution on the right is derived using the posterior density of b and c conditioned on the fact that they are the parameters of a heat for which a sample size n was taken resulting in accepting the heat, i.e. $\theta > A$. This posterior density is just (3.6). We now consider the computation of (4.1). Let this probability be referred to as I. Then

$$I = \iint_{R_c} f(b,h|\theta>A;\ \mu',n',v',r')\,db\,dh \tag{4.2}$$

$$= \iint_{R_c} \frac{P(\theta>A|b,h;\ \mu',n',v',r')\,g(b,h|\mu',n',v',r')}{P(\theta>A|\mu',\ n',\ v',\ r')}\,db\,dh$$

where $R_c = \left\{b,h|b+z_\alpha/h^{1/2} > W_{\alpha,\beta}\right\}$

We note that as before, given b and h, θ can be expressed in terms of a non-central t-variate. Thus, dropping the notation μ',n',v',r' from each expression,

$$I = [1/P(\theta > A)] \iint_{R_c} (1-T(\sqrt{n}A|n-1,-(nh)^{1/2}(L-b))\cdot g(b,h)\,db\,dh \tag{4.3}$$

Through appropriate changes of variables we find that I can be expressed as

$$I = [1/P(\theta > A)] \tag{4.4}$$

$$\cdot\int_{\bar{h}=o}^{\infty}\left[\int_{u=u(h_R)}^{1} (1-T(\sqrt{n}A\ n-1,\delta(\bar{h},u))\,du\right]$$

$$\frac{1}{\Gamma(r'/2)}\ \bar{h}^{\,r'/2-1}\ \exp(-\bar{h})\ d\bar{h}$$

where

$$u(h_R) = \Phi\left[(W_{\alpha,\beta}-\mu')\left(\frac{2\bar{h}n'}{r'v'}\right)^{1/2} - z_\alpha n'^{1/2}\right] \tag{4.5}$$

$$\delta(\bar{h},u) = -\left(\frac{2n\bar{h}}{r'v'}\right)^{1/2}(L-\mu') + (n/n')^{1/2}\ z_u \tag{4.6}$$

and z_u is the u'th percentile of the standard normal distribution.

The expression $P(\theta>A)$ is computed from

$$P(\theta>A) = \int_{\bar{h}=o}^{\infty}\int_{u=o}^{\infty} (1-T(\sqrt{n}\ A|n-1,\ \delta(\bar{h},u))\,du \tag{4.7}$$

$$\cdot\ \frac{1}{\Gamma(r'/2)}\ \bar{h}^{\,r'/2-1}\ \exp(-\bar{h})\ d\bar{h}$$

A computer code for calculating I using (4.4) through (4.7) was written using Gaussian quadrature for the inner (u-integral) and Gauss-Laguerre quadrature for the outer $\bar{h}$-integral. Also written was a routine that used this code to determine $W_{\alpha,\beta}$ for a given α, β, A, L, μ', n', v', and r'. These subroutines are described in Fertig and Mann [2].

5.0 *Numerical Example*. Consider the data discussed in Fertig and Mann [1] and [3]. These data are given in Table I below. As noted in the above reference, they represent strength measurements made on a material in use at Rocketdyne.

The maximum likelihood estimates of μ', n', v', and r' are given in Fertig and Mann [3] as $\hat{\mu}' = 194.72$, $\hat{n}' = 0.2574$, $\hat{v}' = 7.159$ and $\hat{r}' = 5.197$. Percentile estimates as the 0.01, 0.05, and 0.10 percentile are given in Table II. They are computed from $\hat{q}_\alpha = \hat{\mu}' + \frac{\hat{v}'(1+\hat{n}')}{\hat{n}'} t_{\alpha}, \hat{r}'$ where $t_{\alpha,r}$ is the 100α th percentile of a Student-t variate with r degrees of freedom. Given in the same table are the strength percentiles after screening given by solving (3.2) for A = 2,3 and L = 175, 180, 185. One can see that the lower percentiles are markedly affected by the screening process.

TABLE I

Batch (k)	n_k	x_k	s_k^2
1	28	193.75	21.10
2	18	202.22	6.83
3	27	189.78	2.18
4	101	186.19	4.64
5	7	194.14	8.47
6	13	195.62	5.59
7	34	201.15	4.61
8	21	199.64	8.34
9	12	197.20	16.87
10	28	197.35	19.72
11	29	196.07	23.35
	318		

TABLE II

A	L	$q_{0.01}$	$q_{0.05}$	$q_{0.10}$	$W_{0.01,0.90}$	$W_{0.05,0.90}$	$W_{0.10,0.90}$
		175	183	186			
3	185	188	191	192	187	188	189
3	180	185	188	190	184	186	187
3	175	182	186	188	182	184	185
2	185	187	189	191	185	187	188
2	180	183	187	189	182	184	186
2	175	181	185	187	180	183	184

REFERENCES

[1] Fertig, Kenneth W. and Mann, Nancy R., Population Percentile Estimation for the Unbalanced Random-Effect Model: Estimation of the Strength of a Material when Sampling from Various Batches of the Material. Appearing in, Reliability and Biometry, Statistical Analyses of Life Data, edited by Frank Proscham and R. J. Serfling, SIAM Series in Applied Mathematics, (1974a).

[2] Fertig, Kenneth W. and Mann, Nancy R., Computer Program for Computing Design Allowables: Screened Batches, Rocketdyne Research Report, RR 74-02, (1974b).

[3] Fertig, Kenneth W. and Mann, Nancy R., Unbalanced Random-Effect Model with Unequal Group Variances: Maximum Likelihood Estimation of Prior Parameters and Test Design. Rocketdyne Research Report, RR 74-03, Canoga Park, California, submitted for publication, (1974c).

[4] Harville, D. A., Variance Component Estimation for the Unbalanced One-Way Random Classification - A critique Aerospace Research Laboratories 69-0180, Wright Patterson Air Force Base, Ohio, (1969).

[5] Raiffa, Howard and Schlaifer, Robert, Applied Statistical Decision Theory. Division of Research, Graduate School of Business Administration, Harvard University, Boston, (1961).

SOME STOCHASTIC CHARACTERIZATIONS OF MULTIVARIATE SURVIVAL†

by WILLIAM B. BUCHANAN*
and NOZER D. SINGPURWALLA**
The George Washington University

Abstract. In this paper we attempt to obtain some definitions for classes of multivariate life distributions based upon the notions of wear. Our rationale for considering these multivariate definitions is dictated by the following criteria: (i) conditions shall be imposed upon the joint survival function rather than on the corresponding random variables; (ii) a preservation of the established univariate definition in the case of a single variable; (iii) the existence of a chain of implications between the various definitions which is analogous to the known chain of implications between the various univariate classes; (iv) the definitions should be prompted by arguments which are natural generalizations of those used to define the various univariate classes.

1.0 *Introduction*. In this paper we present several definitions for classes of multivariate life distributions based upon the notion of wear. We emphasize our rationale for choosing these definitions, and present a chain of implications between them. In the interest of brevity, we merely outline the implications and omit details about their proofs. These have been delegated to another paper, where, in addition to such details, properties of survival times which satisfy the definitions given here are discussed (cf. Buchanan and Singpurwalla [5]).

---

†This work was supported in part by the Air Force Flight Dynamics Laboratory, Air Force Systems Command, USAF, under Contract F33615-74-C-4040, and in part by the Office of Naval Research under Contract N00014-75-C-0729, Project NR 347 020.

*Center for Naval Analyses, Arlington, Virginia.

**Department of Operations Research, The George Washington University, Washington, D.C. 20052.

Previous research towards defining classes of univariate distributions based upon the notion of wear has led to five different classes. These classes have been discussed by Barlow and Proschan [1], who also give results concerning the implications between these classes. In this paper we seek multivariate analogues to these five classes and obtain implications between them. Our main consideration behind this effort is that many biological and engineering systems consist of more than one component, and that it is more realistic to assume some form of dependence among the various components.

If $X_1, X_2, \ldots, X_m$ are the survival times of m devices, then $P(X_1 > x_1, \ldots, X_m > x_m)$ will be denoted by $\bar{F}_m(x_1, \ldots, x_m)$ or simply by $\bar{F}_m$. We shall say that a function of $m \geq 1$ variables, $\Gamma_m(x_1, \ldots, x_m)$, is increasing (decreasing) in $x_1, \ldots, x_m$ if it is non-decreasing (non-increasing) when any or all of the x_i's, $1 \leq i \leq m$, increase. We shall denote this as

$$\Gamma_m(x_1, \ldots, x_m) \uparrow (\downarrow) \; x_1, \ldots, x_m.$$

In what follows, $\underset{\sim}{x}$ and $\underset{\sim}{t}$ will be vectors, so that $\bar{F}_m(\underset{\sim}{x})$ denotes $\bar{F}_m(x_1, \ldots, x_m)$, and x^* and t^* denote vectors whose elements are identical to x and t, respectively; thus $x^* = (x, \ldots, x)$.

In the univariate case, a distribution F_1 is said to be

$$(1.1) \quad \text{IFR(DFR) iff } \frac{\bar{F}_1(x+t)}{\bar{F}_1(x)} \downarrow (\uparrow) \; x, \quad \forall x, t \geq 0, \; \bar{F}_1(x) > 0;$$

$$(1.2) \quad \text{IFRA(DFRA) iff } [\bar{F}_1(x)]^{1/x} \downarrow (\uparrow) \; x, \quad x > 0;$$

$$(1.3) \quad \text{NBU(NWU) iff } \bar{F}_1(x)\bar{F}_1(t) \underset{(\leq)}{\geq} \bar{F}_1(x+t), \; \forall x, t \geq 0, \; \bar{F}_1(x), \bar{F}_1(t) > 0;$$

$$(1.4) \quad \text{NBUE(NWUE) iff } \int_0^\infty \bar{F}_1(t)\,dt \underset{(\leq)}{\geq} \frac{\int_0^\infty \bar{F}_1(x+t)\,dt}{\bar{F}_1(x)}, \quad \forall x \geq 0, \; \bar{F}_1(x) > 0;$$

(1.5) DMRL(IMRL) iff $\frac{\int_0^\infty \overline{F}_1(x+t)dt}{\overline{F}_1(x)} \downarrow(\uparrow)\ x, \quad x \geq 0,\ \overline{F}_1(x) > 0.$

The above definitions have been motivated by an aging or a wearout characteristic of the item whose life length is of interest. We note here that the above definitions are conditions on the survival function $\overline{F}_1(x)$.

The following chain of implications between the classes defined above is well known:

$$\begin{matrix} \text{NBUE} & \Longleftarrow & \text{DMRL} & \Longleftarrow & \text{IFR} & \Longrightarrow & \text{IFRA} & \Longrightarrow & \text{NBU} & \Longrightarrow & \text{NBUE} . \\ \text{(NWUE)} & & \text{(IMRL)} & & \text{(DFR)} & & \text{(DFRA)} & & \text{(NWU)} & & \text{(NWUE)} \end{matrix}$$

2.0 *Requirements for the Multivariate Definitions.* We shall now define some multivariate analogues to the Conditions (1.1) through (1.5) given above. Our multivariate definitions shall be dictated by the following requirements:

(i) The multivariate definitions should be based upon conditions imposed upon the joint survival function rather than on the corresponding random variables.

(ii) The definition should be valid for the established definition in the case of a single variable.

(iii) The multivariate ($m \geq 2$) definitions should lead to a chain of implications which is analogous to the known chain of implications between the univariate definitions given in Section 1.0.

(iv) The arguments which prompt the multivariate definitions should be natural extensions of those used to define the various univariate classes.

The above four requirements are very reasonable, and should collectively be taken as a motivation for choosing the multivariate definitions given in this paper. We could regard Requirement (i) as being the one which does not restrict the resulting definitions to coherent life functions, Requirement (ii) as being the one which ensures consistency and conformity of the system of definitions, Requirement (iii) as the one which retains the esthetic feature of the univariate definitions, and Requirement (iv) as the one which interprets these definitions in the light of physical situations. We shall formalize Requirement (i) by

Definition 1.0: Survival times $X_1, X_2, \ldots, X_m$ are jointly □ if the distribution function of each subset of them is □.

Multivariate extensions of the basic IFR(DFR) condition and its equivalents have been discussed by Harris [8], Brindley and Thompson [3], Block [2], and Marshall [9], each in their own way. Of direct analogy to us here is the work by Brindley and Thompson and that by Marshall.

Multivariate extensions of the basic IFRA property have recently been discussed by Esary and Marshall [7], but these have been stated in terms of the behavior of the coherent life functions. An exception to this is Condition A of Esary and Marshall, but as pointed out by them, it suffers from a lack of intuitive content. During the course of this paper, we shall point out the equivalence between one of the conditions of Esary and Marshall (which is in terms of coherent life functions) and our definition of multivariate IFRA (which is in terms of the survival function).

With the exception of the multivariate IFR(DFR) and the multivariate IFRA, no literature dealing with the multivariate analogues of the other classes has come to our attention.

3.0 *Classes of Multivariate Life Distributions*. In what follows, the prefix M before the notations for the univariate class definitions given in Equations (1.1) through (1.5) will denote the multivariate extension of the class. Unlike the univariate definitions, each multivariate definition can take several variations which are quite natural. The variations between each class form a natural hierarchy, and we shall denote the "very strong" version of this hierarchy by adding a suffix VS after the corresponding multivariate class definition. Similarly, S, W, and VS will denote the "strong," the "weak," and the "very weak" versions of the hierarchy, respectively. We feel that this manner of notation, though a bit cumbersome, is indicative of the behavior of our system of definitions.

In all cases, without much difficulty, we can verify that

$$(\cdot)\text{-VS} \Longrightarrow (\cdot)\text{-S} \Longrightarrow (\cdot)\text{-W} \Longrightarrow (\cdot)\text{-VW}\,. \tag{3.0}$$

We now proceed to give our definitions.

Definition 3.1: For $\overline{F}_m > 0$, a multivariate distribution F_m defined on the positive orthant is

(3.1.1) MIFR-VS (MDFR-VS) if $\dfrac{\overline{F}_m(\underset{\sim}{x}+\underset{\sim}{t})}{\overline{F}_m(\underset{\sim}{x})} \downarrow(\uparrow)\ x, \quad \forall \underset{\sim}{t}, \underset{\sim}{x} \geq 0;$

(3.1.2) MIFR-S (MDFR-S) if $\dfrac{\overline{F}_m(\underset{\sim}{x}+t^*)}{\overline{F}_m(\underset{\sim}{x})} \downarrow(\uparrow)\ \underset{\sim}{x}, \quad \forall \underset{\sim}{t}, \underset{\sim}{x} \geq 0;$

(3.1.3) MIFR-W (MDFR-W) if $\dfrac{\overline{F}_m(x^*+\underset{\sim}{t})}{\overline{F}_m(x^*)} \downarrow(\uparrow)\ x, \quad \forall \underset{\sim}{t}, x \geq 0;$

(3.1.4) MIFR-VW (MDFR-VW) if $\dfrac{\overline{F}_m(x^*+t^*)}{\overline{F}_m(x^*)} \downarrow(\uparrow)\ x, \quad \forall t, x \geq 0.$

We note here that (3.1.1) through (3.1.4) are identical to Conditions I(I'), IV(IV'), III(III') and II(II') of Marshall [9]. Also, Definition (3.1.2) is identical to the Brindley-Thompson [3] definition of a multivariate increasing (decreasing) failure rate.

We can easily verify that the above definitions satisfy requirements (i) and (ii) of Section 2.0.

The conditional probabilities in the above definitions can be nicely interpreted in the light of physical situations (cf. Marshall [9]); in what follows, we shall discuss these briefly.

From the point of view of a systematic exposition, it is most convenient to start with the weakest case, namely the MIFR-VS (MDFR-VW). We recall that the defining property for this case requires that

$$\frac{\overline{F}_m(x+t,\ldots,x+t)}{\overline{F}_m(x,\ldots,x)} \downarrow(\uparrow)\ x, \quad x,t \geq 0.$$

The above condition is a natural extension of Definition (1.1), which implies that the residual life of an unfailed component is stochastically decreasing in the age of the device. In the multivariate case, we think of $m \geq 2$ components, all of which are new at inception, and given the survival of these to time x, we are interested in their survival to a common horizon of time, say t. The fact that time moves at the same rate for all components is reflected in writing the conditional probability.

A strengthening of the above definition leads us to the MIFR-W (MDFR-W) case, for which the defining property requires that

$$\frac{\bar{F}_m(x+t_1,\ldots,x+t_m)}{\bar{F}_m(x,\ldots,x)} \downarrow(\uparrow)\ \underset{\sim}{x}, \quad \forall x,\underset{\sim}{t} \geq 0.$$

The situation here is identical to the one discussed above, except that we are now interested in the survival of the m components to distinct horizon times $t_1,\ldots,t_m$.

A further strengthening of the above condition leads us to the MIFR-S (MDFR-S) case which requires that

$$\frac{\bar{F}_m(x_1+t,\ldots,x_m+t)}{\bar{F}_m(x_1,\ldots,x_m)} \downarrow(\uparrow)\ \underset{\sim}{x}, \quad \forall t,\underset{\sim}{x} \geq 0.$$

This condition is of interest when the m components in the system have different ages, say $x_1,x_2,\ldots,x_m$, and we are interested in their survival to a common time horizon, say t. The different ages of the components could arise due to a replacement policy, and the survival of the components to a common horizon time t is of particular interest when the system is a series system.

The strongest condition is the MIFR-VS (MDFR-VS), and this requires that

$$\frac{\bar{F}_m(x_1+t_1,\ldots,x_m+t_m)}{\bar{F}_m(x_1,\ldots,x_m)} \downarrow(\uparrow)\ \underset{\sim}{x}, \quad \forall \underset{\sim}{t},\underset{\sim}{x} \geq 0.$$

Clearly, this condition is appropriate when the components in a system with different ages, say $x_1,x_2,\ldots,x_m$, are required to survive to distinct horizon times $t_1, t_2, \ldots, t_m$, respectively.

In the strong and the very strong versions of the MIFR (MDFR) definitions, we had motivated the distinct ages, $x_1, x_2, \ldots, x_m$, of the components by visualizing a replacement policy. Another angle for motivating these is by considering new components at inception, but then assuming that the time moves at different rates for different components. This consideration is appropriate under accelerated life test conditions, wherein different stresses are imposed upon the different components. The effect of imposing elevated stresses on a component is equivalent to transforming the time.

We shall now define multivariate analogues of the basic IFRA (DFRA) property given by Equation (1.2). We bear in mind that mathematically, our definitions should satisfy the first three requirements of Section 2.0. In this case, we note that the multivariate analogue of the basic IFRA (DFRA) property suggests that we define two variations of the very weak version, the "very weak conditional" denoted by VWC and the "very weak unconditional" denoted by VWU. We shall elaborate upon this later.

Definition 3.2: For $\overline{F}_m > 0$, a multivariate distribution F_m defined on the positive orthant is

(3.2.1) MIFRA-VS (MDFRA-VS) if $\left[\dfrac{\overline{F}_m(\underset{\sim}{x}+\underset{\sim}{t})}{\overline{F}_m(\underset{\sim}{x})}\right]^{\left(\sqrt{\sum_{i=1}^{m} t_i^2/m}\right)^{-1}} \downarrow (\uparrow)\ \underset{\sim}{x},$

$\forall \underset{\sim}{t}, \underset{\sim}{x} \geq 0$, some $t_i > 0$;

(3.2.2) MIFRA-S (MDFRA-S) if $\left[\dfrac{\overline{F}_m(\underset{\sim}{x}+t^*)}{\overline{F}_m(\underset{\sim}{x})}\right]^{\frac{1}{t}} \downarrow (\uparrow)\ \underset{\sim}{x},\quad \forall t > 0,\ \forall \underset{\sim}{x} \geq 0;$

$$\text{(3.2.3)}\quad \text{MIFRA-W(MDFRA-W) if } \left[\frac{\overline{F}_m(x^*+\underset{\sim}{t})}{\overline{F}_m(x^*)}\right]^{\left(\sqrt{\sum_{i=1}^{m} t_i^2/m}\right)^{-1}} \downarrow(\uparrow)\ x,$$

$\forall \underset{\sim}{t}, x \geq 0$, some $t_i > 0$;

$$\text{(3.2.4)}\quad \text{MIFRA-VWC(MDFRA-VWC) if } \left[\frac{\overline{F}_m(x^*+t^*)}{\overline{F}_m(x^*)}\right]^{\frac{1}{t}} \downarrow(\uparrow)\ x,\quad \forall t, x > 0;$$

$$\text{(3.2.4a)}\quad \text{MIFRA-VWU(MDFRA-VWU) if } \left[\overline{F}_m(x^*)\right]^{1/x} \downarrow(\uparrow)\ x,\quad x > 0.$$

A motivation for considering the root-mean-square as the average in (3.2.1) and (3.2.3) follows from the fact that the distance between the points $(\underset{\sim}{x}+\underset{\sim}{t})$ and $\underset{\sim}{x}$ in the m-dimensional Euclidean space is $\left(\sum t_i^2\right)^{1/2}$. Dividing by m^2, we obtain a root-mean-square value for the average time over which we calculate the continued probability of survival, given survival to $\underset{\sim}{x}$. From a geometrical point of view, the above argument is a natural extension of the failure rate average concept. Conditions (3.2.2) and (3.2.4) represent a weakening of Conditions (3.2.1) and (3.2.3), respectively.

Condition (3.2.4a), the only unconditional version of the MIFRA definition, is a direct extension of the univariate IFRA definition (1.2). A justification for Condition (3.2.4a) also arises out of Condition E of Esary and Marshall [7], who define a multivariate IFRA by imposing conditions on the corresponding random variables. Specifically, their Condition E states that random variables $X_1, X_2, \ldots, X_m$ have a joint distribution which is IFRA if

$\min_{i \in S} X_i$ is IFRA for all nonempty subsets S of $\{1,2,\ldots,m\}$.

Clearly, if a set of random variables satisfies Condition E, then $\min_i X_i$ is IFRA, and thus

$$\left[P\left(\min_i X_i \geq x\right)\right]^{\frac{1}{x}} \downarrow x \implies \left[\bar{F}_m(x^*)\right]^{\frac{1}{x}} \downarrow x,$$

and this is precisely our Definition (3.2.4a).

We can verify that the various versions of the MIFRA condition satisfy the following chain of implications:

MIFRA-VS $\implies$ MIFRA-S $\implies$ MIFRA-W $\implies$ MIFRA-VWC $\implies$ MIFRA-VWU.

An analogous chain holds for the MDFRA class.

There are some additional features about Definitions (3.2.1) to (3.2.4) which need to be discussed. In the interest of brevity, we shall discuss these only in the context of the MIFRA versions, keeping in mind that these also apply to the MDFRA versions.

After some algebra, we can verify that MIFR($\cdot$) $\implies$ MIFRA($\cdot$), and this is nice because it represents a satisfaction of Requirement (iii) of Section 2.0. However, a consequence of Definitions (3.2.1) through (3.2.4) is that MIFRA(I) $\implies$ MIFR(I), for I = VS, S, W, and VWC, and this is not in the spirit of the univariate chain of implications given in Section 1.0. An exception to this occurs when we consider the very weak unconditional version, for now we can show that MIFRA(VWU) $\not\Longrightarrow$ MIFR($\cdot$), for any of the MIFR versions. Thus, the MIFRA-VWU (MDFRA-VWU) version appears to be the most appropriate one.

In order to ensure that MIFRA($\cdot$) $\not\Longrightarrow$ MIFR($\cdot$) for all its versions, one is tempted to consider alternate ways to define MIFRA. One possibility is to replace the t_i's in the exponent terms of Conditions (3.2.1) through (3.2.4) by x_i's. However, if this is done one cannot show that Requirement (ii) of Section 2.0 is true. Alternatively, if one chooses to delete the denominator term in Conditions (3.2.1) through (3.2.4), then one is not able to show that MIFR($\cdot$) $\implies$ MIFRA($\cdot$).

In view of the above discussion, one may wonder about the purpose of Definitions (3.2.1) through (3.2.4). The reason for their inclusion in our paper becomes convincing if one looks at Figure 1, wherein the chain of implications between our system of definitions is shown. Definitions (3.2.1) through (3.2.4) are not only logical extensions of the geometric property of the univariate IFRA, but more importantly, they serve as intermediate steps (or stops) used to show that MIFR(•) ⟹ MNBU(•), etc., and complete the chain of implications shown in Figure 1. In Figure 1, we draw the reader's attention to the position of the MIFRA-VWU version with respect to its relationship to the MIFR(•) and the MIFRA(I), I = VS, S, W, and VWC classes.

Multivariate versions of the NBU(NWU), NBUE(NWUE), and the DMRL(IMRL) classes are quite straightforward, being based upon generalizations of the arguments used to motivate the univariate versions. Motivations for the univariate version are discussed by Barlow and Proschan [1], and these will not be repeated here. In the definitions which will be given soon, the motivations for considering the various conditional probabilities are the same as those that were used to motivate the multivariate versions of the IFR(DFR). These were given in this paper following Definition (3.1).

Definition 3.3: For $\overline{F}_m > 0$, a multivariate distribution F_m defined on the positive orthant is

(3.3.1) MNBU-VS(MNWU-VS) if $\overline{F}_m(\underset{\sim}{x}+\underset{\sim}{t}) \underset{(\geq)}{\leq} \overline{F}_m(\underset{\sim}{x})\overline{F}_m(\underset{\sim}{t})$, $\forall \underset{\sim}{x},\underset{\sim}{t} \geq 0$;

(3.3.2) MNBU-S(MNWU-S) if $\overline{F}_m(\underset{\sim}{x}+t^*) \underset{(\geq)}{\leq} \overline{F}_m(\underset{\sim}{x})\overline{F}_m(t^*)$, $\forall t,\underset{\sim}{x} \geq 0$;

(3.3.3) MNBU-W(MNWU-W) if $\overline{F}_m(x^*+\underset{\sim}{t}) \underset{(\geq)}{\leq} \overline{F}_m(x^*)\overline{F}_m(\underset{\sim}{t})$, $\forall \underset{\sim}{t},x \geq 0$;

(3.3.4) MNBU-VW(MNWU-VW) if $\overline{F}_m(x^*+t^*) \underset{(\geq)}{\leq} \overline{F}_m(x^*)\overline{F}_m(t^*)$, $\forall x,t \geq 0$.

Definition 3.4: For $\overline{F}_m > 0$, a multivariate distribution F_m defined on the positive orthant is

(3.4.1) MNBUE-VS (MNWUE-VS) if $\int_0^\infty \bar{F}_m(\underset{\sim}{t})d\underset{\sim}{t} \underset{(\leq)}{\geq} \frac{\int_0^\infty \bar{F}_m(\underset{\sim}{x}+\underset{\sim}{t})d\underset{\sim}{t}}{\bar{F}_m(\underset{\sim}{x})}$, $\forall \underset{\sim}{x} \geq 0$;

(3.4.2) MNBUE-S (MNWUE-S) if $\int_0^\infty \bar{F}_m(t^*)dt \underset{(\leq)}{\geq} \frac{\int_0^\infty \bar{F}_m(t^*+\underset{\sim}{x})dt}{\bar{F}_m(\underset{\sim}{x})}$, $\forall \underset{\sim}{x} \geq 0$;

(3.4.3) MNBUE-W (MNWUE-W) if $\int_0^\infty \bar{F}_m(\underset{\sim}{t})d\underset{\sim}{t} \underset{(\leq)}{\geq} \frac{\int_0^\infty \bar{F}_m(x^*+\underset{\sim}{t})d\underset{\sim}{t}}{\bar{F}_m(x^*)}$, $\forall x \geq 0$;

(3.4.4) MNBUE-VW (MNWUE-VW) if $\int_0^\infty \bar{F}_m(t^*)dt \underset{(\leq)}{\geq} \frac{\int_0^\infty \bar{F}_m(x^*+t^*)dt}{\bar{F}_m(x^*)}$, $\forall x \geq 0$.

Definition 3.5: For $\bar{F}_m > 0$, a multivariate distribution F_m defined on the positive orthant is

(3.5.1) MDMRL-VS (MIMRL-VS) if $\frac{\int_0^\infty \bar{F}_m(\underset{\sim}{x}+\underset{\sim}{t})d\underset{\sim}{t}}{\bar{F}_m(\underset{\sim}{x})} \downarrow(\uparrow)\ \underset{\sim}{x}$, $\forall \underset{\sim}{x} \geq 0$;

(3.5.2) MDMRL-S (MIMRL-S) if $\frac{\int_0^\infty \bar{F}_m(\underset{\sim}{x}+t^*)dt}{\bar{F}_m(\underset{\sim}{x})} \downarrow(\uparrow)\ \underset{\sim}{x}$, $\forall \underset{\sim}{x} \geq 0$;

(3.5.3) MDMRL-W (MIMRL-W) if $\frac{\int_0^\infty \bar{F}_m(x^*+\underset{\sim}{t})d\underset{\sim}{t}}{\bar{F}_m(x^*)} \downarrow(\uparrow)\ x$, $\forall x \geq 0$;

(3.5.4) MDMRL-VW (MIMRL-VW) if $\frac{\int_0^\infty \bar{F}_m(x^*+t^*)dt}{\bar{F}_m(x^*)} \downarrow(\uparrow)\ x$, $\forall x \geq 0$.

All the versions given under Definitions (3.3), (3.4) and (3.5) satisfy the chain of implications given by Equation (3.0).

4.0 *Implications Between Classes.* The implications between the various variations of a class were discussed in Section 3.0. In this section we shall establish a chain of implications between the various classes.

In most cases, the proofs which establish these implications are a generalization of those given by Bryson and Siddiqui [4]. Whenever there is a non-implication between the various univariate classes, such as for example DMRL $\not\Longrightarrow$ IFR, this non-implication holds for all the corresponding multivariate analogues. This is because all the variations of a multivariate class are equivalent for a single variable. Non-implication is generally established by a counter-example, and these would be the same as the ones given by Bryson and Siddiqui [4].

Theorem 4.1: If F_m is a multivariate distribution function defined on the positive orthant, then $\bar{F}_m$

(4.1.1) MIFR(I) $\Longleftrightarrow$ MIFRA(I) $\Longrightarrow$ MNBU(I) $\Longrightarrow$ MNBUE(I),

(4.1.2) MIFR(I) $\Longrightarrow$ MIFRA-VWU,

(4.1.3) MIFR(I) $\Longrightarrow$ MDMRL(I) $\Longrightarrow$ MNBUE(I),

where the index (I) denotes VS, S, W, or VW (VWU in the IFRA case).

Proof: We shall prove this theorem only when the index (I) is VS. An analogous line of proof can be used to prove this theorem for the other variations of (I). An exception to this is the implication between MIFRA-VWU and MNBU-VW, and this will be shown.

- $\bar{F}_m$ MIFR-VS $\Longrightarrow \ell n\bar{F}_m(\underset{\sim}{x}+\underset{\sim}{t}) - \ell n\bar{F}_m(\underset{\sim}{x}) \downarrow \underset{\sim}{x}$

$$\Longleftrightarrow \frac{\ell n\bar{F}_m(\underset{\sim}{x}+\underset{\sim}{t}) - \ell n\bar{F}_m(\underset{\sim}{x})}{\left(\sum_{i=1}^{m} t_i^2/m\right)^{1/2}} \downarrow x \Longrightarrow \bar{F}_m \text{ IFRA-VS.}$$

- $\bar{F}_m$ MIFRA-VS $\Longrightarrow \ell n\bar{F}_m(\underset{\sim}{x}+\underset{\sim}{t}+\underset{\sim}{\Delta}) - \ell n\bar{F}_m(\underset{\sim}{x}+\underset{\sim}{\Delta})$

$$\leq \ell n\bar{F}_m(\underset{\sim}{x}+\underset{\sim}{t}) - \ell n\bar{F}_m(\underset{\sim}{x}), \quad \forall \underset{\sim}{\Delta} \geq 0,$$

where $\underset{\sim}{\Delta}$ is a vector whose elements are Δ_i, $i=1,2,\ldots,m$. If we now set $\underset{\sim}{x} \equiv 0$, then

$$\overline{F}_m \text{ MIFRA-VS} \Longrightarrow \overline{F}_m \text{ MNBU-VS}.$$

- $\overline{F}_m$ MNBU-VS $\Longrightarrow \overline{F}_m(\underset{\sim}{x}+\underset{\sim}{t}) \leq \overline{F}_m(\underset{\sim}{t})\overline{F}_m(\underset{\sim}{x})$, and integrating both sides with respect to $\underset{\sim}{t}$ we have

$$\overline{F}_m \text{ MNBU-VS} \Longrightarrow \overline{F}_m \text{ MNBUE-VS};$$

this completes the proof for (4.1.1).

- $\overline{F}_m$ MIFR-VS $\Longrightarrow \dfrac{\overline{F}_m(\underset{\sim}{x}+\underset{\sim}{t})}{\overline{F}_m(\underset{\sim}{x})} \downarrow \underset{\sim}{x}$; integrating with respect to

$\underset{\sim}{t}$ we have

$$\overline{F}_m \text{ MIFR-VS} \Longrightarrow \overline{F}_m \text{ MDMRL-VS}.$$

- $\overline{F}_m$ MDMRL-VS $\Longrightarrow \dfrac{\int_0^\infty \overline{F}_m(\underset{\sim}{x}+\underset{\sim}{t})dt}{\overline{F}_m(\underset{\sim}{x})} \geq \dfrac{\int_0^\infty \overline{F}_m(\underset{\sim}{x}+\underset{\sim}{t}+\underset{\sim}{\Delta})dt}{\overline{F}_m(\underset{\sim}{x}+\underset{\sim}{\Delta})}$, for $\overline{F}_m(\cdot) >$

0. If we set $\underset{\sim}{x} \equiv 0$, we have

$$\overline{F}_m \text{ MDMRL-VS} \Longrightarrow \overline{F}_m \text{ MNBUE-VS};$$

this completes the proof for (4.1.2).

- $\overline{F}_m$ MIFRA-VWU $\Longrightarrow \dfrac{\ln\overline{F}_m(x^\star)}{x} \geq \dfrac{\ln\overline{F}_m(t^\star)}{t}$, for all $t > x > 0$.

We shall prove that $\overline{F}_m$ MIFRA-VWU $\Longrightarrow \overline{F}_m$ MNBU-VW by contradiction.

Assume that (3.3.4) does not hold when (3.2.4a) is true; that that is,

$$\ln\overline{F}_m(x^\star+t^\star) \geq \ln\overline{F}_m(x^\star) + \ln\overline{F}_m(t^\star).$$

Since $\overline{F}_m$ MIFRA-VWU $\Longrightarrow \ln\overline{F}_m(x^\star) \geq \dfrac{x}{t} \ln\overline{F}_m(t^\star)$, we have

$$\ln\overline{F}_m(x^\star+t^\star) \geq \dfrac{x}{t} \ln\overline{F}_m(t) + \ln\overline{F}_m(t^\star),$$

or that

$$\frac{\ell n\overline{F}_m(x^*+t^*)}{x + t} \geq \frac{\ell n\overline{F}_m(t^*)}{t},$$

a contradiction for $\overline{F}_m$ MIFRA-VWU. //.

In Figure 1, we present a schematic for out chain of implications between the classes, and this is appealing.

5.0 *Properties of Joint Survival Times.* The following properties are reasonable requirements for any multivariate conditions on joint survival times, $X_1, X_2, \ldots, X_m$ (cf. Brindley and Thompson [3], Marshall [9]).

(P_1) $(X_1, \ldots, X_m)$ satisfies Condition □ $\Longrightarrow$ each nonempty subset of $X_1, \ldots, X_m$ satisfies Condition □.

(P_2 $(X_1, \ldots, X_m)$ satisfies Condition □; $(Y_1, \ldots, Y_n)$ satisfies Condition □; and $(X_1, \ldots, X_m)$, $(Y_1, \ldots, Y_n)$ independent $\Longrightarrow$ $(X_1, \ldots, X_m; Y_1, \ldots, Y_n)$ satisfies Condition □.

(P_3) $(X_1, \ldots, X_m)$ satisfies Condition □, and J_i is a subset of $(1, 2, \ldots, m)$, then $(Y_1, \ldots, Y_n)$ satisfies Condition □, where $Y_i = \min(X_j | j \varepsilon J_i)$, for $i = 1, 2, \ldots, n$.

(P_4) $(X_1, \ldots, X_m)$ satisfies both Condition □ and its dual, Condition □^{D} $\Longleftrightarrow$ $(X_1, \ldots, X_m)$ have a multivariate distribution with exponential marginals. For example, if Condition □ denotes MIFR-VS, then the dual Condition □^{D} is MDFR-VS. An example of a multivariate distribution with exponential marginals is a multivariate distribution with exponential minimums (Esary and Marshall [6]).

(P_5) $(X_1, \ldots, X_m)$ satisfies Condition □ $\Longrightarrow$ $(aX_1, \ldots, aX_m)$ satisfies Condition □, for all $a > 0$.

We shall now verify if the Definitions 3.1 through 3.5 satisfy properties P_1 through P_5. We have summarized our results in Table 1.

Because of Definition 1.0, it is immediate that P_1 is satisfied by all the definitions given in Section 3.0.

In order to verify that P_2 is satisfied by 3.1 through 3.5, the only result that we need to use is that if $X_1,\ldots,X_m$ and $Y_1,\ldots,Y_n$ are mutually independent, then

$$\bar{F}_{m+n}(\underset{\sim}{x}+\underset{\sim}{y}) = \bar{F}_m(\underset{\sim}{x})\bar{F}_n(\underset{\sim}{y}).$$

In order to verify if P_3 is satisfied by 3.1 through 3.5, we need to use the fact that $Y_i > y_i$, for $i=1,2,\ldots,n$ is equivalent to $X_j > x_j$, for $j=1,2,\ldots,m$, where $x_j = \max\left(y_i \mid j\varepsilon J_i\right)$; hence

$$P\left(Y_1 > y_1,\ldots,Y_n > y_n\right) \equiv P\left(X_1 > x_1,\ldots,X_m > x_m\right).$$

Thus, for example, $\bar{F}_m$ MIFR-S $\Longrightarrow$

$$\frac{\bar{F}_m(\underset{\sim}{x}+t^*)}{\bar{F}_m(\underset{\sim}{x})} \downarrow \underset{\sim}{x},\ \forall t \geq 0 \quad \Longrightarrow \quad \frac{\bar{F}_n(\underset{\sim}{y}+t^*)}{\bar{F}_n(\underset{\sim}{y})} \downarrow \underset{\sim}{y},\ \forall t \geq 0,$$

and this implies that $Y_1,\ldots,Y_n$ are MIFR-S. Similar arguments lead us to conclude that P_3 is satisfied by all the definitions given in Section 4.0, except the very steong versions and their duals. The fact that the very strong versions do not satisfy P_3 is demonstrated by a counter-example. For this purpose, the Gumbel's distribution $\bar{F}_2(x_1,x_2) = \exp[-x_1-x_2-\delta x_1 x_2]$, $0 \leq \delta \leq 1$, $x_1 \geq 0$, $x_2 \geq 0$, which does satisfy the very strong versions of our definitions, but fails to satisfy P_3 is useful.

In order to investigate the nature of the distributions at the boundaries of the various class definitions and their duals, we note that in the univariate case only exponential distributions are both IFR and DFR. In keeping with this, property P_4 is specified; it requires that multivariate distributions which simultaneously satisfy a condition and its dual have exponential marginals. For the discussion which follows, we shall need

Definition 5.0: (Esary and Marshall [6]). Non-negative random variables $X_1, X_2, \ldots, X_m$ have a joint distribution with exponential <u>minimums</u> if

$$P\left(\min_{i \in I} X_i > x\right) = e^{-\mu_I x}$$

for some $\mu_I > 0$ and all non-empty subsets $I \subset \{1,2,\ldots,m\}$.

We also need to introduce a set of non-negative random variables $T_1, T_2, \ldots, T_m$ having a joint distribution which satisfies one of the following conditions:

(C_1) $T_1, \ldots, T_m$ are independent and each T_i has an exponential distribution;

(C_2) $T_1, \ldots, T_m$ have a multivariate exponential distribution;

(C_3) $\min_{i \in I} a_i T_i$ has an exponential distribution for all $a_i > 0$, $i=1,2,\ldots,m$, and all non-empty subsets $I \subset \{1,2,\ldots,m\}$;

(C_4) $T_1, \ldots, T_m$ have a joint distribution with exponential minimums;

(C_5) Each T_i, $i=1,\ldots,m$ has an exponential distribution.

Esary and Marshall [6] show that the classes of joint distributions defined by the above conditions are distinct, and that the following implication is valid

$$C_1 \Longrightarrow C_2 \Longrightarrow C_3 \Longrightarrow C_4 \Longrightarrow C_5.$$

The functional form $\overline{F}_m(x+t) = \overline{F}_m(x)\overline{F}_m(t)$ characterizes the class of joint distributions which are both MIFR-VS and MDFR-VS. This functional form holds if and only if $\overline{F}_m$ has independent exponential marginals. The functional forms characterizing the class of joint distributions which are both MIFR-S and MDFR-S, and the class of joint distributions which are both MIFR-W and MDFR-W hold if and only if F_m is MVE. The functional form which characterizes the very weak version of this class $\bar{F}_m(x^*+t^*) = \bar{F}_m(x^*)\bar{F}_m(t^*)$ is satisfied if and only if F_m has exponential minimums (cf. Marshall [9]).

Analogous results follow for the various versions of the MIFRA(MDFRA) class. For the MIFRA-VWU(MDFRA-VWU) class, we note that

$$\frac{1}{x} \ln \bar{F}_m(x^*) = \frac{1}{x} \ln P\left[X_1 > x,\ldots,X_m > x\right] \quad \text{constant in } x$$

implies that $P\left[\min_{1 \leq i \leq m} X_i > x\right]^{1/x}$ is constant in x, and this condition is satisfied if and only if F_m has exponential minimums.

Results for the MNBU(MNWU) case are quite straightforward, and these are given in Table 1.

For the MNBUE-VS, we have the result that

$$\int_0^\infty \left[\bar{F}_m(\underset{\sim}{t}) - \frac{\bar{F}_m(\underset{\sim}{x}+\underset{\sim}{t})}{\bar{F}_m(\underset{\sim}{x})}\right] d\underset{\sim}{t} \geq 0.$$

Because a multivariate distribution obtained as a product of independent exponential marginals will satisfy an equality for the expression given above, we shall say that the boundary distributions for this definition form a class which contains the independent exponential marginals. Based upon a similar reasoning, we conclude that the boundary distributions for the strong and the weak versions of the MNBUE(MNWUE) condition form a class which includes the MVE distribution. However, based upon arguments analogous to the VWU version of the MIFRA(MDFRA) condition, we can conclude that a multivariate distribution function is both MNBUE-VW and MNWUE-VW if and only if it has exponential minimums.

Results for the MDMRL(MIMRL) case are analogous to those for the MNBUE(MNWUE) case, and these are summarized in Table 1.

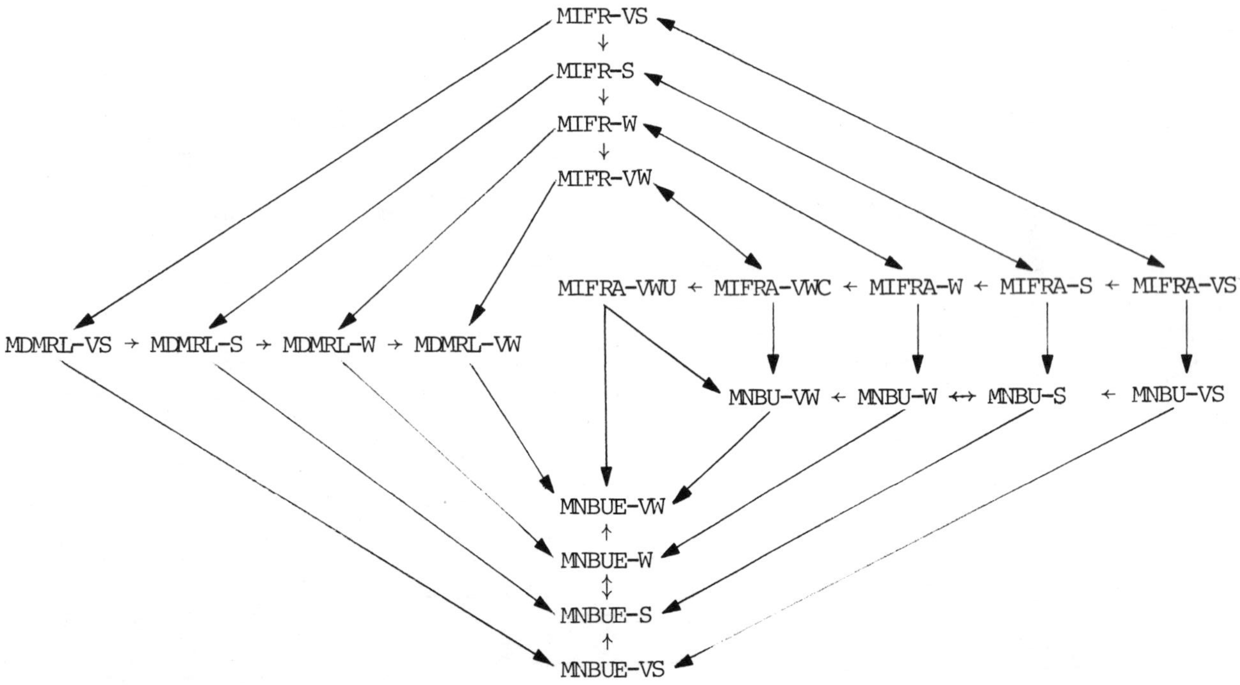

Figure 1.--Schematic of implications between classes of multivariate distributions.

TABLE 1

PROPERTIES OF JOINT SURVIVAL TIMES FOR THE FIVE CLASSES OF MULTIVARIATE DISTRIBUTION FUNCTIONS

Distribution	Survival Time Properties					
	(i) Uni-variate Property	(ii) Union Property	(iii) Subset Property	(iv) Sets of Minimums Property	(v) Boundary Distribution	(vi) Scale Property
MIFR-VS	yes	yes	yes	no	Independent Expon. Marginals	yes
MIFR-S	"	"	"	yes	MVE	"
MIFR-W	"	"	"	"	"	"
MIFR-VW	"	"	"	"	Exponential Minimums	"
MIFRA-VS	yes	yes	yes	no	Independent Expon. Marginals	yes
MIFRA-S	"	"	"	yes	MVE	"
MIFRA-W	"	"	"	"	"	"
MIFRA-VWC	"	"	"	"	"	"
MIFRA-VWU	"	"	"	"	Exponential Minimums	"
MNBU-VS	yes	yes	yes	no	Independent Expon. Marginals	yes
MNBU-S	"	"	"	yes	MVE	"
MNBU-W	"	"	"	"	"	"
MNBU-VW	"	"	"	"	Exponential Minimums	"
MNBUE-VS	yes	yes	yes	no	Contains Indep. Expon. Marginals	yes
MNBUE-S	"	"	"	yes	Contains MVE	"
MNBUE-W	"	"	"	"	"	"
MNBUE-VW	"	"	"	"	Exponential Minimums	"
MDMRL-VS	yes	yes	yes	no	Contains Indep. Expon. Marginals	yes
MDMRL-S	"	"	"	yes	Contains MVE	"
MDMRL-W	"	"	"	"	"	"
MDMRL-VW	"	"	"	"	Exponential Minimums	"

Acknowledgments. We would like to acknowledge the several helpful comments of Professor Henry W. Block.

REFERENCES

[1] Barlow, R. E. and Proschan, F., Statistical Theory of Reliability and Life Testing. Holt, Rinehart and Winston, Inc., New York, (1974).

[2] Block, H. W., Monotone Hazard and Failure Rates for Absolutely Continuous Multivariate Distributions. Research Report #73-20, Department of Mathematics, University of Pittsburgh.

[3] Brindley, E. C., Jr. and Thompson, W. A., Jr., "Dependence and Aging Aspects of Multivariate Survival," *J. Amer. Statist. Assoc.*, *67*, (December, 1972), pp. 822-830.

[4] Bryson, M. C. and Siddiqui, M. M., "Some Criteria for Aging," *J. Amer. Statist. Assoc.*, *64*, (1969), pp. 1472-1483.

[5] Buchanan, W. B. and Singpurwalla, N. D., Some Stochastic Characterizations of Multivariate Survival. Technical Paper Serial T-319, Institute for Management Science and Engineering, The George Washington University, Washington, D.C., (1975).

[6] Esary, J. D. and Marshall, A. W., "Multivariate Distributions with Exponential Minimums," *Ann. Statist.*, *2*, *(1)*, (1974), pp. 84-98.

[7] Esary, J. D. and Marshall, A. W., Multivariate IHRA Distributions. Unpublished manuscript (July, 1975).

[8] Harris, R., "A Multivariate Definition for Increasing Hazard Rate Distribution Functions," *Ann. Math. Statist.*, *41*, (1970), pp. 713-717.

[9] Marshall, A. W., "Multivariate Distributions with Monotone Hazard Rate," in Reliability and Fault Tree Analysis (R. E. Barlow, J. B. Fussell, and N. D. Singpurwalla, eds.), SIAM, Philadelphia, (1975).

A FAMILY OF BIVARIATE LIFE DISTRIBUTIONS

by HENRY W. BLOCK
University of California, Berkeley and University of Pittsburgh

Abstract. A characteristic function equation and an equivalent compounding scheme are studied. The equation arises from a model for a compartmental system and the compounding scheme arises from a shock model. The class of distributions satisfying the equation and the scheme contain the bivariate exponential distributions of Marshall and Olkin, Downton, Hawkes, and Paulson, as well as the bivariate geometric distributions of Hawkes, Paulson and Uppuluri, Esary and Marshall, and Arnold. The scheme is shown to generalize the univariate shock model of Esary, Marshall and Proschan. Properties of the distributions are derived by studying the equation and the compounding scheme which they satisfy.

1.0 Introduction. Unlike the multivariate normal distribution there are many multivariate geometric and exponential distributions. Multivariate geometric distributions have been introduced by Bates and Neyman [4], Paulson and Uppuluri [20], Hawkes [14], Esary and Marshall [10] and Arnold [1]. Multivariate exponential distributions have been considered by Gumbel [13], Marshall and Olkin [16], Downton [9], Hawkes [14], and Paulson [18]. Some multivariate exponential extensions related to the Marshall and Olkin distribution have been given by Block and Basu [6] and Block [5]. Many of these distributions, their relations and their characterizations have been discussed by Basu and Block [3].

In this paper a class of distributions is studied which arises in two ways: first, as a consequence of a random shock model due to Arnold [2]; second, from a characteristic function equation due to Paulson and Uppuluri [20]. It is shown that these two methods are equivalent via a bivariate geometric compounding mechanism and that most of the bivariate geometric and exponential distributions of the authors listed above are in this class. These distributions can then be interpreted as arising from this random shock model and many of their properties can be derived through the use of the characteristic function equation. Furthermore it can be shown that the random shock model generalizes the univariate shock model of Esary, Marshall and Proschan [12]. The bivariate shock models of Marshall and Olkin [16] and Buchanan and Singpurwalla [8] are also seen to be special cases of the random shock model.

In Section 2.0 the bivariate geometric compounding mechanism which connects the two different approaches is introduced. The generalization of the univariate shock model of Esary, Marshall and Proschan is discussed in Section 3.0. In Section 4.0 properties of the bivariate family are discussed.

2.0 The Bivariate Geometric Compounding Mechanism Arnold [2] has given heirarchies of multivariate geometric and exponential distributions. Paulson and Uppuluri [20] and Paulson [18] have used a characteristic function equation to derive particular multivariate geometric and exponential distributions. This equation comes from a model for a compartmental system which is discussed in Paulson and Uppuluri [21]. In the approaches of Arnold and of Paulson and Uppuluri a simple bivariate geometric distribution can be isolated. Using this distribution, in conjunction with either the compounding scheme of Arnold or the equation of Paulson and Uppuluri, many of the bivariate geometric and exponential distributions in the literature can be obtained. These distributions can then be viewed as arising from shock models in the sense of Arnold [2] and their properties can be studied using the characteristic function equation of Paulson and Uppuluri [20].

The fundamental bivariate geometric distribution referred to above will be designated BVG. The bivariate random variable (N_1,N_2) has the BVG with nonnegative parameters $p_{00},p_{01},p_{10},p_{11}$

if it has survival function

$$\bar{F}(n_1,n_2)=P\{N_1>n_1,N_2>n_2\}=\begin{cases} p_{11}^{\,n_1}(p_{01}+p_{11})^{n_2-n_1} & \text{if } n_1 \le n_2 \\ \\ p_{11}^{\,n_2}(p_{10}+p_{11})^{n_1-n_2} & \text{if } n_2 \le n_1 \end{cases} \quad (2.1)$$

where $p_{00}+p_{01}+p_{10}+p_{11}=1$, $p_{10}+p_{11}<1$, and $p_{01}+p_{11}<1$.

The generating function of (N_1,N_2) is then found to be

$$g(s_1,s_2) = \frac{s_1s_2}{1-p_{11}s_1s_2} \quad p_{00} + \frac{p_{01}(p_{00}+p_{10})s_2}{1-(p_{01}+p_{11})s_2} +$$

$$\frac{p_{10}(p_{00}+p_{01})s_1}{1-(p_{10}+p_{11}s_1} \quad . \quad (2.2)$$

From the form of the survival function this is seen to be the BVG-W of Esary and Marshall [10] who have derived this distribution from a shock model. This shock model concerns two devices which are exposed to failure which occurs in discrete cycles. On each cycle there is a shock to both devices and with probability p_{11} both devices survive, with probability p_{10} the first survives and the second does not, with probability p_{01} the first fails and the second survives and with probability p_{01} the first fails and the second survives and with probability p_{00} both fail. The events of surviving the shocks are independent from cycle to cycle. Following Arnold [2] and conditioning on the first cycle, the generating function has the form

$$g(s_1,s_2) = s_1s_2\{p_{00}+p_{01}g(1,s_2)+p_{10}g(s_1,1)+p_{11}g(s_1,s_2)\} \quad (2.3)$$

and this can be solved to give (2.2). This is also done in Hawkes [14] and the same distribution is obtained in slightly different notation. It should be noticed that (2.3) is a special case of the characteristic function equation

$$\phi(t_1,t_2) = \psi(t_1,t_2)[p_{00}+p_{01}\phi(0,t_2)+p_{10}\phi(t_1,0)+p_{11}\phi(t_1,t_2)] \quad (2.4)$$

where $p_{00}, p_{01}, p_{10}, p_{11}$ are probabilities such that $p_{01}+p_{11} < 1$ and $p_{10} + p_{11} < 1$. This equation is only a slightly more general version of (8) of Paulson and Uppuluri [20] or (15) of Paulson [18]. Both of these equations are obtained from the above by taking $\psi(t_1,t_2)$ to be a product of univariate characteristic functions and $p_{00} = a$, $p_{01} = c$, $p_{10} = b$ and $p_{11} = d$.

Equation (2.4) gives the characteristic equation of the bivariate random variable

$$\left(\sum_{i=1}^{N_1} X_{i1}, \sum_{i=1}^{N_2} X_{i2}\right) \quad (2.5)$$

where (N_1,N_2) is BVG and is independent of (X_{i1},X_{i2}), $i = 1,2, \ldots$ which are i.i.d. bivariate r.v.'s with characteristic function $\psi(t_1,t_2)$. Thus (2.4) corresponds to bivariate geometric compounding of the distribution with c.f. $\psi(t_1,t_2)$ with respect to the BVG.

It is also clear that if $\psi(t_1,t_2)$ has geometric (exponential) marginals, then $\phi(t_1,t_2)$ will have geometric (exponential) marginals. This is even more apparent from (2.5) since univariate random geometric sums of i.i.d. geometric (exponential random variables are geometric (exponential).

The concept of univariate geometric compounding has a connection with renewal processes. This connection was observed by Renyi and is called thinning of renewal point processes or rerefaction. In the simplest situation this procedure consists of observing a renewal process and recording an event of the process with probability p and not recording it with probability $1 - p$. The resulting process is called a rare-faction or a thinned process. The new interarrival times are random geometric sums and so can be given by the univariate characteristic function equation analog of (2.4) (see (5) of Paulson [18]). A similar bivariate interpretation can be made. For example, let two identical Poisson processes $Z_1(t)$ and $Z_2(t)$ (i.e., $Z_1(t) \equiv Z_2(t)$) describe the behavior of two components and assume an event of the $Z_1(t)$ process (which necessarily occurs at the same time for $Z_2(t)$) is recorded for the first component and not for the second with probability p_{10}, for the second component and not the first with probability p_{01}, for neither component with probability p_{00}, and for both components with probability p_{11}. The resulting bivariate Poisson process has interarrival times which follow the Marshall and Olkin [16] bivariate exponential distribution. This approach corresponds to a special case of these author's nonfatal shock model and will be discussed in more detail in Section 4.0.

The great generality of the equation (2.4) and the compounding scheme should now be evident. Table 2.1 summarizes the ways in which the various distributions satisfy the equation and hence fit into the compounding scheme. The distributions are arranged in order from simplest to most complex. The ranges of the values of the parameters are as given in the article in which the distribution originally appeared.

The table shows very clearly the relationships among the various distributions. The BVG discussed earlier is seen to be the geometric distribution of Hawkes and the BVG-W of Esary and Marshall. The generalized Arnold distribution appears to be more general but by the alternate form given, it is seen to be a reparameterized version of the BVG. Furthermore the BVG-N of Esary and Marshall and the ordinary geometric of Arnold are special cases of the BVG. The Paulson and Uppuluri distribution

Table 2.1

Distributions from Paulson-Uppuluri Equation (2.4)

Distribution	$\psi(t_1,t_2)$	Parameters	References In Source
A. Bivariate Geometric 1. Independent Marginals	$e^{i(t_1+t_2)}$	$p_{11}=p_1p_2,\ p_{01}=(1-p_1)p_2$ $p_{10}=p_1(1-p_2), p_{00}=(1-p_1)(1-p_2)$	
2. Esary & Marshall (1973) a. BVG-N	$e^{i(t_1+t_2)}$	(given in Fig. 1 of source)	(2.1)
b. BVG-W	$e^{i(t_1+t_2)}$	$p_{ij}=p_{ij};\ i,j=0,1$	(2.2)
3. Hawkes (1972)	$e^{i(t_1+t_2)}$	$p_{ij}=p_{ij};\ i,j=0,1$	p. 130 $(\pi(z_1,z_2))$
4. Arnold (1974) a. Ordinary	$e^{i(t_1+t_2)}$	$p_{00}=0, p_{01}=p_1, p_{10}=p_2, p_{11}=p_0$	(2)
b. Generalized	$\dfrac{e^{i(t_1+t_2)}}{1+\frac{p}{1-p}(1-e^{i(t_1+t_2)})}$	$p_{ij}=p'_{ij};\ i,j=0,1$	Described in Section 4 of Arnold (1974)
(Alternately)	$e^{i(t_1+t_2)}$	$p_{00}=(1-p)p'_{00}, p_{01}=(1-p)p'_{01}$ $p_{10}=(1-p)p'_{10}, p_{11}=p+(1-p)p'_{11}$	

Table 2.1 (continued)

Distribution	$\psi(t_1,t_2)$	Parameters	References In Source
5. Paulson and Uppuluri (1972)	$[1+\frac{p}{1-p}(1-e^{it_1})]^{-1} \times [1+\frac{q}{1-q}(1-e^{it_2})]^{-1}$	$p_{00}=a, p_{01}=c, p_{10}=b, p_{11}=d$	(8)
(Readjusted version so that geometric variables start at 1 instead of zero.)	$\frac{e^{it_1}}{[1+\frac{q}{1-q}(1-e^{it_1})]} \times \frac{e^{it_2}}{[1+\frac{q}{1-q}(1-e^{it_2})]}$	$p_{00}=a, p_{01}=c, p_{10}=b, p_{11}=d$	
6. Arnold (1975)			
a. $G_0^{(2)}=\{\phi(t_1+t_2) \mid \phi(t) = \frac{e^{it}}{[1+\frac{p}{1-p}(1-e^{it})]}, 0 \leq p < 1\}$			Described in Section 4 of Arnold (1975)

Table 2.1 (continued)

Distribution	$\psi(t_1,t_2)$	Parameters	References In Source
b. $G_n^{(2)}=\{\phi(t_1,t_2)\mid\phi$ satisfies equation with ψ and p_{ij} as follows$\}$	$\psi \varepsilon G_{n-1}^{(2)}$	$p_{ij}=p_{ij};i,j=0,1$	Described in Section 4 of Arnold (1975)

Table 2.1 (continued)

Distribution	$\psi(t_1,t_2)$	Parameters	References In Source
B. Bivariate Exponential			
1. Independent Marginals $\phi(t_1,t_2) = [1-i\theta_1 t_1]^{-1} \times [1-i\theta_2 t_2]^{-1}$	$\frac{1}{1-i\theta(t_1+t_2)}$	$p_{00}=0, p_{01}=\frac{\theta}{\theta_1}, p_{10}=\frac{\theta}{\theta_2},$ $p_{11}=1-\frac{\theta}{\theta_1}-\frac{\theta}{\theta_2}$	
2. Marshall and Olkin (1967)	$\frac{1}{1-i\theta(t_1+t_2)}$	$p_{00}=\theta\lambda_{12}, p_{01}=\theta\lambda,$ $p_{10}=\theta\lambda_2, p_{11}=1-\theta\lambda$ $(\lambda=\lambda_1+\lambda_2+\lambda_{12})$	Section 3.2
3. Downton (1970)	$[1-i\frac{t_1}{\mu_1(1+\gamma)}]^{-1}$ $[1-i\frac{t_2}{\mu_2(1+\gamma)}]^{-1}$	$p_{00}=\frac{1}{1+\gamma}, p_{01}=0, p_{10}=0,$ $p_{11}=\frac{\gamma}{1+\gamma}$	(2.10)

Table 2.1 (continued)

Distribution	$\psi(t_1,t_2)$	Parameters	References In Source
4. Hawkes (1972)	$[1=i(\frac{P_1}{\mu_1})t_1]^{-1}\times$ $[1-i(\frac{P_2}{\mu_2})t_2]^{-1}$	$p_{ij}=p_{ji};i=0,1$ $(P_1=p_{11}+p_{10},P_2=p_{11}+p_{01})$	(3)
5. Paulson (1973)	$[1-i\theta_1 t_1]^{-1}\times$ $[1-i\theta_2 t_2]^{-1}$	$p_{00}=a,p_{01}=c,p_{10}=b,p_{11}=d$	(15)
6. Arnold (1975)			
a. $E_0^{(2)}=\{\phi(t_1+t_2)\mid\phi(t)=\frac{1}{1-i\theta t},\ 0<\theta\}$	$\psi(t_1,t_2)\ \varepsilon\ E_{n-1}^{(2)}$	$p_{ij}=p_{ij};i,j=0,1$	Section 2
b. $E_n^{(2)}=\{\phi(t_1,t_2)\mid\phi(t_1,t_2)$ satisfies equation with ψ and p_{ij} as follows$\}$			Section 2

is a more general distribution, reducing to the BVG for $p = 0$ in the readjusted version. The Arnold families given in 6 were derived using geometric compounding via the distribution in 4b. It is no loss of generality to view this compounding with respect to the BVG. It is clear that these families contain all of the distributions listed above. Certainly the Hawkes, Arnold, and Esary and Marshall distributions are in $G_1^{(2)}$. Moreover by examining (2.1) it can be seen that $G_1^{(2)}$ contains the distribution with independent geometric marginals. Thus it follows that the Paulson and Uppuluri distribution is in $G_2^{(2)}$.

Similar observations about the bivariate exponential distributions can be made from the table. First it is clear that the Downton distribution is a special case of the Hawkes and the Paulson distributions. The latter two distributions can be seen to be the same. The Arnold classes given in 6 contain all of the exponential distributions given in the table. First 1 and 2 are in $E_1^{(2)}$ and since 1 is in $E_1^{(2)}$, it follows that 3, 4, and 5 are in $E_2^{(2)}$.

The duality between the two sets of distributions is evident. The Marshall and Olkin distribution is dual to the BVG, while the Hawkes-Paulson distribution is dual to the Paulson and Uppuluri distribution. A distribution which is dual to the Downton distribution can be obtained by taking $b = 0$ and $c = 0$ in the Paulson and Uppuluri distribution. There is also complete duality between the Arnold classes.

The only other major bivariate geometric distribution which does not fit into this scheme is the Bates-Neyman distribution. This does not satisfy the Paulson-Uppuluri characteristic function equation. However using a model given in Downton [9] a similar equation can be derived. Assume two components are affected by shocks and with probability p_1 the first component survives, with probability p_2 the second component survives and with probability $1 - p_1 - p_2$ both components fial. If these shocks are independent, then by conditioning on the first occurrence, the joint characteristic function satisfies

$$\phi(t_1,t_2)=p_1 e^{it_1}\phi(t_1,t_2)+p_2 e^{it_2}\phi(t_1,t_2)+(1-p_1-p_2)e^{i(t_1+t_2)}$$

and solving for ϕ gives

$$\phi(t_1,t_2) = \frac{(1 - p_1 - p_2)e^{i(t_1+t_2)}}{1 - p_1e^{it_1} - p_2e^{it_2}}$$

which is the characteristic function of the Bates-Neyman [4] distribution. If we compound independent exponentials with respect to this distribution the Downton distribution is obtained. Many other analogs of the previous system are also possible.

3.0 Bivariate Shock Models and Wear Processes Esary, Marshall and Proschan [12] have discussed certain univariate shock models and wear processes. These authors have introduced models which can be described as follows. A device is affected by a sequence of shocks occurring randomly in time according to a Poisson process $N(t)$ with rate λ. These shocks occur at times X_1, $X_1 + X_2, \ldots, \sum_{i=1}^{n} X_i, \ldots$ where the X_i are the interarrival times for the Poisson process and consequently are independent exponentials with mean λ^{-1}. The device fails when a random number N of shocks occurs. The time to failure of the device is then $T = \sum_{i=1}^{N} X_i$. If N is a nonnegative integer valued random variable independent of $N(t)$, then T has survival function

$$\overline{H}(t) = \sum_{k=0}^{\infty} \frac{e^{-\lambda t}(\lambda t)^k}{k!} P\{N > k\} . \tag{3.1}$$

If the random number of shocks is caused when the sum of successive amounts of damage given by $Z_1, Z_2, \ldots$ exceeds a critical threshold Y, i.e. $N = \min \{k: \sum_{i=1}^{k} Z_i > Y\}$, then T has survival function

$$\overline{H}(t) = \sum_{k=0}^{\infty} \frac{e^{-\lambda t}(\lambda t)^k}{k!} P\{\sum_{i=1}^{k} Z_i \leq Y\} . \tag{3.2}$$

In the special case that the critical threshold is a constant x and $Z_1, Z_2, \ldots$ are independent with the same distribution $F(x)$, (3.2) becomes

$$\overline{H}(t) = \sum_{k=0}^{\infty} \frac{e^{-\lambda t}(\lambda t)^k}{k!} F^{(k)}(x) . \tag{3.3}$$

More generally assuming that $X_1, X_2, \ldots$ are i.i.d. nonnegative random variables and $N(t)$ is the renewal process defined by

$$N(t) = \max \{k \mid \sum_{i=1}^{k} X_i \leq t\}$$

it follows that T has survival function

$$\overline{H}(t) = \sum_{k=0}^{\infty} P\{N(t) = k\} P\{N > k\} . \tag{3.4}$$

A bivariate generalization will now be described. A two component device is subjected to shocks and the times until shocks n_1 and n_2 in the first and second component respectively are

$$\left(\sum_{i=1}^{n_1} X_{i1}, \sum_{i=1}^{n_2} X_{i2} \right)$$

where for $j = 1, 2$ $X_{1j}, X_{2j}, \ldots$ are independent and identically distributed exponentials. Assuming the number of shocks until failure of these components are N_1 and N_2 respectively (and N_1 and N_2 are independent of the X_{ij}) gives that the times until failure are

$$(T_1,T_2) = \left(\sum_{i=1}^{N_1} X_{i1}, \sum_{i=1}^{N_2} X_{i2} \right) . \tag{3.5}$$

These times have bivariate survival function

$$\overline{H}(t_1,t_2) = \sum_{k_1=0}^{\infty} \sum_{k_2=0}^{\infty} P\{N_1(t_1)=k_1,N_2(t_2)=k_2\}P\{N_1>k_1,N_2>k_2\} \tag{3.6}$$

where

$$N_1(t_1) = \max\left\{ k_1 : \sum_{i=1}^{k_1} X_{i1} \le t_1 \right\}$$

$$N_2(t_2) = \max\left\{ k_2 : \sum_{i=1}^{k_2} X_{i2} \le t_2 \right\} .$$

The quantities $N_1(t_1)$ and $N_2(t_2)$ are univariate Poisson processes which are dependent if X_{i1} and X_{i2} are dependent.

Various special cases of this shock model in different guises have appeared in the literature. These will now be given along with some generalizations which remain within the framework of (3.6) above. Results similar to those in Esary, Marshall and Proschan [12] will be discussed elsewhere.

3.1 Marshall and Olkin Type Shock Models A shock model is said to be of Marshall and Olkin type if the survival functions of the times to failure are given by (3.6) where $N_1(t) \equiv N_2(t)$ with rate θ^{-1}. It then follows that

$$P\{N_1(t_1) = k_1, N_2(t_2) = k_2\} = \begin{cases} \dfrac{\exp(-\theta^{-1}(t_1-t_2))(\theta^{-1}(t_1-t_2))^{k_1-k_2}}{(k_1-k_2)!} \dfrac{\exp(-\theta^{-1}t_2)(\theta^{-1}t_2)^{k_2}}{k_2!} & \text{if } t_2 \le t_1 \text{ and } k_2 \le k_1, \\ \dfrac{\exp(-\theta^{-1}(t_2-t_1))(\theta^{-1}(t_2-t_1))^{k_2-k_1}}{(k_2-k_1)!} \dfrac{\exp(-\theta^{-1}t_1)(\theta^{-1}t_1)^{k_1}}{k_1!} & \text{if } t_1 \le t_2 \text{ and } k_1 \le k_2, \\ 0 \text{ otherwise.} \end{cases} \tag{3.1}$$

The reason that these are said to be of Marshall and Olkin type is that if (N_1,N_2) has the BVG with $p_{00} = \theta\lambda_{12}$, $p_{01} = \theta\lambda_1$, $p_{10} = \theta\lambda_2$ and $p_{11} = 1 - p_{00} - p_{01} - p_{10}$, then (T_1,T_2) has the BVE of Marshall and Olkin. As observed in the paragraph following (2.5), this is a special case of the nonfatal shock model of Marshall and Olkin. Furthermore if N_1 and N_2 are independent geometrics, (T_1,T_2) still has the BVE. The general shock model of this type is obtained if (N_1,N_2) has an arbitrary bivariate distribution on the lattice of nonnegative integers.

The models of Esary and Marshall [11] where the distributions of (N_1,N_2) and (T_1,T_2) are obtained from a bivariate **threshold** random variable (Y_1,Y_2), which is assumed to have the BVE, are special cases of this type of shock model.

3.2 Downton-Hawkes-Paulson Type Shock Models In this case

$$P\{N_1(t_1)=k_1,N_2(t_2)=k_2\}= \frac{\exp(-\theta_1^{-1}t_1)(\theta_1^{-1}t_1)^{k_1}}{k_1!} \frac{\exp(-\theta_2^{-1}t_2)(\theta_2^{-1}t_2)^{k_2}}{k_2!} \tag{3.2}$$

where θ_1 and θ_2 are the rates of the independent Poisson processes $N_1(t_1)$ and $N_2(t_2)$. If (N_1, N_2) is BVG, then for various choices of the parameters (see Table 2.1B) (T_1,T_2) has the Downton, Hawkes or Paulson distributions. The general model of this type occurs when (N_1,N_2) has an arbitrary distribution on the negative integers.

3.3 Buchanan and Singpurwalla Type Shock Models Here $N_1(t_1)$ and $N_2(t_2)$ are defined by

$$(N_1(t_1), N_2(t_2)) = (Z_1(t_1) + Z_{12}(t_1), Z_2(t_2) + Z_{12}(t_2)) \tag{3.3}$$

where $Z_1(t)$, $Z_2(t)$, $Z_{12}(t)$ are independent Poisson processes with rates λ_1, λ_2 and λ_{12} respectively. The particular case where N_1 and N_2 are independent nonnegative integer valued random variables gives the specific model proposed by Buchanan and Singpurwalla [8]. The general case occurs if N_1 and N_2 are not necessarily independent. Also notice that for $\lambda_1 = \lambda_2 = 0$, the Marshall and Olkin type model is obtained.

4.0 Properties of the Class The form of the characteristic function equation (2.4) can be utilized to derive properties of the distributions which satisfy it. One property of interest is that of infinite divisibility. Of all the bivariate exponential distributions given in Table 2.1, only the distribution of Downton and the distribution with independent exponential marginals are known to be infinitely divisible. Through the use of the equation, the infinite divisibility of the Marshall and Olkin, the Paulson-Hawkes and many of the distributions in the Arnold classes can be completely determined. These results have been proved in the paper of Block, Paulson and Kohlberger [7] and are summarized in Table 4.1.

The infinite divisibility of the distributions discussed above is important because of their use in an estimation scheme due to Paulson. This scheme (see Paulson, Holcomb and Leich [19] and Leitch and Paulson [15] involves minimizing

$$I(\underset{\sim}{\theta}) = \int |\hat{\phi}(\underset{\sim}{t}) - \phi_{\underset{\sim}{\theta}}(\underset{\sim}{t})|^2 h(\underset{\sim}{t}) d\underset{\sim}{t} \tag{4.1}$$

with respect to the parameter $\underset{\sim}{\theta} = (\theta_1, \theta_2, \ldots, \theta_k)$ where $\underset{\sim}{t} = (t_1, t_2, \ldots, t_n)$, $\hat{\phi}(\underset{\sim}{t})$ is an empirical characteristic function and $h(\underset{\sim}{t})$ is a weight function. Certain cases are currently being studied by Paulson and his students where $\phi_{\underset{\sim}{\theta}}(t_1, t_2) = [\phi_{\underset{\sim}{\theta}'}(t_1, t_2)]^{\gamma}$ and $\phi_{\underset{\sim}{\theta}'}(t_1, t_2)$ is the characteristic function of one of the distributions from Table 2.1. Since γ

Table 4.1

Infinite Divisibility of Bivariate Distribution

Form of $\psi(t_1,t_2)$ (from (2.4))	Restriction on Parameters (in addition to those following (2.4))	Specific Distributions	Infinite Divisibility
1. Arbitrary c.f.	$p_{01}=p_{10}=0$		Yes
2. Exponential marginals	$p_{01}=0$ or $p_{10}=0$	Members of Arnold Classes	Yes
3. $\psi(t_1,t_2) = [1-i\theta(t_1+t_2)]^{-1}$	$p_{00},p_{01},p_{10},p_{11}$ from Table 2.1 (B.2)	Marshall and Olkin	
	a) $p_{01}=0$ or $p_{10}=0$	a) $\lambda_1=0$ or $\lambda_2=0$	Yes
	b) $p_{00}=0$	b) $\lambda_{12}=0$	Yes
	c) $p_{00}>0,p_{01}>0,p_{10}>0$	c) $\lambda_1>0,\lambda_2>0,\lambda_{12}>0$	No
4. $\psi(t_1,t_2) = (1-i\theta_1 t_1)^{-1}(1-i\theta_2 t_2)^{-1}$	$p_{00}=a,p_{01}=c,p_{10}=b,p_{11}=d$	Paulson-Hawkes	
	a) $p_{01}=0$ or $p_{10}=0$	a) b=0 or c=0	Yes
	b) $p_{01}>0,p_{10}>0,p_{11}=0$	b) b>0,c>0,d=0	No
	c) $p_{00}=0,p_{01}>0,p_{10}>0,p_{11}>0$	c) a=0,b>0,c>0,d>0	No
	d) $p_{00}>0,p_{01}>0,p_{10}>0,p_{11}>0$	d) a>0,b>0,c>0,d>0	No (except if ad=bc)

is one of the parameters to be estimated and it will not in general be an integer, it is important to know that $\phi_{\underset{\sim}{\theta}}(t_1,t_2)$ is infinitely divisible so that $\phi_{\underset{\sim}{\theta}}(t_1,t_2)$ will be a characteristic function. Mitchell [17] has used the above procedure to define a negative binomial distribution where $\phi_{\underset{\sim}{\theta}}(t_1,t_2)$ is a bivariate geometric distribution obtained from equation (2.4). He has fitted this distribution to bivariate data for 1) diseases for industrial workers, 2) shift workers' absenteeism, 3) aircraft aborts, and 4) demand for aircraft parts.

Another use of the characteristic function equation is in studying the families of distributions of Arnold [2]. These arise through multiple compounding of the random geometric sums as discussed in Section 2.0 or equivalently through successive iterations of the characteristic function equation (2.4).

The equation (2.4) can be rewritten as

$$\phi(T) = \psi(T)E(\phi(TV)) \tag{4.2}$$

where $T = (t_1,t_2)$ and the expectation is with respect to the random matrix V which takes values in the set

$$V = \left\{ \begin{bmatrix} 0 & 0 \\ 0 & 0 \end{bmatrix}, \begin{bmatrix} 1 & 0 \\ 0 & 0 \end{bmatrix}, \begin{bmatrix} 0 & 0 \\ 0 & 1 \end{bmatrix}, \begin{bmatrix} 1 & 0 \\ 0 & 1 \end{bmatrix} \right\}$$

with probabilities $P\left\{V = \begin{bmatrix} 0 & 0 \\ 0 & 0 \end{bmatrix}\right\} = p_{00}$, $P\left\{V = \begin{bmatrix} 1 & 0 \\ 0 & 0 \end{bmatrix}\right\} = p_{10}$, $P\ V = \begin{matrix} 0 & 0 \\ 0 & 1 \end{matrix} = p_{01}$, $P\ V = \begin{matrix} 1 & 0 \\ 0 & 1 \end{matrix} = p_{11}$. Arnold's exponential families can then be given by

$$\begin{aligned} E_0^{(2)} &= \{\psi_0(T) : \psi_0(T) = [1 - i\theta(t_1 + t_2)]^{-1},\ 0 < \theta\} \\ E_n^{(2)} &= \{\psi_n(T) : \psi_n(T) = \psi_{n-1}(T)E(\psi_n(TV_n))\} \quad \text{for } n \geq 1 \end{aligned} \tag{4.3}$$

where each V_n has probabilities $p_{00}^{(n)}$, $p_{01}^{(n)}$, $p_{10}^{(n)}$ and $p_{11}^{(n)}$ satisfying $p_{01}^{(n)} + p_{11}^{(n)} < 1$ and $p_{10}^{(n)} + p_{11}^{(n)} < 1$ for $n \geq 1$. The exponential marginal distributions at any stage can be given for $n \geq 1$ by the characteristic function

$$\psi_n(t_1, 0) = \left[1 - i \frac{\mu_{1,n-1}}{p_{00}^{(n)} + p_{01}^{(n)}} t_1\right]^{-1}$$

$$\psi_n(0, t_2) = \left[1 - i \frac{\mu_{2,n-1}}{p_{00}^{(n)} + p^{(n)}} t_1\right]^{-1} \tag{4.4}$$

where $\mu_{1,n-1}$, $\mu_{2,n-1}$ are the marginal first moments of $\psi_{n-1}(T)$. The moments are thus given by

$$\mu_{1,n} = \theta \left[\prod_{k=1}^{n} (p_{00}^{(k)} + p_{01}^{(k)})\right]^{-1}$$

$$\mu_{2,n} = \theta \left[\prod_{k=1}^{n} (p_{00}^{(k)} + p_{10}^{(k)})\right]^{-1} . \tag{4.5}$$

The correlation coefficients can be expressed iteratively for $n \geq 1$ by

$$\rho_n = \frac{\{p_{00}^{(n)} p_{11}^{(n)} - p_{01}^{(n)} p_{10}^{(n)} (1 - \rho_n)^{-1}\} + p_{00}^{(n)} \rho_{n-1}}{1 = p_{11}^{(n)}} \tag{4.6}$$

and it can be shown that if $\rho = \lim_{n\to\infty} \rho_n$ exists, then ρ satisfies $-\frac{1}{3} \leq \rho \leq 1$.

Conditions under which the family of limiting distributions $\lim_{n\to\infty} \psi_n(T)$ exist have been studied by Block, Paulson and Kohlberger [7]. One result which these authors obtain is that compounding of the Downton type (i.e. $p_{01}^{(n)} = p_{10}^{(n)} = 0$) leads to a Downton distribution in the limit if $\prod_{k=2}^{\infty} (1 - p_{11}^{(k)})$ converges.

References

[1] Arnold, B. C., "A Characterization of the Exponential Distribution by Geometric Compounding," *Sankhya*, to appear, (1974).

[2] ____________,"Multivariate Exponential Distributions Based on Hierarchical Successive Damage," *J. Appl. Prob.*,*Vol. 12*, (1975), pp. 142-147.

[3] Basu, A. P. and Block, H. W.,"On Characterizing Univariate and Multivariate Exponential Distributions with Applications", Statistical Distributions In Scientific Work, Eds. G. P. Patil, S. Kotz and J. K. Ord, Vol. 3, D. Reidel, Dordrecht, (1975), pp. 399-422.

[4] Bates, G. E. and J. Neyman,"Contributions to the Theory of Accident Proneness, I, II", *Univ. Calif. Publications in Statistics*, 1, 215-54, 255-75, (1952).

[5] Block, Henry W.,"Continuous Multivariate Exponential Extensions", Reliability and Fault Tree Analysis, (Eds. R. Barlow, J. Fussell, N. Singpurwalla), SIAM, pp. 285-306, (1975).

[6] ______________, and A. P. Basu,"A Continuous Bivariate Exponential Extension", *J. Amer. Statist. Assoc.*, Vol. 69, (1974), pp. 1031-1037.

[7] ______________, A. S. Paulson and R. C. Kohlberger, Some Bivariate Exponential Distributions: Syntheses and Properties", to appear, (1975).

[8] Buchanan, W. B. and N. D. Singpurwalla, "Some Stochastic Characterizations of Multivariate Survival", George Washington University, Serial T-319, (1975).

[9] Downton, F., "Bivariate Exponential Distributions in Reliability Theory," *J. Roy. Statist. Soc.*, Ser. B., Vol. 32, (1970), pp. 822-830.

[10] Esary, J. D. and A. W. Marshall, "Multivariate Geometric Distributions Generated by a Cumulative Damage Process," Naval Postgraduate School Report NP555EY73041A, (1973).

[11] ______________________________, "Families of Components and Systems, Exposed to a Compound Poisson Damage Process," Reliability and Biometry, (Eds. F. Proschan and R. Serfling), SIAM Philadelphia, (1975), pp. 31-46.

[12] ______________________________ and F. Proschan, "Shock Models and Wear Processes," *Ann. Probability*, *1*, (1973), pp. 627-649.

[13] Gumbel, E. J., "Bivariate Exponential Distributions," *J. Amer. Statist. Assoc.*, *56*, (1960), pp. 335-349.

[14] Hawkes, A. G., "A Bivariate Exponential Distribution with Applications to Reliability," *J. Roy. Statist. Soc.*, Ser. B, Vol. 34, (1972) pp. 129-131.

[15] Leitch, R. A. and A. S. Paulson, "Estimation of Stable Law Parameters: Stock Price Behavior Applications, *J. Amer. Statist. Assoc.*, Vol. 70, (1975) pp. 690-697.

[16] Marshall, A. W. and I. Olkin, "A Multivariate Exponential Distribution," *J. Amer. Statist. Assoc.*, Vol. 62, (1967), pp. 30-44.

[17] Mitchell, R., "On a Bivariate Negative Binominal Distribution," unpublished notes, (1975).

[18] Paulson, A. S., "A Characterization of the Exponential Distribution and a Bivariate Exponential Distribution," *Sankhya*, Ser. A, Vol. 35, (1973), pp. 69-78.

[19] ______________, E. W. Holcomb and R. A. Leitch, "The Estimation of the Parameters of the Stable Laws," *Biometrika*, Vol. 62, (1975), pp. 163-170.

[20] Paulson, A. S. and V. R. R. Uppuluri, "A Characterization of the Geometric Distribution and a Bivariate Geometric Distribution," *Sankhya*, Ser. A, Vol. 34, (1972), pp. 88-91.

[21] ________________________________," Limits Laws of a Sequence Determined by a Random Difference Equation Governing a One-Compartment System," *Mathematical Bio-Science*, Vol. 13, (1972a), pp. 325-333.

TECHNIQUES FOR ANALYZING MULTIVARIATE FAILURE DATA

by RICHARD E. BARLOW*
University of California, Berkeley

and FRANK PROSCHAN**
Florida State University

Abstract. New multivariate techniques for analyzing multivariate life test data are introduced. The hazard gradient is estimated using the multivariate piecewise exponential distribution model. An example based on caterpillar tractor data is analyzed. Isoprobability contours are used to analyze joint distribution structure.

1.0 *Introduction*. In this paper, we introduce new multivariate techniques for analyzing life test data. Using the multivariate piecewise exponential distribution model, we show how to estimate the hazard gradient. Through an example based on caterpillar tractor data, we show how to analyze the marginal distributions and the joint distribution structure. Using asymptotic results for conditional maximum likelihood estimators of multivariate piecewise exponential distribution parameters, we show how to construct asymptotic simultaneous confidence limits on univariate marginal failure rate functions.

Since the multivariate piecewise exponential distribution model is in general discontinuous on cell boundaries (see Section 3), it may not provide a good approximation to continuous multivariate life distributions. It does however suggest natural

---

* Research partially supported by the Office of Naval Research under Contract N00014-75-C-0781, the U.S. Nuclear Regulatory Commission under Contract AT(49-24)-0147 and the National Science Foundation under Grant MPS75-14194 with the University of California. Reproduction in whole or in part is permitted for special purposes within the United States Government.

** Research sponsored by the Air Force Office of Scientific Research, AFSC, USAF, under Grant AFOSR 74-2581B.

estimators for the hazard gradient. (The hazard gradient in the multivariate case is analogous to the well-known failure rate in the univariate case.) These hazard gradient estimators can be used to investigate the structure of the joint distribution and also to estimate conditional distributions. Continuous bivariate distributions with piecewise linear hazard gradient coordinates are also considered. Estimation of parameters for this model is more difficult.

2.0 *Preliminaries*. Let F be a d-variate right continuous (i.e., upper semicontinuous life distribution and let $R(\underline{x}) = -\log \overline{F}(\underline{x})$ be the corresponding *hazard function*, where $\underline{x} = (x_1, x_2, \ldots, x_d)$ and $\overline{F}(\underline{x}) = P[X_1 > x_1, X_2 > x_2, \ldots, X_d > x_d]$. Then, assuming $\frac{\partial R}{\partial x_i}$, $1 \leq i \leq d$, exist,

$$\underline{r}(\underline{x}) = \left(\frac{\partial}{\partial x_1} R(\underline{x}), \frac{\partial}{\partial x_2} R(\underline{x}), \ldots, \frac{\partial}{\partial x_d} R(\underline{x})\right)$$

is called the *hazard gradient*. Marshall (1975) shows that, if F is absolutely continuous, then

$$\overline{F}(\underline{x}) = \exp\left[-\int_{\underline{0}}^{\underline{x}} \underline{r}(\underline{z}) \cdot d\underline{z}\right], \tag{2.1}$$

where the integral is a line integral over any sufficiently smooth continuous path beginning at $\underline{0}$ and terminating at $\underline{x}$. Because the integral is independent of the path, this integral can be written as above without ambiguity. It follows that the hazard gradient function, $\underline{r}$, determines F and vice versa.

Given independent multivariate observations $\underline{X}_i = (X_{1i}, X_{2i}, \ldots, X_{di})$, $i = 1, 2, \ldots, n$, from F, we wish to estimate $\underline{r}(\underline{x})$. A generalization of the usual life table approach would consist of constructing a d-dimensional grid and counting the number of random vector observations in each of the d-dimensional cells. Figure 3.1 displays such a grid for the bivariate case. Let $N(\underline{x})$

be the number of $\underline{X}_i$'s such that $X_{si} > x_s$ for $1 \le s \le d$. F can be estimated nonparametrically using the multivariate survival distribution, $\bar{F}_n(\underline{x}) = N(\underline{x})/n$. This estimator is not absolutely continuous, and hence provides no immediate clue to estimation of the hazard gradient. [See Ahmad and Lin [1] for a nonparametric approach.] We seek additional model smoothness assumptions to obtain useful estimators of the hazard gradient.

3.0 *The Multivariate Piecewise Exponential Distribution.* For clarity of exposition, we consider the bivariate case and let (X, Y) have joint bivariate distribution F. Let $0 \equiv x_0 < x_1 < \cdots < x_j < \cdots$ and $0 \equiv y_0 < y_1 < \cdots < y_i < \cdots$ be partitions of the positive x and y axes respectively. [See Figure 3.1.] Note that i always labels rows and j always labels columns.

Definition 1: F is *bivariate piecewise exponential* iff $r_1(u, v) = \lambda_{ij} \ge 0$ and $r_2(u, v) = \nu_{ij} \ge 0$ for $x_j \le u < x_{j+1}$ and $y_i \le v < y_{i+1}$.

Actually, $r_1(x, y) = \frac{\partial}{\partial x} R(x, y)$ and $r_2(x, y) = \frac{\partial}{\partial y} R(x, y)$ only for (x, y) in the interior of cells. F is, in general, discontinuous on cell boundaries and $\underline{r}$ is defined on cell boundaries by upper semicontinuity.

It follows from the definition and (2.1), that for $x_j \le x < x_{j+1}$ and $y_i \le y < y_{i+1}$

$$\frac{\bar{F}(x,y)}{\bar{F}(x_j,y_i)} = \exp\{-\lambda_{ij}(x - x_j) - \nu_{ij}(y - y_i)\}. \qquad (3.1)$$

Let $\bar{F}(x_j, y_i) = e^{-\delta_{ij}}$. Then F is determined by λ_{ij}, ν_{ij}, and δ_{ij}, i, j = 0, 1, 2, ... Since $\bar{F}$ is coordinatewise nonincreasing, the δ_{ij} must also satisfy restrictions imposed by the inequalities

$$\bar{F}(x, y_i^-) \ge \bar{F}(x, y_i) \text{ for } x_j \le x < x_{j+1} \qquad (3.2)$$

$$\bar{F}(x_j^-, y) \ge \bar{F}(x_j, y) \text{ for } y_i \le y < y_{i+1} \qquad (3.3)$$

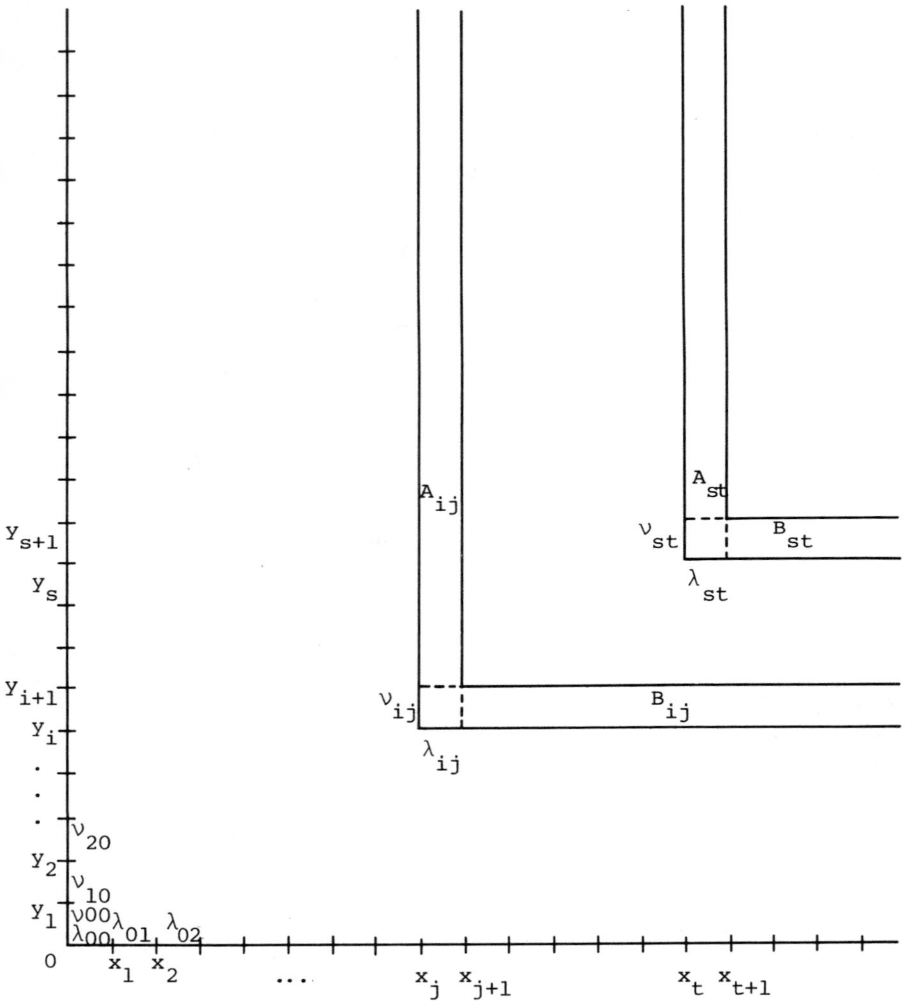

FIGURE 3.1

GRID FOR BIVARIATE PIECEWISE EXPONENTIAL DISTRIBUTION

$$\overline{F}(x_j^-, y_j^-) \geq \overline{F}(x_j, y_i). \tag{3.4}$$

It follows that

$$\delta_{ij} \geq \delta_{i-1,j} + (\lambda_{i-1,j} - \lambda_{ij})(x - x_j) + \nu_{i-1,j}(y_i - y_{i-1}) \tag{3.5}$$

for $x_j \leq x < x_{j+1}$;

$$\delta_{ij} \geq \delta_{i,j-1} + \lambda_{i,j-1}(x_j - x_{j-1}) + (\nu_{i,j-1} - \nu_{i,j})(y - y_i \tag{3.6}$$

for $y_i \leq y < y_{i+1}$; and

$$\delta_{ij} \geq \delta_{i-1,j-1} + \lambda_{i-1,j-1}(x_j - x_{j-1}) + \nu_{i-1,j-1}(y_i - y_{i-1}). \tag{3.7}$$

If we require F to be continuous, then (3.5), (3.6) and (3.7) must be equalities which in turn imply that $\lambda_{ij} = \lambda_{0j}$ and $\nu_{ij} = \nu_{i0}$ for i, j = 0, 1, 2, ... Hence, F is continuous iff the marginals are independent.

By choosing parameters appropriately and letting cell volumes shrink to zero, we can approximate any bivariate life distribution. For example, the bivariate exponential distribution (BVE)

$$\overline{F}(x,y) = \exp\{-\lambda x - \nu y - \Delta \max(x,y)\}$$

for x, y $\geq$ 0, has hazard gradient

$$\underline{r}(x,y) = \begin{cases} (\lambda\ ,\ \nu + \Delta) & \text{if } x < y \\ (\lambda + \Delta\ ,\ \nu) & \text{if } x > y\ . \end{cases}$$

If we choose

$$(\lambda_{ij}, \nu_{ij}) = \begin{cases} (\lambda\ ,\ \nu + \Delta) & \text{if } x_j < y_i \\ (\lambda + \Delta\ ,\ \nu) & \text{if } x_j > y_i\ , \end{cases}$$

then the bivariate piecewise exponential will converge to the BVE as cell volumes shrink to zero. The BVE is continuous but not absolutely continuous. [See Marshall and Olkin [12].]

The Generalized Bivariate Piecewise Exponential Distribution.
By suitable generalization of the bivariate piecewise exponential

distribution, we can obtain continuous bivariate distributions with dependent piecewise exponential marginals. However, the resulting continuous distribution is negative quadrant dependent. Hence, the continuous model is of more limited interest.

Definition 2: F is *generalized bivariate piecewise exponential* iff for $x_j \le x < x_{j+1}$ and $y_i \le y < y_{i+1}$

$$r_1(x, y) = \lambda_{ij}(y) \tag{3.8}$$

$$r_2(x, y) = \nu_{ij}(x). \tag{3.9}$$

Johnson and Kotz [9], Page 61, show that (3.8) and (3.9) imply that r_1 is linear in y and r_2 is linear in x for $x_j \le x < x_{j+1}$ and $y_i \le y < y_{i+1}$. Hence,

$$r_1(x, y) = \lambda_{ij} + \theta_{ij}(y - y_i) \tag{3.8^1}$$

$$r_2(x, y) = \nu_{ij} + \theta_{ij}(x - x_j). \tag{3.9^1}$$

It follows that in this case

$$\frac{\bar{F}(x,y)}{\bar{F}(x_j,y_i)} = \exp\{-\lambda_{ij}(x-x_j) - \nu_{ij}(y-y_i) - \theta_{ij}(x-x_j)(y-y_i)\}. \tag{3.10}$$

If we require F to be continuous, then we must have equality in (3.2), (3.3) and (3.4).

Theorem 3.1: If F is continuous and satisfies Definition 2, then

$$\text{(a)} \quad \lambda_{ij} = \lambda_{0j} + \sum_{r=0}^{i-1} \theta_{rj}(y_{r+1} - y_r), \quad (j \ge 1) \tag{3.11}$$

$$\text{(b)} \quad \nu_{ij} = \nu_{i0} + \sum_{c=0}^{j-1} \theta_{ic}(x_{c+1} - x_c), \quad (i \ge 1). \tag{3.12}$$

Note that if $\theta_{ij} = 0$ for all i and j, then F is determined by the marginal distributions; i.e., the marginals are independent.

Proof: Assume $R(x, y) = -\log \bar{F}(x, y)$ is continuous so that

$$R(x_j^-, y_{i+1}) = R(x_j, y_{i+1}). \tag{3.13}$$

Now $R(x_j^-, y_{i+1}) = R(x_{j-1}, y_i) + \lambda_{i,j-1}(x_j - x_{j-1}) + \nu_{i,j-1}(y_{i+1} - y_i) + \theta_{i,j-1}(x_j - x_{j-1})(y_{i+1} - y_i)$ and for $x \geq x_j$,

$$R(x, y_{i+1}) = R(x_{j-1}, y_i) + \lambda_{i,j-1}(x_j - x_{j-1}) + \lambda_{ij}(x - x_j) + \nu_{ij}(y_{i+1} - y_i) + \theta_{ij}(x - x_j)(y_{i+1} - y_i) .$$

Letting $x \downarrow x_j$, (3.13) implies

$$\lambda_{i,j-1}(x_j - x_{j-1}) + \nu_{i,j-1}(y_{i+1} - y_i) + \theta_{i,j-1}(x_j - x_{j-1})(y_{i+1} - y_i) = \lambda_{i,j-1}(x_j - x_{j-1}) + \nu_{ij}(y_{i+1} - y_i)$$

or

$$\nu_{ij} = \nu_{i,j-1} + \theta_{i,j-1}(x_j - x_{j-1}). \tag{3.14}$$

Similarly

$$\lambda_{ij} = \lambda_{i-1,j} + \theta_{i-1,j}(y_i - y_{i-1}) . \tag{3.15}$$

Applying (3.14) and (3.15) repeatedly, we obtain (3.11) and (3.12).||

Definition 3: [Lehmann [10], p. 1137]. F is positive (negative) quadrant dependent if

$$\overline{F}(x,y) \geq (\leq) \overline{F}_1(x)\overline{F}_2(x)$$

where F_1 and F_2 are the respective marginal distributions.

Theorem 3.2: If F is continuous, and $\theta_{ij} \geq 0$ for all i and j and satisfies Definition 2, then F is negative quadrant dependent.

Proof: From (3.10), (3.11) and (3.12),

$$\overline{F}(x,y) \leq e^{-\lambda_{0j}(x-x_j)-\nu_{i0}(y-y_i)}\overline{F}(x_j, y_i)$$

for $x_j \leq x < x_{j+1}$ and $y_i \leq y < y_{i+1}$. By repeated applications of this argument, we see that

$$\overline{F}(x,y) \leq \exp\left\{- \sum_{c=0}^{j-1} \lambda_{0c}(x_{c+1} - x_c) - \lambda_{0j}(x - x_j) - \sum_{r=0}^{i-1} \nu_{r0}(y_{r+1} - y_r) - \nu_{i0}(y - y_i)\right\} .$$

Hence,

$$\bar{F}(x,y) \le \bar{F}_1(x)\bar{F}_2(y) ,$$

as stated.||

Since *associated* random variables are positive quadrant dependent, it follows that the above continuous distributions are not associated when $\theta_{ij} \ge 0$ for all i and j. [Cf. Theorem 4.2, Page 143, Barlow and Proschan [3].

4.0 *Conditional Maximum Likelihood Estimators of Failure Rate Parameters*. To find the estimators of λ_{ij} and ν_{ij} which maximize the joint likelihood, we would have to do so under the restrictions imposed by (3.5), (3.6), and (3.7). For mathematical convenience, we consider the estimators λ_{ij} (ν_{ij}) which maximize the conditional likelihood, conditional on $Y > y_i$ ($X > x_j$).

The conditional survival distribution

$$\bar{F}(x \mid Y > y_i) = P[X > x \mid Y > y_i] \tag{4.1}$$

is piecewise exponential with possible jumps at x_j, j = 0, 1, ... Let $p_{ij} = \bar{F}(x_j^- \mid Y > y_i) - \bar{F}(x_j \mid Y > y_i)$. Then, for $x_j \le x < x_{j+1}$,

$$\bar{F}(x \mid y_i) = \left[\prod_{r=0}^{j-1} p_{ir} e^{-\lambda_{ir}(x_{r+1}-x_r)}\right] p_{ij} e^{-\lambda_{ij}(x-x_j)}$$

and the conditional density for $x_j < x < x_{j+1}$ is

$$f(x \mid Y > y_i) = -\frac{\partial}{\partial x} \bar{F}(x \mid Y > y_i) \tag{4.2}$$

$$= \lambda_{ij} e^{-\lambda_{ij}(x-x_j)} \left[\prod_{r=0}^{j-1} p_{ir} e^{-\lambda_{ir}(x_{r+1}-x_r)}\right] .$$

Let (X_r, Y_r), r = 1, 2, ..., n be independent observations from F. Let n_{ij} be the number of pairs (X_r, Y_r) such that $x_i < X_r < x_{i+1}$ and $Y_r > y_j$. Then n_{ij} is the number of pairs in the interior of A_{ij}. [See Figure 3.1.] From (4.2), it is easy to see that λ_{ij} will contribute the following terms to the log likelihood.

$$n_{ij} \log \lambda_{ij} - \lambda_{ij} \int_{x_j}^{x_{j+1}} N(u, y_i)du ,$$

where $N(u, y_i)$ is the number of pairs (X_r, Y_r) such that $X_r > u$ and $Y_r > y_i$. It follows that the conditional maximum likelihood estimate of λ_{ij} is

$$\hat{\lambda}_{ij}(n) = \frac{n_{ij}}{\int_{x_j}^{x_{j+1}} N(u,y_i)du} . \tag{4.3}$$

Similarly, if we let m_{ij} be the number of pairs (X_r, Y_r) such that $X_r > x_i$ and $y_j < Y_r < y_{j+1}$, then m_{ij} is the number of pairs in B_{ij}. [See Figure 3.1.] It follows that the conditional maximum likelihood estimator of ν_{ij} is

$$\hat{\nu}_{ij}(n) = \frac{m_{ij}}{\int_{y_i}^{y_{i+1}} N(x_j,v)dv} . \tag{4.4}$$

If both $n_{ij} = 0$ and

$$\int_{x_j}^{x_{j+1}} N(u,y_i)du = 0,$$

then $\hat{\lambda}_{ij}$ is indeterminate. Similarly for $\hat{\nu}_{ij}$. [If $F(x \mid Y > y_i)$ is IFR, we would set $\hat{\lambda}_{ij} = \infty$. If $F(x \mid Y > y_i)$ is DFR, we would set $\hat{\lambda}_{ij} = 0$.]

From Figure 3.1, we see that $\hat{\lambda}_{ij}$ is the number of X failures in the strip A_{ij} divided by the total time on test of X observations in the strip; a similar relationship holds for $\hat{\nu}_{ij}$. In the case of complete observations,

$$\hat{\lambda}_{ij}(n) = F_n[A_{ij}] \Bigg/ \int_{x_j}^{x_{j+1}} \bar{F}_n(u, y_i)\,du \tag{4.5}$$

and

$$\hat{\nu}_{ij}(n) = F_n[B_{ij}] \Bigg/ \int_{y_i}^{y_{i+1}} \bar{F}_n(x_j, v)\,dv. \tag{4.6}$$

In our notation, $F_n[A]$ is the F_n measure of A, i.e., the proportion of (X, Y) pairs which fall in A.

It follows that in the case of complete observations,

$$\lim_{n\to\infty} \hat{\lambda}_{ij}(n) = F(A_{ij}) \Bigg/ \int_{x_j}^{x_{j+1}} \bar{F}(u, y_i)\,du \tag{4.7}$$

$$\lim_{n\to\infty} \hat{\nu}_{ij}(n) = F(B_{ij}) \Bigg/ \int_{y_i}^{y_{i+1}} \bar{F}(x_j, v)\,dv , \tag{4.8}$$

almost surely. Note that $\hat{\lambda}_{ij}(n)$ and $\hat{\nu}_{ij}(n)$ are defined for observations from arbitrary F. In Barlow and Proschan [4], these estimators were called *interval failure rate* estimators (in the univariate case).

Since $F[A_{ij}] = \bar{F}(x_j, y_i) - \bar{F}(x_{j+1}, y_i)$, we see from the definition of F in (3.1) that

$$\frac{F[A_{ij}]}{\int_{x_j}^{x_{j+1}} \bar{F}(u,y_i)\,du} = \frac{1 - e^{-\lambda_{ij}(x_{j+1}-x_j)}}{\int_{x_j}^{x_{j+1}} e^{-\lambda_{ij}(u-x_j)}\,du} = \lambda_{ij} .$$

Hence, $\hat{\lambda}_{ij}(n)$ is a consistent estimator of λ_{ij} when F is bivariate piecewise exponential. A similar result holds for $\hat{\nu}_{ij}(n)$.

For the case of $d > 2$ dimensions, let $\underline{a}$ be a grid point and

$\Delta_i(\underline{a})$ the increment in the ith coordinate as we move to an adjacent "higher" grid point along the ith coordinate axis. Let

$$I_i(\underline{a}) = \{\underline{x} \mid a_i \leq x_i < a_i + \Delta_i(\underline{a}),\ x_j \geq a_j,\ j \neq i\}$$

by the strip (or slab) orthogonal to the ith coordinate axis corresponding to grid point $\underline{a}$. Let $n[I_i(\underline{a})]$ be the number of random vectors, $\underline{X}$, such that $a_i \leq X_i < a_i + \Delta_i(\underline{a})$ and $X_j \geq a_j$, $j \neq i$.

Then

$$r_i(\underline{a}) = n[I_i(\underline{a})] \Bigg/ \int_{a_i}^{a_i+\Delta_i(\underline{a})} N(u_i,\underline{a})\,du_i$$

is the conditional maximum likelihood estimator for $r_i(u_i,\underline{a})$, where $a_i \leq u_i < a_i + \Delta_i(\underline{a})$ and $(u_i,\underline{a}) = (a_1, a_2, \ldots, u_i, a_{i+1}, \ldots, a_d)$. For the case of complete observations

$$\hat{r}_i(\underline{a}) = F_n[I_i(\underline{a})] \Bigg/ \int_{a_i}^{a_i+\Delta_i(a)} \overline{F}_n(u_i,\underline{a})\,du_i \ .$$

5.0 *Application to Failure Data.* To illustrate the application of the previous model and also to introduce some new techniques for analyzing failure data, we consider some specific data. Table 5.1 is a list of paired first failure times (in hours) of the transmission and the transmission pump on 15 caterpillar tractors. A scatter plot diagram is given in Figure 5.1.

Our first step in analyzing the data is to draw total time on test plots for the sample marginal distributions and also for the sample distribution of minima and maxima (Figures 5.2, 5.3, and 5.4). All plots indicate that sample distributions have strongly increasing failure rate (IFR) or increasing failure rate average (IFRA). [Cf. Barlow and Campo [2] for details on plot construction.] Rays from the origin to successive points tend to have decreasing slope, which indicates sample distributions are IFRA [cf. Barlow and Campo [2], Theorem 2.1]. The coefficients of variation are, in each case, less than 1, corroborating the IFR or IFRA conclusion. Comparison of Figure 5.2 with transforms of standard distributions, e.g., Weibull, gamma, lognormal, truncated normal [see Barlow and Campo [2]] does not indicate that any of these models are appropriate. Since the scaled total time on test plots are essentially linear up to the 25th percentile, a Makeham distribution with failure rate: $r(t) = \alpha + \beta e^{\gamma t}$ for $t \geq 0$ would likely provide a good univariate model for the marginal distributions.

Since the sample distribution of maxima and minima are also IFRA, we may tentatively conclude that the joint distribution is probably multivariate IFRA in the sense of Esary and Marshall [7]; namely, $-\log \overline{F}(\alpha \underline{x})/\alpha$ is nondecreasing in $\alpha > 0$ for all $\underline{x}$.

Tractor Number	X	Y
1	1641	850
2	5556	1607
3	5421	2225
4	3168	3223
5	1534	3379
6	6367	3832
7	9460	3871
8	6679	4142
9	6142	4300
10	5995	4789
11	3953	6310
12	6922	6310
13	4210	6378
14	5161	6449
15	4732	6949

TABLE 5.1

PAIRED FIRST FAILURE TIMES OF TRANSMISSIONS (X) AND TRANSMISSION PUMPS (Y) ON D9G-66A CATERPILLAR TRACTORS

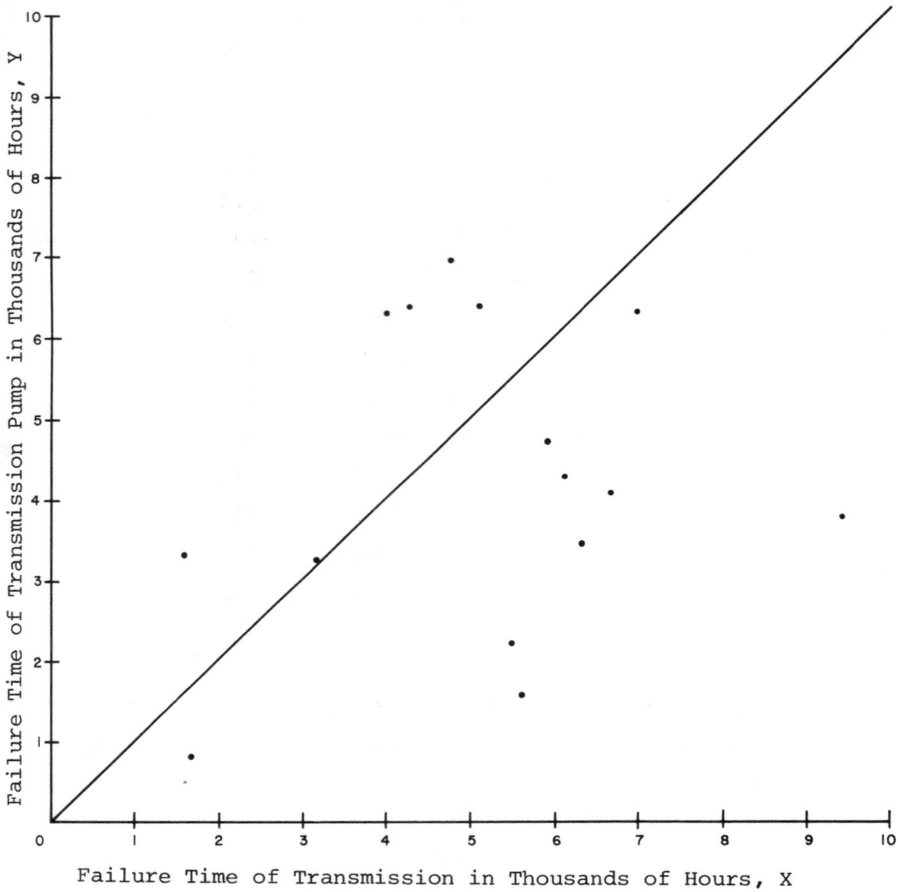

FIGURE 5.1

SCATTER PLOT OF TRANSMISSION (X) AND TRANSMISSION PUMP (Y) FAILURE TIMES

$\overline{X}$ = 5129 Hours, $\overline{Y}$ = 4307 Hours, $\hat{\rho}_{XY}$ = .216 (n = 15)

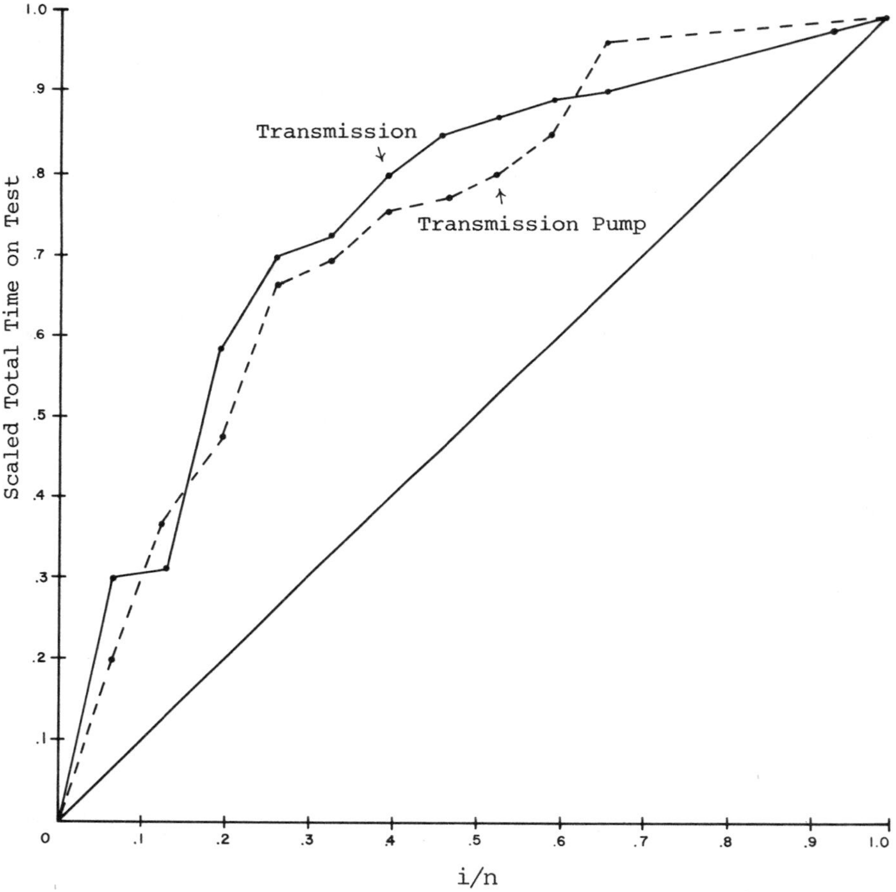

FIGURE 5.2

SCALED TOTAL TIME ON TEST PLOTS FOR MARGINAL DISTRIBUTIONS (TRANSMISSION AND TRANSMISSION PUMP FAILURES)

$$\frac{\hat{\sigma}(X)}{\overline{X}} = .387 , \quad \frac{\hat{\sigma}(Y)}{\overline{Y}} = .424$$

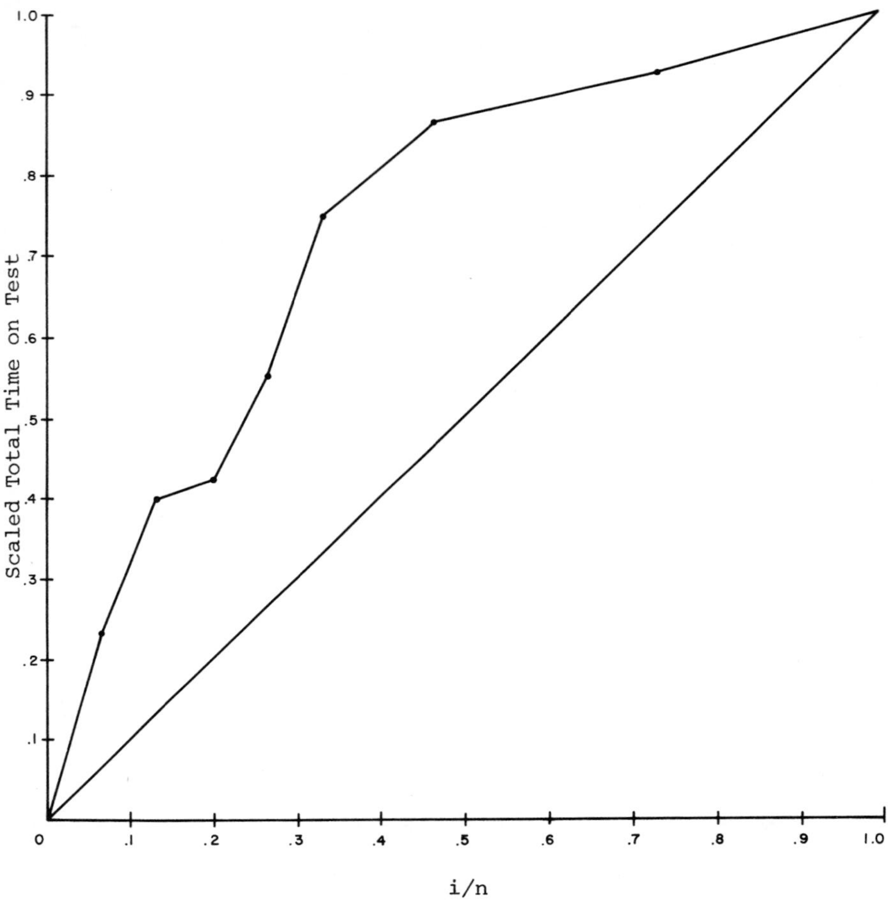

FIGURE 5.3

SCALED TOTAL TIME ON TEST PLOT OF MINIMUM (X,Y)

(n = 15)

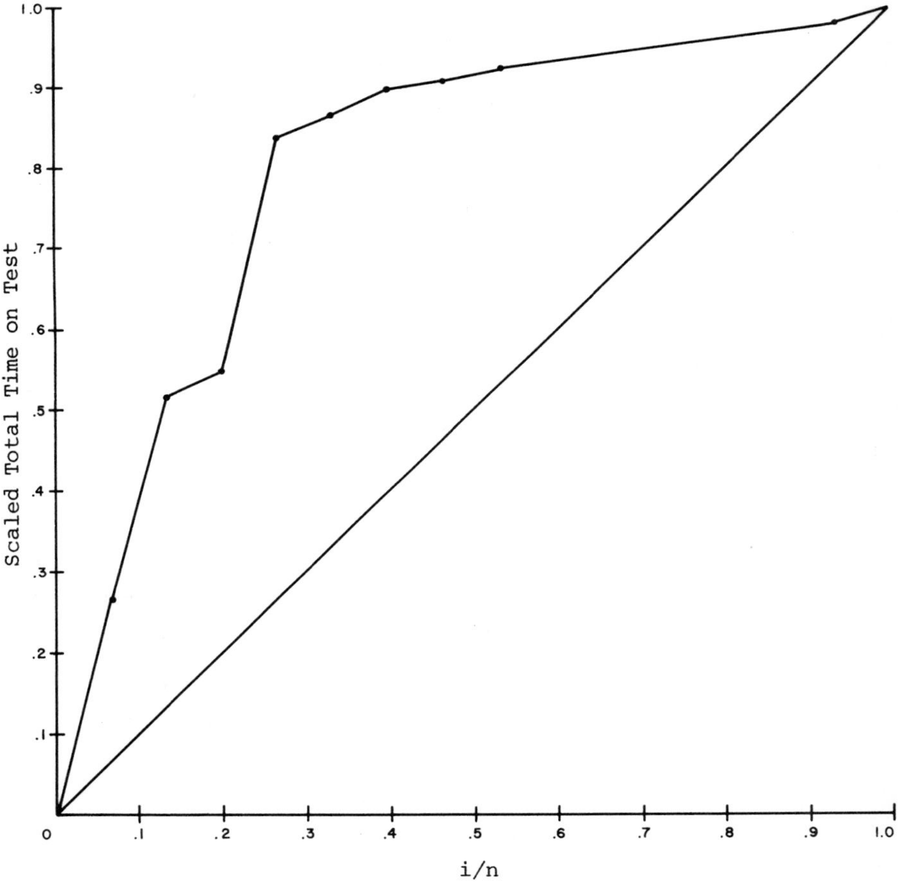

FIGURE 5.4

SCALED TOTAL TIME ON TEST PLOT OF MAXIMUM (X,Y)

(n = 15)

On the basis of Figure 5.1 and the relatively small sample size, n = 15, we have selected 3000 hour intervals as the basis for our grid. Figure 5.5 shows the conditional maximum likelihood estimates $\hat{\lambda}_{ij}$ and $\hat{\nu}_{ij}$ computed according to formulas (4.3) and (4.4). Note that marginal failure rates are increasing even though $\hat{\lambda}_{ij}$ and $\hat{\nu}_{ij}$ are not everywhere increasing in both i and j.

Iso-Probability Contours. To further investigate the dependency between X and Y, we plot empirical iso-probability contours in Figure 5.6. Points having equal empirical survival probability are joined by dashed lines.

Definition: $C = \{\underline{x} \mid R(\underline{x}) = c\}$ is an *iso-probability contour* corresponding to hazard level c, or equivalently, to probability level $p = e^{-c}$.

An iso-probability contour, C, is *concave (convex)* if and only if $R\left[\sum_{j=1}^{K} \lambda_j \underline{x}_j\right] \leq (\geq)\ c$ for all $\underline{x}_j \in C$, $\lambda_j \geq 0$, $\sum_{j=1}^{K} \lambda_j = 1$, and $K \geq 1$, where c defines the contour C. Note that the empirical iso-probability contours in Figure 5.6 are concave. In the case of the Marshall-Olkin bivariate exponential distribution

$$\overline{F}(x,y) = \exp\,[-\lambda_1 x - \lambda_2 y - \lambda_{12} \max\,(x,y)],$$

the iso-probability contours are linear for $\lambda_{12} = 0$ and concave for $\lambda_{12} > 0$. As $\lambda_{12} \uparrow \infty$, corresponding to complete dependency, the contours become rectangular.

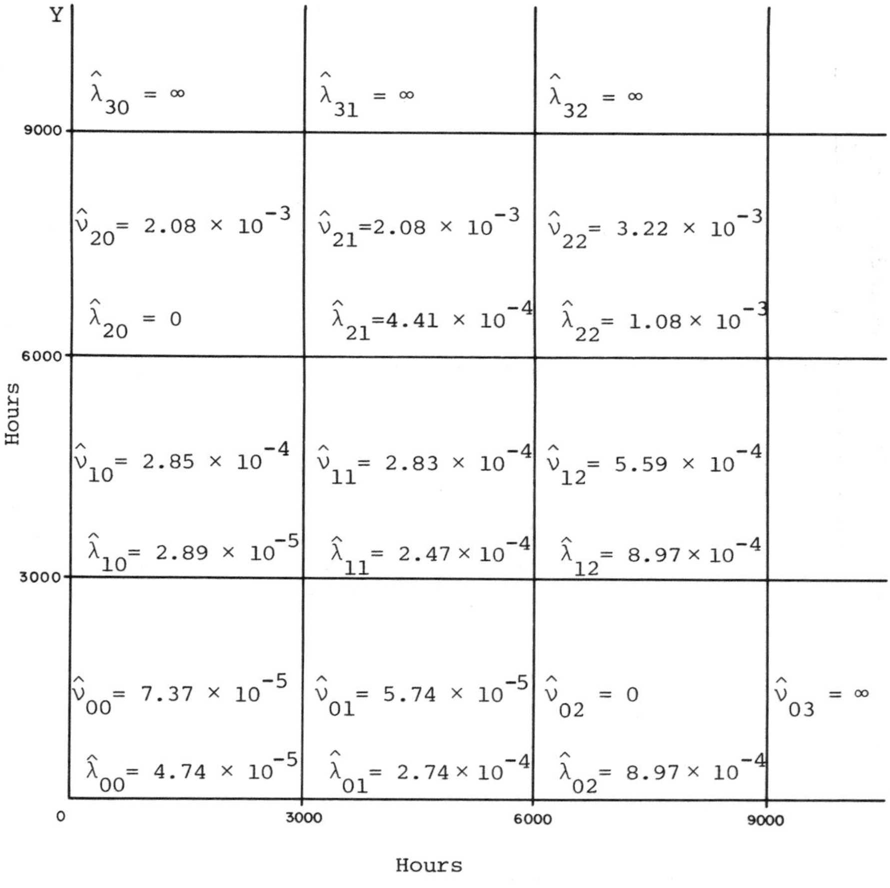

FIGURE 5.5

BIVARIATE PIECEWISE EXPONENTIAL HAZARD GRADIENT ESTIMATORS
(TRANSMISSION AND TRANSMISSION PUMP)

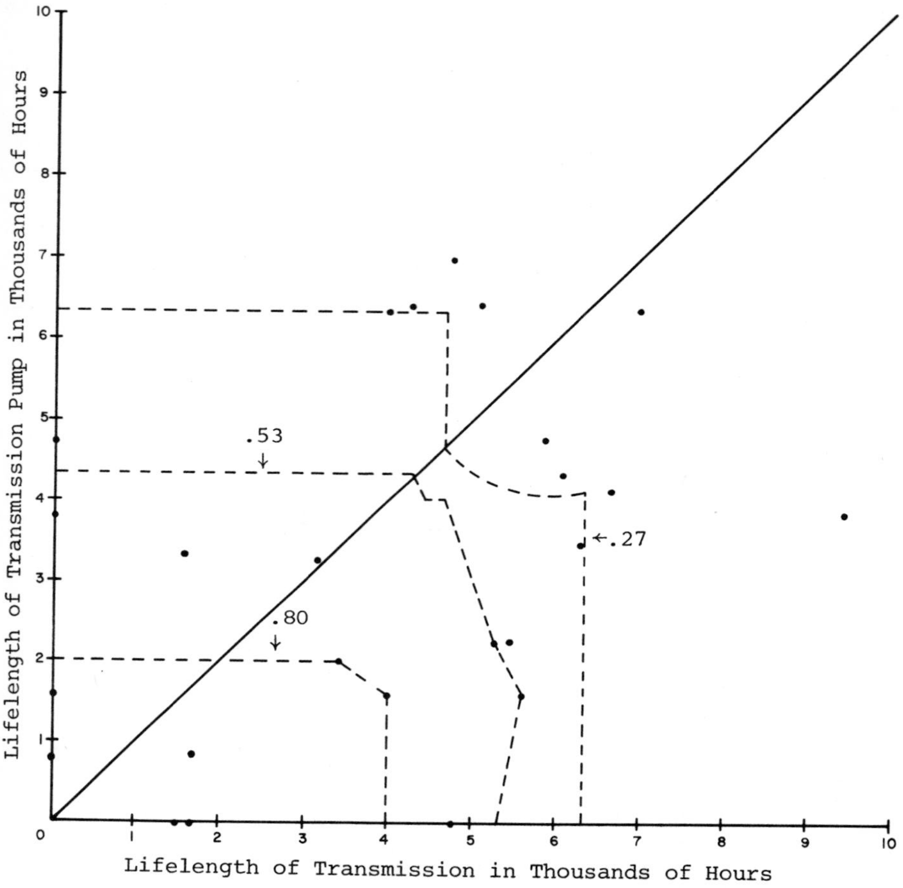

FIGURE 5.6

EMPIRICAL ISO-PROBABILITY CONTOURS

Lemma 5.1: If R is convex (concave), then all iso-probability contours are concave (convex).

Proof: R convex implies

$R\left[\sum_{j=1}^{K} \lambda_j \underline{x}_j\right] \le \sum_{j=1}^{K} \lambda_j R(\underline{x}_j)$. If each $\underline{x}_j$ belongs to iso-probability contour $C = \{\underline{x} \mid R(\underline{x}) = c\}$, then $R\left[\sum_{j=1}^{K} \lambda_j \underline{x}_j\right] \le c$ and C is concave.

The case of concave R is similar.||

Lemma 5.2: If F has independent IFR (DFR) univariate marginals, then F has concave (convex) iso-probability contours.

Proof: Let F_i (i = 1, 2, ..., d) be the univariate marginal distributions and let $R_i(x_i) = -\log \overline{F}_i(x_i)$. Recall that F_i IFR implies R_i convex. From independence, we have $R(\underline{x}) = \sum_{i=1}^{d} R_i(x_i)$, which clearly is convex. The result follows from Lemma 5.1.

A similar proof holds in the DFR case.||

Lemma 5.3: In the bivariate case, if $r_1(x, y)/r_2(x, y)$ is increasing in $x \ge 0$ and decreasing in $y \ge 0$, then iso-probability contours are concave.

Proof: $R(x, y) = c$ implies $\frac{d}{dx} R(x, y) = r_1(x, y) + r_2(x, y) \frac{dy}{dx} = 0$ or $\frac{dy}{dx} = -r_1(x, y)/r_2(x, y)$. Since on an iso-probability contour, the graph of (x, y) is concave if and only if $\frac{dy}{dx}$ is a decreasing function of x, we see that the assumptions on $r_1(x, y)/r_2(x, y)$ imply that the iso-probability contours are concave.||

It follows that λ_{ij}/ν_{ij} increasing in i and decreasing in j suggest that the true distribution may have concave iso-probability contours. However, since F(x, y) is discontinuous for the bivariate piecewise exponential distribution except for the case of independent marginals, F for this model cannot in general have concave or convex iso-probability contours.

In our illustrative example, the sample correlation $\hat{\rho}_{xy} = .216$ indicates that X and Y may be positively dependent. A stronger form of positive dependence is *positive quadrant dependence, i.e.,*

$$\overline{F}(\underline{x}) \geq \prod_{i=1}^{d} \overline{F}_i(\underline{x}) \overset{\text{def}}{=} \overline{F}^*(\underline{x}) \text{ say, for all } \underline{x}.$$ [See Definition 3.]

Lemma 5.4: F is positively quadrant dependent if and only if iso-probability contours for F dominate corresponding iso-probability contours for F^*.

The proof is obvious from the definition of positive quadrant dependence.

A natural rank test for bivariate positive quadrant dependence is based on

$$T_n = \iint [F_n(x,y) - F_{1n}(x)F_{2n}(y)]dF_n(x,y) ,$$

where F_{1n} and F_{2n} are the sample marginal distributions. T_n can also be written as

$$[(n - 1)r_k/4n] - \left[(n^2 - 1)r_s/12n^2\right],$$

where r_k and r_s are respectively the rank correlation coefficients of Kendall and Spearman [cf. Crouse [6]]. Positive values of T_n indicate positive quadrant dependence. The maximum value of T_n is $M(n) = (2n - 1)(n - 1)/(12n^2)$. Then $r_M \overset{\text{def}}{=} T_n/M(n)$ has variance

$$\text{var } r_M = (n^2 + 4n - 3)/\{(2n - 1)^2(n - 1)\}$$

[cf. Crouse [6], p. 107]. In our example, $r_M = .071$ and $\sqrt{\text{var } r_M} = .154$. Since r_M is asymptotically normal, $r_M < \sqrt{\text{var } r_M}$ indicates X and Y are *not* significantly positively quadrant dependent.

On the basis of our previous analysis, we would recommend modeling the joint life distribution in this illustrative example in terms of independent Makeham distributed random variables or independent piecewise exponential random variables.

6.0 *Asymptotic Confidence Limits on Univariate Marginal Failure Rate Functions*. Using the notation at the end of Section 4,

$$\hat{r}_i(\underline{a}) = F_n[I_i(\underline{a})] \Bigg/ \int_{a_i}^{a_i+\Delta_i(\underline{a})} \overline{F}_n(u_i, \underline{a})\, du_i$$

is the maximum likelihood estimate for $r_i(u_i, \underline{a})$, $a_i \leq u_i < a_i + \Delta_i(\underline{a})$, assuming the multivariate piecewise exponential model. If $\underline{a}_1 \leq \underline{a}_2 \leq \cdots \leq \underline{a}_m$ are distinct, coordinatewise increasing grid points, then Barlow and Proschan [5] show that $\sqrt{n}\,[\hat{\underline{r}}(\underline{a}_1) - \underline{r}(\underline{a}_1)]$, ..., $\sqrt{n}\,[\hat{\underline{r}}(\underline{a}_m) - \underline{r}(\underline{a}_m)]$ are asymptotically independent, normally distributed random variables. It follows that for any specified coordinate, i, we can construct asymptotic simultaneous confidence limits for $r_i(\underline{a}_1)$, $r_i(\underline{a}_2)$, ..., $r_i(\underline{a}_m)$ using an estimate of the asymptotic variance, $\sigma_i^2(\underline{a}_j)$, of $\sqrt{n}\,[\hat{r}_i(\underline{a}_j) - r_i(\underline{a}_j)]$ for $1 \leq j \leq m$. For the multivariate piecewise exponential model, the variances depend on F in a relatively simple way. Using the multivariate empirical distribution, F_n, for F we can approximate each $\sigma_i^2(\underline{a}_j)$. Confidence limits may then be calculated using both the asymptotic normality and independence results in the usual way.

Acknowledgment. We would like to acknowledge the help of Myles Hollander with respect to tests for positive quadrant dependence.

REFERENCES

[1] Ahmad, I. A. and Pi-Erh Lin, "Abstract: Nonparametric Estimation of a Vector-Valued Bivariate Failure Rate Function," *Bulletin of the Institute of Mathematical Statistics*, vol. 4, no. 4, (1975), p. 177.

[2] Barlow, R. E. and R. Campo, "Total Time on Test Processes and Applications to Failure Data Analysis," in RELIABILITY AND FAULT TREE ANALYSIS, edited by R. E. Barlow, J. B.

Fussell and N. D. Singpurwalla, SIAM, Philadelphia, Pennsylvania, (1975), pp. 451-481.

[3] Barlow, R. E. and F. Proschan, STATISTICAL THEORY OF RELIABILITY AND LIFE TESTING, Holt, Rinehart and Winston, (1975).

[4] Barlow, R. E. and F. Proschan, "Asymptotic Theory of Total Time on Test Processes, with Applications to Life Testing," to appear in MULTIVARIATE ANALYSIS, vol. 4, edited by P. R. Krishnaiah, North-Holland Publishing Company, (1976a).

[5] Barlow, R. E. and F. Proschan, "Multivariate Life Table Analysis," ORC 76-9, Operations Research Center, University of California, Berkeley, (April, 1976b).

[6] Crouse, C. F., "Distribution Free Tests Based on the Sample Distribution Function," *Biometrika*, vol. 53, (1966), pp. 99-108.

[7] Esary, J. D. and A. W. Marshall, "Multivariate IHRA Distributions," unpublished technical report, (1975).

[8] Gumbel, E. J., "Bivariate Exponential Distributions," *Journal of the American Statistical Association*, vol. 55, (1960), pp. 698-707.

[9] Johnson, N. L. and S. Kotz, "A Vector Multivariate Hazard Rate," *Journal of Multivariate Analysis*, vol. 5, (1975), pp. 53-66.

[10] Lehmann, E. L., "Some Concepts of Dependence," *Annals of Mathematical Statistics*, vol. 37, (1966), pp. 1137-1153.

[11] Marshall, A. W., "Some Comments on the Hazard Gradient," *Journal of Stochastic Processes and Their Applications*, vol. 3, (1975), pp. 293-300.

[12] Marshall, A. W. and I. Olkin, "A Multivariate Exponential Distribution," *Journal of the American Statistical Association*, vol. 62, (1967), pp. 30-44.

OPTIMAL REPLACEMENT POLICIES FOR DEVICES SUBJECT TO A GAMMA WEAR PROCESS*

by MOHAMED ABDEL-HAMEED
The University of North Carolina at Charlotte

Abstract. Suppose that a device is subject to wear occurring randomly in time as a gamma process $\{X(t), t \geq 0\}$. Let ζ be the life length of the device. The device can be replaced before or at failure. The cost of replacing the device, h, is of the form

$$h(X(t)) = \begin{cases} g(X(t)), & t < \zeta \\ c & t = \zeta \end{cases}$$

for some g. In this paper we give a form of an optimal replacement policy that minimizes the expected replacement cost per unit time. Discrete version of the above results is also considered.

1.0 *Introduction and Summary.* Abdel-Hameed [2] studied life distribution properties of a device subject to wear occurring randomly in time as a gamma wear process with density

$f(x, t) = \lambda e^{-\lambda x} \dfrac{(\lambda x)^{t-1}}{\Gamma(t)}$, $t \geq 0$. If $\overline{G}(x)$ is the probability that the device survives x units of wear, $x \geq 0$. Then, for each $t \geq 0$, the probability that the device survives till time t without failure (the reliability) is given by

$$\overline{F}(t) = \lambda\int_0^\infty e^{-\lambda x} \frac{(\lambda x)^{t-1}}{\Gamma(t)} \overline{G}(x)dx. \tag{1.1}$$

Theorems proved in [2] show that life distribution properties of $\overline{G}(x)$ (such as increasing failure rate, increasing failure rate average, new better than used) are inherited as corresponding properties of the reliability $\overline{F}(t)$ given in (1.1). These results have many applications in inventory, biometry, and dam flooding theory. In this paper we study optimal replacement policies for such devices. Let ζ be the failure time of the device; for every $t < \zeta$ the wear at time t

*Research supported in part by the Air Force Office of Scientific Research, AFSC,USAF, under Grant AFOSR-76-2999 and Office of Naval Research Contract N00014-76-C-0840.

is observed and the decision to replace the device or not is based on, and only on, its current value. Upon failure the device is replaced with a new device identical with the original, or repaired to a like-new condition, and the process repeats. Throughout we assume that replacement, or repairs, are done instantaneously. The cost of replacement at time t, before failure, depends only on the total wear at time t and is denoted by h. The cost of replacement at failure is assumed to be a constant $c > 0$. Throughout we assume that $h \in L_\infty(R_+)$ (the space of bounded Lebesgue measurable functions on $R_+ \overset{\text{def}}{=} [0, \infty)$ with the supremum norm $\|h\| = \sup_{x \in R_+} |h(x)|$).

Throughout we use E_x to denote the expectation conditioning on the fact that the device starts with an x amount of wear.

Definition 1.1. A replacement time τ^* is said to be optimal if it minimizes the expected cost per unit time

$\dfrac{E_x[h(X(\tau))I(\tau < \zeta)] + c\,P_x(\tau = \zeta)}{E_x(\tau)}$ over the class of all replacement times $\{\tau\}$, for each $x \geq 0$.

In Section 2 we show that the optimal replacement time is the first time the gamma wear process $\{X(t), t \geq 0\}$ exceeds a prespecified threshold ξ^*, and we give an equation which defines ξ^* in terms of the cost function and the wear survival probability $\overline{G}(x)$. Explicitly, we show that if b is the minimum expected cost per unit time, $b > 0$, and $w(x) \overset{\text{def}}{=} E_x(\zeta)$, then

$$\xi^* = \inf\{x \geq 0: \int_0^\infty \frac{e^{-\lambda y}}{y} [\frac{\{c - h(x+y) - bw(x+y)\}\overline{G}(x+y)}{\overline{G}(x)} - \{c - h(x) - bw(x)\}]dy \leq 0\}.$$

We illustrate the results by considering special forms of h and $\overline{G}$.

In Section 3 we consider a discrete version of the results in Section 2, assuming that the damage or wear occurs only at time $n = 1, 2, \ldots$, damages accumulate and the distribution of the accumulated damage at time $n = 1, 2, \ldots$ is a gamma distribution.

2.0 An Optimal Replacement Policy for a Device Subject to a Gamma Wear Process-Continuous Time Case. Let $\{X(t), t \geq 0\}$ be the gamma wear process described in Section 1. The cost of replacement is given by

$$g(X(t)) = \begin{cases} h(X(t)) \ , & t < \zeta \\ c & t = \zeta \ . \end{cases}$$

If τ is any replacement time then the expected cost per unit time $\psi_\tau \overset{\text{def}}{=} \dfrac{E_x g(X(\tau))}{E_x(\tau)}$. We assume throughout that $E_x(\zeta)$ is finite. The following theorem provides a sufficient condition for the finiteness of $E_x(\zeta)$.

Theorem 2.1. Let the wear survival probability $\overline{G}$ has increasing failure rate average (IFRA) i.e., $[\overline{G}(x)]^{1/x}$ **is decreasing in** $x \geq 0$**, then** $E_x(\zeta) < \infty$.

Proof. We first note that $E_x(\zeta) \leq E_0(\zeta)$ and to show that $E_x(\zeta) < \infty$ it suffices to show that $E_0(\zeta) < \infty$. Since $\overline{G}(x)$ is IFRA, it then follows that for every $\gamma \geq 0$, $\overline{G}(x) - e^{-\gamma x}$ changes sign at most once, and if once, in the order +, -. By the variation diminishing property [7] of the totally positive gamma kernel $\ell(t, x) = e^{-\lambda x}(\lambda x)^{t-1}/\Gamma(t)$ we have $\overline{F}(t) - (\frac{\lambda}{\lambda+\gamma})^t =$

$\lambda\int_0^\infty e^{-\lambda x} \frac{(\lambda x)^{t-1}}{\Gamma(t)} (\overline{G}(x) - e^{-\gamma x})dx$ changes sign at most once, and in $t \geqslant 0$ if once, in the order +, -. It follows immediately that $\overline{F}(t)$ is IFRA. Hence, for every $\lambda \geq 0$, $\overline{F}(t)$ crosses $e^{-\lambda t}$, from above, at most once. Thus for every $\lambda \geq 0$, there exists a point $t_\lambda \geq 0$, depending on λ, such that $\bar{F}(t) \geq e^{-\lambda t}$ for every $t \leq t_\lambda$ and $\bar{F}(t) \leq e^{-\lambda t_\lambda}$ for every $t \geq t_\lambda$. Thus

$$E_0(\zeta) = \int_0^\infty \bar{F}(t)dt \leq \int_0^{t_\lambda} \bar{F}(t)dt + \int_{t_\lambda}^\infty e^{-\lambda t}dt = \int_0^{t_\lambda} \bar{F}(t)dt + (1/\lambda)e^{-\lambda t_\lambda} < \infty.$$

In this section we give a form of an optimal replacement policy that maximizes ψ_τ. Let $b = \inf \psi_\tau > 0$ be the minimum average cost, $\phi(x) = c - h(x) - bw(x)$, and $I(A)$ denotes the indicator function of the set A. The following lemma follows from Theorem 4 of Taylor [9].

Lemma 2.1. The optimal replacement policy that minimizes ψ_τ is the one maximizing $\theta_\tau \overset{\text{def}}{=} bE_x(\tau) - E_x g(X(\tau))$ and the maximum value of θ_τ is zero.

The following lemma of interest plays an important role in establishing the form of the optimal replacement policy.

Lemma 2.2. Let θ_τ be as defined in Lemma 2.1, then

$$\theta_\tau = E_x(\phi(X(\tau))I(\tau < \zeta)) - (c - bw(x)).$$

Proof. Write $E_x(\tau) = E_x(\zeta) - E_x[(\zeta - \tau)I(\tau < \zeta)]$. Conditioning on $\tau = t$, we have that $E_x[(\zeta - t)I(t < \zeta)] = \int_0^\infty P_x(\zeta > t + u)du =$

$$\lambda \int_0^\infty \int_0^\infty e^{-\lambda z} \frac{(\lambda z)^{t+u-1}}{\Gamma(t+u)} \bar{G}(z + x)dzdu =$$

$$\lambda^2 \int_0^\infty \int_0^\infty \int_0^z e^{-\lambda z} \frac{(\lambda y)^{t-1}}{\Gamma(t)} \frac{(\lambda(z-y))^{u-1}}{\Gamma(u)} \bar{G}(z + x)dy\,dz\,du.$$ Let

$v = z - y$, by interchanging the order of integration it follows that the right-hand side of the last equation equals

$$\lambda\int_0^\infty (\lambda\int_0^\infty \int_0^\infty e^{-\lambda v} \frac{(\lambda v)^{t-1}}{\Gamma(t)} \bar{G}(v + y + x)dv\, du)\, e^{-\lambda y} \frac{(\lambda y)^{t-1}}{\Gamma(t)} dy$$

$$= \lambda\int_0^\infty e^{-\lambda y} \frac{(\lambda y)^{t-1}}{\Gamma(t)} w(x + y)dy = E_x W(X(t)).$$ Thus we have,

$E_x(\tau) = w(x) - E_x W(X(\tau))$. Also, note that $E_x(gX(\tau)) = E_x(h(X(\tau))I(\tau < \zeta)) + cP_x(\tau = \zeta) = E_x[\{h(X(\tau)) - c\}I(\tau < \zeta)] + c$. Hence, $\theta_\tau = bw(x) - bE_x(w(X(\tau)I(\tau < \zeta)) - E_x[(h(X(\tau)) -c)I(\tau < \zeta)] - c = E_x[\{c - h(X(\tau)) - bw(X(\tau))\}I(\tau < \zeta)] - (c - bw(x)) = E_x\{\phi(X(\tau))I(\tau < \zeta)\} + (c - bw(x))$.

The following theorem follows from Lemmas 2.1 and 2.2.

Theorem 2.2. The optimal replacement time that minimizes ψ_τ is the one that maximizes $E_x(\phi(X(\tau))I(\tau < \zeta))$ and the maximum equals $c - bw(x)$.

Let $L_\infty(R_+)$ be the space of all bounded functions mapping $R_+ \overset{\text{def}}{=} [0, \infty)$ to R_+. Set $B(L_\infty(R_+))$ equal to the algebra of all (bounded, linear) operators from $L_\infty(R_+)$ to $L_\infty(R_+)$. For any time $t < \zeta$ and any $g \in L_\infty(R_+)$ we let the transformation T_t to be defined by $T_t g(x)=\lambda\int_0^\infty e^{-\lambda y} \frac{(\lambda y)^{t-1}}{\Gamma(t)} (\bar{G}(x+y)/\bar{G}(x))g(x+y)dy$.

Then $\{T_t, t \in R_+\}$ is a <u>semigroup</u> in $B(L_\infty(R_+))$, i.e., $S_0 = I$, the identity operator, and for all t and s in R_+, $T_t T_s = T_{t+s}$. For every $f \in L_\infty(R_+)$ we define the infinitesimal generator of f (denoted by Af) by the relation $Af = \lim_{t\downarrow 0} \frac{T_t f - f}{t}$,

whenever the limit exists. We let $\mathcal{D}(A) \overset{\text{def}}{=} \{f \in L_\infty(R_+) : \lim_{t\downarrow 0} \frac{T_t f - f}{t}$

exists in the sup norm}.

The following lemma is due to Dynkin [6, p. 133].

Lemma 2.3. Let f be any function in $\mathcal{D}(A)$, then for any replacement time τ for which $E_x(\tau) < \infty$, we have that

$$T_\tau f(x) - f(x) = E_x\left(\int_0^\tau Af(X(t))I(t < \zeta)dt\right).$$

Lemma 2.4. Suppose that $\overline{G}$ is differentiable on $[0,\infty)$ and $\frac{\overline{G}'}{\overline{G}} \varepsilon L_\infty$. Then $\mathcal{D}(A) = \{f \varepsilon L_\infty : f$ is absolutely continuous and $f' \varepsilon L_\infty\}$ and $Af(x) = \int_0^\infty e^{-y}[\{f(x+y)\overline{G}(x+y) - f(x)\overline{G}(x)\}/y\overline{G}(x)]dy$.

Proof. Take an arbitrary f in $\{f \varepsilon L_\infty : f$ is absolutely continuous and $f' \varepsilon L_\infty\}$. Note that

$$\left|\left|\frac{T_t f - f}{t} - \int_0^\infty e^{-y}[\{f(x+y)\overline{G}(x+y) - f(x)\overline{G}(x)\}/G(x)y]dy\right|\right|$$

$$= \left|\left|\int_0^\infty [\{f(x+y)\overline{G}(x+y) - f(x)\overline{G}(x)\}/\overline{G}(x)y][e^{-y}\{(y^t/\Gamma(t+1)) - 1\}]dy\right|\right|$$

$$= \left|\left|\int_0^\infty [\{(f(x+y) - f(x))/y\}(\overline{G}(x+y)/\overline{G}(x)) + f(x)\{(\overline{G}(x+y)\right.$$

$$\left. - \overline{G}(x))/\overline{G}(x)y\}] \times [e^{-y}\{(y^t/\Gamma(t+1)) - 1\}]dy\right|\right|$$

By Lagrange's Theorem the last equation equals to

$$\left|\left|\int_0^\infty [f'(x+\theta_1 y)(\overline{G}(x+y)/\overline{G}(x)) + f(x)(\overline{G}'(x+\theta_2 y)/\overline{G}(x))] \times\right.\right.$$

$$\left.\left.[e^{-y}(y^t/\Gamma(t+1)-1)]dy\right|\right|,$$

where $0 \leq \theta_1 \leq 1$ and $0 \leq \theta_2 \leq 1$. This last expression is less than or equal to

$$(||f'|| + ||f|| \; ||\frac{\overline{G}'}{\overline{G}}||) \lim_{t\downarrow 0} \int_0^\infty e^{-y}[(y^t/\Gamma(t+1))-1]dy.$$

Since $||e^{-y}y^t/\Gamma(t+1)|| < 1$ and all $t > 0$, then applying the Bounded Convergence Theorem we have that

$$(||f'|| + ||f|| \; ||\frac{\overline{G}'}{\overline{G}}||)\lim_{t\downarrow 0} \int_0^\infty e^{-y}[(y^t/\Gamma(t+1)) - 1]dy = 0.$$

It follows immediately that $\{f \in L_\infty : f$ is absolutely continuous and $f' \in L_\infty\} \subset \mathcal{D}(A)$.

To show that $\mathcal{D}(A) \subset \{f \in L_\infty : f$ is absolutely continuous and $f' \in L_\infty\}$. Take an $f \in \mathcal{D}(A)$ and note that $||T_t f - f|| \to 0$ as $t \to 0$. This is equivalent to

$$||\int_0^\infty e^{-y}(y^t/\Gamma(t))[f(x+y)\overline{G}(x+y) - f(x)G(x))/G(x)y]dy|| \to 0 \text{ as } t \to 0.$$

Since $f \in L_\infty$ and $||\overline{G}|| < 1$, then f must satisfy the Lipschitz condition $\sup_y|(f(x+y)\overline{G}(x+y) - f(x)\overline{G}(x))/\overline{G}(x)y| < M$, for all $X \in R_+$, for some M in $(0,\infty)$. This together with the assumption that $\frac{\overline{G}'}{\overline{G}} \in L_\infty$ implies that $f' \in L_\infty$. Since

$$\frac{T_t f(x) - f(x)}{t} = \int_0^\infty e^{-y}(y^t/\Gamma(t+1))[\{f(x+y)\overline{G}(x+y)$$

$$- f(x)\overline{G}(x)\}/\overline{G}(x)y]dy,$$

applying the Bounded Convergence Theorem we have that

$$\lim_{t\downarrow 0} \frac{T_t f - f}{t} = \int_0^\infty e^{-y}[\{f(x+y)\overline{G}(x+y) - f(x)\overline{G}(x)\}/\overline{G}(x)y]dy.$$

For any $f \in \mathcal{D}(A)$, the pointwise limit exists for a sequence $t_n \to 0$ as $n \to \infty$ and for each $X \in R_+$, thus we must have that

$$Af(x) = \lim_{t_n \downarrow 0} [\{T_{t_n} f(x) - f(x)\}/t_n]$$

$$= \int_0^\infty e^{-y}[\{f(x+y)\overline{G}(x+y) - f(x)\overline{G}(x)\}/\overline{G}(x)y]dy .$$

Thus $\mathcal{D}(A) \subset \{f \in L_\infty : f$ is absolutely continuous and $f' \in L_\infty\}$. This finishes the proof of the Lemma.

Let $B = \{x: \int_0^\infty \frac{e^{-\lambda y}}{y} [\frac{\phi(x+y)\overline{G}(x+y)}{G(x)} - \phi(x)]dy \leq 0\}$, and ξ^* be the infimum of B. The following theorem gives the form of the optimal replacement policy.

Theorem 2.3. Let $\phi \in \mathcal{D}(A)$ and assume that B is closed in the sense that $P_x(\exists t: X(t) \notin B) = 0$ for all $x \in B$. Then, the replacement time $\tau^* = \inf\{t: X(t) \geq \xi^*\}$ is optimal in the sense that it minimizes ψ_τ.

Proof. Let τ be any arbitrary replacement time. Then

$\theta_{\tau^*} - \theta_\tau = E_x[\phi(X(\tau^*))I(\tau^* < \zeta)] - E_x[\phi(X(\tau))I(\tau < \zeta)] =$

$E_x \int_0^{\tau^*} \{A\phi(X(t))I(t < \zeta)\, dt - E_x \int_0^{\tau} \{A\phi(X(t))I(t < \zeta)\}dt =$

$E_x \int_\tau^{\tau^*} \{A\phi(X(t))I(\tau < t < \tau^* < \zeta)\}dt -$

$E_x \int_{\tau^*}^{\tau} \{A\phi(X(t))I(\tau^* < t < \tau < \zeta)\}dt$. By definition of τ^* we have that $A\phi(X(t)) \geq 0$ on the set $\{\tau < t < \tau^* < \zeta\}$ which in turn implies that $E_x \int_\tau^{\tau^*} A\phi(X(t))I(\tau < t < \tau^* < \zeta)dt$ is nonnegative. Moreover, the definition of τ^* and the fact that B is closed implies that $A\phi(X(t))$ is nonpositive on the set $\{\tau^* < t < \tau < \zeta\}$. Thus, $\int_{\tau^*}^{\tau} A\phi(X(t))I(\tau^* < t < \tau < \zeta)dt$ is nonpositive and hence the second quantity in the last equation is nonnegative. It follows at once that $\theta_{\tau^*} - \theta_\tau \geq 0$. Since τ was arbitrary we must have that $\theta_{\tau^*} = \sup_\tau \theta_\tau$. Using Lemma 2.1 it follows that τ^* is optimal.

Note. The assumptions that B is closed is specially true whenever the infinitesimal generator $A\phi(x)$ is decreasing in $x \geq 0$.

We note that in order to determine ξ^* one must know $b = \inf_\tau \psi_\tau$. Clearly h and w will also play an important role. We illustrate this point by considering the following two special cases.

Special Case I. (Exponential wear survival probability and exponential cost function.) In this case we assume that the wear survival probability is of the form $\overline{G}(x) = e^{-\nu x}$, $x \geq 0$ and $\nu > 0$. Moreover, we let $h(x) = c(1 - e^{-\gamma x})$, $\gamma > 0$, then

$$w(x) = \int_0^\infty P_x(\tau > t)dt =$$

$$\int_0^\infty E_x\overline{G}(X(t))dt = \lambda\int_0^\infty \int_0^\infty e^{-\lambda y}\, \frac{(\lambda y)^{t-1}}{\Gamma(t)}\, e^{-\nu(x+y)}dy\, dt$$

$$= e^{-\nu x}\int_0^\infty \int_0^\infty e^{-(\lambda+\nu)y}\, \frac{y^{t-1}}{\Gamma(t)}\, dy = e^{-\nu x}\int_0^\infty \left(\frac{\lambda}{\lambda+\nu}\right)^t dt$$

$$= e^{-\nu x}/\log\left(\frac{\lambda+\nu}{\lambda}\right). \quad \text{Thus, } A\phi(x) \overset{\text{def}}{=} \int_0^\infty \frac{e^{-\lambda y}}{y}\left[\left(\frac{\phi(x+y)\overline{G}(x+y)}{\overline{G}(x)}\right) - \phi(x)\right] dy$$

$$= \int_0^\infty \frac{e^{-\lambda y}}{y} [\{(b\, e^{-\nu(x+y)}/\log(\tfrac{\lambda+\nu}{\lambda})) - c\, e^{-\gamma(x+y)}\}e^{-\nu y} -$$

$$\{(b\, e^{-\nu x}/\log(\tfrac{\lambda+\nu}{\lambda})) - c\, e^{-\gamma x}\}]dy$$

$$= (b/\log(\tfrac{\lambda+\nu}{\lambda}))e^{-\nu x} \int_0^\infty \frac{e^{-\lambda y}}{y} (e^{-2\nu y} - 1)dy$$

$$- c\, e^{-\gamma x} \int_0^\infty \frac{e^{-\lambda y}}{y} (e^{-(\gamma+\nu)y} - 1)dy$$

$$= (b/\log(\tfrac{\lambda+\nu}{\lambda}))\log(\tfrac{\lambda}{\lambda+2\nu})e^{-\nu x} - c\ \log(\tfrac{\lambda}{\lambda+\gamma+\nu})e^{-\gamma x}$$ [See 8, p. 449]

$$= c\ \log(\tfrac{\lambda+\gamma+\nu}{\lambda})e^{-\gamma x} - b(\log(\tfrac{\lambda+2\nu}{\lambda})/\log(\tfrac{\lambda+\nu}{\lambda}))e^{-\nu x}.$$

If $\log(b\nu\log(\frac{\lambda+2\nu}{\lambda})/c\gamma\log(\frac{\lambda+\gamma+\nu}{\lambda})\log(\frac{\lambda+\nu}{\lambda}))/(\nu-\gamma) \leq 0$, then it follows that $A\phi(x)$ is decreasing and B is closed. We also have that $\xi^* = \log(b\ \log(\frac{\lambda+2\nu}{\lambda})/c\ \log(\frac{\lambda+\nu}{\lambda})\log(\frac{\lambda+\gamma+\nu}{\lambda}))/(\nu-\gamma)$. By Theorem 2.3 it follows that

$$\tau^* = \inf\{t:X(t) \geq \log(b\ \log(\tfrac{\lambda+2\nu}{\lambda})/c\ \log(\tfrac{\lambda+\nu}{\lambda})\log(\tfrac{\lambda+\gamma+\nu}{\lambda}))/(\nu-\gamma)\}$$

is optimal in the sense defined in Theorem 2.2.

From the definition of τ^* we note that the larger the value of c (the cost of replacement at failure) the smaller we expect τ^* to be, a result that is consistent with one's intuition.

Special Case II. (Exponential wear survival probability and constant cost function.)

In this case we assume that $\overline{G}(x) = e^{-\nu x}$, while $h(x) = c - k$, $0 < k < c$. Thus, $\phi(x) = (be^{-\nu x}/\log(\frac{\lambda+\nu}{\lambda})) - k$ and

$$A\phi(x)=\int_0^\infty \frac{e^{-\lambda y}}{y}\{[(be^{-\nu(x+y)}/\log(\frac{\lambda+\nu}{\lambda}))-k]e^{-\nu y}-[(be^{-\nu x}/\log(\frac{\lambda+\nu}{\lambda}))-k]\}dy$$

$$= e^{-\nu x}/\log(\frac{\lambda+\nu}{\lambda}) \int_0^\infty \frac{e^{-\lambda y}}{y}(e^{-2\nu y}-1)dy - k\int_0^\infty \frac{e^{-(\lambda+\nu)y}-e^{-\lambda y}}{y}dy$$

$$= \{e^{-\nu x}/\log(\frac{\lambda+\nu}{\lambda})\}\log(\frac{\lambda}{\lambda+2\nu}) - k\log(\frac{\lambda}{\lambda+\nu}) .$$

In this case $\xi^* = \frac{1}{\nu}\log(\frac{k(\log(\frac{\lambda+\nu}{\lambda}))^2}{\log(\frac{\lambda+\nu}{\lambda})})$. Since $A\phi(x)$ is decreasing in x, then immediate applications of Theorem 2.3 implies that the optimal replacement policy is of the form

$$\tau^* = \inf\{t:X(t) \geq \frac{1}{\nu}\log(\frac{k(\log(\frac{\lambda+\nu}{\lambda}))^2}{\log(\frac{\lambda+\nu}{\lambda})})\}$$

3.0 An Optimal Replacement Policy for a Device Subject to a Gamma Wear Process-Discrete Case. In this section we consider a discrete version of the results discussed in Section 2. We assume that a device is subject to damage occurring randomly at time $n = 1, 2, \ldots$. The distribution of the damage is assumed to be exponential with mean $1/\lambda$, $\lambda > 0$, and damages accumulate additively. Since the convolution of exponentials is a gamma distribution, it then follows that the density function of the accumulated damage at time $n = 1, 2, \ldots$ will be of the form

$$f(x) = \lambda e^{-\lambda x}\frac{(\lambda x)^{n-1}}{\Gamma(n)}, \quad x \geq 0,\ \lambda > 0$$

$$= 0 \qquad \text{otherwise.}$$

Upon failure or replacement the device is replaced with a new one, and the process repeats. Accumulated damage is observed by a controller and a decision to replace the device is based only on its current value. As we assumed in Section 2, the cost of replacement at failure is identically equal to a constant $c > 0$. Cost of replacement at any time before failure is a function depending only on the accumulated damage at that time and is denoted by $h(\cdot)$. Throughout we assume that $h(\cdot)$ is bounded. We assume that failure is only due to the damage the device suffers, thus the life length of the device depends only on the damage it suffers and is independent of any other internal or external causes. Let $\overline{G}(x)$ be the probability that the device survives an x amount of damage, $x \geq 0$, and denote the life length of the device by ζ. The problem is to find an optimal replacement time that minimizes the expected cost per unit time.

We now give a discrete version of Dynkin's lemma. To fully exploit the lemma it behooves us to consider the following general model: For any bounded cost function f on the half line $[0, \infty)$ and for any time $n < \zeta$ we define the expected cost, conditioning on the fact that the device begins with an x amount of damage, $x \geq 0$, by the relationship

$$E_x(f(X(n))I(n < \zeta)) = \lambda \int_0^\infty e^{-\lambda y} \frac{(\lambda y)^{n-1}}{\Gamma(n)} (\overline{G}(x + y)f(x + y)/\overline{G}(x))dy.$$

Define the infinitesimal generator of f (denoted by Af) by the following equation $Af(x) = E_x(f(X(1))I(1 < \zeta)) - f(x)$. Let $F_n = \sigma(X(1),\ldots,X(n))$. The triple $(X(n), \mathcal{F}_n, P_x)$, then describe the gamma process $\{X(n)\}$ and the increasing sequence of σ-algebras $\{\mathcal{F}_n\}$, where P_x is the probability measure that induces $\{X(n)\}$.

Lemma 3.1. Let f be any bounded function, then for every replacement time τ for which $E_x(\tau) < \infty$, we have that

$$E_x(f(X(\tau))I(\tau < \zeta)) - f(x) = (1/\overline{G}(x))E_x \sum_{n=0}^{\tau-1} Af(X(n))\overline{G}(X(n)).$$

Proof. We note that for every n in $\{1, 2,\ldots\}$

$$Af(X(n))\overline{G}(X(n)) = [\,E_{X(n)}\left(\frac{f(X(n+1))\overline{G}(X(n+1))}{\overline{G}(X(n))}\right) - f(X(n))]\overline{G}(X(n))$$

$= E_{X(n)}f(X(n+1))\overline{G}(X(n+1)) - f(X(n))\overline{G}(X(n))$. The assumptions that f is bounded implies that Af is bounded. This together with the assumptions that $E_x(\tau) < \infty$ implies that

$E_x \sum_{n=0}^{\tau-1} Af(X(n))\overline{G}(X(n))$ exists and is finite. Thus we can write

$$E_x(\sum_{n=0}^{\tau-1} Af(X(n))\overline{G}(X(n))) = \int_{(\tau<\infty)} \sum_{n=0}^{\tau-1}[\,E_{X(n)}(f(X(n+1))\overline{G}(X(n+1)) -$$

$$f(X(n))\overline{G}(X(n))]\,dP_x = \sum_{k=1}^{\infty}\sum_{n=0}^{k-1}\int_{(\tau=k)}[\,E_{X(n)}(f(X(n+1))\overline{G}(X(n+1))) -$$

$$f(X(n))\overline{G}(X(n))]\,dP_x = \sum_{n=0}^{\infty}\int_{(\tau>n)}[\,E_{X(n)}(f(X(n+1))\overline{G}(X(n+1))) -$$

$$f(x(n))\overline{G}(X(n))]\,dP_x = \sum_{n=0}^{\infty}\int_{(\tau>n)}[\,f(X(n+1))\overline{G}(X(n+1)) - f(X(n))\overline{G}(x(n))\,dP_x,$$

since $(\tau > n) \in \mathcal{F}_n$ and $E_{X(n)}f(X(n+1))\overline{G}(X(n+1)) \in \mathcal{F}_n$.

$$= \sum_{n=1}^{\infty}\int_{(\tau\geq n)} f(X(n))\overline{G}(X(N))\,dP_x - \sum_{n=0}^{\infty}\int_{(\tau>n)} f(X(n))\overline{G}(X(n))\,dP_x$$

$$= \sum_{n=0}^{\infty}\int_{(\tau=n)} f(X(n))\overline{G}(X(n))\,dP_x - f(x)\overline{G}(x)$$

$$= \overline{G}(x)E_x(f(X(\tau))I(\tau < \zeta)) - f(x)\overline{G}(x).$$

Hence, $(1/\overline{G}(x))[\,E_x \sum_{n=0}^{\tau-1} Af(X(n))\overline{G}(X(n))] = E_x(f(X(\tau))I(\tau<\zeta)) - f(x)$.

If τ is any replacement time then the optimal replacement time minimizing the expected cost per unit time $\psi_\tau \overset{\text{def}}{=} E_x g(X(\tau))/E_x(\tau)$. Adopting the same notations as those used in sections 1 and 2, we let $w(x) = E_x(\zeta)$, $b = \inf_\tau \psi_\tau > 0$ be the minimum average cost,

The cost function g is defined by the following relation

$$g(X(n)) = \begin{cases} h(X(n)), & n < \zeta \\ c & n = \zeta. \end{cases}$$

and $\phi(x) = c - h(x) - bw(x)$.

Let $B = \{x:\lambda\int_0^\infty \phi(x+y)\overline{G}(x+y)e^{-\lambda y}dy - \phi(x)\overline{G}(x) \leq 0\}$, and let $\xi^* = \inf B$. The following theorem gives the form of the optimal replacement policy and is the discrete analogue of Theorem 2.3.

Theorem 2.3′. If B is closed in the sense that $P_x\{\exists n:X(n) \notin B\} = 0$ for all $x \in B$, then the replacement time $\tau^* = \inf\{n:X(n) \geq \xi^*\}$ is optimal in the sense that it minimizes ψ_τ.

As an applications of the results in this section one can consider a discrete version of the exponential wear survival probability with exponential cost function, and exponential wear survival probability with constant cost function case discussed in Section 2. We now give a different application of the results.

Special Case III. Suppose that the device fails the first time the damage exceeds a given threshold x_0, in this case we have that

$$\overline{G}(x) = \begin{cases} 1 & \text{if } x < x_0 \\ 0 & \text{if } x \geq x_0. \end{cases}$$

We let, $h(X(n)) = c - k$ for all $n < \zeta$, and $h(X(\zeta)) = c$ where,

$0 < k < c$. Thus, $w(x) = \sum_{n=1}^{\infty} E_x \overline{G}(X(n))$

$$= \begin{cases} \sum_{n=1}^{\infty} \int_0^{x_0 - x} e^{-\lambda y} \frac{(\lambda y)^{n-1}}{\Gamma(n)} d(\lambda y), & x \le x_0 \\ 0 & x \ge x_0. \end{cases}$$

$$= \begin{cases} \sum_{n=1}^{\infty} \sum_{k=n}^{\infty} e^{-\lambda(x_0 - x)} \frac{[\lambda(x_0 - x)]^k}{k!}, & x \le x_0 \\ 0 & x \ge x_0. \end{cases}$$

$$= \begin{cases} \sum_{k=1}^{\infty} k\, e^{-\lambda(x_0 - x)} \frac{[\lambda(x_0 - x)]^k}{k!}, & x \le x_0 \\ 0 & x \ge x_0. \end{cases}$$

$$= \begin{cases} \lambda(x_0 - x), & x \le x_0 \\ 0 & x \ge x_0. \end{cases}$$

Hence, we have that $\phi(x) = k - b\lambda(x_0 - x)$, and

$$A\phi(x) = \int_0^{x_0 - x} [k - b\lambda(x_0 - x - y)] e^{-\lambda y} d(\lambda y) - k + b\lambda(x_0 - x).$$

Simple calculations yield, $A\phi(x) = b - (b + k)e^{-\lambda(x_0 - x)}$. Thus

$\tau^* = \inf\{n : b - (b + k)e^{-\lambda(x_0 - X(n))} \le 0\} =$

$\inf\{n : X(n) \ge x_0 + \frac{1}{\lambda} \log(b/b + k)\}$.

REFERENCES

[1] Abdel-Hameed, M. S. and Proschan, F., Non-stationary shock models. *Stoch. Proc. and Applic.* *1*, (1973), pp. 383-404.

[2] Abdel-Hameed, M. S., A Gamma Wear Process. (a) *IEEE Trans. Rel.* R-24 #2, (1975), pp. 152-154.
(b) NAPS Document #02574-B. Order from ASIS-NAPS, c/o Microfiche Publications, 440 Park Ave. South, New York, N. Y. 10016.

[3] Abdel-Hameed, M. S. and Proschan, F., Shock models with underlying birth process. *J. Appl. Probability*, *12*, (1975) pp. 18-28.

[4] Abdel-Hameed, M. S. and Proschan, F., Total positivity properties of generating functions, *SIAM J. MATH. ANAL.* *6*, (1975), pp. 81-84.

[5] Abdel-Hameed, M. S. and Proschan, F., Preservation of geometric properties under an integral transformation. *SIAM J. Math. Anal.*, (1976), to appear.

[6] Dynkin, E. G., Markov Processes. English transl., Springer-Verlag, Berlin, (1965).

[7] Karlin, S., Total Positivity, Vol. 1. Stanford University Press, Stanford, (1968).

[8] Standard Mathematical Tables, 19th Edition. The Chemical Rubber Co., Cleveland, Ohio 44128, (1971).

[9] Taylor, H. M., Optimal Stopping in a Markov Process. *Ann. Math. Statist.* *39*, (1968), pp. 1333-1344.

SOME LIMIT THEOREMS IN STANDBY REDUNDANCY WITH RENEWALS*

by Z. KHALIL
Concordia University

Abstract. A class of possible limit distributions is derived for the distribution of the first time system-down of a system consisting of $n + 1$ $(n = 1,2, \ldots)$ exactly similar units where only one unit is in operation at any given time while the other n units are in unloaded standby or are being repaired.

1.0 Introduction and Notation. Consider a system of $n + 1$ identical units where only one unit is operative and n units are in cold standby. The failure time of the operating unit has an arbitrary distribution $F(t)$. The repair time of a failed unit has the exponential distribution $G(t) = 1 - e^{-\nu t}$. There are $r (1 \leq r \leq n)$ repair facilities and one switching device which, upon failure of the working unit, will switch the available standby unit into operation. The switchover time from standby state to operative is random and has an arbitrary distribution $L(t)$. When the working unit fails, a standby will be switched to operate and the failed unit undergoes repair if a repair facility is free if not it lines up in a queue waiting for repair. All random variables involved are independent and non negative. We introduce the following notation:

$$H(x) = \int_0^x L(x - t)dF(t) \quad , \quad G_K(x) = \int_0^\infty G(x - z)dG_{K-1}(z) =$$

$$= 1-e^{-\nu x} \sum_{i=0}^{k-1} \frac{(\nu x)^i}{i!} , \; G_0(x) = G(x)$$

$$\overline{H}(x) = 1 - H(x)$$

---

* Research partially supported by NRC grant #A9095 and Quebec FCAC grant.

$$f(s) = \int_0^\infty e^{-sx} dF(x) \qquad \overline{G}(x) = 1 - G(x)$$

$$l(s) = \int_0^\infty e^{-sx} dL(x) \quad , \quad h(s) = \int_0^\infty e^{-sx} dH(x) = l(s) \cdot f(s)$$

Let $A(t)$ denote the probability that the system with n standby units is operative at time t, given that the system starts at $t = 0$ with all units operative. Let $\Gamma_{K,i}(t)$ denote the probability that in the repair system there are K failed units at $t = 0$ and within time t, i elements are repaired. The function $\Gamma_{K,i}(t)$ has been calculated, see [2]. Furthermore, let

$$a(s) = \int_0^\infty e^{-sx} d[1 - A(x)]$$

and

$$\gamma_{K,i}(s) = \int_0^\infty e^{-sx} \, \Gamma_{K,i}(x) dH(x) \tag{1.1}$$

Considering an expanding backup system, i.e., first finding a solution for the system with K standbys and then proceeding to a solution for $K + 1$ standby units the Laplace-Stieltjes transform $a(s)$ of $1 - A(t)$ was found to be: see [2]

$$a(s) = f(s) \; \frac{h^{n-r}(s + r\nu) \prod_{k=1}^{r} h(s + k\nu)}{\Delta_n(s)}$$

where

$$\Delta_n(s) = \begin{vmatrix} 1-\gamma_{11} & 1-h & 1-h & \cdots & 1-h \\ -\gamma_{22} & 1-\gamma_{21}-\gamma_{22} & 1-h & \cdots & \\ -\gamma_{33} & -\gamma_{32}-\gamma_{33} & 1-\gamma_{31}-\gamma_{32}-\gamma_{33} & & \\ \cdots & \cdots & \cdots & & \\ \cdots & \cdots & \cdots & & \\ \cdots & \cdots & \cdots & & 1-h \\ -\gamma_{nn} & -\gamma_{n,n-1}-\gamma_{n,n} & & & 1-\sum_{i=0}^{n} \gamma_{n,i} \end{vmatrix}$$

From the definition of the functions $\Gamma_{k,i}(\nu)$ we have for any K

$$\sum_{i=0}^{K} \gamma_{K,i}(s) = h(s)$$

but $$\gamma_{k,0}(s) = \int_0^{\infty} e^{-sx}\Gamma_{k,0}(x)\, dH(x) = \begin{cases} h(s+K\nu) & \text{for } k \le r \\ h(s+r\nu) & \text{for } k > r \end{cases}$$

and we have

$$1 - \sum_{i=1}^{k} \gamma_{k,i}(s) = 1 - h(s) + \gamma_{k,0}(s)$$

Using these equations we can rewrite $\Delta_n(s)$ in the following form:

$$\Delta_n(s) = \begin{vmatrix} 1-h(s)+h(s+\nu) & 1-h(s) & 1-h(s) & \cdots & 1-h(s) \\ -1+h(s+\nu)-h(s+2\nu) & h(s+2\nu) & 0 & \cdots & \\ \cdots\cdots\cdots & & & & \\ & & h[s+(r-1)\nu] & & \\ & & & h(s+r\nu) & \\ & & & \cdots & 0 \\ \cdots\cdots\cdots & & & \cdots & h(s+r\nu) \end{vmatrix}$$

$\Delta_n(s)$ could now be written in the form

$$\Delta_n(s) = [1 - h(s) + h(s + \nu)]\ \delta_1(s) + [1 - h(s)] \sum_{i=2}^{n} (-1)^{i-1} \delta_i(s)$$

where $\delta_i(s)$ are determinants of the $(n - 1)^{st}$ order resulting from $\Delta_n(s)$ by deleting the 1st row and the ith column.

since
$$\delta_1(s) = h^{n-r}(s + r\nu) \prod_{k=2}^{r} h(s + k\nu)$$

we have

$$\Delta_n(s) = h^{n-r}(s + r\nu) \prod_{k=1}^{r} h(s + k\nu) + [1 - h(s)] \sum_{i=1}^{n} (-1)^{i-1} \delta_i(s)$$

and

$$a(s) = f(s) \frac{h^{n-r}(s + r\nu) \prod_{k=1}^{r} h(s + k\nu)}{h^{n-r}(s + r\nu) \prod_{k=1}^{r} h(s + k\nu) + [1 - h(s)] \sum_{i=1} (-1)^{i-1} \delta_i(s)}$$

2.0 Limit Theorems. In work [2] the following theorem has been proved for $L(x) = 1, x > 0$ i.e. the instantanious switchover time.

Theorem 1. Let τ denote the failure-free operation time of the system. Let $F(x)$ and $L(x)$ be such that $\int_0^\infty x d\, H(x) < \infty$.

Then as $\nu \to \infty$

$$a(\alpha_\nu s) \to \frac{1}{1+s}$$

i.e.

$$P(\alpha_\nu \tau > t) \to e^{-t}$$

where $\alpha_\nu = \frac{1}{a} h^{n-r}(r\nu)$ and **tends** to zero as $\nu \to \infty$.

This theorem puts on the initial distributions $F(t)$ and $L(t)$ a strong condition namely the existence of its first moments. It is of interest to investigate the extent to which this assumption limits the class of possible limit distributions. Here many problems formulations are possible. Let $F(t)$ and $L(t)$ be fixed. One would ask the following question: under a suitable normalization and as $\nu \to \infty$ what are the possible asymptotic distributions for the system's failure - free operation time distribution. In other words what are the possible limiting functions $\Psi(s)$ for the function

$$a(\beta_\nu s) = f(\beta_\nu s)\,\frac{h^{n-r}(\beta_\nu s+r\nu)\prod_{k=1}^{r} h(\beta_\nu s+k\nu)}{h^{n-r}(\beta_\nu s+r\nu)\prod_{k=1}^{r} h(\beta_\nu s+k\nu)+1-h(\beta_\nu s)\sum_{i=1}^{n}(-1)^{i-1}\delta_i(\beta_\nu s)}$$

if the repair time is small as compared to the failure time of the working unit, i.e. as $\nu \to \infty$ and a suitable choice of the normalizing factors β_ν. Before we proceed to the main result of this work we state without proof the following lemma, see [2].

Lemma. Let $F(x)$ be the distribution function of a non-negative random variable, let $\phi_1(x) \geq 0$ and $\phi_2(x) \geq 0$ where $\phi_1(x)$ is a monotone increasing function and $\phi_2(x)$ a monotone decreasing function, then

$$\int_0^\infty \phi_1(x)\,\phi_2(x)\,dF(x) \leq \int_0^\infty \phi_1(x)\,dF(x) \cdot \int_0^\infty \phi_2(x)\,dF(x).$$

Theorem 2. Let $F(x)$ and $L(x)$ be fixed. As $\nu \to \infty$ the class of limit functions for $a(\beta_\nu s)$ is given by the formula

$$a_o(s) = \lim a(\beta_\nu s) = \frac{1}{1 + cs^\gamma}$$

where $0 \leq \gamma \leq 1$, c is a constant and $\beta_\nu \to 0$ as $\nu \to \infty$.

Proof. Let $\beta_\nu \to 0$ as $\nu \to \infty$. Then for fixed s, $f(\beta_\nu s) \to 1$ and

$$a_o(s)=\lim a(\beta_\nu s)=\left[1+\lim_{\beta_\nu\to 0} \frac{1-h(\beta_\nu s)}{\varepsilon(\beta_\nu)} \lim_{\beta_\nu\to 0} \frac{\varepsilon(\beta_\nu)\sum_{i=1}^{n} (-1)\delta_i(\beta_\nu s)}{h^{n-r}(\beta_\nu s+r\ )\prod_{k=1}^{r} h(\beta_\nu s+k\nu)}\right]^{-1} \tag{2.1}$$

We take $\varepsilon(\beta_\nu) = h^{n-r}(r\nu) \prod_{k=1}^{r} h(k\nu)$ and consider separately each of these limits.

a) As

$$\beta_\nu \to 0$$

$$\sum(-1)^{i-1}\delta_i(\beta_\nu s) \to 1$$

as we have shown in [2].

b)

$$\lim_{\beta_\nu\to 0} \frac{h^{n-r}(\beta_\nu s+r\nu)\prod_{k=1}^{r} h(\beta_\nu n+k\nu)}{\varepsilon(\beta_\nu)} = 1$$

since

$$\frac{h^{n-r}(\beta_\nu s+r_\nu)\prod_{k=1}^{r} h(\beta_\nu s+k\nu)}{\varepsilon(\beta_\nu)} = \frac{h(\beta_\nu s+r\nu)}{h(r\nu)}^{n-r} \prod_{k=1}^{r} \frac{h(\beta_\nu s+k\nu)}{h(k\nu)}$$

and each factor $\dfrac{h(\beta_\nu s + k\nu)}{h(k\nu)}$ for $k = 1,2,\ldots,r$

tends to one by the lemma, see [1].

c) consider now the limit:

$$\theta(s) = \lim_{\beta_\nu \to 0} \frac{1 - h(\beta_\nu s)}{\varepsilon(\beta_\nu)} \quad .$$

Here we use a method due to Kovalenko, see [3].

Let $\theta(s) \neq 0$ and $s > 0$ (non degenerate distribution). Then

$$\theta(s') = \lim_{\beta_\nu \to 0} \frac{1 - h(\beta_\nu s')}{\varepsilon(\beta_\nu)} = \lim_{\beta_\nu \to 0} \frac{1 - h(\beta_\nu \frac{s'}{s} s)}{\varepsilon(\beta_\nu \frac{s'}{s})} \frac{\varepsilon(\beta_\nu \frac{s'}{s})}{\varepsilon(\beta_\nu)}$$

$$= \theta(s) \lim_{\beta_\nu \to 0} \frac{\varepsilon(\beta_\nu \frac{s'}{s})}{\varepsilon(\beta_\nu)} = \theta(s)\; K(\frac{s'}{s}) \tag{2.2}$$

Now since $\theta(s)$ for $s > 0$ is obviously an analytic function we can use Taylor's expansion and we have

$$\theta(s') = \theta(s) + \theta'(s)(s' - s) + o(s' - s)$$

substituting in (2.2) we get

$$\theta'(s)(s' - s) = \theta(s)[K(\frac{s'}{s}) - 1] + o(s' - s)$$

Now dividing this equation by $\theta(s)(s' - s)$ and letting $s' \to s$ we find

$$\frac{\theta'(s)}{\theta(s)} = \frac{\gamma}{s}$$

where

$$\gamma = K'(s)\Big|_{s = 1}$$

Integrating we get

$$\theta(s) = cs^\gamma \tag{2.3}$$

where c is a constant. We remark that $a_0(s)$ is the Laplace-Stieltjes transform of a non-negative nondegenerate random variable, for $s > 0$ $a_0(s)$ must be a decreasing function.

From (a) and (b) and equation (2.3) we get:

$$a_0(s) = \frac{1}{1 + cs^\gamma} \tag{2.4}$$

For this it is necessary that $cs^\gamma > 0$. But for $c < 0$ the function (2.4) will have a pole at the point $s = (-\frac{1}{c})^{\frac{1}{\gamma}} > 0$ which is impossible. Now we have $c > 0$ and $\gamma > 0$. Furthermore since for $s > 0$, $a_0(s)$ is a convex function then it follows that never $\gamma > 1$, otherwise $a_0'(s) = 0 \leq a_0'(s) \leq 0$ and consequently $a_0(s) = 1$.

It follows that

$$c > 0 \quad \text{and} \quad o < \gamma \leq 1$$

Substituting (2.3) in (2.2) we get

$$s'^\gamma = s^\gamma K(\frac{s'}{s})$$

from which

$$K(z) = z^\gamma$$

or in other words

$$\lim_{\beta_\nu \to 0} \frac{\varepsilon(\beta_\nu z)}{\varepsilon(\beta_\nu)} = z^\gamma$$

and the theorem is proved.

REFERENCES

[1] Gnedenko, B. V., "Limit theorems for sums of a random number of positive independent random variables", *Proceedings of the Sixth Berkeley Symposium on Math. Statistics and Probability V. 2*, (1970), p. 537.

[2] Khalil, Z. S., "A redundancy Problem with renewal, IMS and AMS selected translations" in *Math. Statist. and Probability V. 13*, (1973).

[3] Kovalenko, I. N., "On the class of limit distributions for thinning streams of homogeneous events". Selected translation in *Math. Statist. and Prob. V. 9*, (1971).

SOME ASPECTS OF THE USE OF THE MANN-FERTIG STATISTIC TO OBTAIN CONFIDENCE INTERVAL ESTIMATES FOR THE THRESHOLD PARAMETER OF THE WEIBULL

by PAUL N. SOMERVILLE
*Florida Technological University**

Abstract. Mann and Fertig [1] proposed a statistic P for a goodness-of-fit test for the two parameter vs three parameter Weibull. Use of this statistic to obtain confidence interval estimates for the threshold parameter was described. This article discusses use of the statistic for this purpose.

1.0 *Introduction*. A statistic P was proposed by Mann and Fertig [1] for a goodness-of-fit test for the two-parameter Weibull distribution. The alternative was a three-parameter Weibull distribution. The P statistic was a modification of a statistic S previously proposed by Mann, Scheuer and Fertig [3]. The S statistic was shown to have a higher power than any of the analogues of the classical tests for testing the null hypothesis that a set of data was from a two-parameter Weibull distribution against both lognormal and three-parameter Weibull alternatives. A number of additional very desirable properties were also demonstrated. The P statistic is a modification of the S statistic, applicable specifically to tests with three-parameter Weibull alternatives. For positive values of the location parameter, the P statistic was shown to have a higher power than that of the S statistic.

The cumulative distribution of the variable T having the three-parameter Weibull is given by

---

*The research documented in this paper was initiated and completed while the author was a summer employee of the Laboratories Division, Support Operations, NASA, Kennedy Space Center.

$$P(T < t) = \begin{cases} 1 - \exp[-\{(t - \lambda)/\delta\}^{\beta}], & t > \lambda \quad \delta, \beta > 0 \\ 0 & , \quad t \leq \lambda \end{cases}$$

If $\lambda = 0$, and $X = \ln T$, then it is not difficult to show that

$$P(X \leq x) = 1 - \exp(-\exp\,(x - \eta)/\xi\}), \quad \xi > 0 \qquad (1.1)$$

where $\eta = \ln \delta$ and $\xi = 1/\beta$ are the location and scale parameters respectively of the random variable X.

A sample of n ordered observations of the random variable T is given by $t_{1,n} \leq t_{2,n} \leq \cdots \leq t_{n,n}$. (In what follows we will omit the second subscript). In a typical situation, we might have available only the first m ($\leq$ n) ordered observations. The Mann-Fertig statistic is given by

$$P_{k,m} = \sum_{i=k+1}^{m-1} \ell_i \Bigg/ \sum_{i=1}^{m-1} \ell_i \qquad (1.2)$$

where $\ell_i = (X_{i+1} - X_i)/E(X_{i+1} - X_i)$. (1.3)

$$\text{Now } E(X_{i+1} - X_i) = \xi E\{(X_{i+1} - \eta)/\xi\} - \xi E\{(X_i - \eta)/\xi\}$$
$$= \xi E(Z_{i+1} - Z_i)$$

where $Z_i = (X_i - \eta)/\xi$. (1.4)

Values of $E(Z_i)$ are tabulated in White [4] for n = 1(1)50(5)100. Values of $E(Z_{i+1} - Z_i)$ are given in Mann, Schafer and Singpurwalla [2] for i = 1(1)15 and in Mann, Scheuer and Fertig [3] for i = 1(1)25.

The value of k for which the test is most powerful is a function of m, and the size of the test, and is given in Table 3 of Mann and Fertig [1]. Mann and Fertig show that $P_{k,m}$ is approximately a Beta variate with parameters $\nu_1 = m - k - 1$ and $\nu_2 = k$.

Suppose that the data in question have a Weibull distribution with a non-zero location parameter λ. Then subtracting the constant λ from each data point would result in data with a two-parameter Weibull distribution.

Put $X_i^* = \ln(T_i - \lambda)$ (1.5)

$$\ell_i^* = (X_{i+1}^* - X_i^*)/E(X_{i+1} - X_i). \tag{1.6}$$

Define $$P_{k,m}^*(\lambda) = \sum_{i=k+1}^{m-1} \ell_i^* / \sum_{i=1}^{m-1} \ell_i^*. \tag{1.7}$$

Then $P_{k,m}^*(\lambda)$, as $P_{k,m}$ under the null hypothesis that $\lambda = 0$, is approximately a Beta variate with parameters ν_1 and ν_2. Mann and Fertig [1] have shown that $P_{k,m}^*(\lambda)$ is monotonically decreasing in λ for $\lambda < t_1$.

Thus, for a given set of data, it should be possible to obtain a lower confidence bound $\underline{\lambda}$ for λ at a level $1-\alpha$ by solving for the value of λ such that $P_{k,m}^*(\lambda)$ is equal to the $100(1-\alpha)$th percentile of a Beta variate with the appropriate combination of values of k and m.

The statistic $P_{k,m}^*(\lambda)$ approaches the lower bound of 0 as λ approaches t_1. However, it does not in general approach 1 as λ decreases without limit, but approaches a constant (say ρ), which is a function of the data points, and indeed it is possible for ρ to be arbitrarily small. We show in the following section that the value of ρ, for a given set of data $t_1, t_2, \ldots, t_m$ is given by

$$\rho = \sum_{i=k+1}^{m-1} \ell_i' / \sum_{i=1}^{m-1} \ell_i' \tag{1.8}$$

where $\ell_i' = (t_{i+1} - t_i)/[E(Z_{i+1}) - E(Z_i)]$. (1.9)

2.0 *Upper Bound for Statistic $P(\lambda)$*. It should be noted that $P_{k,m}^*(\lambda)$ is independent of the value of ξ, since the values appearing in the numerator and denominator cancel. For convenience, then, set $\xi = 1$.

Put $e_i = (t_{i+1} - t_i)/(t_i - \lambda)$ (2.1)

Then

$$(x^*_{i+1} - x^*_i) = \ln[(t_{i+1} - \lambda)/(t_i - \lambda)] \tag{2.2}$$
$$= \ln(1 + e_i)$$

For $\lambda < 0$ and $|\lambda|$ sufficiently large, we may write

$$e_i = [(t_{i+1} - t_i)/(-\lambda)][1 + t_i/\lambda + (t_i/\lambda)^2 + \cdots] \tag{2.3}$$
$$= [(t_{i+1} - t_i)/(-\lambda)][1 + 0(t_i/\lambda)].$$

Then from (2.2) and (2.3), we have for $\lambda < 0$ and $|\lambda|$ large,

$$x^*_{i+1} - x^*_i = [(t_{i+1} - t_i)/(-\lambda)][1 + 0(t_i/\lambda)] \tag{2.4}$$

and

$$\lim_{\lambda \to -\infty} P^*_{k,m}(\lambda) = \sum_{i=k+1}^{m-1} \ell'_i \Big/ \sum_{i=1}^{m-1} \ell'_i \tag{2.5}$$

where ℓ'_i is given by (1.9).

3.0 *Proportion of Time Finite Confidence Intervals Can Be Obtained.* A study was made, using Monte Carlo methods, of the proportion of the time that various finite confidence interval estimates could be made. The results suggest that the proportion is nearly independent of n and for sufficiently large values of $m \leq n$, approaches 1. However, for small m, with all values of $(1 - \alpha)$ greater than .5 the proportion was considerably less than 1, and decreased with increasing values of β, the shape parameter. This result was independent of the value of k used. For $(1-\alpha)$ equal to .5, the "best" value of k appeared to be the closest integer to m/5, although the results were relatively insensitive to small changes in k. Here "best" is defined to mean the value of k for which the proportion is largest.

Table I gives the proportion of the time that finite confidence intervals are obtained for $(1-\alpha)$ equal to .50, .90 and .95 and for m = 3, 6, 10, 15, and 25 for certain values of n. For $(1-\alpha)$ equal to .50 the value of k used was the integer nearest to m/5. For (1-α) equal to .90 and .95, the values of k used are those given in Table 3 of Mann and Fertig [1]. The values of k used are given in Table II. The proportions were obtained by Monte Carlo simulations, with samples of size 20,000 in all cases. It should be noted that for (1-α) equal to .90 and β=.5, that the proportions would be increased by approximately .05 when m = 6 and approximately .03 when m = 10, if the values of k used had been 1 and 2 respectively. For (1-α) equal to .95 and β=.5, for the same values of m, the use of k = 1 and 2, as above, would have resulted in an increase of the proportions by amounts varying from .01 to .02. These small gains for β=.5 are far outweighed by the losses that would be incurred for $\beta \geq 1$, and unless the value of β is known to be approximately .5 or less the values of k given in the table are recommended.

TABLE I

Value of β for Alternatives

		.5			1			2			4		
		CONFIDENCE			CONFIDENCE			CONFIDENCE			CONFIDENCE		
m	n	.50	.90	.95	.50	.90	.95	.50	.90	.95	.50	.90	.95
3	3	.83	.43	.30	.74	.24	.13	.65	.15	.07	.58	.12	.06
	6	.78	.35	.22	.69	.20	.11	.61	.14	.07	.56	.12	.06
	10	.77	.32	.21	.68	.19	.10	.60	.14	.07	.55	.11	.06
	15	.77	.32	.20	.68	.19	.10	.60	.14	.07	.56	.12	.06
	25	.77	.31	.19	.67	.18	.10	.60	.13	.07	.55	.11	.06
6	6	.99	.82	.68	.94	.52	.39	.79	.28	.16	.64	.16	.08
	10	.98	.71	.54	.90	.41	.28	.74	.23	.13	.61	.15	.08
	15	.98	.68	.50	.89	.39	.25	.73	.22	.12	.61	.15	.08
	25	.97	.65	.47	.88	.37	.25	.72	.21	.11	.61	.15	.07

TABLE I (Con't)

Value of β for Alternatives

		.5			1			2			4		
		CONFIDENCE			CONFIDENCE			CONFIDENCE			CONFIDENCE		
m	n	.50	.90	.95	.50	.90	.95	.50	.90	.95	.50	.90	.95
10	10	1.00	.99	.96	.99	.81	.68	.92	.43	.27	.74	.22	.11
	15	1.00	.96	.91	.98	.69	.54	.87	.36	.22	.71	.20	.11
	25	1.00	.94	.87	.97	.63	.48	.85	.33	.20	.69	.19	.10
15	15	1.00	1.00	1.00	1.00	.95	.89	.97	.60	.42	.82	.28	.15
	25	1.00	1.00	.99	1.00	.86	.75	.94	.49	.33	.77	.24	.13
25	25	1.00	1.00	1.00	1.00	1.00	.99	1.00	.83	.70	.90	.40	.24

Proportion of Time that Finite Confidence Intervals are Obtained

TABLE II

Values of k Used in Obtaining the Results Of Table I

CONFIDENCE $(1-\alpha)$	m				
	3	6	10	15	25
.50	1	1	1	1	1
.90	1	3	4	5	8
.95	1	3	4	5	8

4.0 *Use of $P^*_{k,n}(\lambda)$ for Confidence Interval Estimates.* Based on the results of Table I, unless the value of the shape parameter β is rather large, then it would appear that the use of $P^*_{k,n}(\lambda)$ will usually give finite 50 percent confidence interval estimates for λ unless m is very small. For 90 percent or higher confidence interval estimates, even if the shape parameter β is small (say 1 or less) a much larger value of m (say at least 15) is required for the technique to given a large percentage of finite confidence interval estimates.

It should be noted that the confidence interval estimates include the possibility of negative values of the threshold parameter λ. There is of course no theoretical reason why the value of λ cannot be thought of as negative. From an applied point of view, a negative value of λ, could be thought of as resulting from a translation of variables from the original (positive) values of the data.

The suggested method for attempting to obtain a finite $(1-\alpha)$ lower confidence bound for λ is as follows:

1. Calculate ρ as given by (1.8). This is the largest value attainable for $P^*_{k,n}(\lambda)$ for any finite value of $\lambda \leq t_1$.

2. Obtain the $100(1-\alpha)$th percentile P of the Beta variate

with parameters $\nu_1 = m - k - 1$ and $\nu_2 = k$. If this value is greater than or equal to the value of ρ in 1., then the method will not give a finite lower confidence bound for λ.

3. If $P < \rho$, then compute $P^*_{k,n}(0)$. If this value is greater than P, then the required lower confidence bound $\underline{\lambda}$ is between 0 and t_1 and may be obtained by the bisection technique outlined in Mann and Fertig [1] or by other more sophisticated iterative techniques.

4. If $P^*_{k,n}(0) < \rho$, then the required lower confidence bound $\underline{\lambda}$ is negative. One method of obtaining the bound is to successively calculate $P^*_{k,n}(-it_1)$ for $i = 1, 2, \ldots$ until the first value of i such that $P^*_{k,n}(-it_1) < \rho$ is found, and then to use the bisection technique for values between $-it_1$ and $(1-i)t_i$. There are of course more sophisticated iterative techniques available.

5.0 *Some General Considerations*. The previous sections have indicated that it is not always possible to use $P^*_{k,n}(\lambda)$ to obtain finite confidence interval estimates for λ. This was shown to be especially true when m is small, or when $\beta > 1$. Two important considerations have so far not been stated. First, the statistic $P_{k,n}$ was developed to test the hypothesis that a given set of data was from a two-parameter Weibull distribution, with the alternate hypothesis being that the data was from a three-parameter Weibull distribution (no parameters known in either case). Whenever the null hypothesis is rejected at a level α, then it is always possible to obtain an interval estimate of the Weibull threshold parameter with confidence γ, where γ is less than or equal to $1-\alpha$. When the null hypothesis is accepted (i.e., the data gives no reason, at the significance level α, to doubt that the data is from a two-parameter Weibull) then there is usually no strong motivation to construct an interval estimate for a "threshold parameter."

Second, the ability to estimate the value of the threshold paramater is most important when the data is from a Weibull distribution with small β (say $\leq 1.$). We have shown that the statistic P succeeds with a fairly high probability, in providing useful interval estimates, when $\beta \leq 1$, unless m is small.

It may be noted that the entries in Table I can be interpreted as the maximum values of power for testing the null hypothesis that $\lambda = \lambda_0$ vs the alternate hypothesis that $\lambda = \lambda_1 > \lambda_0$. As $\lambda_1 \to \infty$, the power is bounded above by the values in Table I.

Acknowledgment. I am indebted to Nancy Mann for critically reading an earlier version of this paper, and for many valuable comments.

REFERENCES

[1] Mann, N. R. and Fertig, K. W., "A Goodness-of-Fit Test for the Two Parameter vs. Three Parameter Weibull; Confidence Bounds for Threshold," *Technometrics*, *17*, (1975), pp. 237-245.

[2] Mann, N. R., Schafer, R. E. and Singpurwalla, N.D., Methods for Statistical Analysis of Reliability and Life Data, John Wiley, New York, (1975).

[3] Mann, N. R., Scheuer, E. M., and Fertig, K. W., "A New Goodness-of-Fit Test for the Two Parameter Weibull or Extreme Value Distribution with Unknown Parameters," *Communication in Statistics*, *2*, (1973), pp. 282-400.

[4] White, J. S., "The Moments of Log-Weibull Order Statistics," General Motors Research Publication G. M. R-717, General Motors Research Laboratories, Warren, Michigan, (1967).

OPPORTUNISTIC REPLACEMENT POLICIES

by DAVINDER P. S. SETHI
University of California, Berkeley

Abstract. Consider a unit in continuous operation. When this unit fails it is immediately replaced. In addition, opportunities arise according to a renewal sequence at which times we can replace this unit at a reduced cost. When the unit has increasing failure rate, the replacement policy that minimizes the expected total discounted cost and the average cost of maintenance is determined.

A series system of two units where the failure of either unit is an opportunity to replace the other unit at a reduced cost is considered. When the units have increasing failure rate, the structure of the optimal policy is again determined.

1.0 *Background*. Opportunistic policies pertain to situations where it costs less to replace two or more units concurrently than to replace them at different times. These cost advantages may be due to lower overhead costs and the economies of scale. Thus the necessary replacement of a unit upon its failure may also justify the replacement of some other units whose failure seems imminent. Radner and Jorgenson have considered the inspection and replacement policies for a unit in series with several monitored units having exponential life (see [3] and [5]). These and other related models have been surveyed by John McCall in [4].

All the replacement models considered in this paper are modeled as Markov Decision Processes. An extensive treatment of the subject can be found in (Derman [2]). Let random vector X_t be the state of the system at time t = 0, 1, 2, ... We observe $X_t = \underline{i}$, and choose an action a_t from a finite set of choices. In doing so, we incur a cost $C(\underline{i}, a_t)$ and our next state $X_{t+1} = \underline{j}$ with probability $P_{\underline{ij}}(a_t)$. Let π be the policy or rule for choosing the actions. We wish to determine a π that minimizes the "cost of maintaining the system." One measure of our objective is the total expected discounted cost V. Given initial state $\underline{i}$, and policy π we define

$$V_\pi(\underline{i}) = E_\pi \{C(X_0, a_0) + \alpha C(X_1, a_1) + \alpha^2 C(X_2, a_2) + \cdots \mid X_0 = \underline{i}\} \tag{1.1}$$

where α is the rate at which future costs are discounted. For $\alpha < 1$ and $C(\ ,\ ) < \infty$, $V_\pi(\underline{i})$ is finite and a meaningful objective. Let

$$V(\underline{i}) = \operatorname*{Min}_\pi V_\pi(\underline{i}).$$

Under our assumptions, an optimal policy that is stationary (i.e. depends only on the state of the system) exists (for proof see Ross [6], Theorem 6.3). Let it be π^*. Under policy π^*, $V(\cdot)$ satisfies the functional equation

$$V(\underline{i}) = \operatorname*{Min}_a \{C(\underline{i}, a) + \alpha \sum_{\underline{j}} P_{\underline{ij}}(a)\, V(\underline{j})\} \quad \text{for all } \underline{i}. \tag{1.2}$$

Another objective of interest is the expected average cost of maintaining the system, $\phi(\cdot)$. Given initial state $\underline{i}$, and policy π, we define

$$\phi_\pi(\underline{i}) = \lim_{n\to\infty} E_\pi \frac{\{\sum_{t=0}^{n} C(X_t, a_t) \mid X_0 = \underline{i}\}}{(n+1)}. \tag{1.3}$$

We say policy π^* is optimal if

$$\phi_{\pi^*}(\underline{i}) = \operatorname*{Min}_\pi \phi_\pi(\underline{i}) \quad \text{for all } \underline{i}.$$

In the replacement models to follow, we minimize the two cost objectives defined above and deduce the nature of the optimal policy π^*.

2.0 *Opportunistic Replacements of a Single Unit*. Example (1): Consider a unit that is in continuous operation. When this unit fails, an emergency crew is called to replace it immediately at a cost (K+C). In addition, a repairman visits the facility periodically and offers to replace the (working) unit at a reduced cost C. If we choose to replace the unit, we save K dollars but sacrifice the remaining life of the unit in operation. Clearly, every visit of the repairman is a decision state where we may replace the working unit or forego the chance.

Suppose the unit has a discrete failure probability density g_i, $i = 0, 1, 2, \ldots$ The repairman's visits are also random with probability density f_i, $i = 0, 1, 2, \ldots$ and $f_0 < 1$. Suppose that replacements, whether opportunistic or upon failure, take 1 unit of time. First, we minimize the total expected discounted cost V_π of maintaining the unit. If future costs are discounted at rate $\alpha < 1$ and $(K+C) < \infty$, then V_π is finite. This follows from the upper bound

$$V_\pi \leq (K+C) + \alpha\ (K+C) + \alpha^2(K+C) + \cdots \text{ for any policy } \pi$$
$$< \infty$$

which is obvious if we note that (K+C) is the maximum we spend in any period.

<u>Example (2)</u>: The above problem is equivalently defined by the following system. Unit 1 is in series with sub-system 2. When 1 fails we replace it at cost (K+C) where K is the cost of system failure. Failure epochs of 2 constitute a renewal process and are potential opportunities to replace 1 at reduced cost C, and, are equivalent to the repairman's visits in Example (1). As before, let the failure densities of 1 and 2 be g_i and f_i. Suppose replacements of 1 or 2 or both take one unit of time. V_π is now the cost associated with maintaining unit 1 under some replacement policy π. For convenience, we shall follow Example (2) for further developments. Define the system state by a pair

(i, j) where i: age of unit 1

j: is the time left to the failure of 2.

Note that the option to replace 1 or not arises only when subsystem 2 fails, i.e. $j = 0$. So (i, 0) is the only decision state. When $j \neq 0$, we do not know its value, nor do we need to know it since (i, j), $j \neq 0$, is not a decision state. Let R_i be the conditional probability of failure at age i of unit 1 given survival to age i-1. R_i is the discrete analog of the failure rate (see Barlow & Proschan [1]) and,

$$R_i = \frac{g_i}{\sum_{k \geq i} g_k} .$$

If $V(i, j)$ is the minimum cost objective given we are in state (i, j), it obeys the functional equations

$$V(i, j) = R_i \{K+C + \alpha V(0,j-1)\} + (1-R_i) \cdot \alpha V(i + 1, j - 1) \text{ for } j > 0 \quad (2.1a)$$

and,

$$V(i, 0) = R_i \{C + \alpha \sum_j f_j V(0, j)\} + (1-R_i) \cdot \text{Min} \{C + \alpha \sum_j f_j V(0, j);\ \alpha \sum_j f_j V(i + 1, j)\} \quad (2.1b)$$

In equation (2.1a), the first term is the failure rate multiplied by the cost if 1 fails. The second term is the survival probability of 1 multiplied by the cost if 1 does not fail. When $j = 0$ (i.e. 2 fails) and unit 1 fails also, we incur only C to replace unit 1. Our next stage is $(0, j)$ with probability f_j. So, the total expected cost is $\{C + \alpha \sum_j f_j V(0, j)\}$. This accounts for the first term in (2.1b). If 1 does not fail, we either replace it and expect to spend $C + \alpha \sum_j f_j V(0, j)$, or, do not replace it and expect to spend $\alpha \sum_j f_j (i + 1, j)$ depending on which is less.

We now turn our attention to evaluating the optimal objective $V(i, j)$, and, the structure of the associated optimal policy π^*. If unit 1 has increasing failure rate ($R_i \uparrow i$), we show that $V(i, j)$ is monotonic in i, and π^* has the following simple structure:

There exists $i^* \in (0, \infty]$ such that whenever we are in a decision state (i, 0), replace unit 1 only if $i \geq i^*$.

Successive Approximation Technique to Evaluate V. Define $V_0(i, j) = 0$ for all i, j. Let

$$V_{k+1}(i, j) = R_i \{K + C + \alpha V_k(0, j - 1)\} + (1-R_i)\cdot \alpha V_k(i + 1, j - 1) \text{ for } j > 0 \tag{2.2a}$$

and

$$V_{k+1}(i, 0) = R_i \{C + \alpha \sum_j f_j V_k(0, j)\} + (1 - R_i)\cdot \text{Min} \{C + \alpha \sum_j f_j V_k(0, j);\ \alpha \sum_j f_j V_k(i + 1, j)\}. \tag{2.2b}$$

For example, when k = 0 equations (2.2a) and (2.2b) would yield

$$V_1(i, j) = R_i(K + C) \quad \text{for} \quad j > 0$$

and,

$$V_1(i, 0) = R_i \cdot C.$$

Intuitively, $V_k(i, j)$ is the cost if we follow policy π^* for k periods and incur a terminal cost of zero, given we start in state (i, j). Given $\alpha < 1$, it follows that

$$\lim_{k\to\infty} V_k(i, j) = V(i, j) \quad \text{for all } i, j.$$

Equations (2.2a) and (2.2b) give an easy technique to evaluate V in terms of K, C, R, g, and α. In addition, they will be helpful in establishing the monotonicity of V in i, the age of unit 1.

Lemma 2.1. If unit 1 has increasing failure rate ($R_i \uparrow i$), and K, C > 0, then for all k

(a) $V_k(i, j) \uparrow i$ for all j

(b) $K + C + \alpha V_k(0, j) \geq \alpha V_k(i, j)$ for all i, j.

Proof: For k = 0, $V_0(i, j) = 0$ for all i, j by definition. Therefore, (a) and (b) are trivially true. The proof proceeds by induction on k. Suppose (a) and (b) are true for some k. First, we show $V_{k+1}(i, j) \uparrow i$ for all j.

Case (1): j > 0. Rewriting the recursive equation (2.2a)

$$\begin{aligned}
V_{k+1}(i, j) &= R_i \{K + C + \alpha V_k(0, j-1)\} + (1-R_i)\alpha V_k(i + 1, j - 1) \\
&= R_i \{K + C + \alpha V_k(0, j-1) - \alpha V_k(i + 1, j - 1)\} \\
&\qquad + \alpha V_k(i + 1, j - 1) \\
&\le R_{i+1} \{K + C + \alpha V_k(0, j-1) - \alpha V_k(i + 1, j - 1)\} \\
&\qquad + \alpha V_k(i + 1, j - 1) \\
&= R_{i+1} \{K + C + \alpha V_k(0, j-1)\} + (1 - R_{i+1})\cdot \\
&\qquad \alpha V_k(i + 1, j - 1) \\
&\le R_{i+1} \{K + C + \alpha V_k(0, j-1)\} + (1 - R_{i+1})\cdot \\
&\qquad \alpha V_k(i + 2, j - 1) \\
&= V_{k+1}(i + 1, j).
\end{aligned}$$

The first inequality is true since $R_{i+1} \ge R_i$ and the coefficient of R_i is non-negative by the induction hypothesis (b). The second inequality follows from induction hypothesis (a), namely $V_k \uparrow i$.

Case (ii): $j = 0$. First note that

$$\begin{aligned}
&V_k(i, j) \uparrow i \quad \text{for all } j \\
&\Longrightarrow \sum_j f_j V_k(i + 1, j) \uparrow i \\
&\Longrightarrow \text{Min } \{C + \alpha \sum_j f_j V_k(0, j);\ \alpha \sum_j f_j V_k(i + 1, j)\} \uparrow i.
\end{aligned} \tag{2.3}$$

Now,

$$\begin{aligned}
V_{k+1}(i, 0) &= R_i \{C + \alpha \sum_j f_j V_k(0, j)\} + (1 - R_i)\cdot \text{Min} \\
&\qquad \{C + \alpha \sum_j f_j V_k(0, j);\ \alpha \sum_j f_j V_k(i + 1, j)\} \\
&\le R_{i+1} \{C + \alpha \sum_j f_j V_k(0, j)\} + (1 - R_{i+1})\cdot \text{Min} \\
&\qquad \{C + \alpha \sum_j f_j V_k(0, j);\ \alpha \sum_j f_j V_k(i + 1, j)\} \\
&\le R_{i+1} \{C + \alpha \sum_j f_j V_k(0, j)\} + (1 - R_{i+1})\cdot \text{Min} \\
&\qquad \{C + \alpha \sum_j f_j V_k(0, j);\ \alpha \sum_j f_j V_k(i + 2, j)\} \\
&= V_{k+1}(i + 1, 0).
\end{aligned}$$

The first inequality follows from $R_{i+1} \geq R_i$ and

$$C + \alpha \sum_j f_j V_k(0, j) \geq \text{Min}\,\{C + \alpha \sum_j f_j V_k(0, j);\ \alpha \sum_j f_j V_k(i + 1, j)\}.$$

The second inequality follows from (2.3). We have shown $V_{k+1}(i, j) \uparrow i$ for all j. It still remains to show that part (b) is true for (k+1). That is,

$$K + C + \alpha V_{k+1}(0, j) \geq \alpha V_{k+1}(i, j) \text{ for all } i, j. \quad (2.4)$$

For $j > 0$,

$$\begin{aligned}
\text{L.H.S.} &= (K + C) + \alpha R_0 \{K + C + \alpha V_k(0, j - 1)\} \\
&\quad + \alpha^2 (1 - R_0) \cdot V_k(i + 1, j - 1) \\
&= (K + C) + \alpha R_0 (K + C) + \alpha^2 [R_0 V_k(0, j - 1) \\
&\quad + (1 - R_0) V_k(i + 1, j - 1)] \\
&\geq (K + C) + \alpha R_0 (K + \ ) + \alpha^2 [R_0 V_k(0, j - 1) \\
&\quad + (1 - R_0) V_k(0, j - 1)] \\
&\quad \text{(by the induction hypothesis } V_k \uparrow i) \\
&= (K + C) + \alpha R_0 (K + C) + \alpha^2 V_k(0, j - 1) = A \\
\text{R.H.S.} &= \alpha [R_i \{K + C + \alpha V_k(0, j - 1)\} + (1 - R_i) \cdot \\
&\quad \alpha V_k(i + 1, j - 1)] \\
&\leq \alpha [R_i \{K + C + \alpha V_k(0, j - 1)\} + (1 - R_i) \cdot \\
&\quad \{K + C + \alpha V_k(0, j - 1)\}] \\
&\quad \text{(by the induction hypothesis (b))} \\
&= \alpha (K + C) + \alpha^2 V_k(0, j - 1) \\
&< A, \text{ (by inspection)}.
\end{aligned}$$

When $j = 0$ Equation (2.4) follows similarly. When we are in state $(i, 0)$ the optimal policy π^* chooses the action (replace Unit 1 or not) that minimizes the total expected cost from thereon. The next theorem determines its structure.

<u>*Theorem* 2.2.</u> If Unit 1 is IFR, and $K, C > 0$, then there exists an $i^* \in (0, \infty)$ such that when in state $(i, 0)$,

(a) replace Unit 1 if $i \geq i^*$

(b) do not replace Unit 1 if $i < i^*$

(c) $i^* = \text{Min}\{i : \alpha\sum_j f_j V(i, j) > C + \alpha\sum_j f_j V(0, j)\}$.

Proof: As $k \to \infty$, Lemma 2.1 implies $V(i, j) \uparrow i$ for all j. Therefore, $\alpha \sum_j f_j V(i, j) \uparrow i$.

Define $i^* = \text{Min}\{i : \alpha \sum_j f_j V(i, j) > C + \alpha \sum_j f_j V(0, j)\}$. Then, for all states (i, 0) where $i \geq i^*$,

$$\alpha \sum_j f_j V(i, j) > C + \alpha \sum_j f_j V(0, j)$$

and replacing Unit 1 minimizes the cost objective. The above structure of π^* follows.

Extensions of the Model. A stationary policy that minimizes the expected average cost of maintaining Unit 1 exists when $V(i, j) - V(0, 0) < \infty$ for all i, j and when K and C are finite. Furthermore, the average cost optimal policy has the same structure as the discounted cost optimal policy when Unit 1 is IFR (see Ross [6], Section 6.7). When Units 1 and 2 have continuous density functions, the above results are true for any discrete approximation of the failure rate function of Unit 1 and density function of Unit 2, as well as for the limiting continuous functions.

The preceding model also adapts itself to a series system of many components. Suppose we want a replacement policy for some critical component in a system. We regard this component as Unit 1 and the rest of the system as Unit 2 and note that Example 2 describes this situation. Furthermore, the failure distribution for Unit 2 tends to the exponential as its complexity and the time of operation increase (see Barlow and Proschan [1], Section 2.3). Similarly, we may consider another critical component as Unit 1 and determine a replacement policy for it. This gives us an approximate technique to introduce an opportunistic policy for components with greatest benefit in a complex system.

Another situation of interest arises when the failure distribution for repaired components is different than for original components. Let m = 0 if a new component is in service and m = 1 if a used component is in service. Let (i, j, m) describe the state of the system, and, $R_{i,m}$ and V(i, j, m) be the associated failute rate and the cost objective respectively. The functional equations for this case are

$$V(i, j, m) = R_{i,m}\{K + C + \alpha V(0, j-1, 1)\} + (1-R_{i,m})\ \alpha\ V(i+1, j-1,m)$$
$$\text{for } m = 0, 1 \text{ and } j > 0$$

and,

$$V(i, 0, m) = R_{i,m}\{C + \alpha \sum_j f_j V(0, j, 1)\} + \text{Min}$$
$$\{C + \alpha \sum_j f_j V(0, j, 1);\ \alpha \sum_j f_j V(i + 1, j, m)\}$$
$$\text{for } m = 0, 1 \text{ and } j = 0.$$

If the new and the repaired units have increasing failure rates and, $R_{i,1} \geq R_{i,0}$ for all i, there is a critical replacement age i_0^* (i_1^*) for the new (repaired) unit and the optimal policy again has a simple structure as given in Theorem 2.2. The successive approximation technique can be used to evaluate the cost and approximate i_m^*.

3.0 *Simultaneous Replacement of Two Components*. Consider a series system of two units, 1 and 2. The system fails when either unit fails and is repaired by replacing the failed unit immediately. If unit n fails we incur a cost of $(K + C_n)$ where C_n is replacement cost of component n and K is the overhead cost associated with system failure. When the system is down, we have the option to replace the non-failed unit at its marginal cost C_n. In replacing the non-failed unit, we sacrifice its remaining life but hope to forestall the next system failure.

Clearly, failure epochs of Unit 1 (Unit 2) are potential opportunities to replace Unit 2 (Unit 1). As in Section 2, we wish to determine an optimal stationary replacement policy that minimizes the total expected discounted cost of maintaining the entire system. Let V denote the minimum cost objective and π^* the "best" policy. Let α be the discount rate.

Suppose the system operates in discrete time $t = 0, 1, 2, \ldots$. Let (i, j) denote the state of the system where $i = 0, 1, 2, \ldots$ is the age of Unit 1 and $j = 0, 1, 2, \ldots$ is the age of Unit 2. Suppose that if Unit n fails (or is opportunistically replaced) at time t, it returns to state 0 at time $t + 1$. Let R_i (P_j) denote the discrete failure rate of Unit 1 (Unit 2) at age $i(j)$. In this discrete time space we allow the simultaneous failure of both units. Given state (i, j), the minimum cost function $V(i, j)$ obeys the functional equation

$$\begin{aligned} V(i, j) = {} & R_iP_j\{K + C_1 + C_2 + \alpha V(0, 0)\} + R_i\overline{P}_j \\ & \{K + C_1 + S(0, j + 1)\} \\ & + \overline{R}_iP_j\{K + C_2 + T(i+1, 0\} + \overline{R}_i\overline{P}_j \\ & \{\alpha V(i+1, j+1)\} \end{aligned} \tag{3.1}$$

where $\overline{R}_i \equiv 1 - R_i$ and $\overline{P}_j \equiv 1 - P_j$, and where

$$S(0, j+1) = \text{Min } \{C_2 + \alpha V(0, 0); \alpha V(0, j+1)\} \tag{3.2}$$

$$T(i+1, 0) = \text{Min } \{C_1 + \alpha V(0, 0); \alpha V(i+1, 0)\} \tag{3.3}$$

If Unit 1 fails and Unit 2 does not, we would take the opportunity and replace 2 only if $S(0, j+1) = C_2 + \alpha V(0, 0)$. If 2 fails and 1 does not, we would take the opportunity and replace 1 only if $T(i+1, 0) = C_1 + \alpha V(0, 0)$.

Given the costs K, C_1, and C_2, and failure rate functions R_i and P_j, we can calculate $V(i, j)$ by using the successive approximation technique again:

$$\begin{aligned} V_{k+1}(i, j) = {} & R_iP_j\{K + C_1 + C_2 + \alpha V_k(0, 0)\} \\ & + R_i\overline{P}_j\{K + C_1 + S_k(0, j+1)\} \\ & + \overline{R}_iP_j\{K + C_2 + T_k(i+1, 0)\} \\ & + \overline{R}_i\overline{P}_j\{\alpha V_k(i+1, j+1)\} \end{aligned} \tag{3.4}$$

where

$$S_k(0, j+1) = \text{Min } \{C_2 + \alpha V_k(0, 0); \alpha V_k(0, j+1)\} \tag{3.5}$$

$$T_k(i+k, 0) = \text{Min } \{C_1 + \alpha V_k(0, 0); \alpha V_k(i+1, 0)\} \tag{3.6}$$

Define $V_0(i, j) = 0$ for all (i, j). Then $S_0(0, j+1) = 0$ for all j, and $T_0(i+1, 0) = 0$ for all i.

Intuitively, $V_k(i, j)$ is the expected discounted cost if we follow policy π^* for k periods and end with zero terminal cost. For K, C_1, and $C_2 < \infty$, and $\alpha < 1$,

$$\lim_{k\to\infty} V_k(i, j) = V(i, j)$$

$$\lim_{k\to\infty} S_k(0, j+1) = S(0, j+1)$$

$$\lim_{k\to\infty} T_k(i+1, 0) = T(i+1, 0).$$

Also note that $V_k(i, j) \uparrow k$. Besides its computational value, this technique helps us to establish the monotonicity of V in i and j when Unit 1 and Unit 2 are IFR. This ultimately leads to a simple structure for the best replacement policy π^*.

Lemma 3.1. If Unit 1 and Unit 2 are both IFR ($R_i \uparrow i$ and $P_j \uparrow j$), and K, C_1, and $C_2 > 0$, then for all k

(a) $V_k(i, j) \uparrow i$ for all j

(b) $V_k(i, j) \uparrow j$ for all i.

Proof. For $k = 0$, (a) and (b) are trivially true since $V_k(i, j) = 0$ for all i, j. Also observe that for $k = 0$,

(a') $K + C_2 + T_k(i+1, 0) \geq \alpha V_k(i+1, j+1)$, and

(b') $K + C_1 + S_k(0, j+1) \geq \alpha V_k(i+1, j+1)$.

Intuitively, (a') and (b') imply that no failure entails no more cost than failure of either unit for a k period problem.

The proof proceeds by induction on k. Suppose (a), (b), (a'), and (b') are true for some k. First, we show that $V_{k+1}(i, j) \uparrow j$ for all i. Rewriting the iterative equation (3.4),

$$\begin{aligned} V_{k+1}(i, j) = R_i[&P_j\{K + C_1 + C_2 + \alpha V_k(0, 0)\} \\ &+ \overline{P}_j\{K + C_1 + S_k(0, j+1)\}] \\ &+ \overline{R}_i[P_j\{K + C_2 + T_k(i+1, 0)\} \\ &+ \overline{P}_j\{\alpha V_k(i+1, j+1)\}] = R_i A(j) + \overline{R}_i B(j), \end{aligned}$$

where A(j) and B(j) are functions of j, given some value of i. We proceed to show that both A(j) and B(j) ↑ j for all i.

$$A(j) = P_j\{K + C_1 + C_2 + \alpha V_k(0, 0)\}$$
$$+ \overline{P}_j \cdot \text{Min}\ \{K + C_1 + C_2 + \alpha V_k(0, 0);\ K + C_1 + \alpha V_k(0, j+1)\}.$$

Since P_j ↑ j, and

$$K + C_1 + C_2 + \alpha V_k(0, 0)$$
$$\geq \text{Min}\ \{K + C_1 + C_2 + \alpha V_k(0, 0);\ K + C_1 + \alpha V_k(0, j+1)\},$$

therefore

$$A(j) \leq P_{j+1}\{K + C_1 + C_2 + \alpha V_k(0, 0)\}$$
$$+ \overline{P}_{j+1} \cdot \text{Min}\ \{K + C_1 + C_2 + \alpha V_k(0, 0);\ K + C_1 + \alpha V_k(0, j+1)\}$$
$$\leq P_{j+1}\{K + C_1 + C_2 + \alpha V_k(0, 0)\}$$
$$+ \overline{P}_{j+1} \cdot \text{Min}\ \{K + C_1 + C_2 + \alpha V_k(0, 0);\ K + C_1 + \alpha V_k(0, j+2)\}$$

(since V_k ↑ j for all i by the induction hypothesis (b))

$$= A(j + 1).$$

Now,

$$B(j) = P_j\{K + C_2 + T_k(i+1, 0)\} + \overline{P}_j\{\alpha V_k(i+1, j+1)\}.$$

Since P_j ↑ j, and by the induction hypothesis (a')

$$K + C_2 + T_k(i+1, 0) \geq \alpha V_k(i+1, j+1),$$

therefore

$$B(j) \leq P_{j+1}\{K + C_2 + T_k(i+1, 0)\} + \overline{P}_{j+1}\{\alpha V_k(i+1, j+1)\}$$
$$\leq P_{j+1}\{K + C_2 + T_k(i+1, 0)\} + \overline{P}_{j+1}\{\alpha V_k(i+1, j+2)\}$$

(since V_k ↑ j for all i)

$$= B(j + 1).$$

Hence, $V_{k+1}(i, j) = [R_i A(j) + \overline{R}_i B(j)]$ ↑ j for all i. Similarly, we can show that $V_{k+1}(i, j)$ ↑ i for all j.

It still remains to show that assumptions (a') and (b') extend to k + 1. That is,

(a') $K + C_2 + T_{k+1}(i+1, 0) \geq \alpha V_{k+1}(i+1, j+1)$ for all i, j (3.7) (b') $K + C_1 + S_{k+1}(0, j+1) \geq \alpha V_{k+1}(i+1, j+1)$ for all i, j. (3.8)

Equation (3.7) restated is

$$K + C_2 + \text{Min}\ \{C_1 + \alpha V_{k+1}(0, 0);\ \alpha V_{k+1}(i+1, 0)\} \geq \alpha V_{k+1}(i+1, j+1),$$

which is equivalent to

$$K + C_1 + C_2 + \alpha V_{k+1}(0, 0) \geq \alpha V_{k+1}(i+1, j+1), \quad \text{and} \tag{3.9a}$$

$$K + C_2 + \alpha V_{k+1}(i+1, 0) \geq \alpha V_{k+1}(i+1, j+1). \tag{3.9b}$$

To see (3.9a) note that

$$\begin{aligned}
\alpha V_{k+1}(i+1, j+1) &= \alpha[R_{i+1}P_{j+1}\{K + C_1 + C_2 + \alpha V_k(0, 0)\} \\
&+ R_{i+1}\overline{P}_{j+1}\{K + C_1 + S_k(0, j+2)\} \\
&+ \overline{R}_{i+1}P_{j+1}\{K + C_2 + T_k(i+2, 0)\} + \overline{R}_{i+1}\overline{P}_{j+1}\{\alpha V_k(i+2,j+2)\}] \\
&\leq \alpha[K + C_1 + C_2 + \alpha V_k(0, 0)] \\
&\leq K + C_1 + C_2 + \alpha V_k(0, 0), \text{ since } \alpha < 1 \\
&\leq K + C_1 + C_2 + \alpha V_{k+1}(0, 0), \text{ since } V_k \uparrow k.
\end{aligned}$$

For Equation (3.9b), the left hand side is

$$\begin{aligned}
&(K + C_2) + \alpha V_{k+1}(i+1, 0) \\
&\quad = (K + C_2) + \alpha\{R_{i+1}P_0\alpha V_k(0, 0) + R_{i+1}\overline{P}_0S_k(0, 1) \\
&\qquad\qquad + \overline{R}_{i+1}P_0T_k(i+2, 0) \\
&\qquad + \overline{R}_{i+1}\overline{P}_0\alpha V_k(i+2, 1) + R_{i+1}C_1 + P_0C_2 \\
&\qquad\qquad + (R_{i+1} + P_0 - R_{i+1}P_0)K\} \\
&\quad \geq (K + C_2) + \alpha\{R_{i+1}\alpha V_k(0, 0) + \overline{R}_{i+1}T_k(i+2, 0) \\
&\qquad\qquad + R_{i+1}C_1 + P_0C_2 + (R_{i+1} + P_0 - R_{i+1}P_0)K\} \\
&\quad = \beta .
\end{aligned}$$

The inequality follows from $\alpha V_k(0, 0) \leq S_k(0, 1)$ and $T_k(i+2, 0) \leq \alpha V_k(i+2, 1)$. The right hand side of (3.9b) is

$$\alpha V_{k+1}(i+1,\ j+1) \le \alpha\{R_{i+1}(K + C_1 + C_2 + \alpha V_k(0,\ 0))$$
$$+ \overline{R}_{i+1}(K + C_2 + T_k(i+2,\ 0))\}$$
$$\le \beta \text{ (by inspection).}$$

The first inequality follows from $K + C_1 + C_2 + V_k(0,\ 0) \ge K + C_1 + S_k(0,\ j+2)$ and by induction hypothesis (a'). Therefore, the L.H.S. $\ge \beta \ge$ R.H.S. of equation (3.9b). The proof for equation (3.8) is analogous.

If we let $k \to \infty$ in Lemma 3.1, we get $V(i,\ j) \uparrow i$ for all j, and $V(i,\ j) \uparrow j$ for all i. This leads us to the simple replacement policy given in the next theorem.

Theorem 3.2. If Units 1 and 2 are IFR, and K, C_1, and $C_2 > 0$, then there exist i^* and j^* such that

(a) if Unit 1 fails and $j \ge j^*$, replace both units. Otherwise, replace only the failed Unit 1.

(b) If Unit 2 fails and $i \ge i^*$, replace both units. Otherwise replace only the failed Unit 2.

(c) $j^* = \text{Min }\{j\colon V(0,\ j) > C_2 + \alpha V(0,\ 0)\}$.

(d) $i^* = \text{Min }\{i\colon V(i,\ 0) > C_1 + \alpha V(0,\ 0)\}$.

Proof. Define j^* and i^* as above. From Lemma 3.1, $V(0,\ j) \uparrow j$ and $V(i,\ 0) \uparrow i$. Therefore for all $j \ge j^*$, $V(0,\ j) > C_2 + \alpha V(0,\ 0)$ and Unit 2 should be opportunistically replaced. Similarly, for all $i \ge i^*$, $V(i,\ 0) > C_1 + \alpha V(0,\ 0)$ which implies Unit 1 should be opportunistically replaced. The above structure of the optimal policy π^* follows.

In practice the critical replacement ages i^* and j^* can be approximated by using the successive approximation technique for a large number of iterations.

Extensions to the Model. For the expected average cost criterion, a sufficient condition for a stationary optimal policy to exist is that K, C_1, $C_2 < \infty$. Furthermore, when Units 1 and 2 are IFR, the optimal policy inherits the structure from the expected discounted cost case (see Ross [6], Sect. 6.7). The structure of the optimal policy also extends to the case where Units 1 and 2 have continuous distributions.

For series systems with more than two components, the optimal policy has no simple structure. To see this, let (i, j, ℓ) be the age vector for a system of three units in series. If Unit 1 fails, the decision to replace Unit 2 (Unit 3) would also depend on the age of Unit 3 (Unit 2) and the associated cost advantages thereof. Models for a parallel system of two components are discussed in Sethi [7].

REFERENCES

[1] Barlow, R. E. and F. Proschan, MATHEMATICAL THEORY OF RELIABILITY, John Wiley & Sons, (1965).

[2] Derman, Cyrus, FINITE STATE MARKOVIAN DECISION PROCESSES, Academic Press, (1970).

[3] Jorgenson, D., J. McCall, and R. Radner, OPTIMAL REPLACEMENT POLICIES, Rand McNally, (1967).

[4] McCall, John J. "Maintenance Policies for Stochastically Failing Equipment: A Survey," Management Science, vol. 11, no. 5, (1965).

[5] Radner, R. and D. W. Jorgenson, "Optimal Replacement of a Single Part in the Presence of Several Monitored Parts," Management Science, vol. 10, no. 1, (1963).

[6] Ross, Sheldon M., Applied Probability Models with Optimization Applications, Holden-Day, (1969).

[7] Sethi, Davinder, "Opportunistic Replacement Policies for Maintained Systems," Operations Research Center Report, University of California, Berkeley, (1976).

ON SELECTING WHICH OF k POPULATIONS EXCEED A STANDARD

by RAY E. SCHAFER
Hughes Aircraft Co.

Abstract. In this paper we investigate a procedure for selecting from k populations, $\{\Pi_i\}$, those, and only those, which have a corresponding parameter θ_i that equals or exceeds a minimum standard, say θ_*. A lower bound on the probability of correct selection is provided when a pivotal function, $h(\theta, \overline{\theta}, n)$, exists such that h has a Chi-square distribution. This bound is tabulated for a large choice of degrees of freedom and for $k \geq 1$ when θ_* is known and for k = 1, 2 when θ_* is unknown.

1.0 *Introduction*. A great deal of work on selection and multiple decision procedures has been completed in the relatively recent past. Gupta and Panchapakesan [4] give a comprehensive survey. Also, Barr and Rizvi [1] give a good introduction to the methods used. Rizvi and Saxena [8] give a procedure for constructing a confidence interval for a certain ordered location or scale parameter and for simultaneously selecting all populations having parameters equal to or larger than the certain ordered parameter with preassigned minimal probability. However, here we will consider the problem of identifying those (of $k \geq 1$) populations having parameters equal to or greater than a fixed standard. Desu [2], Gupta [3], and Gupta and Sobel [5] have considered this latter problem: a procedure is given which, with guaranteed minimal probability, selects a subset of the k populations which contains all the populations having parameters equal to or greater than a standard. This procedure is not entirely satisfactory since the subset selected usually contains populations that do not equal or exceed the standard. The guarantee means only that the complement of the selected subset contains only "eliminated" populations but not necessarily all of them that should have been eliminated. Thus the procedure of [2], [3], and [5] has the effect of eliminating some of the k populations from consideration as having parameters "better" than the standard. Lehmann [6] considers a similar problem but uses a different criterion. Finally, Paulson [7] treats the problem of selecting the single best population (of k populations) exceeding a standard.

The procedure we give selects, with guaranteed minimal probability, those, and only those populations having parameters equal to or exceeding a standard. Problems of this sort arise repeatedly in practice. For example, machine tool capability is often best measured by variability. If the random variable of interest is normally distributed and k suppliers are available each with variance σ_i^2 it is useful to be able to state which, and only which, of the σ_i^2 satisfy $\sigma_i^2 < \sigma_*^2$ (this is the same as identifying those σ_i^2 that are $\geq \sigma_*^2$). In the next section we give a formal description of the problem and the assumptions to be made.

2.0 *Formulation of the Problem and Assumptions.* Suppose there is a family F of continuous probability distributions indexed by an m-component vector of parameters $(\theta, \phi_1, \phi_2, \ldots, \phi_{m-1})$. Suppose also that $\theta > 0$ is the parameter of interest (the set $\{\phi_j\}$ is a (vector) nuisance parameter) and that for $k \geq 1$ it is desired to identify all the θ_i's satisfying $\theta_i \geq \theta_*$, $i = 1, \ldots, k$. We assume that independent random samples of common size n from the k populations are available and denote $\overline{\theta}_i \equiv \overline{\theta}_i(X_i, \ldots, X_n)$ as an estimator of θ_i for the

i^{th} population Π_i, $i = 1, \ldots, k$. The ordered θ_i's are denoted as $\theta_{[1]} \leq \theta_{[2]} \leq \cdots \leq \theta_{[k]}$. Further, let $\underset{\sim}{\theta} = (\theta_1, \theta_2, \ldots, \theta_k)$ and let $\Omega = \{\underset{\sim}{\theta}\}$ be the parameter space.

Following the usual <u>indifference zone</u> approach we seek a procedure such that the probability of selecting those (and only those) $\theta_i \geq \theta_*$ exceeds a minimal probability $P^*(k)$ when $\underset{\sim}{\theta} \varepsilon S \subset \Omega$. Letting t denote the unknown number of $\theta_i < \theta_*$ then $t = 0, 1, \ldots, k$ and $S = \{\underset{\sim}{\theta} | \theta_{[t]} \leq \psi(\theta_*)$ and $\theta_{[t+1]} \geq \psi'(\theta_*)$, $0 < t < k$; $\theta_{[1]} \geq \psi'(\theta_*)$, $t = 0$; $\theta_{[k]} \leq \psi(\theta_*)$, $t = k\}$. The complement of S with respect to Ω is called the indifference zone and ψ and ψ' are functions such that $\psi(v) < v$ and $\psi'(v) > v$.

For any given selection procedure the probability of the correct selection (CS) of those $\theta_i \geq \theta_*$, say $P_{\underset{\sim}{\theta}}[CS]$, depends on the particular (unknown) $\underset{\sim}{\theta} \varepsilon S$ which obtains. For $t = t(\underset{\sim}{\theta})$ assumed known the desired lower bound on $P_{\underset{\sim}{\theta}}[CS]$ is

$$P''_t = \inf_{\{\underset{\sim}{\theta} | t(\underset{\sim}{\theta}) = t\} \cap S} P_{\underset{\sim}{\theta}}[CS].$$

where the infimum is taken over the region indicated. However, since t is generally unknown, it is clear that

$$P^*(k) = \min_{0 \leq t \leq k} P''_t$$

The demand that we identify those, and only those, $\theta_i \geq \theta_*$ is entirely realistic from a practical point of view but generally it makes the determination of $P^*(k)$ very difficult. In the balance of this paper we solve the problem of finding $P^*(k)$ in a restricted but important setting.

In particular we assume that there exists a (pivotal) function $h(\theta, \overline{\theta}, n) = g(n)\overline{\theta}/\theta$ which has a χ^2 distribution with $g(n)$ degrees of freedom; $g(n)$ being an increasing function of n. The following properties of h will be needed:

(i) $h(\theta, \bar{\theta}, n)$ is a continuous, increasing function of $\bar{\theta}$.

(ii) $h(\theta/c, b\bar{\theta}, n) = bch(\theta, \bar{\theta}, n)$.

(iii) $h(\theta_*, \theta_*, n) = g(n)$.

Also, the set $\{h(\theta_i, \bar{\theta}_i, n)\}$ is mutually independent. The form of h insures that the c.d.f. of $\bar{\theta}$ is stochastically increasing in θ.

There are many situations in which such a function h exists. The following examples give two of the more important applications.

Example 1. The two-parameter exponential distribution.

$$f(x; \theta, \mu) = (1/\theta) \exp[-(x-\mu)/\theta], \ x \geq \mu, \ \theta > 0$$
$$= 0 \text{ elsewhere.}$$

Here $m = 2$ and $\phi_1 = \mu$. The μ known case is trivial. For μ unknown and (possibly) different for each population Π_i, $h(\theta, \bar{\theta}, n) = 2(n-1)\bar{\theta}/\theta$ where $\bar{\theta} = (n-1)^{-1}[\Sigma x_j - nx_{(1)}]$, $x_{(1)}$ the first order statistic.

Example 2. The normal distribution with the means μ_i unknown. Then $m = 2$, $\theta = \sigma^2$, $\phi_1 = \mu$ and $h(\theta, \bar{\theta}, n) = (n-1)\bar{\theta}/\theta$ where $\bar{\theta} = \Sigma(x_j - \bar{x})^2/(n-1)$.

There is a large number of choices for the pair of functions (ψ, ψ'). We will consider two quite natural choices.

Case I. $\psi(\theta_*) = \theta_*/c$, $\psi'(\theta_*) = c\theta_*$, $c > 1$. Although it is not required that θ be a scale parameter it invariably will be and this choice is natural in the scale parameter setting.

Case II. In Case I the minimum distance between $\theta_{[t+1]}$ and θ_* is not equal to the minimum distance between θ_* and $\theta_{[t]}$. A choice which makes these minimum distances equal is $\psi(\theta_*) = (2-c)\theta_*$, $\psi'(\theta_*) = c\theta_*$. Here $1<c<2$.

In both cases $c = 1$ is the trivial case. In Case II $c < 2$ is required so that $\psi(\theta_*)$ is not negative; i.e., since $\theta, \theta_* > 0$ then $\theta_{[t]}$ negative is absurd. Cases I and II are considered for the standard known and unknown.

3.0 *Selection Procedure and Probability of Correct Selection.*

3.1 Standard Known. When θ_* is known the procedure is:

All populations Π_i which have $\bar{\theta}_i \geq B\theta_*$ should be selected as having $\theta_i \geq \theta_*$.

The constant B, which depends on $g(n)$, c and the Case considered (but not on k) is chosen to maximize $P^*(k)$. It turns out that $P^*(k)$ depends on B and that generally $B \neq 1$.

The probability of correct selection is, for given $\underset{\sim}{\theta}$ with $t(\underset{\sim}{\theta}) = t$,

$$P_{\underset{\sim}{\theta}}[CS] = P_{\underset{\sim}{\theta}}[\bar{\theta}_i < B\theta_* \text{ for the t populations with } \theta_i < \theta_* \text{ and } \bar{\theta}_i \geq B\theta_* \text{ for the other k-t populations}]. \quad (3.1)$$

Because of the properties of $h(\theta, \bar{\theta}, n)$ and the independence of the $\bar{\theta}_i$ the infimum of $P_{\underset{\sim}{\theta}}[CS]$ over $\underset{\sim}{\theta}$ in S with $t(\underset{\sim}{\theta}) = t$ is obtained (for proof see Appendix):

Case I. When every $\theta_{[i]}$ which is less than θ_* satisfies $\theta_{[i]} = \theta_*/c$ and every $\theta_{[j]}$ which is greater than θ_* satisfies $\theta_{[j]} = c\theta_*$, $c > 1$.

Case II. When every $\theta_{[i]}$ which is less than θ_* satisfies $\theta_{[i]} = (2-c)\theta_*$ and every $\theta_{[j]}$ greater than θ_* satisfies $\theta_{[j]} = c\theta_*$, $1 < c < 2$.

Hence,

$$P_t'' = \inf. P_{\underset{\sim}{\theta}}[CS] = Q^t(Q')^{k-t}, \quad t = 0, 1, \ldots, k. \quad (3.2)$$

In equation (3.2) for Case I

$$Q = \int_0^{Bcg(n)} f(x)\, dx, \quad Q' = \int_{Bg(n)/c}^{\infty} f(x)\, dx, \tag{3.3}$$

and for Case II

$$Q = \int_0^{Bg(n)/(2-c)} f(x)\, dx \text{ and } Q' = \int_{Bg(n)/c}^{\infty} f(x)\, dx,$$

where f(x) represents the p.d.f. of a χ^2 variate with g(n) degrees of freedom.

Since t is unknown, $P^*(k) = \min_{0 \le t \le k} P''_t$. It is clear that when $Q > Q'$, $P^*(k) = P''_0$; i.e., the lower bound is at t = 0. Conversely, when $Q < Q'$, $P^*(k) = P''_k$. Thus to maximize $P^*(k)$ B should be chosen so that $Q = Q'$. Then from equation (3.2)

$$P^*(k) = Q^k \tag{3.4}$$

and this is the largest $P^*(k)$ can be in this formulation. It is important to note that

$$P^*(k) = [P^*(1)]^k. \tag{3.5}$$

The constant B was found by a simple search routine and $Q = P^*(1)$ was computed using the IBM CDTR subroutine. For Case I $P^*(1)$ and B are given in Tables 1 and 2 respectively for c = 1.1(0.1)2.0 and g(n) = d.f. = 1(1)20(2)40(4)100. For Case II $P^*(1)$ and B are given in Tables 3 and 4 respectively for c = 1.1(0.1)1.5 and the same degrees of freedom as above.

It should be noted that as $g(n) \to \infty$, $P^*(k) \to 1$; $P^*(k)$ is monotone increasing (decreasing) in c(k). Also $P^*(k)$ (Case II) $\ge P^*(k)$ (Case I) and for both cases $B \to 1$ as $g(n) \to \infty$. Although $P^*(k)$ is a variable depending on B, c and g(n) in practical situations it may be treated as a preassigned constant (along with c) in order to determine g(n).

3.2 Standard Unknown. It may happen that the standard, θ_*, is unknown. In particular it might be that there is a population, Π_*, with an associated (unknown) θ_* and that it is desired to determine which populations Π_i have $\theta_i \geq \theta_*$. We will assume that $\overline{\theta}_*$ is based on the common random sample size n and that $\overline{\theta}_*$ is independent of the $\overline{\theta}_i$'s. The selection procedure is:

All populations Π_i which have $\overline{\theta}_i \geq B\overline{\theta}_*$ should be selected as having $\theta_i \geq \theta_*$.

Retaining the assumptions, formulation and the two cases of Section 2 the infimum of $P_{\underset{\sim}{\theta}}[CS]$, namely P_t'', occurs at the same $\underset{\sim}{\theta}$ as in Section 3.1. It is straightforward to show that

Case I.

$$P_t'' = \underbrace{\int_{B/c}^{\infty} \cdots \int_{B/c}^{\infty}}_{k-t} \underbrace{\int_{0}^{Bc} \cdots \int_{0}^{Bc}}_{t} \ell(u_1, \ldots, u_k)du_1 \ldots du_k, \quad t = 0, 1, \ldots, k \qquad (3.6)$$

where

$$\ell(u_1, \ldots, u_k) = \frac{\Gamma\left[\frac{g(n)(k+1)}{2}\right]\left(\prod^{k} u_i\right)^{\frac{g(n)}{2} - 1}}{\Gamma(k+1)\left[\frac{g(n)}{2}\right]\left(1 + \sum^{k} u_i\right)^{\frac{g(n)(k+1)}{2}}}$$ is the

joint p.d.f. of $U_i = \dfrac{h(\theta_{[i]}, \overline{\theta}_{[i]}, n)}{h(\theta_*, \overline{\theta}_*, n)}$ and $\overline{\theta}_{[i]}$ is, as before,

the estimator corresponding to $\theta_{[i]}$.

Case II.

$$p_t'' = \underbrace{\int_{B/c}^{\infty} \cdots \int_{B/c}^{\infty}}_{k-t} \underbrace{\int_{0}^{B/(2-c)} \cdots \int_{0}^{B/(2-c)}}_{t} \ell(u_1, \ldots, u_k)du_1, \ldots du_k, \quad t = 0, 1, \ldots, k \ (3.7)$$

with the same definitions as in Case I above. It should be noted that while the U_i have identical (marginal) F distributions they are not independently distributed.

We will discuss Case I first. Denoting the cumulative F distribution with equal degrees of freedom as G(x) it is clear that for k = 1 equation (3.6) becomes G(c) because G(c) = 1 - G(1/c) and hence $P_0'' = P_1''$. We also note that k = 1 is the same as completely ordering two unknown parameters and P_0'' may be found in [9]. In any event B = 1 for k = 1, 2 (we conjecture but have not proved B = 1 for all k) and for $k \geq 2$ the multiple integral in equation (3.6) may be evaluated by numerical integration or by Monte Carlo methods. We have been unable to find an analytic expression for the value of t at which the minimum occurs but for k = 2 $P^*(2) = P_1''$ and the results are given in Table 5.

For Case II $B \neq 1$ and since the value of t at which min P_t'' occurs is unknown finding $P^*(k)$ is likely to be expensive even on a high speed digital computer. However, we have worked out $P^*(1)$ and it is tabulated in Table 6. In this Case II, B does not depend on g(n) and B = 0.995, 0.980, 0.954, 0.916, 0.866 for c = 1.1(0.1)1.5 respectively.

Details of the calculations for both Cases are available from the Author.

4.0 *Example*. An electronics manufacturer desires to compare two possible sources for a piece of equipment against a minimum standard mean-time-to-failure $\theta_* = 1000$ hrs. Suppose he chooses the Case I formulation, c = 1.4 and $P^*(2) = 0.90$. Then $P^*(1) = (0.90)^{1/2} = 0.949$; Table 1 leads to g(n) = 48 and Table 2 gives B = 0.968. Thus, $B\theta_* = 968$. Assuming a single parameter exponential time-to-failure distribution for each source g(n) = 2n = 48 and n = 24 failures. The results were ($\overline{\theta}_1$ and $\overline{\theta}_2$ are the sample means)

Source 1: $\overline{\theta}_1 = 1120$ hrs

Source 2: $\overline{\theta}_2 = 940$ hrs

Thus Source 1 is identified as the only Source with $\theta > 1000$ hrs.

5.0 *Conclusions.* The methods presented here are somewhat more general than might be thought. There are many pivotal functions, not distributed as χ^2, to which the methods can be applied. For example, consider the Weibull distribution with shape parameter β and scale parameter α. Thoman, Bain and Antle [10] have shown that $\hat{\beta}/\beta$ ($\hat{\beta}$ the m.l.e. of β) is distributed independently of α and β and have obtained, by Monte Carlo means, the distribution $\hat{\beta}/\beta$. Methods identical to those used here would lead to tables for $P^*(k)$ for selecting those $\beta_i \geq \beta_*$.

If the degrees of freedom are not identical for the samples from the k populations the Tables given here can still be used by entering them with $g(n) = \min_{1 \leq i \leq k} g(n_i)$. The $P^*(k)$ so obtained will still be a lower bound since it is clear from the properties of the χ^2 distribution and equations (3.6) and (3.7) that $P^*(k)$ is an increasing function of each $g(n_i)$.

Finally, it should be noted that if $\overline{\theta}$ is consistent then $P^*(k) \to 1$ as $n \to \infty$ whatever be the p.d.f. of the pivotal function.

6.0 *Appendix.* We give a proof that $P_t'' = \inf. P_{\underset{\sim}{\theta}}[CS]$ occurs for the configuration as stated for the Case I situation. The proof for the Case II situation is similar.

For Case I and for arbitrary t we describe a vector $\underset{\sim}{\theta} \in S$ as follows. Choose a set of constants $c_i \geq c > 1$, $i = 1, \ldots, t$ and $c_j \geq c > 1$, $j = t + 1, \ldots, k$ such that for those $\theta_{[i]}$ less than θ_*, $\theta_{[i]} = \theta_*/c_i$ and for those $\theta_{[j]}$ greater than θ_*, $\theta_{[j]} = c_j\theta_*$. The restriction that the constants equal or exceed c insures that $\underset{\sim}{\theta} \in S$. Denoting a χ^2 p.d.f. by f(x) then from equation (3.1)

$$P_{\underset{\sim}{\theta}}[CS] = \left\{ \prod_{i=1}^{t} \left[\int_0^{Bc_i g(n)} f(x)\, dx \right] \right\} \left\{ \left[\prod_{j=t+1}^{k} \int_{Bg(n)/c_j}^{\infty} f(x)\, dx \right] \right\} \tag{6.1}$$

where if t = 0 the left hand product on the right hand side is defined to be one and if t = k the right hand product on the right hand side is defined to be one.

To see how this equation is obtained consider an arbitrary parameter and its corresponding estimator, which for notational convenience, we denote simply by θ and $\overline{\theta}$ respectively. Suppose also $\theta = \theta_*/c$. Then

$$P(\overline{\theta} < B\theta_*) = P(\overline{\theta} < Bc\theta) = P(g(n)\overline{\theta}/\theta < Bcg(n)) = \int_0^{Bcg(n)} f(x)\, dx.$$

The last equation follows because by hypothesis $g(n)\overline{\theta}/\theta$ has a χ^2 p.d.f.. The independence of the k $\overline{\theta}_i$'s and similar reasoning when $\theta = c\theta_*$ leads to equation (6.1).

Clearly, each integral in the left hand product equals or exceeds Q of equation (3.3) and each integral in the right hand product equals or exceeds Q' of equation (3.3) hence $P_{\underset{\sim}{\theta}}[CS] \geq P_t''$.

Table 1

P*(1) for Case I, Standard Known

D.F.	C=1.1	C=1.2	C=1.3	C=1.4	C=1.5	C=1.6	C=1.7	C=1.8	C=1.9	C=2.0
1	0.520	0.539	0.556	0.572	0.586	0.599	0.612	0.623	0.634	0.644
2	0.533	0.563	0.590	0.615	0.637	0.658	0.677	0.694	0.710	0.724
3	0.542	0.580	0.615	0.646	0.674	0.699	0.722	0.742	0.761	0.778
4	0.550	0.595	0.635	0.671	0.703	0.731	0.756	0.779	0.799	0.817
5	0.557	0.607	0.652	0.692	0.727	0.758	0.785	0.808	0.829	0.847
6	0.563	0.618	0.667	0.710	0.748	0.780	0.808	0.832	0.853	0.871
7	0.568	0.628	0.681	0.727	0.766	0.799	0.828	0.852	0.873	0.891
8	0.573	0.637	0.693	0.741	0.782	0.817	0.845	0.870	0.890	0.907
9	0.578	0.646	0.705	0.755	0.797	0.832	0.861	0.885	0.904	0.920
10	0.582	0.654	0.715	0.767	0.810	0.845	0.874	0.897	0.916	0.932
11	0.586	0.661	0.725	0.778	0.822	0.857	0.886	0.909	0.927	0.941
12	0.590	0.669	0.735	0.789	0.833	0.868	0.896	0.918	0.936	0.949
13	0.594	0.675	0.743	0.799	0.843	0.878	0.906	0.927	0.943	0.956
14	0.598	0.682	0.751	0.808	0.852	0.887	0.914	0.935	0.950	0.962
15	0.601	0.688	0.759	0.816	0.861	0.895	0.922	0.941	0.956	0.967
16	0.604	0.694	0.767	0.824	0.869	0.903	0.928	0.947	0.961	0.971
17	0.608	0.699	0.774	0.832	0.876	0.910	0.935	0.953	0.966	0.975
18	0.611	0.705	0.780	0.839	0.883	0.916	0.940	0.957	0.970	0.978

Table 1 - continued

P*(1) for Case I, Standard Known

D.F.	C=1.1	C=1.2	C=1.3	C=1.4	C=1.5	C=1.6	C=1.7	C=1.8	C=1.9	C=2.0
19	0.614	0.710	0.787	0.846	0.890	0.922	0.945	0.962	0.973	0.981
20	0.617	0.715	0.793	0.852	0.896	0.927	0.950	0.965	0.976	0.984
22	0.622	0.725	0.804	0.864	0.907	0.937	0.958	0.972	0.981	0.987
24	0.628	0.734	0.815	0.875	0.917	0.945	0.964	0.977	0.985	0.990
26	0.633	0.742	0.825	0.884	0.925	0.952	0.970	0.981	0.988	0.993
28	0.638	0.750	0.834	0.893	0.933	0.958	0.974	0.985	0.991	0.994
30	0.643	0.758	0.843	0.901	0.939	0.963	0.978	0.987	0.993	0.996
32	0.647	0.765	0.850	0.908	0.945	0.968	0.982	0.990	0.994	0.997
34	0.652	0.772	0.858	0.915	0.951	0.972	0.984	0.991	0.995	0.997
36	0.656	0.778	0.865	0.921	0.955	0.975	0.987	0.993	0.996	0.998
38	0.660	0.785	0.871	0.927	0.960	0.978	0.989	0.994	0.997	0.998
40	0.664	0.791	0.878	0.932	0.963	0.981	0.990	0.995	0.998	0.999
44	0.671	0.802	0.889	0.941	0.970	0.985	0.993	0.997	0.998	0.999
48	0.679	0.812	0.899	0.949	0.975	0.989	0.995	0.998	0.999	1.000
52	0.685	0.822	0.908	0.956	0.980	0.991	0.996	0.998	0.999	1.000
56	0.692	0.831	0.916	0.961	0.983	0.993	0.997	0.999	1.000	1.000
60	0.698	0.840	0.923	0.966	0.986	0.995	0.998	0.999	1.000	1.000
64	0.704	0.847	0.930	0.971	0.988	0.996	0.998	0.999	1.000	1.000

Table 1 - continued

P*(1) for Case I, Standard Known

D.F.	C=1.1	C=1.2	C=1.3	C=1.4	C=1.5	C=1.6	C=1.7	C=1.8	C=1.9	C=2.0
68	0.710	0.855	0.936	0.974	0.990	0.997	0.999	1.000	1.000	1.000
72	0.715	0.862	0.941	0.977	0.992	0.997	0.999	1.000	1.000	1.000
76	0.721	0.868	0.946	0.980	0.993	0.998	0.999	1.000	1.000	1.000
80	0.726	0.874	0.951	0.983	0.995	0.998	1.000	1.000	1.000	1.000
84	0.731	0.880	0.955	0.985	0.995	0.999	1.000	1.000	1.000	1.000
88	0.736	0.886	0.958	0.987	0.996	0.999	1.000	1.000	1.000	1.000
92	0.740	0.891	0.962	0.988	0.997	0.999	1.000	1.000	1.000	1.000
96	0.745	0.896	0.965	0.990	0.997	0.999	1.000	1.000	1.000	1.000
100	0.749	0.900	0.968	0.991	0.998	1.000	1.000	1.000	1.000	1.000

Table 2

B for Case I, Standard Known

D.F.	C=1.1	C=1.2	C=1.3	C=1.4	C=1.5	C=1.6	C=1.7	C=1.8	C=1.9	C=2.0
1	0.454	0.453	0.451	0.448	0.445	0.441	0.438	0.434	0.430	0.426
2	0.692	0.690	0.686	0.681	0.676	0.670	0.664	0.658	0.651	0.644
3	0.787	0.784	0.780	0.775	0.768	0.762	0.754	0.747	0.739	0.732
4	0.838	0.835	0.830	0.824	0.817	0.810	0.802	0.794	0.786	0.778
5	0.869	0.866	0.861	0.854	0.847	0.840	0.832	0.823	0.815	0.806
6	0.890	0.886	0.881	0.875	0.868	0.860	0.852	0.843	0.834	0.825
7	0.905	0.902	0.896	0.890	0.883	0.875	0.866	0.857	0.848	0.839
8	0.917	0.913	0.908	0.901	0.894	0.885	0.877	0.868	0.859	0.849
9	0.925	0.922	0.916	0.910	0.902	0.894	0.885	0.876	0.867	0.858
10	0.933	0.929	0.924	0.917	0.909	0.901	0.892	0.883	0.874	0.864
11	0.939	0.935	0.929	0.923	0.915	0.907	0.898	0.889	0.879	0.870
12	0.944	0.940	0.934	0.928	0.920	0.911	0.902	0.893	0.884	0.874
13	0.948	0.944	0.938	0.932	0.924	0.915	0.906	0.897	0.888	0.878
14	0.951	0.947	0.942	0.935	0.927	0.919	0.910	0.900	0.891	0.881
15	0.954	0.951	0.945	0.938	0.930	0.922	0.913	0.903	0.894	0.884
16	0.957	0.953	0.948	0.941	0.933	0.924	0.915	0.906	0.896	0.886
17	0.960	0.956	0.950	0.943	0.935	0.927	0.918	0.908	0.898	0.889
18	0.962	0.958	0.952	0.945	0.937	0.929	0.920	0.910	0.900	0.891

Table 2 - continued

B for Case I, Standard Known

D.F.	C=1.1	C=1.2	C=1.3	C=1.4	C=1.5	C=1.6	C=1.7	C=1.8	C=1.9	C=2.0
19	0.964	0.960	0.954	0.947	0.939	0.931	0.921	0.912	0.902	0.892
20	0.965	0.961	0.956	0.949	0.941	0.932	0.923	0.914	0.904	0.894
22	0.968	0.964	0.959	0.952	0.944	0.935	0.926	0.916	0.907	0.897
24	0.971	0.967	0.961	0.954	0.946	0.938	0.928	0.919	0.909	0.899
26	0.973	0.969	0.963	0.956	0.948	0.940	0.930	0.921	0.911	0.901
28	0.975	0.971	0.965	0.958	0.950	0.941	0.932	0.922	0.913	0.902
30	0.976	0.972	0.967	0.960	0.952	0.943	0.934	0.924	0.914	0.904
32	0.978	0.974	0.968	0.961	0.953	0.944	0.935	0.925	0.915	0.905
34	0.979	0.975	0.969	0.962	0.954	0.945	0.936	0.926	0.916	0.906
36	0.980	0.976	0.970	0.963	0.955	0.946	0.937	0.927	0.917	0.907
38	0.981	0.977	0.971	0.964	0.956	0.947	0.938	0.928	0.918	0.908
40	0.982	0.978	0.972	0.965	0.957	0.948	0.939	0.929	0.919	0.909
44	0.983	0.979	0.974	0.966	0.958	0.950	0.940	0.930	0.920	0.910
48	0.985	0.981	0.975	0.968	0.960	0.951	0.941	0.932	0.922	0.911
52	0.986	0.982	0.976	0.969	0.961	0.952	0.942	0.933	0.923	0.912
56	0.987	0.983	0.977	0.970	0.962	0.953	0.943	0.933	0.923	0.913
60	0.987	0.983	0.978	0.970	0.962	0.953	0.944	0.934	0.924	0.914
64	0.988	0.984	0.978	0.971	0.963	0.954	0.945	0.935	0.925	0.915

Table 2 - <u>continued</u>

B for Case I, Standard Known

D.F.	C=1.1	C=1.2	C=1.3	C=1.4	C=1.5	C=1.6	C=1.7	C=1.8	C=1.9	C=2.0
68	0.989	0.985	0.979	0.972	0.964	0.955	0.945	0.935	0.925	0.915
72	0.989	0.985	0.979	0.972	0.964	0.955	0.946	0.936	0.926	0.916
76	0.990	0.986	0.980	0.973	0.965	0.956	0.946	0.936	0.926	0.916
80	0.990	0.986	0.980	0.973	0.965	0.956	0.947	0.937	0.927	0.917
84	0.990	0.987	0.981	0.974	0.965	0.956	0.947	0.937	0.927	0.917
88	0.991	0.987	0.981	0.974	0.966	0.957	0.947	0.938	0.927	0.918
92	0.991	0.987	0.981	0.974	0.966	0.957	0.948	0.938	0.928	0.918
96	0.991	0.987	0.982	0.974	0.966	0.957	0.948	0.938	0.928	0.919
100	0.992	0.988	0.982	0.975	0.967	0.958	0.948	0.938	0.928	0.919

Table 3

P*(1) for Case II, Standard Known

D.F.	C=1.1	C=1.2	C=1.3	C=1.4	C=1.5
1	0.521	0.543	0.566	0.590	0.606
2	0.535	0.570	0.606	0.643	0.682
3	0.545	0.589	0.635	0.681	0.728
4	0.553	0.605	0.658	0.711	0.764
5	0.560	0.619	0.678	0.736	0.792
6	0.566	0.631	0.695	0.757	0.816
7	0.571	0.642	0.711	0.776	0.836
8	0.577	0.652	0.724	0.792	0.854
9	0.582	0.661	0.737	0.807	0.869
10	0.586	0.670	0.749	0.820	0.882
11	0.591	0.678	0.760	0.832	0.894
12	0.595	0.686	0.770	0.843	0.904
13	0.599	0.693	0.779	0.854	0.913
14	0.603	0.700	0.788	0.863	0.921
15	0.606	0.707	0.797	0.871	0.929
16	0.610	0.713	0.805	0.879	0.935
17	0.613	0.719	0.812	0.887	0.941
18	0.616	0.725	0.819	0.893	0.946
19	0.620	0.731	0.826	0.900	0.951
20	0.623	0.736	0.832	0.906	0.955
22	0.629	0.746	0.844	0.916	0.963
24	0.634	0.756	0.855	0.926	0.969
26	0.640	0.765	0.865	0.934	0.974
28	0.645	0.773	0.874	0.941	0.978
30	0.650	0.781	0.882	0.947	0.982
32	0.654	0.789	0.890	0.953	0.985
34	0.659	0.796	0.897	0.958	0.987
36	0.664	0.803	0.903	0.962	0.989
38	0.668	0.809	0.909	0.966	0.991
40	0.672	0.816	0.915	0.969	0.992

Table 3 - continued

P*(1) for Case II, Standard Known

D.F.	C=1.1	C=1.2	C=1.3	C=1.4	C=1.5
44	0.680	0.827	0.925	0.975	0.994
48	0.687	0.838	0.934	0.980	0.996
52	0.694	0.848	0.941	0.984	0.997
56	0.701	0.857	0.948	0.987	0.998
60	0.708	0.865	0.954	0.989	0.999
64	0.714	0.873	0.959	0.991	0.999
68	0.720	0.880	0.963	0.993	0.999
72	0.725	0.887	0.967	0.994	0.999
76	0.731	0.893	0.971	0.995	1.000
80	0.736	0.899	0.974	0.996	1.000
84	0.741	0.904	0.977	0.997	1.000
88	0.746	0.910	0.979	0.997	1.000
92	0.751	0.915	0.982	0.998	1.000
96	0.756	0.919	0.984	0.998	1.000
100	0.760	0.923	0.985	0.999	1.000

Table 4

B for Case II, Standard Known

D.F.	C=1.1	C=1.2	C=1.3	C=1.4	C=1.5
1	0.452	0.443	0.428	0.407	0.400
2	0.689	0.675	0.652	0.618	0.573
3	0.783	0.768	0.741	0.703	0.651
4	0.834	0.817	0.788	0.747	0.693
5	0.864	0.847	0.817	0.775	0.718
6	0.885	0.867	0.837	0.793	0.735
7	0.900	0.882	0.851	0.807	0.748
8	0.912	0.893	0.862	0.817	0.757
9	0.921	0.902	0.870	0.825	0.764
10	0.928	0.909	0.877	0.831	0.770
11	0.934	0.915	0.883	0.837	0.775
12	0.939	0.920	0.887	0.841	0.779
13	0.943	0.924	0.891	0.845	0.782
14	0.946	0.927	0.895	0.848	0.785
15	0.949	0.930	0.898	0.851	0.788
16	0.952	0.933	0.900	0.853	0.790
17	0.955	0.935	0.902	0.855	0.792
18	0.957	0.937	0.904	0.857	0.794
19	0.959	0.939	0.906	0.859	0.795
20	0.960	0.941	0.908	0.860	0.797
22	0.963	0.944	0.911	0.863	0.799
24	0.966	0.946	0.913	0.865	0.801
26	0.968	0.948	0.915	0.867	0.803
28	0.970	0.950	0.917	0.869	0.805
30	0.971	0.952	0.918	0.870	0.806
32	0.973	0.953	0.919	0.871	0.807
34	0.974	0.954	0.920	0.872	0.808
36	0.975	0.955	0.921	0.873	0.809
38	0.976	0.956	0.922	0.874	0.810
40	0.977	0.957	0.923	0.875	0.810

Table 4 - continued

B for Case II, Standard Known

D.F.	C=1.1	C=1.2	C=1.3	C=1.4	C=1.5
44	0.978	0.958	0.925	0.876	0.812
48	0.979	0.960	0.926	0.877	0.813
52	0.981	0.961	0.927	0.878	0.813
56	0.981	0.961	0.928	0.879	0.814
60	0.982	0.962	0.928	0.880	0.815
64	0.983	0.963	0.929	0.880	0.815
68	0.984	0.964	0.930	0.881	0.816
72	0.984	0.964	0.930	0.881	0.816
76	0.985	0.965	0.931	0.882	0.817
80	0.985	0.965	0.931	0.882	0.817
84	0.985	0.965	0.931	0.883	0.817
88	0.986	0.966	0.932	0.883	0.818
92	0.986	0.966	0.932	0.883	0.818
96	0.986	0.966	0.932	0.883	0.818
100	0.987	0.967	0.933	0.884	0.818

Table 5

P*(2) for Case I, Standard Unknown

D.F.	C=1.1	C=1.2	C=1.3	C=1.4	C=1.5	C=1.6	C=1.7	C=1.8	C=1.9	C=2.0
1	0.184	0.204	0.212	0.221	0.240	0.250	0.261	0.271	0.283	0.293
2	0.194	0.219	0.239	0.260	0.289	0.309	0.323	0.352	0.365	0.392
3	0.201	0.231	0.261	0.290	0.328	0.351	0.372	0.408	0.425	0.460
4	0.208	0.242	0.281	0.319	0.362	0.388	0.417	0.453	0.477	0.513
5	0.214	0.254	0.302	0.347	0.387	0.423	0.455	0.489	0.520	0.558
6	0.220	0.263	0.319	0.367	0.417	0.452	0.492	0.527	0.574	0.601
7	0.226	0.273	0.333	0.386	0.441	0.480	0.525	0.560	0.605	0.637
8	0.231	0.284	0.349	0.404	0.462	0.504	0.556	0.591	0.638	0.670
9	0.237	0.293	0.362	0.421	0.482	0.530	0.583	0.621	0.664	0.699
10	0.240	0.301	0.375	0.436	0.501	0.550	0.604	0.645	0.689	0.723
11	0.243	0.312	0.389	0.454	0.519	0.572	0.629	0.671	0.712	0.750
12	0.249	0.321	0.400	0.470	0.537	0.592	0.650	0.693	0.732	0.770
13	0.252	0.329	0.412	0.485	0.553	0.611	0.669	0.712	0.751	0.788
14	0.255	0.335	0.423	0.500	0.569	0.627	0.687	0.730	0.771	0.803
15	0.259	0.342	0.434	0.512	0.584	0.643	0.704	0.748	0.788	0.818
16	0.262	0.351	0.444	0.527	0.599	0.659	0.721	0.762	0.802	0.831
17	0.265	0.358	0.454	0.539	0.613	0.672	0.735	0.778	0.817	0.843
18	0.268	0.364	0.463	0.551	0.626	0.686	0.749	0.791	0.829	0.853

Table 5 - continued

P*(2) for Case I, Standard Unknown

D.F.	C=1.1	C=1.2	C=1.3	C=1.4	C=1.5	C=1.6	C=1.7	C=1.8	C=1.9	C=2.0
19	0.271	0.370	0.472	0.562	0.639	0.699	0.761	0.803	0.841	0.863
20	0.273	0.374	0.479	0.573	0.649	0.714	0.771	0.816	0.850	0.875
22	0.279	0.389	0.490	0.592	0.675	0.734	0.794	0.837	0.872	0.889
24	0.282	0.400	0.513	0.612	0.698	0.758	0.824	0.854	0.889	0.903
26	0.287	0.410	0.529	0.630	0.717	0.776	0.832	0.870	0.901	0.918
28	0.290	0.420	0.542	0.646	0.734	0.794	0.850	0.883	0.913	0.930
30	0.292	0.429	0.554	0.661	0.750	0.811	0.863	0.898	0.922	0.940
32	0.296	0.438	0.567	0.673	0.765	0.825	0.878	0.908	0.931	0.952
34	0.298	0.447	0.578	0.687	0.779	0.839	0.889	0.917	0.939	0.958
36	0.299	0.453	0.588	0.698	0.791	0.850	0.899	0.925	0.946	0.964
38	0.301	0.462	0.598	0.708	0.802	0.860	0.909	0.932	0.951	0.969
40	0.302	0.467	0.608	0.719	0.814	0.870	0.916	0.938	0.954	0.971

Table 6

P*(1) for Case II, Standard Unknown

D.F.	C=1.1	C=1.2	C=1.3	C=1.4	C=1.5
1	0.516	0.532	0.549	0.567	0.586
2	0.525	0.551	0.577	0.604	0.634
3	0.532	0.564	0.597	0.632	0.668
4	0.538	0.576	0.614	0.654	0.696
5	0.542	0.585	0.629	0.673	0.719
6	0.547	0.594	0.642	0.690	0.739
7	0.551	0.602	0.653	0.705	0.757
8	0.555	0.609	0.664	0.719	0.773
9	0.558	0.616	0.674	0.731	0.787
10	0.561	0.623	0.683	0.742	0.800
11	0.565	0.629	0.692	0.753	0.812
12	0.568	0.634	0.700	0.763	0.823
13	0.570	0.640	0.708	0.772	0.833
14	0.573	0.645	0.715	0.781	0.842
15	0.576	0.650	0.722	0.789	0.851
16	0.578	0.655	0.728	0.797	0.859
17	0.581	0.660	0.735	0.804	0.866
18	0.583	0.664	0.741	0.811	0.873
19	0.585	0.668	0.747	0.818	0.880
20	0.588	0.673	0.752	0.824	0.886
22	0.592	0.681	0.763	0.836	0.897
24	0.596	0.688	0.773	0.847	0.907
26	0.600	0.695	0.782	0.857	0.916
28	0.604	0.702	0.791	0.866	0.924
30	0.607	0.709	0.799	0.874	0.931
32	0.611	0.715	0.807	0.882	0.937
34	0.614	0.721	0.814	0.889	0.943
36	0.617	0.727	0.821	0.896	0.948
38	0.621	0.732	0.828	0.902	0.953
40	0.624	0.738	0.834	0.908	0.957

Table 6 - continued

P*(1) for Case II, Standard Unknown

D.F.	C=1.1	C=1.2	C=1.3	C=1.4	C=1.5
44	0.630	0.748	0.846	0.918	0.964
48	0.635	0.757	0.856	0.927	0.970
52	0.641	0.766	0.866	0.935	0.975
56	0.646	0.775	0.875	0.942	0.979
60	0.651	0.783	0.883	0.948	0.982
64	0.655	0.790	0.891	0.954	0.985
68	0.660	0.797	0.898	0.959	0.988
72	0.664	0.804	0.904	0.963	0.989
76	0.668	0.811	0.910	0.967	0.991
80	0.673	0.817	0.916	0.970	0.993
84	0.677	0.823	0.921	0.973	0.994
88	0.681	0.828	0.926	0.976	0.995
92	0.684	0.834	0.930	0.978	0.995
96	0.688	0.839	0.934	0.980	0.996
100	0.692	0.844	0.938	0.982	0.997

REFERENCES

[1] Barr, D. R. and Rizvi, M. H., "An Introduction to Ranking and Selection Procedures", *Journal of the American Statistical Association*, *61*, (1966), pp. 640-646.

[2] Desu, M., "A Selection Problem", *Annals of Mathematical Statistics*, *41*, (1970), pp. 1596-1603.

[3] Gupta, S. S., "On Some Multiple Decision (Selection and Ranking) Rules", *Technometrics*, *7*, (1965), pp. 225-245.

[4] Gupta, S. S. and Panchapakesan, S., "Inference for Restricted Families: (A) Multiple Decision Procedures; (B) Order Statistics Inequalities", Reliability and Biometry, ed. Proschan F. and Serfling, R.J., SIAM (1974), pp. 503-596.

[5] Gupta, S. S. and Sobel, M., "On Selecting a Subset which Contains all Populations Better than a Standard", *Annals of Mathematical Statistics*, *29*, (1958), pp. 235-244.

[6] Lehmann, E. L., "Some Model I Problems of Selection", *Annals of Mathematical Statistics*, *32*, (1961), pp. 990-1012.

[7] Paulson, Edward, "On the Comparison of Several Experimental Categories with a Control", *Annals of Mathematical Statistics*, *23*, (1952), pp. 239-246.

[8] Rizvi, M. H. and Saxena, K. M. L., "On Interval Estimation and Simultaneous Selection of Ordered Location or Scale Parameters", *Annals of Statistics*, *2*,(1974), pp. 1340-1345.

[9] Schafer, R. E., "A Single-Sample Complete Ordering Procedure for Certain Populations", Reliability and Biometry, ed. Proschan, F., and Serfling, R. J., SIAM (1974), pp. 597-617.

[10] Thoman, D. R., Bain, L. J., and Antle, C. E., "Inferences on the Parameters of the Weibull Distribution", *Technometrics*, *11*, (1969), pp. 445-460.

ON GENERALIZED MULTIVARIATE GAMMA TYPE DISTRIBUTIONS AND THEIR APPLICATIONS IN RELIABILITY

by P. R. KRISHNAIAH
University of Pittsburg

Abstract. In this paper, the author considers the joint distributions of the linear combinations of correlated gamma variables and their applications in reliability theory. Asymptotic expressions for these distributions are given in terms of the linear combinations of the multivariate normal densities and the derivatives of these densities.

1.1 *Introduction* Multivariate distribution theory plays an important role in studying the reliability of complicated systems since the failures of various components may be correlated. This has motivated some workers in the field to investigate various multivariate exponential distributions. For a review of the literature on these distributions, the reader is referred to Johnson and Kotz [14] and Basu and Block [3]. The object of this paper is to investigate some other multivariate distributions which are useful in studying the reliability and maintainability of complicated systems.

In Section 2 of this paper, we discuss the distributions of the order statistics from a conditional generalized gamma distribution and the usefulness of these distributions in reliability. In Section 3, we discuss the distribution of a linear combination of gamma variates when these variates are not necessarily independent. The applications of this distribution in reliability studies are discussed. In Section 4, we discuss the joint distributions of correlated linear functions of gamma variables and their applications in reliability. These distributions also play an important role in the area of simultaneous test procedures (e.g., see Krishnaiah ([17], [18]), which we discuss in Section 5. In Section 6, we discuss the generation of multivariate distributions useful in reliability theory by using mixtures of distributions.

2.0 *Order Statistics From A Conditional Distribution* Kibble [16] defined a bivariate gamma distribution whose characteristic function is given by

$$\phi(t_1, t_2) = \{(1-it_1)(1-it_2) + \rho^2 t_1 t_2\}^{-\alpha} \tag{2.1}$$

where ρ^2 is the correlation between the gamma variables and α is a half-integer. Since the above characteristic function is infinitely divisible (see Vere-Jones [26]), (2.1) is a characteristic function for any real α greater than zero. Kibble [16] obtained an expression for the bivariate gamma distribution in terms of the Laguerre polynominals. If the characteristic function of (x, y) is given by Equation (2.1), then the distribution of (x, y) is given by the following expression:

$$f(x,y) = (1 - \rho^2)^{\alpha} \sum_{j=0}^{\infty} \frac{\rho^{2j} \exp\{-(x+y)/(1-\rho^2)\}(xy)^{\alpha+j-1}}{\Gamma(\alpha)\; j!\; \Gamma(j+\alpha)(1-\rho^2)^{2\alpha+2j}} \tag{2.2}$$

When α is a half-integer, an expression similar to the above was derived by S. Bose [5] for the joint density of $\sqrt{x}$ and $\sqrt{y}$. Now, let $x = c_1 z_1^{\beta_1}$ and $y = c_2 z_2^{\beta_2}$ where $\beta_1, \beta_2 > 0$. Then, the joint distribution of z_1 and z_2 is given by the following expression:

$$f(z_1,z_2) = (1-\rho^2)^{\alpha}\beta_1\beta_2 \sum_{j=0}^{\infty} \frac{\rho^{2j}(c_1c_2)^{\alpha+j}}{\Gamma(\alpha)\; j!\; \Gamma(j+\alpha)(1-\rho^2)^{2\alpha+2j}}$$

$$\times\, z_1^{\beta_1(\alpha+j)-1}\, z_2^{\beta_2(\alpha+j)-1} \exp\left\{-\frac{1}{(1-\rho^2)}\left(c_1 z_1^{\beta_1} + c_2 z_2^{\beta_2}\right)\right\}. \tag{2.3}$$

The marginal distribution of z_2 is given by

$$f(z_2) = \frac{\beta_2 \exp\left(-c_2 z_2^{\beta_2}\right) z_2^{\beta_2 \alpha - 1} c_2^{\alpha}}{\Gamma(\alpha)} . \tag{2.4}$$

The conditional distribution of z_1 given z_2 is given by

$$f(z_1 | z_2) = \sum_{j=0}^{\infty} d_j \, z_1^{\beta_1(\alpha+j)-1} \exp\left\{-c_1 z_1^{\beta_1}/(1-\rho^2)\right\} , \tag{2.5}$$

where

$$d_j = \frac{\beta_1 \exp\left\{-c_2 z_2^{\beta_2} \rho^2/(1-\rho^2)\right\} \rho^{2j} c_1^{\alpha+j} c_2^{j} z_2^{\beta_2 j}}{j! \, \Gamma(j+\alpha)(1-\rho^2)^{\alpha+2j}} .$$

Now, let $u_1 \leq \cdots \leq u_n$ be the order statistics from the above distribution. Then, the rth moment of u_k is given by

$$\mu_r'(k,n) = \int_0^{\infty} x^r \left[\int_0^x f(z_1|z_2)dz_1\right]^{k-1} \left[\int_x^{\infty} f(z_1|z_2)dz_1\right]^{n-k} f(x|z_2)dx_1. \tag{2.6}$$

After a simple transformation, we can write (2.6) as

$$\mu_r'(k,n) = \int_0^{\infty} \left\{u(1-\rho^2)/c_1\right\}^{r/\beta_1} \left\{\sum_{j=0}^{\infty} d_j \, \eta(u;j)\right\}$$
$$\times \left[\int_0^u \left\{\sum_{j=0}^{\infty} d_j \, \eta(y;j)\right\} dy\right]^{k-1}$$
$$\times \left[\int_u^{\infty} \left\{\sum_{j=0}^{\infty} d_j \, \eta(y;j)\right\} dy\right]^{n-k} du$$

where

$$\eta(y;j) = \frac{\exp(-y)y^{\alpha+j-1}(1-\rho^2)^{\alpha+j}}{\beta_1 c_1^{\alpha+j}} .$$

Using a method similar to the one used by Breiter and Krishnaiah [6], we can evaluate the moments $\mu_r'(k,n)$. The first moments of the order statistics from the gamma and Weibull populations were evaluated by Harter [12]. The first four moments of the order statistics from gamma population were evaluated by Gupta [11] and Breiter and Krishnaiah [6] for various values of the parameters.

If the failure times of two components are correlated, then the performance of one component will be affected by the other. In these situations, it would be desirable to consder the conditional distribution of the failure times of one component given the distribution of the failure times of the second component. Also, in some situations it would be realistic to assume that the joint distribution of the failure times (or intervals between failure times) follow the generalized bivariate gamma distribution discussed in this paper.

3.0 Distributions of Linear Combinations of Gamma Variables We will first discuss the distribution of $z = c_1x_1 + \ldots + c_qx_q$ where $x_1, \ldots, x_q$ are distributed independently as

$$f(x_i) = \frac{\exp(-x_i)x_i^{\alpha_i-1}}{\Gamma(\alpha_i)} .$$

The characteristic function of z is given by

$$\phi(t) = \prod_{j=1}^{q} (1 - c_j it)^{-\alpha_j} . \tag{3.1}$$

Now, let $\alpha^* = c_1\alpha_1 + \ldots + c_q\alpha_q$ and $\alpha^{**} = c_1^2\alpha_1 + \ldots + c_q^2\alpha_q$.

The characteristic function of $y = (z - \alpha^*)/\sqrt{\alpha^{**}}$ is given by

$$\phi_1(t) = \exp(-it\alpha^*/\sqrt{\alpha^{**}}) \prod_{j=1}^{q} \left(1 - \frac{itc_j}{\sqrt{\alpha^{**}}}\right)^{-\alpha_j}. \tag{3.2}$$

But

$$\prod_{j=1}^{q} \left(1 - \frac{itc_j}{\sqrt{\alpha^{**}}}\right)^{-\alpha_j} = \exp -\left[\sum_{j=1}^{q} \alpha_j \log\left(1 - \frac{itc_j}{\sqrt{\alpha^{**}}}\right)\right]$$

$$= \exp\left[\sum_{j=1}^{q} \alpha_j \sum_{r=1}^{\infty} (itc_j/\sqrt{\alpha^{**}})^r/r\right]. \tag{3.3}$$

Using (3.3) in (3.2), we obtain

$$\phi_1(t) = \exp(-t^2/2) \left[1 + \frac{(it)^3}{3\sqrt{\alpha^{**}}} \sum_{j=1}^{q} \mu_j c_j + \frac{1}{\alpha^{**}}\left\{\frac{(it)^4}{4} \sum_{j=1}^{q} \mu_j c_j^2 + \frac{(it)^6}{9}\left(\sum_{j=1}^{q} \mu_j c_j\right)^2\right\} + O(\alpha^{**-3/2})\right] \tag{3.4}$$

where $\mu_j = \alpha_j c_j^2/\alpha^{**}$.

Inverting the above characteristic function, we obtain the following expression for the density of y:

$$f_1(y) = \psi(y) \left[1 - \frac{1}{3\sqrt{\alpha^{**}}} \sum_{j=1}^{q} \mu_j c_j \; H_3(y) + \frac{1}{\alpha^{**}}\left\{\frac{1}{4}\left(\sum_{j=1}^{q} \mu_j c_j^2\right) H_4(y) + \frac{1}{9}\left(\sum_{j=1}^{q} \mu_j c_j\right)^2 H_6(y)\right\} + O(\alpha^{**-3/2})\right] \tag{3.5}$$

where $\psi(y) = \frac{1}{\sqrt{2\pi}} \exp(-y^2/2)$ and $H_j(y)\psi(y) = (-1)^j \frac{d^j}{dy^j} \psi(y)$.

Next, let us assume that $x_1, \ldots, x_q$ are distributed jointly as a multivariate gamma distribution with characteristic function

$$\phi_2(t_1,\ldots,t_q) = |I - iT\Sigma|^{-\alpha} \prod_{j=1}^{q} (1 - id_j t_j)^{-\alpha_j} \tag{3.6}$$

where $T = \text{diag.}(t_1,\ldots,t_q)$, $\Sigma = (\sigma_{ij})$ is the matrix of population parameters, $a_j \geq 0$, $d_j \geq 0$ and α is a half-integer. Krishnamoorthy and Parthasarthy [23] obtained an expression for the density of the multivariate gamma distribution in terms of the Laguerre polynominals. Krishnaiah and Rao [21] discussed some properties of this distribution. It is not known as to whether this distribution is infinitely divisible in the general case. Moran and Vere-Jones [24] showed that the above distribution is infinitely divisible when $\rho_{ij} = (\sigma_{ij}/\sqrt{\sigma_{ii}\sigma_{jj}}) = \rho$ for $i \neq j$; they also showed that this distribution is infinitely divisible when $q = 3$ and $\rho_{ij} = \rho^{|i-j|}$. Griffiths [10] gave the necessaryand sufficient conditions for the infinite divisibility of the trivariate gamma distribution. The characteristic function of z is given by

$$\phi_3(t) = \prod_{j=1}^{q} \left\{(1 - it\lambda_j)^{-\alpha}(1-itc_j d_j)^{-\alpha_j}\right\} \tag{3.6a}$$

where $\lambda_1, \ldots, \lambda_q$ are the eigenvalues of $C\Sigma$, and $C = \text{diag}(c_1, \ldots, c_q)$. Here, we note that $\phi_3(t)$ is of the same form as $\phi(t)$ in Equation (3.1). It is known (see Krishnaiah and Rao [21]) that $\phi_3(t)$ is infinitely divisible. Robbins and

Pitman [25] gave an exact expression for the density of z when $x_1, \ldots, x_q$ are independently distributed. For a discussion of the distribution of z when $x_1, \ldots, x_q$ are correlated, the reader is referred to Krishnaiah and Waikar [22].

We will now discuss the applications of the distributions of linear combinations of gamma variables in reliability problems.

Let $t_1 \le \ldots \le t_q$ denote the times of shock on a component and $\tau_j = t_j - t_{j-1}$ $(j = 1, \ldots, q)$ with the understanding that $t_0 = 0$. Let us assume that the intervals $\tau_1, \ldots, \tau_q$ between the shocks are independently distributed as

$$f(\tau_j) = \frac{\exp(-\tau_j/c_j)\,\tau_j^{\alpha_j-1}}{c_j^{\alpha_j}\Gamma(\alpha_j)}. \tag{3.7}$$

Then the characteristic function of the total time for q shocks is given by (3.1).

Next, let us assume that the time intervals between the replacements (or repairs) of a component are distributed independently as (3.7). Then the characteristic function of the waiting time for q^{th} replacement (or q^{th} repair) is given by (3.1).

Next, let us assume that the intervals $\tau_1, \ldots, \tau_q$ between shocks are distributed independently as in Equation (3.7). Also, let us assume that the component fails after N shocks where N is a random variable. Then, the characteristic function of the failure time is given by

$$\phi_4(t) = \sum_{n=1}^{\infty} p_n \prod_{j=1}^{n} (1 - c_j it)^{-\alpha_j}, \tag{3.8}$$

where p_n denotes the probability that $N = n$.

4.0 Joint Distributions of Several Linear Combinations of Gamma Variables Let $t_{1j} \leq \cdots \leq t_{q_j,j}$ be the times at which the shocks occur on the jth component of a complicated system and let the intervals between shocks be denoted by $\tau_{uj} = t_{uj} - t_{u-1,j}$. We assume that the joint distribution of the intervals $(\tau_{11}, \ldots, \tau_{q_1,1}, \ldots, \tau_{1p}, \ldots, \tau_{q_p,p})$ between the shocks is a multivariate gamma distribution with the characteristic function[1]

$$\phi_5(t_1, \ldots, t_q) = |I - iT\Sigma|^{-\alpha} \tag{4.1}$$

where $\alpha = \frac{n}{2}$, $T = \text{diag.}\ (t_1, \ldots, t_q)$, $q = q_1 + \cdots + q_p$ and $\Sigma = (\sigma_{ij})$ depends upon the parameters of the population. The joint characteristic function of $s_1, \ldots, s_p$ is given by

$$\phi_6(t_1^*, \ldots, t_p^*) = |I - iT^*\Sigma|^{-\alpha} \tag{4.2}$$

where $T^* = \text{diag}(t_1^* I_{q_1}, \ldots, t_p^* I_{q_p})$, $t_1 = \cdots = t_{q_1} = t_1^*, \ldots, t_{q_{p-1}+1} = \cdots = t_{q_p} = t_p^*$ and $s_j = \tau_{1j} + \cdots + \tau_{q_j,j}$. In the above discussion, if t_{uj} denotes the time at which the uth replacement of the jth component occurs, then s_j is the waiting time for the q_j^{th} replacement of the jth component.

Khatri, Krishnaiah and Sen [15] considered the following distribution. Let $\underset{\sim}{x}' = (\underset{\sim}{x}_1', \ldots, \underset{\sim}{x}_p')$ be distributed as the multivariate normal with zero mean vector and covariance matrix Σ

1 For the cases where (4.1) is infinitely divisible, α can be replaced with any real $\alpha^* > 0$.

and let x_j' be of order $1 \times q_j$. Also, let $y_j = \frac{1}{2} \sum_{j=1}^{m} x_j' A_j x_j$ where $A_1, \ldots, A_p$ are positive definite. Then, the characteristic function of $(y_1, \ldots, y_p)$ is given by

$$\phi_7(t_1^*, \ldots, t_p^*) = |I - iT^*\Sigma^*|^{-n/2} \tag{4.3}$$

where $\Sigma^* = A\Sigma$ and $A = \text{diag}(A_1, \ldots, A_p)$. Here we note that the characteristic function given in Equation (4.3) is of the same form as the characteristic function given by Equation (4.2). Khatri, Krishnaiah and Sen [15] gave two expressions for the joint density of $y_1, \ldots, y_p$. The first expression was based on the linear combinations of the products of gamma variables whereas the second expression was based upon the linear combinations of the products of Laguerre polynomials. We will now obtain an asymptotic expression for the joint density of $s_1^*, \ldots, s_p^*$ where $s_j^* = (s_j - n\text{tr}\Sigma_{jj})/\sqrt{n}$, $(j = 1, \ldots, p)$. Here we note that Σ is partitioned as

$$\Sigma = \begin{matrix} \Sigma_{11} & \cdots & \Sigma_{1p} \\ \Sigma_{21} & \cdots & \Sigma_{2p} \\ \vdots & \cdots & \vdots \\ \Sigma_{p1} & \cdots & \Sigma_{pp} \end{matrix}$$

and Σ_{uj} is of order $q_u \times q_j$. The characteristic function of $s_1^*, \ldots, s_p^*$ is given by

$$\phi_8(t_1, \ldots, t_p) = \text{etr}(-i\sqrt{n}T^*\Sigma)\left|I - \frac{i}{\sqrt{n}} T^*\Sigma\right|^{-n/2} \tag{4.4}$$

where etr denotes the exponential of the trace

But

$$|I - \frac{i}{\sqrt{n}} T^*\Sigma|^{-n/2} = \exp\left[-\frac{n}{2}\log|I - \frac{i}{\sqrt{n}} T^*\Sigma|\right]$$
$$= \exp\left[\frac{n}{2}\sum_{r=1}^{\infty}\frac{\text{tr}(iT^*\Sigma/\sqrt{n})^r}{r}\right]. \qquad (4.5)$$

Using (4.4) and (4.5), we obtain

$$\phi_8(t_1, \ldots, t_p) = \text{etr}\left(-\frac{1}{4}(T^*\Sigma)^2\right)\left[1 + \frac{i^3}{6\sqrt{n}}\text{tr}(T^*\Sigma)^3\right.$$
$$+ \frac{1}{n}\left\{\frac{i^4}{8}\text{tr}(T^*\Sigma)^4 + \frac{i^6}{36}\left(\text{tr}(T^*\Sigma)^3\right)^2\right\}$$
$$\left. + 0(n^{-3/2})\right]. \qquad (4.6)$$

We can write $\frac{1}{2}\text{tr}(T^*\Sigma)^2$ as $\underset{\sim}{t}'\Omega\underset{\sim}{t}$ where $\underset{\sim}{t}' = (t_1, \ldots, t_p)$ and $\Omega = (w_{ju})$ where $w_{ju} = \frac{1}{2}\text{tr}\Sigma_{ju}\Sigma_{uj}$. Now, we can write (4.6) as

$$\phi_8(t_1, \ldots, t_p) = \exp\left[-\frac{1}{2}\underset{\sim}{t}'\Omega\underset{\sim}{t}\right]\left[1 + \frac{i^3}{6\sqrt{n}}\sum\nolimits_1 c_{j_1j_2j_3}t_{j_1}t_{j_2}t_{j_3}\right.$$
$$+ \frac{1}{n}\left\{\sum\nolimits_2 c_{j_1j_2j_3j_4}t_{j_1}t_{j_2}t_{j_3}t_{j_4}\right.$$
$$\left. + \sum\nolimits_3 c_{j_1\ldots j_6}t_{j_1}t_{j_2}t_{j_3}t_{j_4}t_{j_5}t_{j_6}\right\}$$
$$\left. + 0(n^{-3/2})\right] \qquad (4.7)$$

where $\sum_1$ denotes the summation over j_1, j_2, j_3, $\sum_2$ denotes the summation over j_1, j_2, j_3, j_4 and $\sum_3$ denotes the summation over $j_1, j_2, \ldots, j_6$. Also, $c_{j_1j_2j_3}$, $c_{j_1j_2j_3j_4}$ and $c_{j_1\ldots j_6}$ are defined by the following equations:

$$\mathrm{tr}(T^*\Sigma)^3 = \sum_1 c_{j_1j_2j_3} t_{j_1} t_{j_2} t_{j_3}$$

$$\mathrm{tr}(T^*\Sigma)^4 = \sum_2 c_{j_1j_2j_3j_4} t_{j_1} t_{j_2} t_{j_3} t_{j_4}$$

$$\mathrm{tr}(T^*\Sigma)^6 = \sum_3 c_{j_1\ldots j_6} t_{j_1} t_{j_2} \cdots t_{j_6} \quad .$$

Now, let $H_{j_1j_2\ldots j_v}(\underset{\sim}{x})$ denote the multivariate Hermite polynomial where

$$\psi(\underset{\sim}{x};c)H_{j_1\ldots j_v}(\underset{\sim}{x}) = (-1)^v \frac{\partial^v}{\partial x_{j_1} \cdots \partial x_{j_v}} \psi(\underset{\sim}{x};c),$$

$$\psi(\underset{\sim}{x};c) = \frac{1}{(2\pi)^{p/2}|C|^{1/2}} \operatorname{etr}\left(-\frac{1}{2}\underset{\sim}{x}'C^{-1}\underset{\sim}{x}\right)$$

and $\underset{\sim}{x}' = (x_1, \ldots, x_p)$. For some details regarding the above polynomials, the reader is referred to Appel and Kampe de Feriet [1]. If we now invert the right side of (4.7), we obtain the following asymptotic expression for the joint density of $s_1^*, \ldots, s_p^*$:

$$f(s_1^*, \ldots, s_p^*) = \psi(s^*;\Omega) \left[1 + \frac{1}{6\sqrt{n}} \sum_1 c_{j_1j_2j_3} H_{j_1j_2j_3}(\underset{\sim}{s}^*) + \frac{1}{n}\left\{\sum_2 c_{j_1j_2j_3j_4} H_{j_1j_2j_3j_4}(\underset{\sim}{s}^*) + \sum_3 c_{j_1\ldots j_6} H_{j_1\ldots j_6}(\underset{\sim}{s}^*)\right\} + 0(n^{-3/2})\right] . \tag{4.8}$$

Next, let us assume that $x_1,\ldots,x_q$ are distributed independently as

$$f(x_j) = \frac{\exp(-x_j)x_j^{\alpha_j-1}}{\Gamma(\alpha_j)} .$$

Also, let $z_j = d_{j1}x_1 + \ldots + d_{jq}x_q$, $(j=1,\ldots,p)$, $d_j^* = d_{j1}\alpha_1 + \ldots + d_{jq}\alpha_q$, $\alpha_j^{**} = d_{j1}^2\alpha_1^2 + \ldots + d_{jq}^2\alpha_q^2$, and $y_j^* = (z_j-\alpha_j^*)/\sqrt{\alpha_j^{**}}$.

Then, the characteristic function of the joint distribution of $y_1^*,\ldots,y_p^*$ is given by

$$\phi(t_1,\ldots,t_p) = \exp\left\{-i \sum_{j=1}^{p} (t_j \alpha_j^*/\sqrt{\alpha_j^{**}})\right\} \times \prod_{j=1}^{q} \left\{1-i \sum_{r=1}^{p} (t_r d_{rj}/\sqrt{\alpha_r^{**}})\right\}^{-\alpha_j} \tag{4.9}$$

$$= \exp\left\{\sum_{j=1}^{q} \sum_{u=2}^{\infty} \alpha_j \, i \sum_{r=1}^{p} \left(t_r d_{rj}/\sqrt{\alpha_r^{**}}\right)^u/u \right\}.$$

So, we can write $\phi(t_1,\ldots t_p)$ as

$$\phi(t_1,\ldots,t_p) = \exp(-\tfrac{1}{2}\, t'\Omega t) \left[1 + \textstyle\sum_1 c_{u_1u_2u_3} \prod_{k=1}^{3} (it_{u_k}/\sqrt{\alpha_{u_k}^{**}}) + \textstyle\sum_2 c_{u_1\ldots u_4} \prod_{k=1}^{4} (it_{u_k}/\sqrt{\alpha_{u_k}^{**}}) + \textstyle\sum_3 c_{u_1\ldots u_6} \prod_{k=1}^{6} (it_{u_k}/\sqrt{\alpha_{u_k}^{**}}) + \ldots\right]$$

where $\Omega = (\omega_{uv})$, $\omega_{uv} = \sum_{j=1}^{v} (\alpha_j d_{uj} d_{vj}/ \alpha_u^{**} \alpha_v^{**})$, $\sum_1$ denotes the summation over values of u_1, u_2, u_3, $\sum_2$ denotes the summation over values of u_1, u_2, u_3, u_4 and $\sum_3$ denotes the summation over $u_1, \ldots, u_6$. Also

$$c_{u_1 u_2 u_3} = \sum_{j=1}^{q} \frac{\alpha_j}{3} d_{u_1 j}\, d_{u_2 j}\, d_{u_3 j}$$

$$c_{u_1 \ldots u_4} = \sum_{j=1}^{v} \frac{\alpha_j}{4} d_{u_1 j}\, d_{u_2 j}\, d_{u_3 j}\, d_{u_4 j}$$

and $c_{u_1 \ldots u_6}$ is given by the relation

$$\frac{1}{2}\left\{\sum_1 \left(c_{u_1 u_2 u_3} \prod_{k=1}^{3} \left(it_{u_k} / \sqrt{\alpha^{**}_{u_k}}\right)\right\}^2 = \sum_3 c_{u_1 \ldots u_6} \prod_{k=1}^{6} \left(it_{u_k} / \sqrt{\alpha^{**}_{u_k}}\right).$$

Inverting (4.10), we obtain the following expression for the joint density of $y_1^*, \ldots y_p^*$:

$$f(y_1^*, \ldots, y_p^*) = \psi(\underset{\sim}{y}^*; \Omega) \left[1 + \sum_1 \frac{c_{u_1 u_2 u_3} H_{u_1 u_2 u_3}(\underset{\sim}{y}^*)}{\sqrt{\alpha^{**}_{u_1} \alpha^{**}_{u_2} \alpha^{**}_{u_3}}} \right.$$

$$+ \sum_2 \frac{c_{u_1 u_2 u_3 u_4} H_{u_1 u_2 u_3 u_4}(\underset{\sim}{y}^*)}{\sqrt{\alpha^{**}_{u_1} \alpha^{**}_{u_2} \alpha^{**}_{u_3} \alpha^{**}_{u_4}}}$$

$$\left. + \sum_3 \frac{c_{u_1 \ldots u_6} H_{u_1 \ldots u_6}(\underset{\sim}{y}^*)}{\sqrt{\alpha^{**}_{u_1} \ldots \alpha^{**}_{u_6}}} + \ldots \right]$$

5.0 Simultaneous Tests for the Equality of the Mean Times Between Failures of Components In the previous section, let us assume that t_{uj} denotes the time at which the jth component is replaced the uth time. Then the joint characteristic function of $s_1, \ldots, s_p$ is given by (4.2) where s_j denotes the waiting time for the q_j^{th} replacement. The mean, variances, and covariances of $s_1, \ldots, s_p$ are given by $E(s_j) = \alpha \mathrm{tr}\Sigma_{jj}$, $\mathrm{var}(s_j) = \frac{\alpha}{2} w_{jj}$, $\mathrm{cov}(s_{jj}, s_{uu}) = \frac{\alpha}{2} w_{ju}$, where the notations are defined in in the preceding section. We now propose simultaneous test procedures for testing the hypothesis of the equality of the mean times between replacements of the components where $\rho_{ij} = \sigma_{ij}/\sqrt{\sigma_{ii}\sigma_{jj}}$ are known.

Let $\theta_j = \mathrm{tr}\Sigma_{jj}/q_j$, $H_{ij}:\theta_i = \theta_j$, $A_{ij}:\theta_i \neq \theta_j$, $H:\theta_1 = \ldots = \theta_p$, $A_1 = \bigcup_{i=1}^{p-1} A_{ip}$, $A_2 = \bigcup_{i=1}^{p-1} A_{i,i+1}$ and $A_3 = \bigcup_{i<j}^{p} A_{ij}$. Also, let $\hat{\theta}_j = s_j/q_j$ for $j = 1, \ldots, p$. Then, the hypothesis H_{ip} is accepted if

$$d_{1\alpha} \leq \hat{\theta}_i/\hat{\theta}_p \leq d_{2\alpha}$$

and rejected otherwise where

$$P[d_{1\alpha} \leq (\hat{\theta}_i/\hat{\theta}_p) \leq d_{2\alpha},\ i = 1, \ldots, (p-1) \,|\, H] = (1 - \alpha). \tag{5.1}$$

The total hypothesis H when tested against A_1 is accepted if all the components hypotheses H_{ip} are accepted. If we test $H_{i,i+1}$ $(i = 1, \ldots, p-1)$ and H simultaneously against $A_{i,i+1}$ and A_2, we accept $H_{i,i+1}$ if

$$d_{1\alpha} \leq (\hat{\theta}_i/\hat{\theta}_{i+1}) \leq d_{2\alpha}$$

and reject it otherwise where

$$P[d_{1\alpha} \leq (\hat{\theta}_i/\hat{\theta}_{i+1}) \leq d_{2\alpha};\ i = 1, \ldots, (p-1) | H] = (1 - \alpha). \quad (5.2)$$

The total hypothesis H is accepted if all the component hypotheses $H_{i,i+1}$ are accepted. We can similarly propose a procedure for testing H_{ij} $(i < j)$ and H simultaneously against A_{ij} $(i < j)$ and A_3. Since the joint distribution of $\hat{\theta}_1, \ldots, \hat{\theta}_p$ is essentially the same as the joint distribution of correlated quadratic forms given by Khatri, Krishnaiah and Sen [15], we can compute the critical values in (5.1) and (5.2). We can obtain bounds on the critical values associated with all the above three tests by using Poincare's exclusion-inclusion formula. Here, we note that if Σ in (4.2) is completely unknown, the probability integrals associated with the above procedures involve certain nuisance parameters.

Krishnaiah [19] proposed simultaneous procedures for testing the hypothesis of the equality of the variances in a multivariate normal population. The procedures proposed in this section are proposed in the same spirit.

6.0 Joint Distribution of the Failure Times Let $t_{1j} \leq \ldots \leq t_{N_j,j}$ denote the times at which shocks occur on j^{th} component in a multicomponent system. Let $\tau_{uj} = t_{uj} - t_{u-1,j}$ for $u = 1, \ldots, N_j$ and $j = 1, \ldots, q$. Also, let us assume that the jth component fails after M_j shocks where M_j is a random variable. Also, let $p_{u_1 \ldots u_q}$ denote the probability that $M_1 = u_1, \ldots, M_q = u_q$. In addition, let $T_{jM_j} = \tau_{1j} + \ldots + \tau_{M_j,j}$ for $j = 1, \ldots, q$. Then, the joint characteristic function of $T_{1M_1}, \ldots, T_{qM_q}$ is given by

$$\phi_{10}(\theta_1, \ldots, \theta_q) = \sum p_{n_1 \ldots n_q} \phi_{11}(\theta_1, \ldots, \theta_q; T_{1n_1}, \ldots, T_{qn_q}) \tag{6.1}$$

where $\phi_{11}(\theta_1, \ldots, \theta_q; T_{1n_1}, \ldots, T_{qn_q})$ is the joint characteristic function of $T_{1n_1}, \ldots, T_{qn_q}$ for fixed values of $n_1, \ldots, n_q$ and the summation is over all possible values of $n_1, \ldots, n_q$. If the joint characteristic function of $T_{1n_1}, \ldots, T_{qn_q}$ is of the form (4.1), then $\phi_{11}(\theta_1, \ldots, \theta_q; T_{1n_1}, \ldots, T_{qn_q})$ is of the form (4.2). We now consider various special cases of (6.1).

Suppose each component requires the same number (M) of shocks for failure. Then (6.1) becomes

$$\phi_{10}(\theta_1, \ldots, \theta_q) = \sum_{n=1}^{\infty} p_n \phi_{11}(\theta_1, \ldots, \theta_q; T_{1n}, \ldots, T_{qn}) \tag{6.2}$$

where p_n denotes the probability that $M = n$. Now, let us assume that the vectors $(\tau_{u1}, \ldots, \tau_{uq})$, $(u = 1, \ldots, n)$, are independently and identically distributed with the common characteristic function $\phi_{12}(\theta_1, \ldots, \theta_q)$. Then, (6.2) becomes

$$\phi_{10}(\theta_1, \ldots, \theta_q) = \sum_{n=1}^{\infty} p_n \{\phi_{12}(\theta_1, \ldots, \theta_q)\}^n. \tag{6.3}$$

If $p_n = p(1-p)^{n-1}$, then the above equation becomes equal to

$$\phi_{10}(\theta_1, \ldots, \theta_q) = p\phi_{12}(\theta_1, \ldots, \theta_q)/1 - q\phi_{12}(\theta_1, \ldots, \theta_q). \tag{6.4}$$

If we assume that the distribution of M is a negative binomial with the probability generating function of the form

$$\Pi(z) = \sum_{n=1}^{\infty} p_n z^{n-1} = \left[\frac{\beta}{1+\beta-z}\right]^{-\nu},$$

then Equation (6.3) becomes equal to

$$\phi_{10}(\theta_1, \ldots, \theta_q) = \phi_{12}(\theta_1, \ldots, \theta_q) \left[\frac{\beta}{1+\beta-\phi_{12}(\theta_1,\ldots,\theta_q)}\right]^{-\nu}. \quad (6.5)$$

Next, let us assume that the number of shocks required for failure of each component need not be the same. Also, let us assume that the intervals τ_{uj} are distributed independently with the characteristic function $\phi(\theta_j;\tau_{uj})$. Then, Equation (6.1) becomes equal to

$$\phi_{10}(\theta_1, \ldots, \theta_q) = \sum p_{n_1\ldots n_q} \prod_{j=1}^{q} \prod_{u=1}^{n_j} \phi(\theta_j;\tau_{uj}) \quad (6.6)$$

In addition, let us assume that $\phi(\theta_j;\tau_{uj})$ is equal to $\phi_{13}(\theta_j)$ for $u = 1, \ldots, n_j$. Then, Equation (6.6) becomes equal to

$$\phi_{10}(\theta_1, \ldots, \theta_q) = \sum p_{n_1\ldots n_q} \{\phi_{13}(\theta_1)\}^{n_1} \ldots \{\phi_{13}(\theta_q)\}^{n_q}. \quad (6.7)$$

We can obtain various multivariate distributions from (6.7) in the reliability context by assuming different joint distribution of $M_1,\ldots,M_q$. For example, we can assume that the joint distribution of M_1 and M_2 to be the bivariate geometric distributions considered by Downton [7] and Hawkes [13] or various bivariate distributions considered by Edwards and Gurland [8] in the context of accident proneness. Similarly, we may assume that the joint distribution of $M_1, \ldots, M_q$ belongs to the class of multivariate geometric distributions discussed by Arnold [2].

Gaver [9] generated a multivariate gamma distribution by mixture. Various bivariate distributions were constructed, by mixtures, in the reliability context by Downton [7], Hawkes [13] and Block [4].

If the joint characteristic function of $T_{1n_1}, \ldots, T_{qn_q}$ is the noncentral multivariate gamma distribution in the sense of Krishnamoorthy and Parthasarathy [23], then $\phi_{11}(\theta_1, \ldots, \theta_q; T_{1n_1}, \ldots, T_{qn_q})$ in (6.1) is the joint characteristic function of correlated quadratic forms. Now, by making different assumptions on the nature of the joint distribution of $M_1, \ldots, M_q$, we can generate a very rich class of multivariate distributions which are useful in reliability studies.

Acknowledgement. This work was done when the author was at the Air Force Flight Dynamics Laboratory. The author wishes to thank Professor C. G. Khatri for going through an earlier version of the paper and for making suggestions for its improvement.

REFERENCES

[1] APPEL, P. and KAMPE DE FERIET, J. Functions Hypergeometriques et Hyperspheriques. Gauthier-Villars, Paris, (1926).

[2] ARNOLD, B. C. Multivariate exponential distributions based on hierarchical successive damage. *J. Applied Probability* 12, (1975), pp. 142-147.

[3] BASU, A. P. and BLOCK, H. W. On characterizing univariate and multivariate exponential distributions with applications. In A Modern Course on Statistical Distribution in Scientific Work (G. P. Patil, et al, editors). Reidel Publishing Company, Holland, (1975).

[4] BLOCK, H. W. A family of bivariate life distributions. *Proceedings of the Conference on the Theory and Applications of Reliability with Emphasis on Bayesian and Nonparametric Methods*, (1976).

[5] BOSE, S. On the distribution of the ratio of variances of two samples drawn from a given normal bivariate correlated population. *Sankhya 2*, (1935), pp. 65-72.

[6] BREITER, M. C. and KRISHNAIAH, P. R. Tables for the moments of gamma order statistics. *Sankhya Ser. B 30*, (1968), pp. 59-72.

[7] DOWNTON, F. Bivariate distribution of reliability theory. *Journal of the Royal Statistical Society Ser. B* 32, (1970), pp. 408-417.

[8] EDWARDS, C. B. and GURLAND, J. A class of distributions applicable to accidents. *Journal of the American Statistical Association 56*, (1961), pp. 503-517.

[9] GAVER, D. P. Multivariate gamma distribution generated by mixture. *Sankhya Ser. A 32*, (1970), pp. 123-126.

[10] GRIFFITHS, R. C. Infinitely divisible multivariate gamma distributions. Sankhya Ser. A 32, (1970), pp. 393-404.

[11] GUPTA, S. S. Order statistics from the gamma distribution. *Technometrics 2*, (1960), pp. 243-262.

[12] HARTER, H. L. Expected values of exponential Weibull and gamma order statistics. ARL 64-31. Aerospace Research Labs., WPAFB, Ohio, (1964).

[13] HAWKES, A. G. A bivariate exponential distribution with application in reliability. *Journal of the Royal Statistical Society Ser. B 34*, (1972), pp. 129-131.

[14] JOHNSON, N. L. and KOTZ, S.. Distributions in Statistics: Continuous Multivariate Distribution. John Wiley & Sons, (1972).

[15] KHATRI, C. G., KRISHNAIAH, P. R. and SEN, P. K. A note on the joint distribution of the correlated quadratic forms. (Abstract). *Annals of Mathematical Statistics 50*, (1970).

[16] KIBBLE, W. F. A two-variate gamma type distribution. *Sankhya 5*, (1941), pp. 137-150.

[17] KRISHNAIAH, P. R. Simultaneous test procedures under general MANOVA models. In Multivariate Analysis-II (Krishnaiah, P. R., editor), Academic Press, Inc., (1969a).

[18] ________________ Further Results on "Simultaneous test procedures under general MANOVA models". *Proceedings of the 39th Session of the International Statistical Institute* held at London, (1969b).

[19] ________________ Tests for the equality of the covariance matrices of correlated multivariate normal populations. In A Survey of Statistical Design and Linear Models (J. N. Srivastava, editor). North-Holland Publishing Company, (1975).

[20] KRISHNAIAH, P. R., HAGIS, P. and STEINBERG, L. A note on the bivariate chi distribution. *SIAM Review 5*, (1963), pp. 140-144.

[21] ________________ and RAO, M. M. Remarks on a multivariate gamma distribution. *American Mathematical Monthly 68*, (1961), pp. 342-346.

[22] ________________ and WAIKAR, V. B. On the distribution of a linear combination of correlated quadratic forms. *Comm. Statist. 1*, (1973), pp. 371-380.

[23] KRISHNAMOORTHY, A. S. and PARTHASARATHY, M. A multivariate gamma type distribution. *Annals of Mathematical Statistics 22*, (1951), pp. 549-557; correction ibid 31, 229.

[24] MORAN, P. A. P. and VERE-JONES, D. The infinite divisibility of multivariate gamma distributions. *Sankhya Ser. A 31*, (1969), pp. 191-194.

[25] ROBBINS, H. and PITMAN, E. J. G. Application of the method of mixtures to quadratic forms in normal variates. *Annals of Mathematical Statistics 20*, (1949), pp. 552-560.

[26] VERE-JONES, D. The infinite divisibility of a bivariate gamma distribution. *Sankhya Ser. A 29*, (1967), pp. 421-422.

ON SOME OPTIMAL SAMPLING PROCEDURES FOR SELECTION PROBLEMS*

by SHANTI S. GUPTA and DENG-YUAN HUANG
Purdue University

Abstract. Let Π_i, $i = 1,\ldots,k$, be k populations with associated continuous distribution functions $F(x,\theta_i) = F(x-\theta_i)$, $i = 1,\ldots,k$, where θ_i are unknown. We select the population associated with the largest θ_i by employing the 'natural' rule which selects the population that yields the largest sample mean based on n independent observations from each population. The problem of optimal allocation of the sample size is studied using the Γ-minimax criterion, which minimizes the expected risk over a class Γ of prior distributions. The loss function is set up taking into account the sampling cost as well as the cost of making a wrong decision. Similar investigation is carried out when $F(x,\theta_i) = F(x/\theta_i)$, $x \geq 0$, $\theta_i > 0$.

1.0 Introduction. An experimenter is asked which of k populations has the largest mean and must decide how large a sample he should take to decide this question. Taking a large sample decreases the probability of an incorrect decision, but at the same time increases the cost of sampling. It seems reasonable that the "optimum" sample size should depend both on the cost sampling and the amount of use to be made of the decision. In this paper, loss functions are set up which take into consideration the amount of use to be made of the result, the cost of making a wrong decision and the cost of sampling. Assume that the parameters are random variables and only partial prior infor-

---

* This research was supported by the Office of Naval Research Contract N00014-67-A-0226-0014 at Purdue University. Reproduction in whole or in part is permitted for special purposes within the United States Government.

mation is available. The Γ-minimax criterion allows one to determine a sample size that minimizes the maximum expected risk over Γ which is a class of prior distributions. Note that if Γ consists of a single prior, then the Γ-minimax criterion is the Bayes criterion for that prior. At the other extreme, if Γ consists of all priors then the Γ-minimax criterion is the usual minimax criterion. Some statements for the development of minimax criterion have been discussed by Gupta and Huang [3]. Dunnett [1] discussed an optimal sampling problem for the normal means problem by the usual minimax and Bayes criterion. Ofosu [4] considered the minimax criterion for gamma populations. In Section 2.0, we discuss the location parameter problem. The scale parameter problem is considered in Section 3.0.

2.0 Location Parameters. Let there be $k(\geq 2)$ independent populations with continuous distribution functions $F(x-\theta_1)$, $F(x-\theta_2),\ldots,F(x-\theta_k)$, where the location parameters θ_i's are unknown. Let $X_{i1},\ldots,X_{in}$ denote n independent observations from the ith population and define $\overline{X}_i = \frac{1}{n}\sum_{j=1}^{n} X_{ij}$, $(1 \leq i \leq k)$. It is well known that the distribution of $X_{ij} - \theta_i$ does not depend on $\theta_i (1 \leq i \leq k)$.

Define

$$G_n(y|F) = P\{\frac{1}{n}[(X_{i1}-\theta_i)+\ldots+(X_{in}-\theta_i)] \leq y\}.$$

Then for any i, $1 \leq i \leq k$,

$$P\{\overline{X}_i \leq x\} = G_n(x-\theta_i|F).$$

We wish to select the population associated with the largest θ_i's using the usual selection procedure. We wish to determine the optimal sampling by Γ-minimax criterion. Let δ_i denote the probability of selecting the ith population. Define the usual procedure as follows:

$$\delta_i(x) = \begin{cases} 1 & \text{if} \quad \overline{x}_i \geq \max\limits_{\substack{1\leq j\leq k \\ j\neq i}} \overline{x}_j, \\ 0 & \qquad\qquad < \end{cases}$$

Then the probability p_i that the ith population is selected is given by

$$p_i = P\{\max_{\substack{1\leq j\leq k \\ j\neq i}} \overline{X}_j \leq \overline{X}_i\}$$

$$= \int_{-\infty}^{\infty} \prod_{\substack{j=1 \\ j\neq i}}^{k} G_n(x+\theta_i-\theta_j|F)\, dG_n(x|F).$$

Let $\Omega = \{\underline{\theta}|\underline{\theta} = (\theta_1,\theta_2,\ldots,\theta_k)\}$ and $\Omega_i = \{\underline{\theta}|\theta_i \geq \max\limits_{\substack{1\leq j\leq k \\ j\neq i}} \theta_j + \Delta\}$,

$i = 1,2,\ldots,k$, and Δ is a given positive constant. Then $\Omega = \Omega_0 \cup \Omega_1 \cup \cdots \cup \Omega_k$, where Ω_0 is that part of Ω usually called indifference zone.

For $\underline{\theta} \in \Omega_i$, $1 \leq i \leq k$, define $L^{(r)}(\underline{\theta},\delta_j) = 0$ for all $r \neq i$, $L^{(i)}(\underline{\theta},\delta_j) = c'(\theta_i - \theta_j)\delta_j$, $j = 1,2,\ldots,k$, where $L^{(i)}(\underline{\theta},\delta_j)$ represents the loss for $\underline{\theta} \in \Omega_i$ when the jth population is selected, c' being a positive constant. For $\underline{\theta} \in \Omega_0$, the loss is zero.

The probability of making a wrong decision can be decreased by increasing the size of the experiment on which the decision is to be based, but this increases the cost of experimentation, which must also be considered. It will be assumed here that the cost of performing an experiment involving n observations from each population is cnk, where c is a positive constant. Let

ρ be a distribution over Ω. Then the risk function, or the expected loss, with experimentation costs included is

$$\gamma_n(\rho) = cnk + c' \sum_{i=1}^{k} \sum_{j=1}^{k} \int_{\Omega_i} \int_{E^k} (\theta_i - \theta_j)\delta_j(\underline{x})dF_{\underline{\theta}}(x)d\rho(\underline{\theta}).$$

Assume that partial information is available in the selection problem, so that we are able to specify $\pi_i = P\{\underline{\theta} \in \Omega_i\}$, $\sum_{i=0}^{k} \pi_i = 1$. Define

$$\Gamma = \{\rho(\underline{\theta}) | \int_{\Omega_i} d\rho(\underline{\theta}) = \pi_i,\ i = 1,2,\ldots,k\}.$$

We know that

$$\gamma_n(\rho) = cnk + c' \sum_{i=1}^{k} \sum_{j=1}^{k} \int_{\Omega_i} (\theta_i - \theta_j)p_j d\rho(\underline{\theta}).$$

For any i, $1 \le i \le k$, let $\underline{\theta}_i^*$ be some point in Ω_i such that

$$\sup_{\underline{\theta}\in\Omega_i} \sum_{j=1}^{k} (\theta_i - \theta_j)p_j = \sum_{j=1}^{k} (\theta_i^* - \theta_j^*)p_j^*,$$

where $\underline{\theta}_i^* = (\theta_1^*, \theta_2^*, \ldots, \theta_k^*)$ and

$$p_j^* = \int_{-\infty}^{\infty} \prod_{\substack{\ell=1\\ \ell\neq j}}^{k} G_n(x + \theta_j^* - \theta_\ell^*)dG_n(x).$$

Then

$$\sup_{\rho\in\Gamma} \gamma_n(\rho) = cnk + c' \sum_{i=1}^{k} \pi_i \sum_{j=1}^{k} (\theta_i^* - \theta_j^*)p_j^*.$$

For $\underline{\theta} \in \Omega_i$, $1 \le i \le k$, let

$$R^{(i)}(\theta_1,\theta_2,\ldots,\theta_k) = \sum_{j=1}^{k} (\theta_i - \theta_j)p_j,$$

and $g_{ij} = \theta_i - \theta_j$, then

$$R^{(i)}(\theta_1,\theta_2,\ldots,\theta_k) = \sum_{j=1}^{k} g_{ij} \int_{-\infty}^{\infty} \prod_{\substack{\ell=1\\ \ell\neq j}}^{k} G_n(x+g_{j\ell})dG_n(x).$$

Note that all the g_{ij} are positive, by definition, and $g_{ii} = 0$.

We first require to determine the values of the parameters in Ω_i for which $R^{(i)}$ is a maximum. Somerville [5] shows that this is achieved when $g_{ij}(j \neq i)$ are positive and equal, while $g_{ii} = 0$. Denote the common value of the positive g_{ij} by g_i, then

$$R^{(i)} = \sum_{\substack{j=1 \\ j \neq i}}^{k} g_i p_j = g_i(1-p_i).$$

If we denote by $R_M^{(i)}$, the maximum of the function $R^{(i)}$, then the maximum risk is given by

$$\sup_{\rho \in \Gamma} \gamma_n(\rho) = cnk + c' \sum_{i=1}^{k} \pi_i R_M^{(i)}.$$

Since

$$R^{(i)} = g_i\{1 - \int_{-\infty}^{\infty} G_n^{k-1}(x + g_i)dG_n(x)\},$$

hence we know that

$$\sup_{g_1 \geq \Delta} R^{(1)} = \ldots = \sup_{g_k \geq \Delta} R^{(k)} = \sup_{g \geq \Delta} R = R_M .$$

Thus, $$\sup_{\rho \in \Gamma} \gamma_n(\rho) = cnk + c'(\sum_{i=1}^{k} \pi_i)R_M$$

$$= cnk + c'(1 - \pi_0)R_M.$$

For the problem of normal distributions $N(\theta_i, 1)$, $i = 1,2,\ldots,k$, so that we are interested in the selection of the means θ_i, the function R becomes

$$R = g[1-\Phi_{k-1,1/2}(g\sqrt{\tfrac{n}{2}},\ldots, g\sqrt{\tfrac{n}{2}})] ,$$

where $\Phi_{k-1,1/2}(\cdot)$ is the cumulative distribution function of the (k-1) - variate normal distribution with zero means, unit variances, and all the correlation **coefficients equal to $\frac{1}{2}$.** If we denote by M_{k-1} the maximum, with respect to $y(\geq\Delta)$ for sufficiently small Δ, of the function

$$y[1-\Phi_{k-1,1/2}(y,\ldots,y)],$$

then the maximum risk is given by

$$\sup_{\rho\varepsilon\Gamma} \gamma_n(\rho) = cnk + \frac{1}{\sqrt{\frac{n}{2}}} c' \ (1-\pi_0)M_{k-1} \ .$$

This can be minimized with respect to n by taking

$n = <(\frac{1}{2c^2k^2} c'^2(1 - \pi_0)^2 M_{k-1}^2)^{\frac{1}{3}}>$, where $<x>$ is the smallest integer greater than or equal to x. The values of M_{k-1} have been discussed by Dunnett [1]. The values of $\Phi_{k-1,1/2}(y,\ldots,y)$ are also tabulated in Gupta [2] for $k = 2(1)13$.

By using Somerville's table [5] for the values of M_{k-1}, $k = 2(1)6$, we compute some n values as follows:

Table I

Minimum Sample Sizes for the Normal Means Problem

k	π_0 \ $\frac{c'}{ck}$	5	10	15	30	50	100
2	0.10	1	2	2	3	4	5
	0.30	1	1	2	2	3	5
	0.50	1	1	1	2	3	4
3	0.10	1	2	2	3	5	7
	0.30	1	2	2	3	4	6
	0.50	1	1	2	2	3	5
4	0.10	2	2	3	4	5	8
	0.30	1	2	2	3	5	7
	0.50	1	2	2	3	4	6
5	0.10	2	2	3	4	6	9
	0.30	1	2	2	4	5	8
	0.50	1	2	2	3	4	6
6	0.10	2	2	3	4	6	9
	0.30	2	2	3	4	5	8
	0.50	1	2	2	3	4	6

Note that we choose $\Delta \leq 0.5$.

3.0 Scale Paramaters We assume that the k independent populations have continuous distribution functions $F(\frac{x}{\theta_1})$, $F(\frac{x}{\theta_2})$, $\ldots, F(\frac{x}{\theta_k})$, respectively, where the scale parameter θ_i is positive and unknown and $x \geq 0$. Let $X_{i1}, \ldots, X_{in}$ denote n independent observations from Π_i and define

$\overline{X}_i = \frac{1}{n} \sum_{j=1}^{n} X_{ij}$, $1 \leq i \leq k$. As before, we know that any i, $1 \leq i \leq k$,

$$P\{\bar{X}_i \le x\} = H_n(\frac{x}{\theta_i}|F).$$

Let $\delta_i(\underline{x})$ denote the same procedure as before. Then the probability q_i that Π_i is selected is given by

$$q_i = P\{\max_{\substack{1\le j\le k \\ j\ne i}} \bar{X}_j \le \bar{X}_i\}$$

$$= \int_0^\infty \prod_{\substack{j=1 \\ j\ne i}}^{k} H_n(x \cdot \frac{\theta_i}{\theta_j}|F)dH_n(x|F).$$

Let $\Omega_i = \{\underline{\theta}|\theta_i \ge \delta \max_{\substack{1\le j\le k \\ j\ne i}} \theta_j\}$, $i = 1,2,\ldots,k$, and $\delta(>1)$ is a given constant. For $\underline{\theta} \in \Omega_i$, $1 \le i \le k$, define $L^{(r)}(\underline{\theta},\delta_j) = 0$ for all $r \ne i$,

$$L^{(i)}(\underline{\theta},\delta_j) = c'\delta_j \log\frac{\theta_i}{\theta_j}, \; j = 1,2,\ldots,k,$$

where $L^{(i)}(\underline{\theta},\delta_j)$ represents the loss for $\underline{\theta} \in \Omega_i$ but the jth population is selected, c' being a positive constant. For $\underline{\theta} \in \Omega_0$, the loss is zero.

By using similar discussion as before, for any i, $1 \le i \le k$, let $\underline{\theta}_i^*$ be some point in Ω_i such that

$$\sup_{\underline{\theta}\in\Omega_i} \sum_{j=1}^{k} q_j \log\frac{\theta_i}{\theta_j} = \sum_{j=1}^{k} q_j^* \log\frac{\theta_i^*}{\theta_j^*},$$

where $\underline{\theta}_i^* = (\theta_1^*,\theta_2^*,\ldots,\theta_k^*)$ and

$$q_j^* = \int_0^\infty \prod_{\substack{\ell=1 \\ \ell\ne j}}^{k} H_n(x \cdot \frac{\theta_j^*}{\theta_\ell^*})dH_n(x).$$

Then

$$\sup_{\rho \in \Gamma} \gamma_n (\rho) = cnk + c' \sum_{i=1}^{k} \pi_i \sum_{j=1}^{k} q_j^* \log \frac{\theta_i^*}{\theta_j^*} ,$$

where Γ is defined as before.

Let

$$Q^{(i)} (\theta_1, \theta_2, \ldots, \theta_k) = \sum_{j=1}^{k} q_j \log \frac{\theta_i}{\theta_j} ,$$

and $h_{ij} = \theta_i/\theta_j$, then

$$Q^{(i)} (\theta_1, \theta_2, \ldots, \theta_k) = \sum_{j=1}^{k} (\log h_{ij}) \int_0^\infty \sum_{\substack{\ell=1 \\ \ell \neq j}}^{k} H_n(x\, h_{j\ell}) dH_n(x) .$$

Note that all the h_{ij} are greater than 1, by definition, and $h_{ii}=1$. By using a discussion similar to that in Section 2, we have

$$\sup_{\rho \in \Gamma} \gamma_n (\rho) = cnk + c' (1 - \pi_0) Q_M ,$$

where $Q_M = \sup_{h \geq \delta} Q = \sup_{h_1 \geq \delta} Q^{(1)} = \ldots = \sup_{h_k \geq \delta} Q^{(k)} = (\log h)$

$$\{1 - \int_0^\infty H_n^{k-1} (xh) dH_n(x)\}.$$

For the selection of the scale parameters of gamma populations with densities

$$\frac{1}{\theta_i} \left(\frac{x}{\theta_i}\right)^{a-1} \frac{1}{\Gamma(a)} \exp\{- \frac{x}{\theta_i}\}, \; x > 0,$$

where θ_i, $1 \leq i \leq k$, are the unknown scale parameters, $\theta_i > 0$ and $a (>0)$ is a known shape parameter. Let

$$A_n = cnk + c' (1 - \pi_0) (\log h) [1 - \int_0^\infty G_\nu^{k-1} (hx) dG_\nu(x)],$$

where $G_\nu(x)$ is the cdf of a standardized gamma random variable with $\nu = na$. Let N_{k-1} be the maximum, with respect to $h (\geq \delta)$, of the function

$$[1 - \int_0^\infty G_\nu^{k-1}(hx)dG_\nu(x)]\log h,$$

then the maximum risk is given by

$$\sup_{\rho\varepsilon\Gamma} \gamma_n(\rho) = cnk + c'(1 - \pi_0)N_{k-1} .$$

It is not analytically feasible to minimize $\sup_{\rho\,\Gamma} \gamma_n(\rho)$ with respect to n by differentiation. We can use the same method as Ofosu [4] to make a numerical study. Some asymptotic Γ-minimax solutions are discussed as follows.

It is known that as ν tends to ∞,

$$\left(\frac{2\nu-1}{2}\right)^{\frac{1}{2}} \log\{\overline{X}_i/(a\theta_i)\}, \quad (i = 1,2,\ldots,k)$$

is asymptotically distributed as $N(0,1)$. Hence, as $\nu \to \infty$, we have

$$\sup_{\rho\varepsilon\Gamma} \gamma_n(\rho) = cnk + c'(1 - \pi_0)\ 2(2\nu-1)^{-\frac{1}{2}} M_{k-1},$$

Where $M_{k-1} = \sup_{h\geq\delta} h\{1 - \Phi_{k-1,1/2}(h,\ldots,h)\}$. Then $\sup_{\rho\varepsilon\Gamma} \gamma_n(\rho)$ can be minimized with respect to n, for large ν, by taking

$$n = \left< \frac{1}{2a}\left\{\left(\frac{2ac'(1-\pi_0)M_{k-1}}{ck}\right)^{2/3} + 1\right\}\right> ,$$

where $\langle x\rangle$ is the smallest integer greater than or equal to x. this result is consistent with the work of Sommerville [5].

Applications to selection of Weibull populations scale parameters and normal variances problems can be obtained in the same way as in Ofosu [4].

References

[1] Dunnett, C. W., On selecting the largest of k normal population means. *J. Roy Statist. Soc. Ser. B, 22,* (1960), pp. 1 - 40.

[2] Gupta, S.S., Probability integrals of multivariate normal and multivariate t. *Ann. Math Statist.* 34, (1963), pp. 792-828.

[3] Gupta, S.S. and Huang, D. Y., On some Γ-minimax subset selection and multiple comparison procedures. Department of Statistics, Purdue University, Mimeo Series #392, W. Laf., IN. (1974).

[4] Ofosu, J. B., A minimax procedure for selecting the population with the largest (smallest) scale parameter. *Calcutta Statist. Assoc. Bulletin 21*, (1972) pp. 143-154.

[5] Somerville, P. N., Some problems of optimum sampling. *Biometrika 41*, (1954), pp. 420-429.

[6] Somerville, P. N., Optimal sample size for choosing the population having the smallest (largest) variance. *J. Amer. Statist. Assoc., 70,* (1975), pp. 852-858.

CLASSIFICATION RULES FOR EXPONENTIAL POPULATIONS: TWO PARAMETER CASE

by A. P. BASU*
University of Missouri

and A. K. GUPTA
The University of Michigan

Abstract. This paper extends many of the results for one-parameter exponential distributions, considered by the authors earlier [4], to the case of the two-parameter exponential distributions. In this paper the classification rules based on the likelihood ratio criterion for the two-parameter exponential populations have been studied. These classification rules have been extended to many populations and to the situation where observations are censored. In cases where the likelihood-ratio criterion does not lead to appealing classification rules, ad hoc rules are proposed. An additional rule is considered from "life testing" point of view. In each case, the probability of misclassification is derived exactly where the parameters are known. The rules considered have been shown to possess a consistency property. An alternate approach based on Bayesian considerations is also explored.

1.0 *Introduction*. Let π_i denote a two-parameter exponential population with probability density function (pdf)

$$f_i(x) = f(x|\mu_i,\theta_i) = \theta_i^{-1} e^{-(x-\mu_i)/\theta_i} I(x \geq \mu_i) \quad (1.1)$$

$$(i = 0, 1, 2),$$

where $\mu_i(-\infty < \mu_i < \infty)$ and $\theta_i(\theta_i > 0)$ are the location and the

*Research supported by the Air Force Office of Scientific Research under Grant No. AFOSR-75-2795.

scale parameters respectively. And $I(x \geq \mu)$ is the indicator function of the set $\{x: x \geq \mu\}$ defined by $I(x \geq \mu) = \begin{matrix} 1 \text{ if } x \geq \mu \\ 0 \text{ if } x < \mu \end{matrix}$. It is known that $(\mu_0, \theta_0) = (\mu_i, \theta_i)$ for exactly one i (i = 1 or 2). Let $\underline{X} = (X_1, X_2, \ldots, X_n)$ be a random sample from π_0. The problem of classification is to classify π_0 as π_1 or π_2 based on the random sample $\underline{X}$. The theoretically best procedures for classification or alternatively, for testing the null hypothesis of one distribution against the alternative hypothesis of the other distribution, are based on the likelihood ratio criterion (see Anderson [1]).

The classification problem for the normal population has been considered by many authors in the statistical literature. For an extensive bibliography the reader is referred to Anderson, et al. [2]. The probability of misclassification in such classification procedures is not necessarily known to the experimenter. Several authors have studied the probability of misclassification problem (e.g., see Smith [13]), John [10], Okamoto [11], Sedransk [12], Hills [9], and Sorum [14]. Gupta and Govindarajulu [8] have studied some heuristic classification rules for m univariate normal populations. A class of admissible decision procedures under a suitable loss function is given by Bhattacharya and Das Gupta [5]. Geisser [7] has studied the classification problem for normal distributions from a Bayesian point of view.

However, in many physical models for example, in problems of life testing and reliability analysis and in many biomedical problems the exponential distribution appears as a natural model (e.g., see Epstein and Sobel [6] and Zelen [16]), where X denotes the life length of a system and the location parameter μ_i (here $0 < \mu_i < \infty$) is the minimal life of a system. Classification of observations from m exponential populations often becomes necessary. In this paper we extend many of the results for one-parameter exponential distributions obtained earlier [4] to the case of the two parameter exponential distributions. In section 2, we study the classification rules based on likelihood-ratio criterion for the exponential populations when (i) (μ_i, θ_i), i = 1, 2 known, (ii) θ_i, i = 1, 2 known and μ_i, i = 1, 2 unknown, and (iii) (μ_i, θ_i), i = 1, 2 unknown. We have also proved a consistency property for these rules. In section 3, exact probability of misclassification has been derived. In sections 4 and 5, the rules have been extended to m populations and to the situation where observations are censored. In section 6, we consider an ad hoc classification procedure, based on physical considerations, and prove that it is consistent. Finally, in section 7, an alternate approach based on Bayesian considerations is explored.

2.0 *Classification Rules*. Let q_i be the prior probability of drawing an observation from population π_i with density $f_i(x)$, (i = 1, 2), and let the cost of misclassifying an observation from π_i as from π_j be $c(j|i)$. Then it is well known that the regions of classification to π_1 and π_2 given by

$$\begin{aligned} S_1&: f_1(\underline{x})/f_2(\underline{x}) \geq c(1|2)q_2/c(2|1)q_1 \\ S_2&: f_1(\underline{x})/f_2(\underline{x}) < c(1|2)q_2/c(2|1)q_1 \end{aligned} \tag{2.1}$$

respectively minimize the expected loss from costs of misclassification which is $q_1\ c(2|1)\ P(2|1) + q_2\ c(1|2)\ P(1|2)$ where $P(i|j)$ is the probability of misclassifying an observation to π_i when actually it is from π_j. Here $f_i(\underline{x}) = \prod_{\ell=1}^{n} f_i(x_\ell)$ is the joint density function of the random sample $\underline{X}$ assuming that $\underline{X}$ is from π_i. Further if

$$P\left\{\frac{f_1(\underline{x})}{f_2(\underline{x})} = \frac{q_2 c(1|2)}{q_1 c(2|1)} \,\Big|\, \pi_i\right\} = 0, \; i = 1, 2 \tag{2.2}$$

then the procedure is unique except for sets of probability zero (see Anderson [1], pp. 131).

For studying the classification rules based on likelihood ratio, it will be assumed that $\theta_1 \neq \theta_2$. However, two ad hoc rules will be studied where this restriction will not be necessary. Also, in what follows it is assumed that the populations are equally likely and the costs of misclassification are the same, that is $q_1 = q_2$ and $c(1|2) = c(2|1)$. In the sequel, to prove some asymptotic properties, we will also need the following definition.

Definition 1. A classification rule R based on n observations is said to be consistent if $P(MC|R) \to 0$ as $n \to \infty$.

2.1. μ_i, θ_i, $(i = 1, 2)$ Known

The following theorem is now a direct consequence of the above results (2.1) and (2.2).

Theorem 1. If π_i has density $f_i(x)$ given by (1.1), then the best regions of classification are given by

S_1: either $U_1(\underline{X}) \leq 0$ and $X_{(1)} \geq \mu$, or $\mu_1 \leq X_{(1)} < \mu_2$,

S_2 either $U_1(\underline{X}) > 0$ and $X_{(1)} \geq \mu$, or $\mu_2 \leq X_{(1)} < \mu_1$,

where

$$U_1(\underline{X}) = \frac{\overline{X} - \mu_1}{\theta_1} - \frac{\overline{X} - \mu_2}{\theta_2} - \ln(\theta_2/\theta_1), \tag{2.3}$$

$$\overline{X} = \frac{1}{n}\sum_{i=1}^{n} X_i, \; X_{(1)} = \min(X_1, \ldots, X_n), \; \mu = \max(\mu_1, \mu_2),$$

and "best" means minimum expected cost.

Hence the classification rule in this case is

R_1: Classify π_0 to π_1 if

$$U_1(\underline{X}) \leq 0 \text{ and } X_{(1)} \geq \mu, \text{ or } \mu_1 \leq X_{(1)} < \mu_2 \quad (2.4)$$

and to π_2 if

$$U_1(\underline{X}) > 0 \text{ and } X_{(1)} \geq \mu, \text{ or } \mu_2 \leq X_{(1)} < \mu_1.$$

The classification rule R_1 has the asymptotic property that the probability of erroneous classification, $P(MC|R_1)$, approaches zero as the sample size n becomes large. This "consistency property" is stated below.

Theorem 2. The classification rule R_1 is consistent, i.e., $P(MC|R_1) \to 0$ as $n \to \infty$.

Proof. We have

$$\begin{aligned} P(MC|R_1) &= 1/2\, P[\{U_1(\underline{X}) \leq 0 \text{ and } X_{(1)} \geq \mu\} \text{ or } \\ &\quad \{\mu_1 \leq X_{(1)} < \mu_2\}|\pi_2] \\ &\quad + 1/2\, P[\{U_1(\underline{X}) > 0 \text{ and } X_{(1)} \geq \mu\} \text{ or } \\ &\quad \{\mu_2 \leq X_{(1)} < \mu_1\}|\pi_1] \\ &= 1/2\,(P_1 + P_2) + 1/2\,(P_3 + P_4), \text{ say} \end{aligned}$$

where

$$\begin{aligned} P_1 &= P(U_1(\underline{X}) \leq 0, X_{(1)} \geq \mu|\pi_2) \\ P_2 &= P(\mu_1 \leq X_{(1)} < \mu_2|\pi_2) = 0 \\ P_3 &= P(U_1(\underline{X}) > 0, X_{(1)} \geq \mu|\pi_1) \\ P_4 &= P(\mu_2 \leq X_{(1)} < \mu_1|\pi_1) = 0. \end{aligned}$$

To prove the theorem it suffices to show that P_1 and $P_3 \to 0$ as $n \to \infty$. Now

$$0 \leq P_1 \leq P(U_1(\underline{X}) \leq 0|\pi_2)$$

and by strong law of large numbers, we have with probability one as $n \to \infty$

$$U_1(\underline{X})/\pi_2 \to \frac{1}{\theta_1}(\mu_2 - \mu_1) - 1 + \frac{\theta_2}{\theta_1} - \ln(\theta_2/\theta_1),$$

which is positive if $\mu = \mu_2$ since $x - 1 - \ln x > 0$ and therefore $P_1 \to 0$ as $n \to \infty$. On the other hand, if $\mu = \mu_1$ we have

$$0 \leq P_1 \leq P\,(X_{(1)} \geq \mu | \pi_2) = e^{-n(\mu - \mu_2)/\theta_2} \to 0 \text{ as } n \to \infty.$$

Hence $P_1 \to 0$ as $n \to \infty$. Similarly, it can be shown that $P_3 \to 0$ as $n \to \infty$ which completes the proof of the theorem.

Remark 1. If $\theta_1 = \theta_2$, the classification rule R_1 simplifies to

R_1^*: classify π_0 to π_1 iff (2.5)

$X_{(1)} \geq \mu_1 > \mu_2$ or $\mu_1 \leq X_{(1)} < \mu_2$.

The rule R_1^* is consistent follows from the proof of Theorem 2.

2.2. $\theta_i\,(i = 1, 2)$ Known and $\mu_i\,(i = 1, 2)$ Unknown

In this case, it is assumed that a random sample X_{ij} $(j = 1, \ldots, n_i)$ is available from population $\pi_i\,(i = 1, 2)$. Then the maximum-likelihood estimator of μ_i is given by $\hat{\mu}_i = X_{i(1)} = \min(X_{i1}, \ldots, X_{in_i})$. Define

$$U_2(\underline{X}) = \frac{\overline{X} - X_{1(1)}}{\theta_1} - \frac{\overline{X} - X_{2(1)}}{\theta_2} - \ln(\theta_2/\theta_1) \text{ and} \tag{2.6}$$

$$\hat{\mu} = \max(\hat{\mu}_1, \hat{\mu}_2).$$

The classification rule proposed is

R_2: Classify π_0 to π_1 if

$$U_2(\underline{X}) \leq 0 \text{ and } X_{(1)} \geq \hat{\mu} \text{ or } \hat{\mu}_1 \leq X_{(1)} < \hat{\mu}_2, \tag{2.7}$$

and to π_2 if

$$U_2(\underline{X}) > 0 \text{ and } X_{(1)} \geq \hat{\mu} \text{ or } \hat{\mu}_2 \leq X_{(1)} < \hat{\mu}_1.$$

Such a rule is sometimes known as a "plug-in" rule. The rule R_2 has the same asymptotic property as R_1.

Theorem 3. The classification rule R_2 is consistent, i.e.,

$$P(MC|R_2) \to 0 \text{ as } n, n_1 \text{ and } n_2 \to \infty.$$

Proof. The proof follows immediately from the arguments of Theorem 2 since

$$U_2(\underline{X}) - U_1(\underline{X}),\ \mu_1 - X_{1(1)} \text{ and } \mu_2 - X_{2(1)} \to 0 \text{ as}$$

$$n_1 \text{ and } n_2 \to \infty$$

in probability.

Remark 2. In case $\theta_1 = \theta_2$, the classification rule R_2 reduces to R_2^*: Classify π_0 to π_1 iff

$$\hat{\mu}_1 \leq X_{(1)} < \hat{\mu}_2 \text{ or } X_{(1)} \geq \hat{\mu}_1 > \hat{\mu}_2.$$

Again the consistency of R_2^* is derived from the proof of Theorem 3.

2.3. μ_i, θ_i, (i = 1, 2) Unknown

As in 2.2 it is assumed that two independent samples are available from π_1 and π_2. The unknown parameters are estimated by $\hat{\theta}_i = \bar{X}_i - X_{i(1)}$, and $\hat{\mu}_i = X_{i(1)}$, i = 1, 2. Then define

$$V(\underline{X}) = \frac{\bar{X} - X_{1(1)}}{\bar{X}_1 - X_{1(1)}} - \frac{\bar{X} - X_{2(1)}}{\bar{X}_2 - X_{2(1)}} - \ln \frac{\bar{X}_2 - X_{2(1)}}{\bar{X}_1 - X_{1(1)}} \tag{2.8}$$

where $\bar{X}_i = \frac{1}{n_i} \sum_{j=1}^{n_i} X_{ij}$, i = 1, 2. We propose the following plug-in classification rule:

R_3: Classify π_0 to π_1 if

$$V(\underline{X}) < 0 \text{ and } X_{(1)} \geq \hat{\mu} \text{ or } \hat{\mu}_1 \leq X_{(1)} < \hat{\mu}_2, \tag{2.9}$$

and to π_2 if

$$V(\underline{X}) > 0 \text{ and } X_{(1)} \geq \hat{\mu} \text{ or } \hat{\mu}_2 \leq X_{(1)} < \hat{\mu}_1.$$

It may now be noted that the classification rule R_3 is consistent, i.e.,

$$P(MC|R_3) \to 0 \text{ as } n, n_1 \text{ and } n_2 \to \infty$$

This can be easily seen since $V \to U_1$ with probability 1 as $n_1 \to \infty$ and $n_2 \to \infty$.

Remark 3. If it is further assumed that $\mu_1 = \mu_2 = \mu$ (say) is known, then since $X - \mu$ will follow a one parameter exponential distribution with scale parameter θ_i if X is from $\pi_i (i = 1, 2)$, the problem in this case reduces to classification between two one-parameter exponential distributions, which has been discussed earlier in detail by the authors [4].

Remark 4. In case $\mu_1 = \mu_2 = \mu_0$ (say) is unknown, it can be estimated by its maximum-likelihood estimator $\hat{\mu}_0 = X_{(1)} = \min(X_1, \ldots, X_n)$ and plugged-in the rule R_1 above to obtain the following studentized classification rule.

R_1^{**}: Classify π_0 to π_1 iff

$$(\overline{X} - X_{(1)})\left(\frac{\theta_2 - \theta_1}{\theta_1\theta_2}\right) - \ln(\theta_2/\theta_1) \leq 0. \qquad (2.10)$$

But $\overline{X} - X_{(1)} = \sum_{i=2}^{n} Y_{in}/n$, where $Y_{in} = (n-i+1)(X_{(i)} - X_{(i-1)})$ $X_{(i)}$ is the ith order statistic from $(X_1, \ldots, X_n)$. Now $Y_{in} (i = 1, \ldots, n)$ are iid with p.d.f. $\theta_i^{-1} \exp(-y/\theta_i)$; $y > 0$ if X is from π_i. Hence, the classification problem, as in the above case, reduces to the one-parameter case discussed by the authors [4].

3.0 *Probability of Misclassification.* As stated earlier, the probability of misclassification is not known to the experimenter and should be evaluated for any given classification procedure. In this section, we evaluate $P(MC|R_1)$ exactly. We have from Theorem 2,

$$\begin{aligned} P(MC|R_1) &= 1/2\ P\{U_1(\underline{X}) \leq 0,\ X_{(1)} \geq \mu|\pi_2\} \\ &\quad + 1/2\ P\{U_1(\underline{X}) > 0,\ X_{(1)} \geq \mu|\pi_1\} \\ &= 1/2\ P\{\overline{X} \leq k,\ X_{(1)} \geq \mu|\pi_2\} \\ &\quad + 1/2\ P\{\overline{X} > k,\ X_{(1)} \geq \mu|\pi_1\} \end{aligned}$$

where $k = [\mu_1\theta_2 - \mu_2\theta_1 + \theta_1\theta_2 \ln(\theta_2/\theta_1)]/(\theta_2 - \theta_1)$ and

loss of generality

$(n - i + 1)(X_{(i)} - X_{(i-1)})$, $i = 1, \ldots, n$ with $X_{(0)} = 2$. Since, given $\pi_0 = \pi_j$, $2u_i/\theta_j$ are i.i.d. χ^2 $(j = 1, 2)$ we have

$$P(MC|R_1) = 1/2 + 1/2\ \{F_{2(n-1)}[\tfrac{2n}{\theta_2}(k-\mu)] \tag{3.1}$$

$$- F_{2(n)}[\tfrac{2n}{\theta_1}(k-\mu)]\} - \sum_{k=0}^{n-1}\sum_{j=0}^{k}\left[\frac{(-1)^j}{k!}\exp\left\{\frac{-n(k-\mu_1)}{\theta_1}\right\}\right.$$

$$\left.\left(\frac{n(k-\mu_1)}{\theta_1}\right)^{k-j} \cdot \frac{(n(\mu_2-\mu_1))^{j+1}}{j+1} \cdot\right]$$

where F_m denotes the cdf of chi-square random variable with m degrees of freedom. Hence, (3.1) gives the exact probability of misclassification of Rule R_1. A similar expression is obtained if $\mu_1 > \mu_2$. This is also the conditional probability of Rule R_2 given (μ_i, θ_i), $(i = 1,2)$ are known and of Rule R_3 given (μ_i, θ_i), $(i = 1,2)$ are known. The unconditional probability of misclassification of Rules R_2 and R_3 seems difficult to obtain since the distribution of the classification statistics $U_2(\underline{X})$ and $V(\underline{X})$ is not known. However, using the asymptotic theory, approximations can be obtained from (3.1).

4.0 *Extension to m Populations*. The classification rule R_1 can be extended to m populations. Suppose there are $(m + 1)$ populations where the density of the population π_i is given by $f_i(x)$ in (1) $(i = 0, 1, \ldots, m)$. The problem is to classify π_0 to one of the m populations $\pi_1, \ldots, \pi_m$. Assuming that the costs of misclassification are the same and that all populations are equally likely, the classification rule is

R_4: Classify π_0 to π_i $(i = 1, \ldots, m)$ iff

$$U_{ij}(\underline{X}) \leq 0,\ j = 1, \ldots, m;\ j \neq i, \text{ and } X_{(1)} > \mu \tag{4.1}$$

$$= \max(\mu_1, \ldots, \mu_m)$$

or $X_{(1)} > \mu_i$ only

or $X_{(1)} > \mu_r$ for $r = i, k_1, \ldots, k_\ell$ only and $U_{ik_j}(\underline{X}) \leq 0$

$k_j = 1, \ldots, m;\ k_j \neq i;\ k_r \neq k_j (r \neq j);\ j = 1, \ldots, \ell,$

for $\ell = 1, 2, \ldots, m;\ \ell \neq i$.

The classification rule R_4 is simple in application. For example, when there are three populations $(m = 3)$, it says: classify π_0 to π_1 iff $U_{12}(\underline{X}) \leq 0$, $U_{13}(\underline{X}) \leq 0$ and $X_{(1)} > \mu$; or $X_{(1)} > \mu_1$ only; or $X_{(1)} < \mu_3$ only and $U_{12}(\underline{X}) \leq 0$ or $X_{(1)} < \mu_2$ only and $U_{13}(\underline{X}) \leq 0$. Also the rule R_4 has the same asymptotic property which is enjoyed by R_1, R_2 and R_3.

Theorem 4. The classification rule R_4 is consistent i.e. $P(MC|R_4) \to 0$ as $n \to \infty$.

Proof. The probability of misclassification to i^{th} population is given by

$$P_i(MC|R_4) = \frac{1}{m} \sum_{\substack{r=1 \\ r \neq i}}^{m} [P_{1r} + P_{2r} + \sum_{\substack{\ell=1 \\ \ell \neq i}}^{m} P_{3\ell r}]$$

where

$$P_{1r} = P(U_{ij}(\underline{X}) \leq 0,\ j = 1, \ldots, m;\ j \neq i;\ X_{(1)} > \mu | \pi_r)$$

$$P_{2r} = P(X_{(1)} > \mu_i \text{ only } | \pi_r)$$

and

$$P_{3\ell r} = P[U_{ik_j}(\underline{X}) \leq 0,\ k_j = 1, \ldots, m;\ k_j \neq i;$$
$$j = 1, \ldots, \ell;\ X_{(1)} > \mu_r,\ r = i, k_1, \ldots, k_\ell$$
$$\text{only } | \pi_r].$$

Now

$$P_{1r} \leq P(X_{(1)} > \mu | \pi_r)$$
$$= e^{-n(\mu_r - \mu)/\theta_r}$$
$$\to 0 \text{ as } n \to \infty \text{ if } \mu \neq \mu_r.$$

However if $\mu = \mu_r$, we have

$$P_{1r} \leq P(U_{ir}(\underline{X}) \leq 0 | \pi_r).$$

But with probability 1, $U_{ir}(\underline{X}) | \pi_r \to \dfrac{\mu_r - \mu_i}{\theta_i} + \dfrac{\theta_r}{\theta_i} - 1 - \dfrac{\theta_r}{\theta_i}$

which is always positive if $\mu_r > \mu_i$. Hence $P_1 \to 0$ as $n \to \infty$.

Also trivially $P_{2r} = 0$. Furthermore, we have

$$P_{3\ell r} \leq P(U_{ir}(\underline{X}) \leq 0 | \pi_r)$$

and as before $P_{3\ell r} \to$ as $n \to \infty$ if

$\mu_r > \mu_i$. In case $\mu_r < \mu_i$, we get

$$P_{3\ell r} \leq P(X_{(1)} > \mu_r,\ r = i, k_1, \ldots, k_\ell \text{ only} | \pi_r)$$
$$= e^{-n\{\max(\mu_i, \mu_{k_1}, \ldots, \mu_{k_\ell}) - \mu_r\}/\theta_r}$$
$$\to 0 \text{ as } n \to \infty.$$

In any case each term of the finite sum approaches zero as $n \to \infty$, which proves that probability of misclassification to π_i approaches zero as $n \to \infty$. Hence the proof.

As in the case of two populations the theorem can be extended to the cases when μ_i's or both μ_i's and θ_i's are unknown.

5.0 *Classification Based on Censored Observations*. So far we have assumed that complete samples are available for classification. However, in many physical situations, for example, in problems of life testing and reliability analysis and in many clinical trials not all the observations are always available, we need to develop a classification rule based on the first few observations only. In this section we point out how the various classification rules can be modified to meet this situation.

Let $f_i(x)$ be as given in (1.1) $(i = 0, 1, 2)$. However, assume that only the first r-ordered observations $X_{(1)} < X_{(2)} < \dots < X_{(r)}$ from π_0 are given out of a sample of size n. Here r is a fixed number and is determined beforehand from non-statistical considerations. Then it is well known (c.f. Basu [3]) that for each n

$$U_{in} = (n - 1 + 1)(X_{(i)} - X_{(i-1)}) \qquad (5.1)$$

$$(i = 1, 2, \dots, r, X_{(0)} = \mu_0)$$

are independent and identically distributed random variables each having the common density function $f_0(x)$ with $\mu_0 = 0$. Thus the classification problem of Section 2 reduces to classifying π_0 to π_1 or π_2 based on a random sample $(U_{1n}, \dots, U_{rn})$ of size r. That is the appropriate classification rule in this case is given by:

R': Classify π_0 to π_1 if

$$W(\underline{U}) = \frac{\overline{U} - \mu_1}{\theta_1} - \frac{\overline{U} - \mu_2}{\theta_2} - \ln(\theta_2/\theta_1) \leq 0 \text{ and } X_{(1)} > \mu \text{ or}$$

$$\mu_1 \leq X_{(1)} < \mu_2 \qquad (5.2)$$

and to π_2 if

$$W(\underline{U}) > 0 \text{ and } X_{(1)} > \mu \text{ or } \mu_2 < X_{(1)} \leq \mu_1.$$

where $\overline{U} = \frac{1}{r}\sum_{i=1}^{r} U_{in} = \frac{1}{r}[\sum_{1}^{r} X_{(i)} + (n-r)X_{(r)}]$.

In this case, following the method described in Section 3, the probability of misclassification can be obtained. Following Theorem 2, one can prove that the classification rule R_1' is consistent. The cases when μ_i's or (μ_i, θ_i)'s are unknown and when the classification has to be made to one of the m populations, are similarly described as before.

6.0 *An Ad Hoc Classification Rule Based on Physical Considerations and Its Properties.* Sometimes it is of interest to classify an observation with respect to a certain function of the parameters, due to the inherent physical importance of such a function, rather than the parameters themselves. Such classification rules with respect to the coefficient of variation for normal populations have been studied by Gupta and Govindarajulu [8]. In the present context of reliability theory and life testing one such function is $\mu + \theta$ which is the average life. In this section we no longer assume that $\theta_1 \neq \theta_2$. Consider two separate cases.

Case (i) (μ_i, θ_i), i = 1, 2 known.

In this case we suggest the ad hoc classification rule

R_5: Classify π_0 to π_1 if

$$|\overline{X} - (\mu_1 + \theta_1)| < |\overline{X} - (\mu_2 + \theta_2)| \text{ and } X_{(1)} \geq \mu \text{ or,}$$

$$\mu_1 \leq X_{(1)} < \mu_2.$$

and classify π_0 to π_2 if

$$|\bar{X} - (\mu_1 + \theta_1| > |\bar{X} - (\mu_2 + \theta_2) \text{ and } X_{(1)} \geq \mu \tag{6.1}$$
$$\text{or } \mu_2 \leq X_{(1)} < \mu_1$$

This rule is equivalent to classifying π_0 to π_1 when $\bar{X} > d$ and $X_{(1)} \geq \mu$ or $\mu_1 \leq X_{(1)} < \mu_2$ and to π_2 if $\bar{X} < d$ and $X_{(1)} \geq \mu$ or $\mu_2 \leq X_{(1)} < \mu_1$ where $d = (\mu_1 + \mu_2 + \theta_1 + \theta_2)/2$ and it is assumed without loss of generality that $\mu_2 + \theta_2 < \mu_1 + \theta_1$. The rule R_5 is consistent as is proved in the following theorem.

Theorem 5. With the above notation, the classification rule R_5 is consistent, i.e.,

$$P(\text{MC } R_5) \to 0 \text{ as } n \to \infty$$

Proof.

$$P(\text{MC}|R_5) = \frac{1}{2} P[\bar{X} > d \text{ and } X_{(1)} \geq \mu|\pi_0 = \pi_2]$$
$$+ \frac{1}{2} P[\mu_1 \leq X_{(1)} < \mu_2|\pi_0 = \pi_2]$$
$$+ \frac{1}{2} P[\bar{X} < d \text{ and } X_{(1)} \geq \mu|\pi_0 = \pi_1]$$
$$+ \frac{1}{2} P[\mu_2 < X_{(1)} < \mu_1|\pi_0 = \pi_1],$$
$$= \frac{1}{2} P[\bar{X} > d \text{ and } X_{(1)} \geq \mu|\pi_0 = \pi_2]$$
$$+ \frac{1}{2} P[\bar{X} < d \text{ and } X_{(1)} \geq \mu|\pi_0 = \pi_1].$$

Now, $P[\bar{X} > d \text{ and } X_{(1)} \geq \mu|\pi_0 = \pi_2] \leq P[\bar{X} > d|\pi_0 = \pi_2]$ and by strong law of large numbers with probability one $\bar{X}|\pi_2 \to \mu_2 + \theta_2$ which is, without any loss of generality $< d$.

Hence

$$P[\bar{X} > d \text{ and } X_{(1)} \geq \mu|\pi_0 = \pi_2] \to 0 \text{ as } n \to \infty.$$

Similarly, $P[\bar{X} < d \text{ and } X_{(1)} \leq \mu|\pi_0 = \pi_1] \to 0$ as $n \to \infty$. Hence $P(\text{MC}|R_5) \to 0$ as $n \to \infty$.

The probability of misclassification here, following the method of Section 3, is given by

$$P(MC|R_5) = \frac{1}{2} + \frac{1}{2}\{F_{2(n-1)}(\frac{2n(d-\mu)}{\theta_2}) - F_{2(n-1)}(\frac{2n(d-\mu)}{\theta_1}) \tag{6.2}$$

<u>Case (ii) (μ_i, θ_i), i = 1, 2 unknown.</u>

As before, assume independent samples X_{ij}, j = 1, ..., n_i, from π_i, i = 1, 2 are available. The plug-in classification rule suggested is R_5^*: Classify π_0 to π_1 if

$$\overline{X} < \frac{\overline{X}_1 + \overline{X}_2}{2} \text{ and } X_{(1)} \geq \max(\hat{\mu}_1, \hat{\mu}_2) \text{ or } \hat{\mu}_1 \leq X_{(1)} < \hat{\mu}_2$$

and to π_2 if

$$\overline{X} > \frac{\overline{X}_1 + \overline{X}_2}{2} \text{ and } X_{(1)} \geq \max(\hat{\mu}_1, \hat{\mu}_2) \text{ or } \hat{\mu}_2 \leq X_{(1)} < \hat{\mu}_1.$$

where $\overline{X}_i = \frac{1}{n_i} \sum_{j=1}^{n_i} X_{ij}$, and $\hat{\mu}_i = X_{i(1)}$.

Again the rule R_5^* is consistent since $\overline{X} - (\overline{X}_1 + \overline{X}_2)/2 \to \overline{X} - d$ with probability one as n_1 and $n_2 \to \infty$. The probability of misclassification in this case can be estimated by (6.2) at least for moderate values of n_1 and n_2.

7.0 *<u>Bayesian Approach</u>*. In this section the problem of classification is considered from the Bayesian point of view. Here the parameters (μ_i, θ_i) are considered to be random variables with given prior density $g(\mu_i, \theta_i)$ when the random sample $\underline{X}_i = (X_{i1}, X_{i2}, \ldots, X_{in_i})$ from π_i is available (i = 1, 2). Using $g(\mu_i, \theta_i)$, the posterior density $p(\mu_i, \theta_i|\underline{X}_i)$ and hence the <u>predictive probability density</u> for classifying a future sample $\underline{Z} \equiv (Z_1, Z_2, \ldots, Z_{n_0})$ into π_i is computed. Finally the predictive odds ratio R_{ij} for classifying Z into π_i as compared with π_j is calculated and this ratio is used to determine from which population $\underline{Z}$ is more likely to have arisen.

Let (μ_1, θ_1) and (μ_2, θ_2) be identically distributed with a common prior distribution. The problem of choosing a suitable prior distribution has been considered by Varde [15] for the two parameter exponential distribution, and is given by the following conjugate prior distribution.

$$g(\mu_i, \theta_i) \propto \frac{1}{\theta_i^{\nu_i+1}} e^{-(\alpha_i - \lambda_i\mu_i)/\theta_i} I(\eta_i \geq \mu_i) \qquad \mu_i > 0,\ \theta_i > 0. \tag{7.1}$$

Here, α_i, λ_i, ν_i and η_i are known constants reflecting the degree of prior knowledge one has about μ_i and θ_i. Note, that for $\lambda_i = \alpha_i = 0$ and $\eta_i \to \infty$ (7.1) reduces to the non-informative prior

$$g(\mu_i, \theta_i) \propto \frac{1}{\theta_i^{\nu_i+1}}, \qquad \mu_i \geq 0,\ \theta_i > 0. \tag{7.2}$$

For a related problem of choosing suitable priors and other details see Basu and Gupta [4].

The posterior density of (μ_i, θ_i), given the random sample $\underline{X}_i$ from Π_i and the prior density (7.1) is given by

$$p(\mu_i, \theta_i | \underline{x}_i) = \frac{f(\underline{x}_i|\mu_i,\theta_i) g(\mu_i,\theta_i)}{\int_0^{m_i}\int_0^{\infty} f(\underline{x}_i|\mu_i,\theta_i) g(\mu_i,\theta_i)\, d\theta_i\, d\mu_i} \tag{7.3}$$

$$= \frac{e^{-(n_i\bar{x}_i+\alpha_i-(n_i+\lambda_i)\mu_i)/\theta_i} (n_i\bar{x}_i+\alpha_i)^{n_i+\nu_i-1} (n_i+\lambda_i)}{\theta_i^{n_i+\nu_i+1}\ \Gamma(n_i+\nu_i-1)\left[1-\left(1-\dfrac{(n_i+\lambda_i)m_i}{n_i\bar{x}_i+\alpha_i}\right)^{1-n_i-\nu_i}\right]}$$

provided $n_i\overline{x}_i + \alpha_i > (n_i + \lambda_i)\mu_i$, $n_i + \nu_i > 1$ and $\eta_i > \mu_i > 0$, $\theta_i > 0$. Here $m_i = \min(\eta_i, X_{i(1)})$, where $X_{i(1)} = \min(X_{i1},\ldots,X_{in_i})$. Here from (7.3), the corresponding predictive probability density for classifying $\underline{Z} = (Z_1, Z_2, \ldots, Z_{n_0})$ into π_i is

$$f(\underline{z}\,\underline{x}_i, \pi_i) = \int_0^{m_i}\int_0^{\infty} f(\underline{z}|\mu_i,\theta_i)p(\mu_i\theta_i|\underline{x}_i)\, d\theta_i\, d\mu_i \tag{7.4}$$

$$= C_i \frac{\Gamma(n_0+n_i+\nu_i-1)}{(n_0+n_i+\lambda_i)\,(n_0\overline{z}+n_i\overline{x}_i+\alpha_i)^{n_0+n_i+\nu_i-1}}$$

$$\cdot \left[1 - \left(1 - \frac{(n_0+n_i+\lambda_i)m_i}{n_0\overline{z}+n_i\overline{x}_i+\alpha_i}\right)^{1-n_0-n_i-\lambda_i}\right]$$

where

$$C_i = \frac{(n_i\overline{x}_i+\alpha_i)^{n_i+\nu_i-1}(n_i+\lambda_i)}{(n_i+\nu_i-1)\left[1 - \left(1 - \frac{(n_i+\lambda_i)m_i}{n_i\overline{x}_i+\alpha_i}\right)^{1-n_i-\eta_i}\right]}$$

Finally, the <u>predictive odds ratio</u> R_{ij} for classifying $\underline{Z}$ into π_i as compared with π_j based on the sample mean $\overline{Z}$, is given by

$$R_{ij} = \frac{f(\underline{z}|\underline{x}_i,\pi_i)}{f(\underline{z}|\underline{x}_j,\pi_j)} \qquad (i, j = 1, 2;\ i \neq j)$$

Thus if $R_{ij} > 1$, $\underline{Z}$ is more likely to have arisen from Π_i, and if $R_{ij} < 1$, $\underline{Z}$ is more likely to have come from Π_j. Similarly, letting $\lambda_i = \alpha_i = 0$ and $\eta_i \to \infty$ in (7.3), the posterior density of (μ_i, θ_i) corresponding to the quasi-prior density (7.2) is obtained as

$$p(\mu_i,\theta_i|\underline{X}_i) = \frac{e^{-n_i(\bar{x}_i-\mu_i)/\theta_i}(n_i\bar{x}_i)^{n_i-1}n_i}{\theta_i^{n_i+\nu_i-1}\Gamma(n_i+\nu_i-1)\left[1-\left(1-\frac{x_{i(1)}}{\bar{x}_i}\right)^{1-n_i-\nu_i}\right]},$$

$$\theta_i > 0, \quad 0 \le \mu_i < x_{i(1)} \tag{7.5}$$

The predictive odds ratio using (7.5) can similarly be computed.

REFERENCES

[1] Anderson, T. W., An Introduction to Multivariate Statistical Analysis, Wiley, New York, (1958).

[2] Anderson, T. W., Das Gupta, S., and Styan, G.P.H., A Bibliography of Multivariate Statistical Analysis, Oliver and Boyd, Edinburg, (1972).

[3] Basu, A. P., "On Some Tests of Hypotheses Relating to the Exponential Distribution When Some Outliers are Present", *Jour. Am. Stat. Assn.*, *60*, pp. 548-559.

[4] Basu, A. P. and Gupta, A. K., "Classification Rules for Exponential Populations", *Proc. Conf. Reliability and Biometry, SIAM Publ.*, (1974), pp. 537-650.

[5] Bhattacharya, P. K. and Das Gupta, S., "Clafficication Between Univariate Exponential Populations", *Sankhya, Ser. A.*, *26*, (1964), pp. 17-24.

[6] Epstein, B. and Sobel, M., "Life Testing", *Jour. Am. Stat. Assn.*, *48*, (1953), pp. 486-502.

[7] Geisser, S., "Posterior Odds for Multivariate Normal Classifications", *Jour. Roy. Stat. Soc. Ser. B.*, *26*, (1964), pp. 69-76.

[8] Gupta, A. K. and Govindarajulu, Z., "Some New Classification Rules for c Univariate Normal Populations", *Canad. J. Statist. 1, No. 2*, (1973), pp. 139-157.

[9] Hills, M., "Allocation Rules and Their Error Rates", *J. Roy. Stat. Soc. Ser. B.*, *28*, (1966), pp. 1-32.

[10] John S., "Errors in Discriminations", *Ann. Math. Statist.*, *32*, (1961), pp. 1125-1144.

[11] Okamoto, M., "An Asymptotic Expansion for the Distribution of the Linear Discriminant Function", *Ann. Math. Statist.*, *34*, (1963), pp. 1286-1301.

[12] Sedransk, N., Contributions to Discriminant Analysis, Ph.D. Dissertation, Iowa State University, (1969).

[13] Smith, C. A. B., "Some Examples of Classification", *Ann. Eugenics*, *13*, (1947), pp. 272-282.

[14] Sorum, M. J., Estimating the Probability of Misclassification, Ph.D. Dissertation, The University of Minnesota, (1968).

[15] Varde, S. D., "Life Testing and Reliability Estimation for the Two Parameter Exponential Distribution", *Jour. Am. Stat. Assn.*, *64*, (1969), pp. 621-631.

[16] Zelen, M., "Application of Exponential Model to Problems in Cancer Research", *J. R. Statist. Soc. A.*, *129*, (1966), pp. 368-398.

A BIVARIATE EXPONENTIAL MODEL WITH APPLICATIONS TO RELIABILITY AND COMPUTER GENERATION OF RANDOM VARIABLES

by D. S. FRIDAY and G. P. PATIL
The Pennsylvania State University

Abstract. A new bivariate life distribution is derived. It is called the BEE distribution and includes as special cases the BVE, ACBVE, Freund, and Proschan-Sullo bivariate life distributions. Models for the BEE distribution and their physical interpretations are given. Properties of this distribution are discussed including the bivariate loss of memory property and the decomposition into absolutely continuous and singular distributions. We show that if Y_1, Y_2 are independent standard exponential rv's and X_1, X_2 have a BEE distribution then there is a linear transformation from Y_1, Y_2 to X_1, X_2. All previous independence transformations for such distributions relate X_1, X_2 to more than 2 independent exponential rv's. The support of X_1, X_2 is partitioned into regions $X_1 > X_2$, $X_1 < X_2$, and $X_1 = X_2$. The distribution on each region is obtained via a linear transformation from a corresponding region in the support of Y_1, Y_2. The transformation is one-to-one for the absolutely continuous component of the BEE distribution. An immediate application of these transformations is an efficient method of generating rv's from the BEE distribution and its special cases on the computer. The computational efficiency is discussed.

1.0 Introduction. The bivariate exponential distribution of Marshall and Olkin [8] (BVE) occupies an important place among bivariate life distributions in that it has the bivariate loss of memory property (BLMP) and its marginals have the loss of memory property (LMP). Also of importance are the absolutely continuous bivariate exponential extensions of Block and Basu [4] (ACBVE) and of Freund [6]. The original derivations of all of these distributions were based on simple and plausible models. Subsequent research [1,2,3,9,11,14,15] has resulted in additional models leading to some or all of these distributions and their multivariate analogs.

We propose a new distribution in Section 2 which includes the BVE, ACBVE and the Freund distribution as special cases and is related to a distribution of Proschan and Sullo [14]. We call this distribution the BEE and present several models for it. Their physical interpretations suggest the names Threshold, Gestation, and Warmup models. A mathematical derivation is given for this distribution and some of its properties are discussed.

In section 3 transformations are given from 2 independent standard exponential random variables (rv's) to rv's with a BEE distribution. As special cases of this piecewise linear transformation we present a transformation from independence to the BVE distribution and also present one-to-one transformations from independence to the Freund and ACBVE distributions.

In section 4 we consider the computer generation of rv's having a BEE distribution or one of its special cases. A comparison is made between generating random observations using the basic models for each distribution and generating random observations using the transformations introduced in section 3. It is shown that a significant reduction in computational effort may be achieved.

2.0 *A Bivariate Distribution Including the Freund, BVE and ACBVE.*

The bivariate exponential type distribution proposed by Freund [5] was derived from a model for the random lifetimes X_1, X_2 of two components C_1, C_2 vulnerable to shocks S_1, S_2 respectively. The time at which the initial shock occurs is determined by two independent exponential rv's Z_1, Z_2 with parameters $\alpha_1, \alpha_2 (> 0)$.

Component C_1 fails at $Z_1 = x_1$ if $Z_1 < Z_2$ or C_2 fails at $Z_2 = x_2$ if $Z_2 < Z_1$. After the initial shock the remaining lifetime until the subsequent shock is Z_1' if $Z_2 < Z_1$ and Z_2' if $Z_1 < Z_2$, where rv's Z_1' and Z_2' are exponential with parameters α_1' and α_2', respectively. The rv's Z_1, Z_2, Z_1', and Z_2' are independent. The component lifetimes are $X_1 = Z_1$, $X_2 = Z_1 + Z_2'$ if $Z_1 < Z_2$ and $X_1 = Z_2 + Z_1'$, $X_2 = Z_2$ if $Z_2 < Z_1$. The density function is:

$$f_{\underset{\sim}{x}}(x_1,x_2) = \begin{cases} \alpha_1\alpha_2'\exp\{-(\alpha_1+\alpha_2-\alpha_2')x_1-\alpha_2'x_2\}; & 0 < x_1 < x_2 \\ \alpha_2\alpha_1'\exp\{-\alpha_1'x_1-(\alpha_1+\alpha_2-\alpha_1')x_2\}; & 0 < x_2 < x_1 \end{cases} \tag{2.1}$$

The marginal densities are of the form (for $x > 0$):

$$f_1(x) = \begin{cases} \{\alpha_1'\alpha_2\exp\{-\alpha_1'x\}+(\alpha_1-\alpha_1')(\alpha_1+\alpha_2)\exp\{-(\alpha_1+\alpha_2)x\}\}(\alpha_1+\alpha_2-\alpha_1')^{-1}; \\ \qquad \alpha_1+\alpha_2-\alpha_1' \neq 0 \\ \alpha_1 \exp\{-\alpha_1'x\}+\alpha_2\alpha_1'x \exp\{-\alpha_1'x\}; \quad \alpha_1+\alpha_2-\alpha_1' = 0 \end{cases} \tag{2.2}$$

$$f_2(x) = \begin{cases} \{\alpha_1\alpha_2'\exp\{-\alpha_2'x\}+(\alpha_2-\alpha_2')(\alpha_1+\alpha_2)\exp\{-(\alpha_1+\alpha_2)x\}(\alpha_1+\alpha_2-\alpha_2')^{-1}1 \\ \qquad \alpha_1+\alpha_2-\alpha_2' \neq 0. \\ \alpha_2 \exp\{-\alpha_2'x\}+\alpha_2'\alpha_1x \exp\{-\alpha_2'x\}; \quad \alpha_1+\alpha_2-\alpha_2' = 0 \end{cases} \tag{2.3}$$

Observe that the marginal distributions allow various representations depending on the parameter values. Consider the density of X_1 (2.2). For values $\alpha_1' < \alpha_1$, $f_1(x)$ is a mixture of two exponential densities and when $\alpha_1' = \alpha_1$ it is exponential. When $\alpha_1 < \alpha_1' < \alpha_1+\alpha_2$, $f_1(x)$ is a negative mixture [12] with the first coefficient greater than unity and when $\alpha_1+\alpha_2 < \alpha_1'$ it is a negative mixture with the second coefficient greater than unity.

When $\alpha_1' = \alpha_1+\alpha_2$, $f_1(x)$ is a mixture of an exponential density and a gamma density. The gamma density may be interpreted as resulting from size biased sampling with weight function $w(x) = x$ [13]. Corresponding properties hold for the marginal density $f_2(x)$ (2.3) and parameter α_2'. When $\alpha_1 = \alpha_1'$ and $\alpha_2 = \alpha_2'$ the marginals X_1 and X_2 are exponential and also independent.

Unlike the Freund distribution the BVE [8] distribution always has exponential marginals and it is not absolutely continuous. We denote the joint survival distribution function of rv's X_1, X_2 by $\bar{F}_{\tilde{X}}(x_1,x_2) = P(X_1 > x_1, X_2 > x_2)$. If X_1, X_2 have a BVE distribution with parameters $\lambda_0, \lambda_1, \lambda_2$ (> 0) then $\bar{F}_{\tilde{X}}(x_1,x_2) = \exp\{-\lambda_1 x_1 - \lambda_2 x_2 - \lambda_0 \max(x_1,x_2)\}$. It contains a singular component which allows $X_1 = X_2$ to occur with positive probability. In fact $\bar{F}_{\tilde{X}}(x_1,x_2) = (\lambda_1+\lambda_2)\lambda^{-1}\,\bar{F}_A(x_1,x_2) + \lambda_0\lambda^{-1}\,\bar{F}_S(x_1,x_2)$ is the decomposition of $\bar{F}_{\tilde{X}}$ into an absolutely continuous distribution $\bar{F}_A(x_1,x_2)$ and a singular distribution $\bar{F}_S(x_1,x_2) = \exp\{-\lambda\max(x_1,x_2)\}$ where $\lambda = \lambda_0 + \lambda_1 + \lambda_2$. The density of the absolutely continuous part is given by:

$$f_A(x_1,x_2) = \begin{cases} \lambda\lambda_1(\lambda_0+\lambda_2)(\lambda_1+\lambda_2)^{-1}\exp\{-\lambda_1 x_1 - (\lambda_0+\lambda_2)x_2\}; & x_1 < x_2 \\ \lambda\lambda_2(\lambda_0+\lambda_1)(\lambda_1+\lambda_2)^{-1}\exp\{-(\lambda_0+\lambda_1)x_1 - \lambda_2 x_2\}; & x_2 < x_1\ . \end{cases} \tag{2.4}$$

The basic derivation of the BVE is the fatal shock model. Consider two components C_1, C_2 and three shocks S_0, S_1, S_2 such that S_1 destroys only C_1, S_2 destroys only C_2 and S_0 destroys both C_1 and C_2 simultaneously. Suppose the shocks occur at random times Z_0, Z_1, Z_2 which have independent exponential distributions with parameters λ_0, λ_1, λ_2. Then X_1, the lifetime of C_1, and X_2, the lifetime of C_2, have a BVE distribution with parameters λ_0, λ_1, λ_2. The BVE has also been derived via more complex models.

The ACBVE [4] distribution is derived from the requirements of absolute continuity, bivariate loss of memory, and marginals which are mixtures of certain exponentials. The ACBVE is related to both the BVE and the Freund distribution. It is identical to the absolutely continuous component of the BVE (but not a special case of the BVE) and it is a special case of the Freund distribution when the parameters are constrained so that $\alpha_1' > \alpha_1$, and $\alpha_2' > \alpha_2$ in a particular way (see (2.16)).

We will construct a distribution which includes each of the above distributions as special cases. The term bivariate exponential extension has been used [2,4,6] to describe bivariate distributions which are derived from univariate exponentials and have non-exponential marginal distributions. These marginals are usually mixtures or negative mixtures of exponentials. The distribution to be derived in Section 2.2 also has marginals of this type. The special case of exponential marginals occurs when the distribution coincides with the BVE. These facts suggest the name BEE (Bivariate Exponential Extension) distribution. Proschan and Sullo [14] introduce a bivariate exponential extension based on a different model which is a reparameterized analog of the BEE on a restricted parameter space $\alpha_1', \alpha_2' \ (>(\alpha_1+\alpha_2)\ (1-\alpha_0))$.

2.1 Interpretation - Threshold, Gestation, and Warmup Models. The mathematical derivation of the BEE distribution which follows in Section 2.2 allows several physical interpretations which we now discuss. The name given to each interpretation is associated with the chance mechanism peculiar to that model.

(i) Threshold Model: Consider two components C_1, C_2 with lifetimes X_1, X_2. We now allow for shocks S_1, S_2 of varying intensity and assume the intensity is random and independent of the time at which the shocks occur. The intensity is always sufficient for S_1 to destroy C_1 and S_2 to destroy C_2. If C_1 and C_2 are in close proximity then for high intensity levels either shock may destroy not only its component but also the other component. Proximity may be interpreted as any type of "closeness" which renders a component vulnerable to the other component's shock. For each component C_1 or C_2 there is a fixed intensity threshold. If the intensity of a shock is below this threshold the other component will not be destroyed. If the intensity exceeds this threshold both components are destroyed simultaneously. The probability of each shock exceeding its threshold is $1 - \alpha_0$, $0 < \alpha_0 \leq 1$. Assume that the underlying shock model is that of the Freund distribution with the additional characteristic that the shocks are of varying intensity and there is an intensity threshold beyond which both components are destroyed. Then components X_1, X_2 have a BEE distribution with parameters α_0, α_1, α_2, α_1', α_2'.

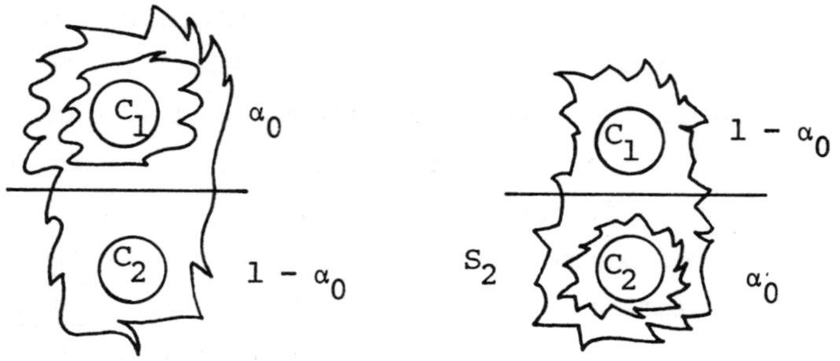

Figure 2.1 Representation of the Threshold Model

(ii) Gestation Model: Consider a system whose component lifetimes are preceded by a gestation interval. If a shock occurs during the gestation period the nature of the vulnerability of the system to the subsequent shock is changed. Let C_1 denote the offspring component and C_2 the parent component. Between the conception time $-\theta$ $(\theta > 0)$ and the instant of birth, time 0, C_1 and C_2 are in close proximity. This closeness may be interpreted not only in terms of time or space but more generally in terms of dependence on the same life-support system, nutrients, environmental factors, etc. Shock S_1 occurs at time $Z_1(> -\theta)$ and S_2 at time $Z_2(> 0)$. If Z_1 is in the interval $(-\theta, 0)$ S_1 prevents birth or separation and then shock S_2, now with an intensity parameter corresponding to the superposition of S_1 and S_2, simultaneously kills both the parent and offspring components at time Z_2. If $Z_1 > 0$ then S_1 kills C_1 and S_2 kills C_2. The Freund model then governs the occurrence times of S_1 and S_2. The death of a parent may affect the lifetime of the offspring and vice-versa. Component lifetimes X_1, X_2 have a BEE distribution.

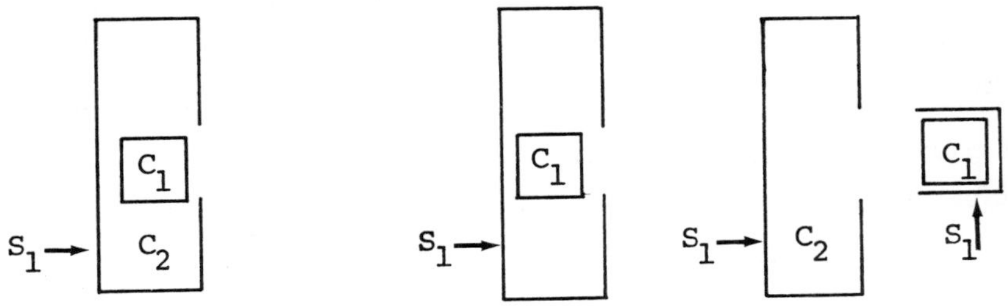

(i) Before time 0 (ii) After time 0

Figure 2.2 A Representation of the Gestation Model

(iii) Warmup Model: A similar interpretation is in terms of the lifetimes X_1, X_2 of two subsystems C_1, C_2 of some large system. Some examples are computer systems, automated industrial processes, power systems, and electronic devices. Such systems often enter a warmup mode immediately prior to operation during which they undergo automatic self-testing and internal self-adjustment. For two output components, random internal structural changes occurring during warmup may force component failures to occur simultaneously.

2.2 Derivation of the BEE Distribution. In Freund's model shock S_1 kills only component C_1, and shock S_2 kills only component C_2. In the BVE fatal shock model S_1 kills only C_1, S_2 kills only C_2, and S_0 kills both C_1 and C_2 simultaneously. Furthermore each shock kills instantly and with probability one. The non-fatal shock model and others [11] allow for shocks which choose and/or kill components randomly. The BVE and Freund models have a fixed correspondence between shocks and components and shocks cannot occur before time zero. We now modify these assumptions relative to the Freund model and obtain the BEE distribution.

We now consider the threshold model. Let Z_1, Z_2, Z_1', Z_2' be as in Section 2.0 and X_1, X_2 the component lifetimes. Suppose X_1, X_2 have the Freund distribution except for the following modification. With probability $1 - \alpha_0$ let the initial shock kill both components. This corresponds to a shock intensity which exceeds the threshold. Then $X_1 = X_2 = \min(Z_1, Z_2)$ is distributed exponentially with parameter $(\alpha_1+\alpha_2)$. With probability α_0 the initial shock kills only its corresponding component and the Freund model remains unchanged. After the initial shock at most one component remains and the intensity of the remaining shock no longer affects the system. The distribution of lifetimes $X_1 = X_2$ is therefore:

$$\bar{F}_S(x_1,x_2) = \exp\{-(\alpha_1+\alpha_2)\max(x_1,x_2)\};\ x_1x_2 > 0 \qquad (2.5)$$

Then with probability $1-\alpha_0$ X_1,X_2 have singular distribution (2.5) and with probability α_0 X_1,X_2 have Freund's absolutely continuous distribution with density (2.1). The joint survival distribution is therefore given by:

$$\bar{F}_X(x_1,x_2) = \alpha_0\,\bar{F}_A(x_1,x_2) + (1-\alpha_0)\,\bar{F}_S(x_1,x_2) \qquad (2.6)$$

where $\bar{F}_A$ corresponds to density (2.1) and $\bar{F}_S$ is given by (2.5). X_1,X_2 then have a BEE distribution with parameters α_0, α_1, α_2, α_1', α_2'.

In the gestation model a location parameter is introduced into one of the shock time distributions in the Freund model. Let $\theta \geq 0$ and let rv Z_1^* have an exponential distribution on support $(-\theta, \infty)$. Let rv Z_2 have an exponential distribution on $(0, \infty)$. Z_1^* and Z_2 are independent with parameters α_1 and α_2, respectively. Given that $Z_1^* > 0$, by the loss of memory property (LMP), $Z_1 = Z_1^* - 0$ is exponential with location parameter 0, scale parameter α_1, and Z_1 remains independent of Z_2. These conditions are identical to the initial part of Freund's derivation and, when $Z_1^* > 0$, we proceed according to that model. If Z_1^* occurs in $(-\theta, 0]$ then the Z_2 parameter increases to $\alpha_1 + \alpha_2$. This choice of parameter is consistent with the threshold model and the BVE distribution. It also follows from the requirement that $\min(X_1,X_2)$ has an exponential distribution or from reasoning that the double kill shock has parameter value corresponding to the minimum of the two single kill shocks. In the gestation model $\alpha_0 = \exp\{-\alpha_1\theta\}$ since

$P(X_1=X_2) = P(Z_1^* < 0) = 1 - \alpha_0$. The distributions (2.5) and (2.6) therefore apply to X_1, X_2. The warmup model results in a BEE distribution in a similar manner, mixing with probability α_0.

2.3 Properties and Special Cases. The joint survival distribution function for BEE distributed rv's X_1, X_2 is:

$$\bar{F}_{\underset{\sim}{X}}(x_1,x_2) = \begin{cases} \phi_1 \exp\{-(\alpha_1+\alpha_2-\alpha_2')x_1-\alpha_2'x_2\} + (1-\phi_1)\exp\{-(\alpha_1+\alpha_2)x_2\}; & x_1 < x_2 \\ \phi_2 \exp\{-\alpha_1'x_1-(\alpha_1+\alpha_2-\alpha_1')x_2\} + (1-\phi_2)\exp\{-(\alpha_1+\alpha_2)x_1\}; & x_2 < x_1 \end{cases} \tag{2.7}$$

where: $\phi_1 = \alpha_0\alpha_1(\alpha_1+\alpha_2-\alpha_2')^{-1}, \phi_2 = \alpha_0\alpha_2(\alpha_1+\alpha_2-\alpha_1')^{-1}$.

This follows from the mixture representation (2.6) with $\bar{F}_A(x_1,x_2)$ as given by Brindley and Thompson[[5], p. 823]. The marginal distributions are:

$$\bar{F}_1(x) = \bar{F}_{\underset{\sim}{X}}(x,0) = \phi_2 \exp\{-\alpha_1'x\} + (1-\phi_2)\exp\{-(\alpha_1+\alpha_2)x\}; \; x \geq 0 \tag{2.8}$$

$$\bar{F}_2(x) = \bar{F}_{\underset{\sim}{X}}(0,x) = \phi_1 \exp\{-\alpha_2'x\} + (1-\phi_1)\exp\{-(\alpha_1+\alpha_2)x\}; \; x \geq 0. \tag{2.9}$$

Observe that each marginal distribution is a mixture of the same exponential distributions as the corresponding Freund marginals, the only difference being that the mixture coefficients ϕ_1 and ϕ_2 are reduced by a multiplicative factor α_0, $0 < \alpha_0 \leq 1$.

The BEE distribution has the bivariate loss of memory property (BLMP) since

$$\bar{F}_{\underset{\sim}{X}}(x_1+t,x_2+t) = \bar{F}_{\underset{\sim}{X}}(x_1,x_2)\bar{F}_{\underset{\sim}{X}}(t,t); \; x_1x_2,t > 0. \tag{2.10}$$

We may also use the representation:

$$\bar{F}_{\underset{\sim}{X}}(x_1,x_2) = \begin{cases} \bar{F}_{\underset{\sim}{X}}(x_1,x_1)\bar{F}_2(x_2-x_1); & x_1 \le x_2 \\ \bar{F}_{\underset{\sim}{X}}(x_2,x_2)\bar{F}_1(x_1-x_2); & x_2 \le x_1 . \end{cases} \tag{2.11}$$

The $\min(X_1,X_2)$ is exponentially distributed with parameter $\alpha_1 + \alpha_2$ since $\bar{F}_{\underset{\sim}{X}}(x,x) = \exp\{-(\alpha_1+\alpha_2)x\}$.

The means, variances and covariances for the BEE distribution are:

$$E(X_1) = \frac{\alpha_0\alpha_2 + \alpha_1'}{\alpha_1'(\alpha_1+\alpha_2)}, \quad E(X_2) = \frac{\alpha_0\alpha_1 + \alpha_2'}{\alpha_2'(\alpha_1+\alpha_2)} \tag{2.12}$$

$$V(X_1) = \frac{\alpha_1'^2 + 2\alpha_0\alpha_2(\alpha_1+\alpha_2) - (\alpha_0\alpha_2)^2}{\alpha_1'^2(\alpha_1+\alpha_2)^2}, \tag{2.13}$$

$$V(X_2) = \frac{\alpha_2'^2 + 2\alpha_0\alpha_1(\alpha_1+\alpha_2) - (\alpha_0\alpha_1)^2}{\alpha_2'^2(\alpha_1+\alpha_2)^2}$$

$$C(X_1,X_2) = \frac{\alpha_1'\alpha_2' - \alpha_0^2\alpha_1\alpha_2}{\alpha_1'\alpha_2'(\alpha_1+\alpha_2)^2}. \tag{2.14}$$

The moment generating function is

$$M_{\underset{\sim}{X}}(t_1,t_2) = E[\exp\{t_1X_1+t_2X_2\}] = \tag{2.15}$$

$$\frac{\alpha_1'\alpha_2'(\alpha_1+\alpha_2) - \alpha_2'(\alpha_1+\alpha_2(1-\alpha_0))t_1 - \alpha_1'(\alpha_2+\alpha_1(1-\alpha_0))t_2 + (1-\alpha_0)(\alpha_1+\alpha_2)t_1t_2}{(\alpha_1+\alpha_2-t_1-t_2)(\alpha_1'-t_1)(\alpha_2'-t_2)}.$$

The BEE distribution reduces to the Freund distribution when the threshold exceedance probability is zero. That is when $\alpha_0 = 1$ or equivalently when gestation period $\theta = 0$. To obtain the BVE distribution we must have $\alpha_0 < 1$ or equivalently $\theta > 0$ for the singular component. In addition, the absolutely continuous component must assume the ACBVE form. This relationship is discussed by Block and Basu [4]. If the Freund parameters are $\alpha_1, \alpha_2, \alpha_1', \alpha_2'$ and the BVE parameters are $\lambda_0, \lambda_1, \lambda_2$ then the relationship between the parameters is:

$$\alpha_1 = \lambda_1(1+\lambda_0(\lambda_1+\lambda_2)^{-1})$$

$$\alpha_2 = \lambda_2(1+\lambda_0(\lambda_1+\lambda_2)^{-1})$$

$$\alpha_1' = \lambda_0 + \lambda_1 \tag{2.16}$$

$$\alpha_2' = \lambda_0 + \lambda_2$$

Let $\gamma_1 = \lambda_0\lambda_1(\lambda_1+\lambda_2)^{-1}$ and $\gamma_2 = \lambda_0\lambda_2(\lambda_1+\lambda_2)^{-1}$. Then $\gamma_1 + \gamma_2 = \lambda_0$, $\alpha_1 = \lambda_1 + \gamma_1$ and $\alpha_2 = \lambda_2 + \gamma_2$. The BVE may be interpreted as the special case of the BEE distribution in which after occurrence of one shock the parameter of the remaining shock increases to $\alpha_1' = \lambda_1 + \gamma_1 + \gamma_2$ or $\alpha_2' = \lambda_2 + \gamma_1 + \gamma_2$. An additional transformation

$$\alpha_0 = (\lambda_1 + \lambda_2)\lambda^{-1} \tag{2.17}$$

is required for the threshold parameter. Then X_1, X_2 have the BVE distribution with parameters $\lambda_0, \lambda_1, \lambda_2$. All of these distributions may be reduced to the case of independent marginals but they are not special cases of one another. Figure 2.3 illustrates the relationships between these distributions.

2.4 A Symmetric Generalization. The gestation model of Section 2.1 lacks symmetry since only one of the shocks can occur before time 0. We consider briefly the effect when both shocks are allowed to occur during the gestation period. Suppose shocks

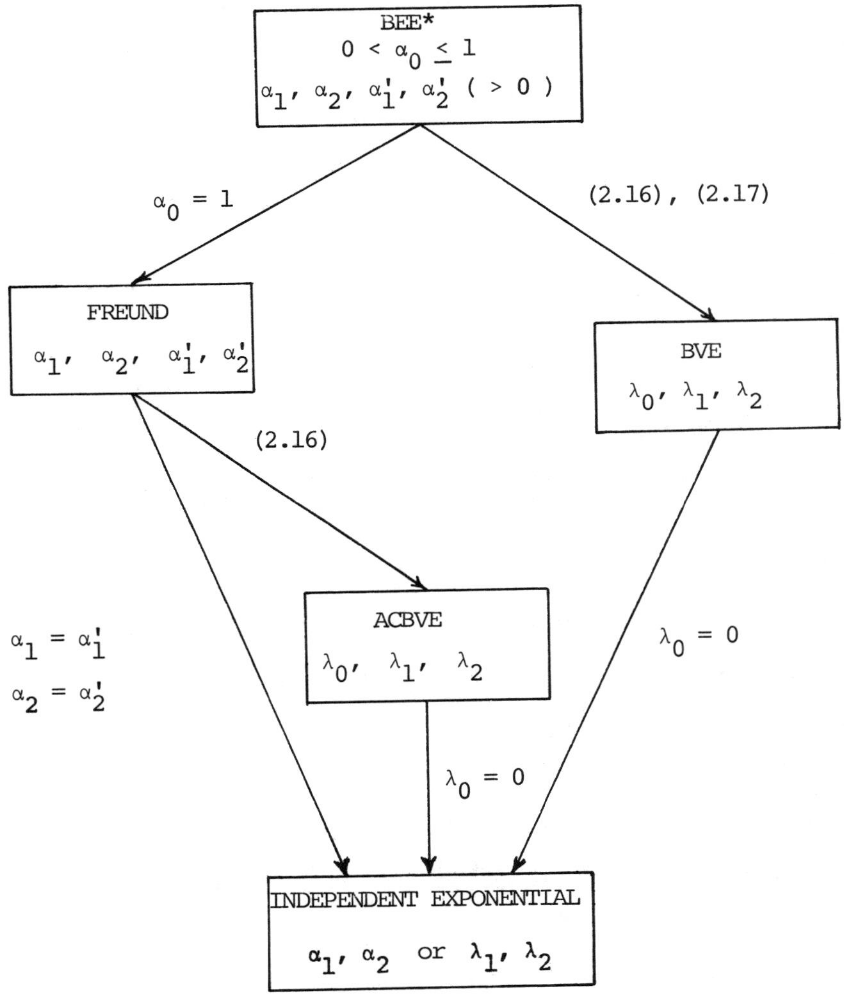

*Reduces to the **extended** Freund Distribution of Proschan and Sollo [14] when α_1', α_2' $> ((\alpha_1 + \alpha_2)(1 - \alpha_0)$.

Figure 2.3. The BEE Distribution and Its Special Cases.

occur at times $Z_1^* > -\theta_1$, $Z_2^* > -\theta_2 (\theta_1 > 0, \theta_2 > 0)$.

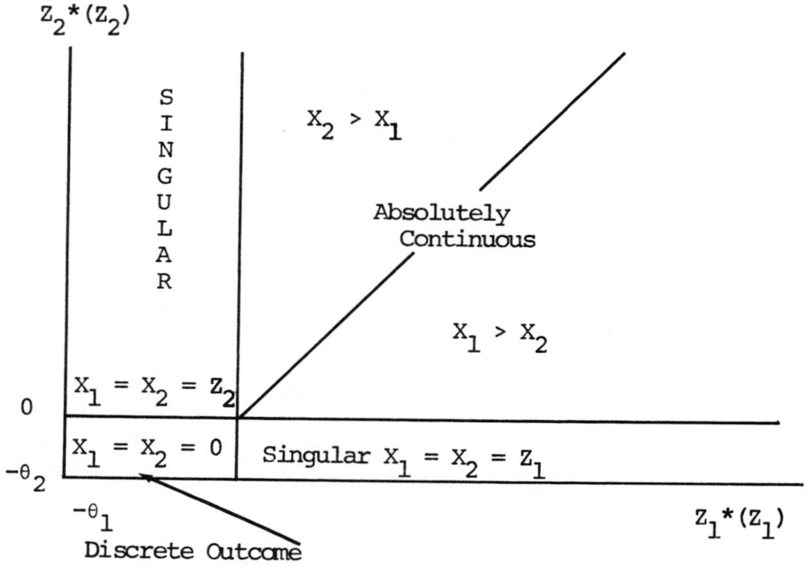

Figure 2.4 Interpretation of Symmetric Model

If both $Z_1^*, Z_2^* > 0$ then X_1, X_2 will have an absolutely continuous distribution and the Freund-type of dependence may be introduced. If both $Z_1^* < 0$, $Z_2^* < 0$ then we define $X_1 = X_2 = 0$, and components C_1, C_2, affected by shocks occurring during gestation, are dead at birth. In this case the distribution contains another singularity, an atom at $(X_1, X_2) = (0,0)$ which occurs with probability $P(Z_1^* < 0, Z_2^* < 0)$. If either of the two events $\{Z_1^* < 0, Z_2^* > 0\}$ or $\{Z_1^* > 0, Z_2^* < 0\}$ occurs then $X_1 = X_2$ and the outcome is from a singular distribution which now is a mixture of two possibly different singular components. Details are analogous to the gestation model.

3.0 Transformation from Independence. A central theme in the research on multivariate exponential models involves relationships between dependent rv's and independent rv's. The fatal shock model of Marshall and Olkin is summarized in the following result [8, Theorem 3.2]: Random variables X_1, X_2 have a BVE distribution if and only if there exist independent exponential random variables Z_o, Z_1, Z_2 such that $X_1 = \min(Z_o, Z_1)$ and $X_2 = \min(Z_0, Z_2)$. The Freund distribution can also be expressed in terms of independent rv's Z_1, Z_2, Z_1', Z_2' as in Section 2.

In each of the above examples a 2-dimensional rv X_1, X_2 is expressed in terms of 3 or more independent random variables. We now show that if X_1, X_2 have a BEE distribution then they can be expressed simply in terms of only 2 independent rv's.

3.1 The BEE and Freund Independence Transformations. Since the absolutely continuous component of the BEE is the Freund distribution we begin by exhibiting a two-piece linear transformation from 2 independent rv's to rv's with a Freund distribution.

Theorem 3.1 Let rv's Y_1, Y_2 be independent with standard exponential distributions. rv's X_1, X_2 are defined as follows:

$$X_1 = \begin{cases} Y_1 \alpha_1^{-1}; & \text{if } \alpha_2 Y_1 < \alpha_1 Y_2 \\ Y_1 (\alpha_1')^{-1} - (\alpha_1 - \alpha_1') Y_2 (\alpha_1' \alpha_2)^{-1}; & \text{if } \alpha_2 Y_1 > \alpha_1 Y_2 \end{cases} \tag{3.1}$$

$$X_2 = \begin{cases} Y_2 (\alpha_2')^{-1} - (\alpha_2 - \alpha_2') Y_1 (\alpha_1 \alpha_2')^{-1}; & \text{if } \alpha_2 Y_1 < \alpha_1 Y_2 \\ Y_2 \alpha_2^{-1}; & \text{if } \alpha_2 Y_1 > \alpha_1 Y_2 \end{cases} \tag{3.2}$$

where $\alpha_1, \alpha_2, \alpha_1', \alpha_2' > 0$.

Then X_1, X_2 have the bivariate exponential distribution given by Freund [6], with density (2.1).

Proof. Case I: $\alpha_2 Y_1 < \alpha_1 Y_2$, $(X_1 < X_2)$

The inverse transformations are $Y_1 = \alpha_1 X_1$ and $Y_2 = (\alpha_2 - \alpha_2')X_1 + \alpha_2' X_2$. The density of X_1, X_2 is given by $f_{\underset{\sim}{X}}(x_1,x_2) = f_{\underset{\sim}{Y}}(\alpha_1 x_1, (\alpha_1 x_1, (\alpha_2-\alpha_2')x_1 + \alpha_2' x_2)|J| = \alpha_1\alpha_2' \exp\{-(\alpha_1+\alpha_2-\alpha_2')x_1 - \alpha_2' x_2\}$; $x_1 < x_2$.

Case II: $\alpha_2 Y_1 > \alpha_1 Y_2$, $(X_1 > X_2)$

Similar to Case I, $Y_1 = \alpha_1' X_1 + (\alpha_1 - \alpha_1')X_2$, $Y_2 = \alpha_2 X_2$ and $f_X(x_1,x_2) = \alpha_1'\alpha_2 \exp\{-\alpha_1' x_1 - (\alpha_1+\alpha_2-\alpha_1')x_2\}$; $x_1 > x_2$. Therefore X_1, X_2 have the Freund bivariate exponential distribution with parameters $\alpha_1, \alpha_2, \alpha_1', \alpha_2'$.

This theorem shows the Freund distribution can be derived from two independent standard exponential random variables. Apart from its theoretical interest an application will be discussed in Section 4. We now use this result to obtain a transformation from independence for BEE distributed random variables.

Theorem 3.2 Let rv's Y_1^*, Y_2 be independent with standard exponential distributions, and $\theta \geq 0$ an arbitrary constant. If $Y_1^* > \theta$ let $Y_1 = Y_1^* - \theta$, Y_2 be transformed to X_1, X_2 by (3.1) and (3.2) in Theorem 3.1.

If $Y_1^* < \theta$ let $X_1 = X_2 = Y_2(\alpha_1+\alpha_2)^{-1}$. Then X_1, X_2 have a BEE distribution with survival function (2.6) and parameters $\alpha_0 = \exp\{-\theta\}$, $\alpha_1, \alpha_2, \alpha_1', \alpha_2'$.

Proof. Given $Y_1^* > \theta$ then by LMP Y_1 is standard exponential, and it is independent of Y_2. Applying Theorem 3.1 shows X_1, X_2 have a Freund distribution.

Given $Y_1^* \leq \theta$ which occurs with probability $1 - \alpha_0$ then X_1, X_2 have the singular exponential distribution given by (2.5).

The singular and absolutely continuous parts are mixed with probability α_0 as in 2.6. Rv's X_1,X_2 then have the BEE distribution.

3.2 Transformation for the BVE and ACBVE A piecewise linear transformation is given in Theorem 3.1 and 3.2 from independent standard exponential rv's Y_1,Y_2 to BEE rv's X_1,X_2. In section 2.3 the BVE and ACBVE were shown to be special cases of the BEE distribution. Transformations from independence for these cases are now given.

Corollary 3.3: Y_1,Y_2 are independent, standard exponentials rv's, $\lambda_0,\lambda_1,\lambda_2$ (> 0) are constants, and $\lambda = \lambda_0 + \lambda_1 \lambda_2$. Define rv's X_1,X_2 as follows:

$$X_1 = \begin{cases} \{Y_1+\lambda_0\lambda^{-1}Y_2\}(\lambda_0+\lambda_1)^{-1}; & \text{if } \lambda_2Y_1 > \lambda_1Y_2 \\ (\lambda_1+\lambda_2)(\lambda\lambda_1)^{-1}Y_1; & \text{if } \lambda_2Y_1 < \lambda_1Y_2 \end{cases} \tag{3.3}$$

$$X_2 = \begin{cases} (\lambda_1+\lambda_2)(\lambda\lambda_2)^{-1}Y_2; & \text{if } \lambda_2Y_1 > \lambda_1Y_2 \\ \{\lambda_0\lambda^{-1}Y_1+Y_2\}(\lambda_0+\lambda_2)^{-1}; & \text{if } \lambda_2Y_1 < \lambda_1Y_2 \end{cases} \tag{3.4}$$

Then X_1,X_2 have the ACBVE distribution [4] with parameters $\lambda_0,\lambda_1,\lambda_2$ and density (2.4).

Proof: Apply the Freund to ACBVE transformations (2.16) to (3.1) and (3.2). The result is (3.3) and (3.4).

We now use Corollary 3.3 to obtain the BVE.

Corollary 3.4: Y_1^*, Y_2 are independent standard exponential rv's, $\lambda_0,\lambda_1,\lambda_2 > 0$ are constants, and $\lambda = \lambda_0 + \lambda_1 + \lambda_2$. Define $\theta \geq 0$ by $P(Y_1^* < \theta) = \lambda_0\lambda^{-1}$. If $Y_1^* > \theta$, let $Y_1 = Y_1^* - \theta$, Y_2

be transformed to X_1, X_2 by (3.3) and (3.4) in Corollary 3.3. If $Y_1^* < \theta$, let $X_1 = X_2 = Y_2\lambda^{-1}$. Then X_1, X_2 have a BVE distribution with parameters $\lambda_0, \lambda_1, \lambda_2$.

Proof. If $Y_1^* > \theta$ use L.M.P. and Corollary 3.3. If $Y_1^* < \theta$ which occurs with probability $\lambda_0\lambda^{-1}$ then X_1, X_2 are distributed singularly along the diagonal with parameter λ. The mixture on $\lambda_0\lambda^{-1}$ of the singular and absolutely continuous parts gives the BVE.

Specific transformations from independence have been exhibited for the BEE, BVE, ACBVE, and Freund distributions. Two wedge and one rectangular shaped regions in $\{Y_1 > 0, Y_2 > 0\}$ map respectively to two wedges and the diagonal line in $\{X_1 > 0, X_2 > 0\}$. An application of these transformations is now presented.

4.0 Computer Generation of rv's From a BEE Distribution A considerable amount of research has been directed toward generating on a digital computer random observations which have specified univariate distributions. A related bibliography for non-uniform distributions is contained in the paper "Chance Mechanisms in Computer Generation of Random Variables" [12], which was an attempt to identify common structures underlying these techniques. In the multivariate case models for the distributions are sometimes unknown and mechanisms particularly adapted to the computer are rare.

4.1 Generating rv's with a BEE Distribution If a model exists for a given distribution then this model may provide the basis for computer generation of random observations from that distribution. For example the gestation model may be used to generate random variables X_1, X_2 with a BEE distribution. A random observation Z_1^* is generated having an exponential distribution with location parameter $-\theta$ $(\theta > 0)$ and intensity parameter α_1. If Z_1^* is < 0 another independent exponential rv Z_0 is generated. Its location parameter is 0 and intensity parameter $(\alpha_1+\alpha_2)$. Given $Z_1^* > 0$ then Z_2 with parameter α_2 is generated and compared with $Z_1 = Z_1^*$. If $Z_1 > Z_2$ $(Z_2 > Z_1)$ another independent exponential Z_3 with parameter $\alpha_1'(\alpha_2')$ is generated. Then either i) $X_1 = X_2 = Z_0$, ii) $X_1 = Z_2 + Z_3$, $X_2 = Z_2$, or iii) $X_1 = Z_1$, $X_2 = Z_1 + Z_3$ depending on the observations. A pair of random observations of X_1,X_2 with a BEE distribution therefore requires either 2 or 3 random numbers with appropriate independent exponential distributions. The absolutely continuous part of the BEE is the Freund distribution; consequently 3 exponential random numbers are required for a Freund observation via the original model. The fatal shock model [8], the simplest model for the BVE requires 3 independent exponentials to obtain an observation.

Using these models 3 (Freund, ACBVE, BVE, BEE-absolutely continuous part) or possibly 2 (BEE-singular part) independent exponentials are required to obtain the jointly distributed pair X_1,X_2. The transformation discussed in Section 3 provide a direct method of generating X_1,X_2 from only 2 independent standard exponential rv's. Theorem 3.1 generates the Freund, Theorem 3.2 the BEE, Corollary 3.3 the ACBVE, and Corollary 3.4 the BVE, each by linear transformations.

4.2 Computational Efficiency We now consider the computational effort required to generate a pair of observations from the BEE distribution and its special cases. A comparison is made between use of the original models and use of the transformations. The real-life computational cost is dependent upon the particular computer used. In order that this discussion be hardware indepen-

we will consider only the number and type of basic operations required in each case. The computational operations to be considered and their abbreviations are as follows: The generation of an independent standard exponential random number (E), the comparison of two numbers to determine which one is smaller (C), multiplication of two real numbers (M), and finally addition or subtraction of two reals (A). To generate a pair of random numbers from a Freund distribution using the original model requires 3E + 3M + 1C + 1A operations. The computational cost using transformations (Theorem 3.1) is 2E + 4M + 1C + 1A. The only difference being that 1E operation is replaced by 1M operation.

In the case of the BEE distribution via the gestation model 1E + 1M + 1C + 1A operations are initially required, then either with probability $1 - \alpha_0$ 1E + 1M additional operations are required or with probability α_0 2E + 2M + 1C + 1A additional operations are required. The number of operations required using transformations (Theorem 3.2) is initially 1E + 1C, then with probability α_0 an additional 1E + 4M + 1C + 2A operations are required or with probability $1 - \alpha_0$ an additional 1E + 1M operations are required. Similarly the BVE via the fatal shock model requires 3E + 3M + 2C operations and via transformations (Corollary 3.3) initially requires 1E + 1C operations, then an additional 1E + 1M operations with probability $\lambda_0\lambda^{-1}$ or 1E + 4M + 1C +2A operations with probability $(\lambda_1+\lambda_2)\lambda^{-1}$. The ACBVE being a special case of the Freund with no unique mechanism of its own therefore has the same computational cost as the Freund.

If we now allow the symbols E, M, C, A to represent the computational cost for the respective operations then the transformation methods can be compared with original models in terms of these parameters. Table 4.1 contains the expected savings in computational effort which results from using the transformations rather than the models.

Distribution	Expected Savings
BEE	$\alpha_0 E + M(1-2\alpha_0) + (1-\alpha_0)A$
Freund/ACBVE	$E - M$
BVE	$E-(1-3\lambda_0\lambda^{-1})M+\lambda_0\lambda^{-1}C-2(1-\lambda_0\lambda^{-1})A$

Table 4.1 Expected Computational Savings Using Transforms

4.3 Generation via Uniform rv's In all of these techniques independent standard exponential rv's were used as the fundamental source of randomness. The models for these distributions are also based on exponential rv's. The transformation $X = -\lambda^{-1}\ln\{U\}$ relates a uniform rv U on the interval (0,1) to an exponential rv X with parameter λ. Other more efficient techniques than the logarithmic transformation may be used to obtain the exponential however [12]. For example Marsaglia's wedge-tail method may be used. Pseudo-random numbers uniform on (0,1) provide the fundamental source of randomness on digital computers. If a uniform rv is used instead of the first exponential rv an additional savings in computational effort results for BEE and BVE random numbers.

In the case of a BEE rv first observe a uniform rv U. If $U < (1-\alpha_0)$ then $\underset{\sim}{X}$ will be singular $(X_1 = X_2)$ and in addition the rv $U(1-\alpha_0)^{-1}$ is distributed uniform on (0,1). This uniform rv can then be transformed to an exponential rv and used to determine the singular observation. Of course numerical errors for the particular computer must be considered whenever one of these procedures is implemented. Observe that using a uniform it is then possible to generate a singular observation from

1U + 2M + 1C + 1L operations (where L is the logarithm) instead of 2E + 1M + 1C. More efficient methods may replace the logarithm as noted above.

REFERENCES

[1] ARNOLD, B. C., Multivariate Exponential Distributions Based on Hierarchical Successive Damage, *J. Applied Probability, Vol. 12*, (1975), pp. 142-147.

[2] BLOCK, H. W., Continuous Multivariate Exponential Extensions, Reliability and Fault Tree Analysis, *SIAM*, **Philadelphia,** (1975), pp. 285-306.

[3] ___________, Physical Models Leading to Multivariate Exponential and Negative Binomial Distributions, Modeling and Simulation, (Vought, W. G. and Miekle, M. H. - Eds.) Inst. Soc. of Am., Pittsburgh, (1975).

[4] ___________ and BASU, A. P., A Continuous Bivariate Exponential Extension, *J. American Statistical Assoc., Vol. 69*, (1974), pp. 1031-1037.

[5] BRINDLEY, E. C. and THOMPSON, W. A., Dependence and Aging Aspects of Multivariate Survival, *J. American Statistical Assoc., Vol. 67*, (1972), pp. 822-830.

[6] FREUND, J. E., A Bivariate Extension of the Exponential Distribution, *J. American Statistical Assoc., Vol. 56*, (1961), pp. 971-977.

[7] KNUTH, D. E., The Art of Computer Programming, Vol. 2: Seminumerical Algorithms. Addison-Wesley Publishing Company, (1969).

[8] MARSHALL, A. W. and OLKIN, I., A Multivariate Exponential Distribution, *J. American Statistical Assoc., Vol. 62*, (1967), pp. 30-44.

[9] ___________________________, A Generalized Bivariate Exponential Distribution, *J. Applied Probability, Vol. 4*, (1967), pp. 291-302.

[10] NEWMAN, T. C. and ODELL, P. L., *The Generation of Random Variates*. Hafner Press/MacMillan, New York, (1967).

[11] SHAKED, M., A Personal Communication, (1975).

[12] PATIL, G. P., BOSWELL, M. T. and FRIDAY, D. S., Chance Mechanisms in Computer Generation of Random Variables, *Statistical Distributions in Scientific Work*, (Patil, G.P.; Kotz, S.; and Ord, J. K. - Eds.) D. Reidel Publishing Company, Dordrecht and Boston, Vol. 2, (1975), pp. 37-50.

[13] ___________ and ORD, J. K., On Size Biased Sampling and Related Form Invariant Weighted Distributions, *Sankhya - Series B*, (1976 - to appear).

[14] PROSCHAN, F. and SULLO, P., Estimating the Parameters of a Bivariate Exponential Distribution in Several Sampling Situations, *Reliability and Biometry*, S.I.A.M., Philadelphia, (Proschan, F. and Serfling, R. J. - Eds.), (1974), pp. 423-440.

[15] WEINMAN, D. G., A Multivariate Extension of the Exponential Distribution. Ph.D. Thesis, Arizona State University, (1966).

A
B 7
C 8
D 9
E 0
F 1
G 2
H 3
I 4
J 5